General Relativity: An Introduction for Physicists provides a clear mathematical introduction to Einstein's theory of general relativity. A wide range of applications of the theory are included, with a concentration on its physical consequences.

After reviewing the basic concepts, the authors present a clear and intuitive discussion of the mathematical background, including the necessary tools of tensor calculus and differential geometry. These tools are used to develop the topic of special relativity and to discuss electromagnetism in Minkowski spacetime. Gravitation as spacetime curvature is then introduced and the field equations of general relativity are derived. A wide range of applications to physical situations follows, and the conclusion gives a brief discussion of classical field theory and the derivation of general relativity from a variational principle.

Written for advanced undergraduate and graduate students, this approachable textbook contains over 300 exercises to illuminate and extend the discussion in the text.

Michael Hobson specialised in theoretical physics as an undergraduate at the University of Cambridge and remained at the Cavendish Laboratory to complete a Ph.D. in the physics of star-formation and radiative transfer. As a Research Fellow at Trinity Hall, Cambridge, and later as an Advanced Fellow of the Particle Physics and Astronomy Research Council, he developed an interest in cosmology, in particular in the study of fluctuations in the cosmic microwave background (CMB) radiation. He is currently a Reader in Astrophysics and Cosmology at the Cavendish Laboratory, where he is the principal investigator for the Very Small Array CMB interferometer. He is also joint project scientist for the Arcminute Microkelvin Imager project and an associate of the European Space Agency Planck Surveyor CMB satellite mission. In addition to observational and theoretical cosmology, his research interests also include Bayesian analysis methods and theoretical optics and he has published over 100 research papers in a wide range of areas. He is a Staff Fellow and Director of Studies in Natural Sciences at Trinity Hall and enjoys an active role in the teaching of undergraduate physics and mathematics. He is a co-author with Ken Riley and Stephen Bence of the well-known undergraduate textbook *Mathematical Methods for Physics and Engineering* (Cambridge, 1998; second edition, 2002; third edition to be published in 2006) and with Ken Riley of the *Student's Solutions Manual* accompanying the third edition.

George Efstathiou is Professor of Astrophysics and Director of the Institute of Astronomy at the University of Cambridge. After studying physics as an undergraduate at Keble College, Oxford, he gained his Ph.D. in astronomy from Durham University. Following some post-doctoral research at the University of

California at Berkeley he returned to work in the UK at the Institute of Astronomy, Cambridge, where he was appointed Assistant Director of Research in 1987. He returned to the Department of Physics at Oxford as Savilian Professor of Astronomy and Head of Astrophysics, before taking on his current posts at the Institute of Astronomy in 1997 and 2004 respectively. He is a Fellow of the Royal Society and the recipient of several awards, including the Maxwell Medal and Prize of the Institute of Physics in 1990 and the Heineman Prize for Astronomy of the American Astronomical Society in 2005.

Anthony Lasenby is Professor of Astrophysics and Cosmology at the University of Cambridge and is currently Head of the Astrophysics Group and the Mullard Radio Astronomy Observatory in the Cavendish Laboratory, as well as being a Deputy Head of the Laboratory. He began his astronomical career with a Ph.D. at Jodrell Bank, specializing in the cosmic microwave background, which has remained a major subject of his research. After a brief period at the National Radio Astronomy Observatory in America, he moved from Manchester to Cambridge in 1984 and has been at the Cavendish since then. He is the author or co-author of over 200 papers spanning a wide range of fields and is the co-author of *Geometric Algebra for Physicists* (Cambridge, 2003) with Chris Doran.

General Relativity

An Introduction for Physicists

M. P. HOBSON, G. P. EFSTATHIOU
and A. N. LASENBY

CAMBRIDGE
UNIVERSITY PRESS

University Printing House, Cambridge CB2 8BS, United Kingdom

Cambridge University Press is part of the University of Cambridge.

It furthers the University's mission by disseminating knowledge in the pursuit of education, learning and research at the highest international levels of excellence.

www.cambridge.org
Information on this title: www.cambridge.org/9780521829519

First published 2006
7th printing 2014

A catalogue record for this publication is available from the British Library

ISBN 978-0-521-82951-9 Hardback

To our families

Contents

Preface

General relativity is one of the cornerstones of classical physics, providing a synthesis of special relativity and gravitation, and is central to our understanding of many areas of astrophysics and cosmology. This book is intended to give an introduction to this important subject, suitable for a one-term course for advanced undergraduate or beginning graduate students in physics or in related disciplines such as astrophysics and applied mathematics. Some of the later chapters should also provide a useful reference for professionals in the fields of astrophysics and cosmology.

It is assumed that the reader has already been exposed to special relativity and Newtonian gravitation at a level typical of early-stage university physics courses. Nevertheless, a summary of special relativity from first principles is given in Chapter 1, and a brief discussion of Newtonian gravity is presented in Chapter 7. No previous experience of 4-vector methods is assumed. Some background in electromagnetism will prove useful, as will some experience of standard vector calculus methods in three-dimensional Euclidean space. The overall level of mathematical expertise assumed is that of a typical university mathematical methods course.

The book begins with a review of the basic concepts underlying special relativity in Chapter 1. The subject is introduced in a way that encourages from the outset a geometrical and transparently four-dimensional viewpoint, which lays the conceptual foundations for discussion of the more complicated spacetime geometries encountered later in general relativity. In Chapters 2–4 we then present a mini-course in basic differential geometry, beginning with the introduction of manifolds, coordinates and non-Euclidean geometry in Chapter 2. The topic of vector calculus on manifolds is developed in Chapter 3, and these ideas are extended to general tensors in Chapter 4. These necessary mathematical preliminaries are presented in such a way as to make them accessible to physics students with a background in standard vector calculus. A reasonable level of mathematical rigour has been maintained throughout, albeit accompanied by the occasional appeal to geometric intuition. The mathematical tools thus developed are then illustrated in Chapter 5 by re-examining the familiar topic of special relativity in a more formal manner, through the use of tensor calculus in Minkowski spacetime. These methods are further illustrated in Chapter 6, in which electromagnetism is described as a field theory in Minkowski spacetime, serving in some respects as a 'prototype' for the later discussion of gravitation. In Chapter 7, the incompatibility of special relativity and Newtonian gravitation is presented and the equivalence principle is introduced. This leads naturally to a discussion of spacetime curvature and the associated mathematics. The field equations of general relativity are then derived in Chapter 8, and a discussion of their general properties is presented.

The physical consequences of general relativity in a wide variety of astrophysical and cosmological applications are discussed in Chapters 9–18. In particular, the Schwarzschild geometry is derived in Chapter 9 and used to discuss the physics outside a massive spherical body. Classic experimental tests of general relativity based on the exterior Schwarzschild geometry are presented in Chapter 10. The interior Schwarzschild geometry and non-rotating black holes are discussed in Chapter 11, together with a brief mention of Kruskal coordinates and wormholes. In Chapter 12 we introduce two non-vacuum spherically symmetric geometries with a discussion of relativistic stars and charged black holes. Rotating objects are discussed in Chapter 13, including an extensive discussion of the Kerr solution. In Chapters 14–16 we describe the application of general relativity to cosmology and present a discussion of the Friedmann–Robertson–Walker geometry, cosmological models and the theory of inflation, including the generation of perturbations in the early universe. In Chapter 17 we describe linearised gravitation and weak gravitational fields, in particular drawing analogies with the theory of electromagnetism. The equations of linearised gravitation are then applied to the generation, propagation and detection of weak gravitational waves in Chapter 18. The book concludes in Chapter 19 with a brief discussion of classical field theory and the derivation of the field equations of electromagnetism and general relativity from variational principles.

Each chapter concludes with a number of exercises that are intended to illuminate and extend the discussion in the main text. It is strongly recommended that the reader attempt as many of these exercises as time permits, as they should give ample opportunity to test his or her understanding. Occasionally chapters have appendices containing material that is not central to the development presented in the main text, but may nevertheless be of interest to the reader. Some appendices provide historical context, some discuss current astronomical observations and some give detailed mathematical derivations that might otherwise interrupt the flow of the main text.

With regard to the presentation of the mathematics, it has to be accepted that equations containing partial and covariant derivatives could be written more compactly by using the comma and semi-colon notation, e.g. $v^a{}_{,b}$ for the partial derivative of a vector and $v^a{}_{;b}$ for its covariant derivative. This would certainly save typographical space, but many students find the labour of mentally unpacking such equations is sufficiently great that it is not possible to think of an equation's physical interpretation at the same time. Consequently, we have decided to write out such expressions in their more obvious but longer form, using $\partial_b v^a$ for partial derivatives and $\nabla_b v^a$ for covariant derivatives.

It is worth mentioning that this book is based, in large part, on lecture notes prepared separately by MPH and GPE for two different relativity courses in the Natural Science Tripos at the University of Cambridge. These courses were first presented in this form in the academic year 1999–2000 and are still ongoing. The course presented by MPH consisted of 16 lectures to fourth-year undergraduates in Part III Physics and Theoretical Physics and covered most of the material in Chapters 1–11 and 13–14, albeit somewhat rapidly on occasion. The course given by GPE consisted of 24 lectures to third-year undergraduates in Part II Astrophysics and covered parts of Chapters 1, 5–11, 14 and 18, with an emphasis on the less mathematical material. The process of combining the two sets of lecture notes into a homogeneous treatment of relativistic gravitation was aided somewhat by the fortuitous choice of a consistent sign convention in the two courses, and numerous sections have been rewritten in the hope that the reader will not encounter any jarring changes in presentational style. For many of the topics covered in the two courses mentioned above, the opportunity has been taken to include in this book a considerable amount of additional material beyond that presented in the lectures, especially in the discussion of black holes. Some of this material draws on lecture notes written by ANL for other courses in Part II and Part III Physics and Theoretical Physics. Some topics that were entirely absent from any of the above lecture courses have also been included in the book, such as relativistic stars, cosmology, inflation, linearised gravity and variational principles. While every care has been taken to describe these topics in a clear and illuminating fashion, the reader should bear in mind that these chapters have not been 'road-tested' to the same extent as the rest of the book.

It is with pleasure that we record here our gratitude to those authors from whose books we ourselves learnt general relativity and who have certainly influenced our own presentation of the subject. In particular, we acknowledge (in their current latest editions) S. Weinberg, *Gravitation and Cosmology*, Wiley, 1972; R. M. Wald, *General Relativity*, University of Chicago Press, 1984; B. Schutz, *A First Course in General Relativity*, Cambridge University Press, 1985; W. Rindler, *Relativity: Special, General and Cosmological*,

Oxford University Press, 2001; and J. Foster & J. D. Nightingale, *A Short Course in General Relativity*, Springer-Verlag, 1995.

During the writing of this book we have received much help and encouragement from many of our colleagues at the University of Cambridge, especially members of the Cavendish Astrophysics Group and the Institute of Astronomy. In particular, we thank Chris Doran, Anthony Challinor, Steve Gull and Paul Alexander for numerous useful discussions on all aspects of relativity theory, and Dave Green for a great deal of advice concerning typesetting in LaTeX. We are also especially grateful to Richard Sword for creating many of the diagrams and figures used in the book and to Michael Bridges for producing the plots of recent measurements of the cosmic microwave background and matter power spectra. We also extend our thanks to the Cavendish and Institute of Astronomy teaching staff, whose examination questions have provided the basis for some of the exercises included. Finally, we thank several years of undergraduate students for their careful reading of sections of the manuscript, for pointing out misprints and for numerous useful comments. Of course, any errors and ambiguities remaining are entirely the responsibility of the authors, and we would be most grateful to have them brought to our attention. At Cambridge University Press, we are very grateful to our editor Vince Higgs for his help and patience and to our copy-editor Susan Parkinson for many useful suggestions that have undoubtedly improved the style of the book.

Finally, on a personal note, MPH thanks his wife, Becky, for patiently enduring many evenings and weekends spent listening to the sound of fingers tapping on a keyboard, and for her unending encouragement. He also thanks his mother, Pat, for her tireless support at every turn. MPH dedicates his contribution to this book to the memory of his father, Ron, and to his daughter, Tabitha, whose early arrival succeeded in delaying completion of the book by at least three months, but equally made him realise how little that mattered. GPE thanks his wife, Yvonne, for her support. ANL thanks all the students who have sat through his various lectures on gravitation and cosmology and provided useful feedback. He would also like to thank his family, and particularly his parents, for the encouragement and support they have offered at all times.

1

The spacetime of special relativity

We begin our discussion of the relativistic theory of gravity by reviewing some basic notions underlying the Newtonian and special-relativistic viewpoints of space and time. In order to specify an *event* uniquely, we must assign it three spatial coordinates and one time coordinate, defined with respect to some frame of reference. For the moment, let us define such a system S by using a set of three mutually orthogonal Cartesian axes, which gives us spatial coordinates x, y and z, and an associated system of synchronised clocks at rest in the system, which gives us a time coordinate t. The four coordinates (t, x, y, z) thus label events in space and time.

1.1 Inertial frames and the principle of relativity

Clearly, one is free to label events not only with respect to a frame S but also with respect to any other frame S', which may be oriented and/or moving with respect to S in an arbitrary manner. Nevertheless, there exists a class of preferred reference systems called *inertial frames*, defined as those in which Newton's first law holds, so that a free particle is at rest or moves with constant velocity, i.e. in a straight line with fixed speed. In Cartesian coordinates this means that

$$\boxed{\frac{d^2x}{dt^2} = \frac{d^2y}{dt^2} = \frac{d^2z}{dt^2} = 0.}$$

It follows that, *in the absence of gravity*, if S and S' are two inertial frames then S' can differ from S only by (i) a translation, and/or (ii) a rotation and/or (iii) a motion of one frame with respect to the other at a constant velocity (for otherwise Newton's first law would no longer be true). The concept of inertial frames is fundamental to the *principle of relativity*, which states that *the laws of physics take the same form in every inertial frame*. No exception has ever been found to this general principle, and it applies equally well in both Newtonian theory and special relativity.

The Newtonian and special-relativistic descriptions differ in how the coordinates of an event P in two inertial frames are related. Let us consider two Cartesian inertial frames S and S' in *standard configuration*, where S' is moving along the x-axis of S at a constant speed v and the axes of S and S' coincide at $t = t' = 0$ (see Figure 1.1). It is clear that the (primed) coordinates of an event P with respect to S' are related to the (unprimed) coordinates in S via a *linear* transformation[1] of the form

$$\begin{aligned} t' &= At + Bx, \\ x' &= Dt + Ex, \\ y' &= y, \\ z' &= z. \end{aligned}$$

Moreover, since we require that $x' = 0$ corresponds to $x = vt$ and that $x = 0$ corresponds to $x' = -vt'$, we find immediately that $D = -Ev$ and $D = -Av$, so that $A = E$. Thus we must have

$$\begin{aligned} t' &= At + Bx, \\ x' &= A(x - vt), \\ y' &= y, \\ z' &= z. \end{aligned} \tag{1.1}$$

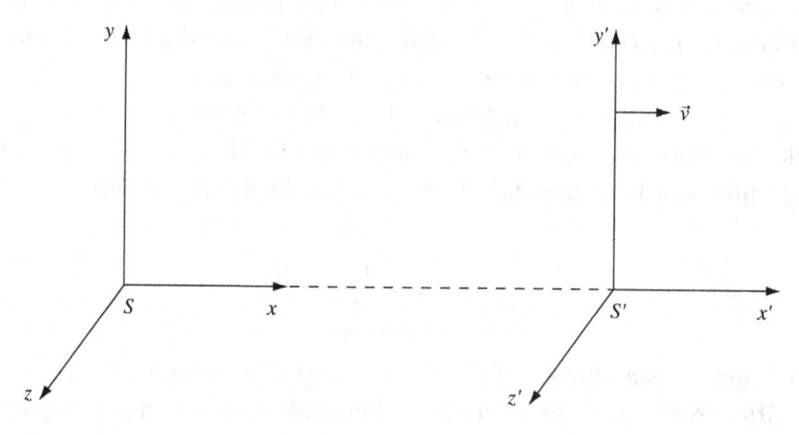

Figure 1.1 Two inertial frames S and S' in standard configuration (the origins of S and S' coincide at $t = t' = 0$).

[1] We will prove this in Chapter 5.

1.2 Newtonian geometry of space and time

Newtonian theory rests on the assumption that there exists an absolute time, which is the same for every observer, so that $t' = t$. Under this assumption $A = 1$ and $B = 0$, and we obtain the *Galilean transformation* relating the coordinates of an event P in the two Cartesian inertial frames S and S':

$$\boxed{\begin{aligned} t' &= t, \\ x' &= x - vt, \\ y' &= y, \\ z' &= z. \end{aligned}} \tag{1.2}$$

By symmetry, the expressions for the unprimed coordinates in terms of the primed ones have the same form but with v replaced by $-v$.

The first equation in (1.2) is clearly valid for any two inertial frames S and S' and shows that the time coordinate of an event P is the same in all inertial frames. The second equation leads to the 'common sense' notion of the addition of velocities. If a particle is moving in the x-direction at a speed u in S then its speed in S' is given by

$$u'_x = \frac{dx'}{dt'} = \frac{dx'}{dt} = \frac{dx}{dt} - v = u_x - v.$$

Differentiating again shows that the acceleration of a particle is the same in both S and S', i.e. $du'_x/dt' = du_x/dt$.

If we consider two events A and B that have coordinates (t_A, x_A, y_A, z_A) and (t_B, x_B, y_B, z_B) respectively, it is straightforward to show that both the time difference $\Delta t = t_B - t_A$ and the quantity

$$\boxed{\Delta r^2 = \Delta x^2 + \Delta y^2 + \Delta z^2}$$

are *separately invariant* under any Galilean transformation. This leads us to consider space and time as separate entities. Moreover, the invariance of Δr^2 suggests that it is a geometric property of space itself. Of course, we recognise Δr^2 as the square of the distance between the events in a three-dimensional Euclidean space. This defines the *geometry* of space and time in the Newtonian picture.

1.3 The spacetime geometry of special relativity

In special relativity, Einstein abandoned the postulate of an absolute time and replaced it by the postulate that *the speed of light c is the same in all inertial frames*.[2] By applying this new postulate, together with the principle of relativity, we may obtain the *Lorentz transformations* connecting the coordinates of an event P in two different Cartesian inertial frames S and S'.

Let us again consider S and S' to be in standard configuration (see Figure 1.1), and consider a photon emitted from the (coincident) origins of S and S' at $t = t' = 0$ and travelling in an arbitrary direction. Subsequently the space and time coordinates of the photon in each frame must satisfy

$$c^2t^2 - x^2 - y^2 - z^2 = c^2t'^2 - x'^2 - y'^2 - z'^2 = 0.$$

Substituting the relations (1.1) into this expression and solving for the constants A and B, we obtain

$$\boxed{\begin{aligned} ct' &= \gamma(ct - \beta x), \\ x' &= \gamma(x - \beta ct), \\ y' &= y, \\ z' &= z, \end{aligned}} \tag{1.3}$$

where $\beta = v/c$ and $\gamma = (1 - \beta^2)^{-1/2}$. This Lorentz transformation, also known as a *boost* in the x-direction, reduces to the Galilean transformation (1.2) when $\beta \ll 1$. Once again, symmetry demands that the unprimed coordinates are given in terms of the primed coordinates by an analogous transformation in which v is replaced by $-v$.

From the equations (1.3), we see that the time and space coordinates are in general mixed by a Lorentz transformation (note, in particular, the symmetry between ct and x). Moreover, as we shall see shortly, if we consider two events A and B with coordinates (t_A, x_A, y_A, z_A) and (t_B, x_B, y_B, z_B) in S, it is straightforward to show that the *interval* (squared)

$$\boxed{\Delta s^2 = c^2\Delta t^2 - \Delta x^2 - \Delta y^2 - \Delta z^2} \tag{1.4}$$

is *invariant* under any Lorentz transformation. As advocated by Minkowski, these observations lead us to consider space and time as united in a four-dimensional continuum called *spacetime*, whose *geometry* is characterised by (1.4). We note that the spacetime of special relativity is non-Euclidean, because of the minus signs in (1.4), and is often called the *pseudo-Euclidean* or *Minkowski geometry*. Nevertheless, for any fixed value of t the spatial part of the geometry remains Euclidean.

[2] The reasoning behind Einstein's proposal is discussed in Appendix 1A.

We have arrived at the familiar viewpoint (to a physicist!) where the physical world is modelled as a four-dimensional spacetime continuum that possesses the Minkowski geometry characterised by (1.4). Indeed, many ideas in special relativity are most simply explained by adopting a four-dimensional point of view.

1.4 Lorentz transformations as four-dimensional 'rotations'

Adopting a particular (Cartesian) inertial frame S corresponds to labelling events in the Minkowski spacetime with a given set of coordinates (t, x, y, z). If we choose instead to describe the world with respect to a different Cartesian inertial frame S' then this corresponds simply to relabelling events in the Minkowski spacetime with a new set of coordinates (t', x', y', z'); the primed and unprimed coordinates are related by the appropriate Lorentz transformation. Thus, describing physics in terms of different inertial frames is equivalent to performing a *coordinate transformation* on the Minkowski spacetime.

Consider, for example, the case where S' is related to S via a spatial rotation through an angle θ about the x-axis. In this case, we have

$$\begin{aligned} ct' &= ct, \\ x' &= x, \\ y' &= y\cos\theta - z\sin\theta, \\ z' &= y\sin\theta + z\cos\theta. \end{aligned}$$

Clearly the inverse transform is obtained on replacing θ by $-\theta$.

The close similarity between the 'boost' (1.3) and an ordinary spatial rotation can be highlighted by introducing the *rapidity* parameter

$$\psi = \tanh^{-1}\beta.$$

As β varies from zero to unity, ψ ranges from 0 to ∞. We also note that $\gamma = \cosh\psi$ and $\gamma\beta = \sinh\psi$. If two inertial frames S and S' are in standard configuration, we therefore have

$$\boxed{\begin{aligned} ct' &= ct\cosh\psi - x\sinh\psi, \\ x' &= -ct\sinh\psi + x\cosh\psi, \\ y' &= y, \\ z' &= z. \end{aligned}} \tag{1.5}$$

This has essentially the same form as a spatial rotation, but with hyperbolic functions replacing trigonometric ones. Once again the inverse transformation is obtained on replacing ψ by $-\psi$.

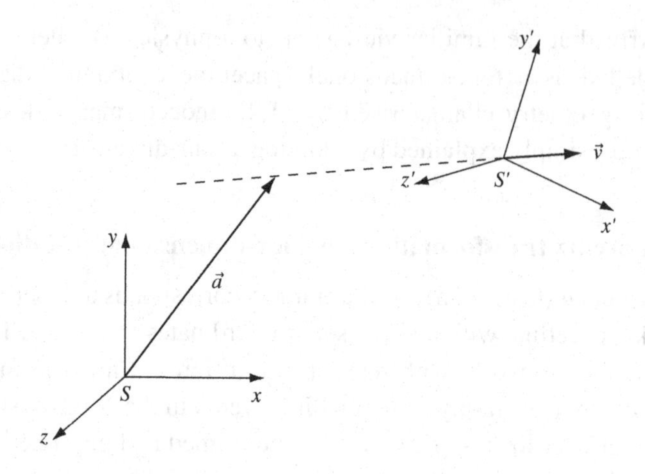

Figure 1.2 Two inertial frames S and S' in general configuration. The broken line shown the trajectory of the origin of S'.

In general, S' is moving with a constant velocity $\vec{v}$ with respect to S in an arbitrary direction[3] and the axes of S' are rotated with respect to those of S. Moreover, at $t = t' = 0$ the origins of S and S' need not be coincident and may be separated by a vector displacement $\vec{a}$, as measured in S (see Figure 1.2).[4] The corresponding transformation connecting the two inertial frames is most easily found by decomposing the transformation into a displacement, followed by a spatial rotation, followed by a boost, followed by a further spatial rotation. Physically, the displacement makes the origins of S and S' coincident at $t = t' = 0$, and the first rotation lines up the x-axis of S with the velocity $\vec{v}$ of S'. Then a boost in this direction with speed v transforms S into a frame that is at rest with respect to S'. A final rotation lines up the coordinate frame with that of S'. The displacement and spatial rotations introduce no new physics, and the only special-relativistic consideration concerns the boost. Thus, without loss of generality, we can restrict our attention to inertial frames S and S' that are in standard configuration, for which the Lorentz transformation is given by (1.3) or (1.5).

1.5 The interval and the lightcone

If we consider two events A and B having coordinates (t'_A, x'_A, y'_A, z'_A) and (t'_B, x'_B, y'_B, z'_B) in S', then, from (1.5), the interval between the events is given by

[3] Throughout this book, the notation $\vec{v}$ is used specifically to denote three-dimensional vectors, whereas $\boldsymbol{v}$ denotes a general vector, which is most often a 4-vector.

[4] If $\vec{a} = \vec{0}$ then the Lorentz transformation connecting the two inertial frames is called *homogeneous*, while if $\vec{a} \neq \vec{0}$ it is called *inhomogeneous*. Inhomogeneous transformations are often referred to as *Poincaré transformations*, in which case homogeneous transformations are referred to simply as Lorentz transformations.

$$\begin{aligned}\Delta s^2 &= c^2\Delta t'^2 - \Delta x'^2 - \Delta y'^2 - \Delta z'^2 \\ &= [(c\Delta t)\cosh\psi - (\Delta x)\sinh\psi]^2 - [-(c\Delta t)\sinh\psi + (\Delta x)\cosh\psi]^2 \\ &\quad - \Delta y^2 - \Delta z^2 \\ &= c^2\Delta t^2 - \Delta x^2 - \Delta y^2 - \Delta z^2.\end{aligned}$$

Thus the interval is invariant under the boost (1.5) and, from the above discussion, we may infer that Δs^2 is in fact invariant under *any* Poincaré transformation. This suggests that the interval is an underlying geometrical property of the spacetime itself, i.e. an invariant 'distance' between events in spacetime. It also follows that the sign of Δs^2 is defined invariantly, as follows:

> for $\Delta s^2 > 0$, the interval is timelike;
> for $\Delta s^2 = 0$, the interval is null or lightlike;
> for $\Delta s^2 < 0$, the interval is spacelike.

This embodies the standard lightcone structure shown in Figure 1.3. Events A and B are separated by a timelike interval, A and C by a lightlike (or null) interval and

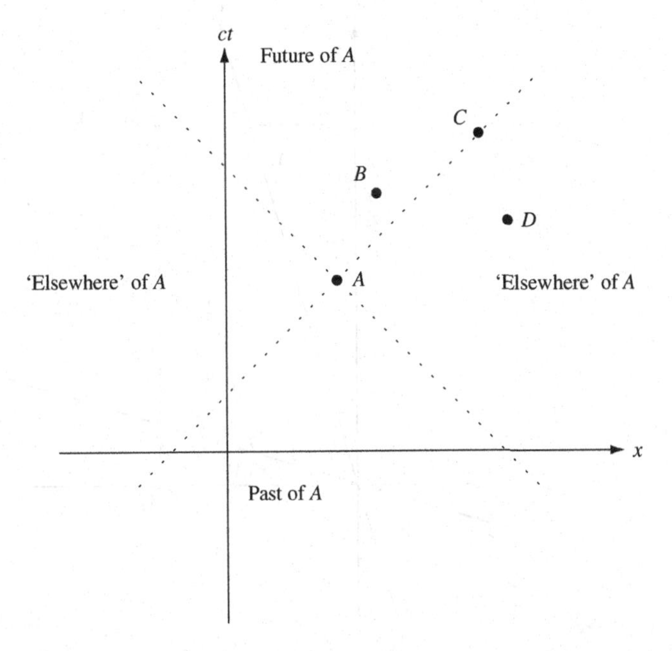

Figure 1.3 Spacetime diagram illustrating the lightcone of an event A (the y- and z- axes have been suppressed). Events A and B are separated by a timelike interval, A and C by a lightlike (or null) interval and A and D by a spacelike interval.

A and D by a spacelike interval. The geometrical distinction between timelike and spacelike intervals corresponds to a physical distinction: if the interval is timelike then we can find an inertial frame in which the events occur at the same spatial coordinates and if the interval is spacelike then we can find an inertial frame in which the events occur at the same time coordinate. This becomes obvious when we consider the spacetime diagram of a Lorentz transformation; we shall do this next.

1.6 Spacetime diagrams

Figure 1.3 is an example of a *spacetime diagram.* Such diagrams are extremely useful in illustrating directly many special-relativistic effects, in particular coordinate transformations on the Minkowski spacetime between different inertial frames. The spacetime diagram in Figure 1.4 shows the change of coordinates of an event A corresponding to the standard-configuration Lorentz transformation (1.5). The x'-axis is simply the line $t' = 0$ and the t'-axis is the line $x' = 0$. From the Lorentz-boost transformation (1.3) we see that the angle between the x- and x'- axes is the same as that between the t- and t'- axes and has the value

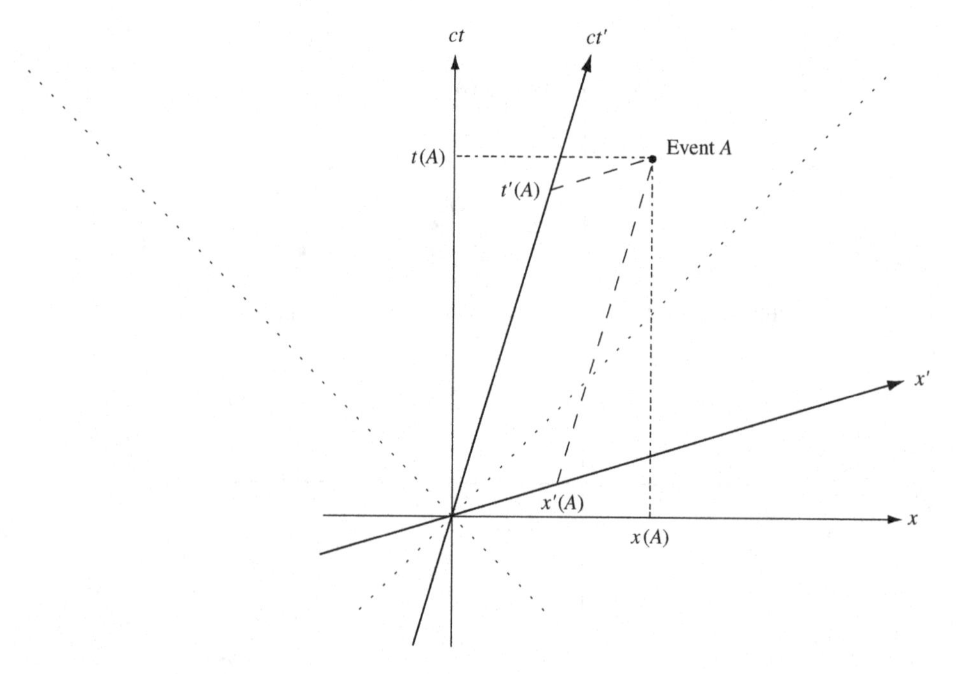

Figure 1.4 Spacetime diagram illustrating the coordinate transformation between two inertial frames S and S' in standard configuration (the y- and z-axes have been suppressed). The worldlines of the origins of S and S' are the axes ct and ct' respectively.

$\tan^{-1}(v/c)$. Moreover, we note that the t- and t'- axes are also the *worldlines* of the origins of S and S' respectively.

It is important to realise that the coordinates of the event A in the frame S' are *not* obtained by extending perpendiculars from A to the x'- and t'- axes. Since the x'-axis is simply the line $t' = 0$, it follows that lines of simultaneity in S' are parallel to the x'-axis. Similarly, lines of constant x' are parallel to the t'-axis. The same reasoning is equally valid for obtaining the coordinates of A in the frame S but, since the x- and t- axes are drawn as orthogonal in the diagram, this is equivalent simply to extending perpendiculars from A to the x- and t- axes in the more familiar manner.

The concept of simultaneity is simply illustrated using a spacetime diagram. For example, in Figure 1.5 we replot the events in Figure 1.3, together with the x'- and t'- axes corresponding to a Lorentz boost in standard configuration at some velocity v. We see that the events A and D, which are separated by a spacelike interval, lie on a line of constant t' and so are *simultaneous* in S'. Evidently, A and D are not simultaneous in S; D occurs at a later time than A. In a similar way, it is straightforward to find a standard-configuration Lorentz boost such that the events A and B, which are separated by a timelike interval, lie on a line of constant x' and hence occur at the same spatial location in S'.

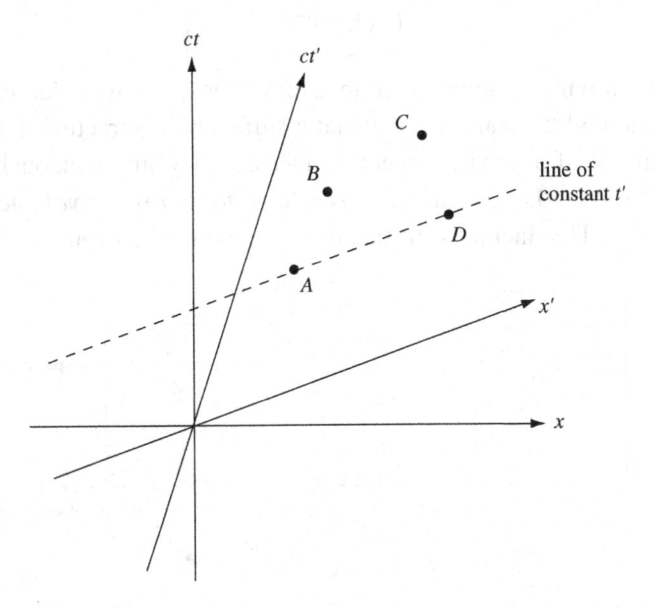

Figure 1.5 The events illustrated in figure 1.3 and a Lorentz boost such that A and D are simultaneous in S'.

1.7 Length contraction and time dilation

Two elementary (but profound) consequences of the Lorentz transformations are length contraction and time dilation. Both these effects are easily derived from (1.3).

Length contraction

Consider a rod of *proper length* ℓ_0 at rest in S' (see Figure 1.6); we have

$$\ell_0 = x'_B - x'_A.$$

We want to apply the Lorentz transformation formulae and so find what length an observer in frame S assigns to the rod. Applying the second formula in (1.3), we obtain

$$x'_A = \gamma\left(x_A - vt_A\right),$$
$$x'_B = \gamma\left(x_B - vt_B\right),$$

relating the coordinates of the ends of the rod in S' to the coordinates in S. The observer in S measures the length of the rod at a *fixed* time $t = t_A = t_B$ as

$$\ell = x_B - x_A = \frac{1}{\gamma}\left(x'_B - x'_A\right) = \frac{\ell_0}{\gamma}.$$

Hence in S the rod appears contracted to the length

$$\boxed{\ell = \ell_0\left(1 - v^2/c^2\right)^{1/2}.}$$

If a rod is moving relative to S in a direction perpendicular to its length, however, it is straightforward to show that it suffers no contraction. It thus follows that the volume V of a moving object, as measured by simultaneously noting the positions of the boundary points in S, is related to its *proper volume* V_0 by $V = V_0(1 - v^2/c^2)^{1/2}$. This fact must be taken into account when considering densities.

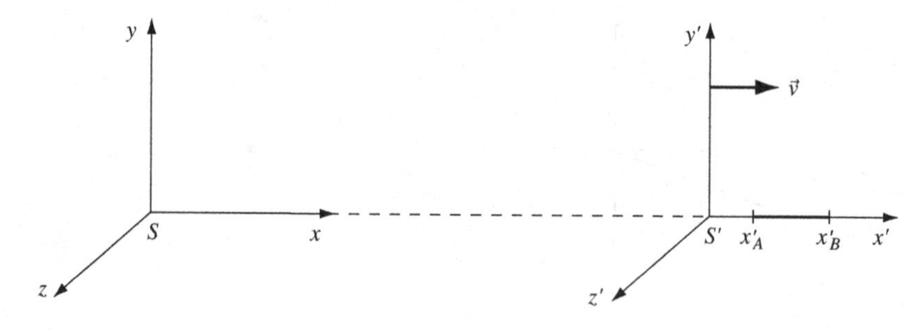

Figure 1.6 Two inertial frames S and S' in standard configuration. A rod of proper length ℓ_0 is at rest in S'.

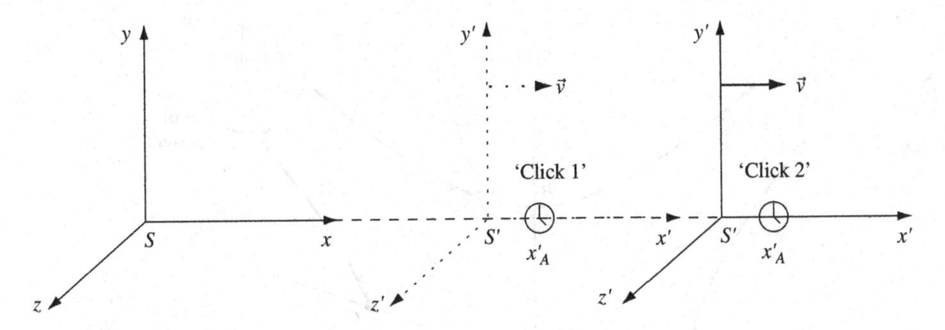

Figure 1.7 Two inertial frames S and S' in standard configuration. A clock is at rest in S'.

Time dilation

Suppose we have a clock at rest in S', in which two successive 'clicks' of the clock (events A and B) are separated by a time interval T_0 (see Figure 1.7). The times of the clicks as recorded in S are

$$t_A = \gamma\left(t'_A + vx'_A/c^2\right),$$
$$t_B = \gamma\left(t'_A + T_0 + vx'_B/c^2\right).$$

Since the clock is at rest in S' we have $x'_A = x'_B$, and so on subtracting we obtain

$$\boxed{T = t_B - t_A = \gamma T_0 = \frac{T_0}{(1 - v^2/c^2)^{1/2}}.}$$

Hence, the moving clock ticks more slowly by a factor of $(1 - v^2/c^2)^{1/2}$ (time dilation).

Note that an *ideal clock* is one that is unaffected by acceleration – external forces act identically on all parts of the clock (an example is a muon).

1.8 Invariant hyperbolae

Length contraction and time dilation are easily illustrated using spacetime diagrams. However, while Figure 1.4 illustrates the positions of the x'- and t'- axes corresponding to a standard Lorentz boost, we have not yet calibrated the length scales along them. To perform this calibration, we make use of the fact that the interval Δs^2 between two events is an invariant, and draw the *invariant hyperbolae*

$$c^2t^2 - x^2 = c^2t'^2 - x'^2 = \pm 1$$

on the spacetime diagram, as shown in Figure 1.8. Then, if we first take the negative sign, setting $ct = 0$, we obtain $x = \pm 1$. It follows that OA is a unit

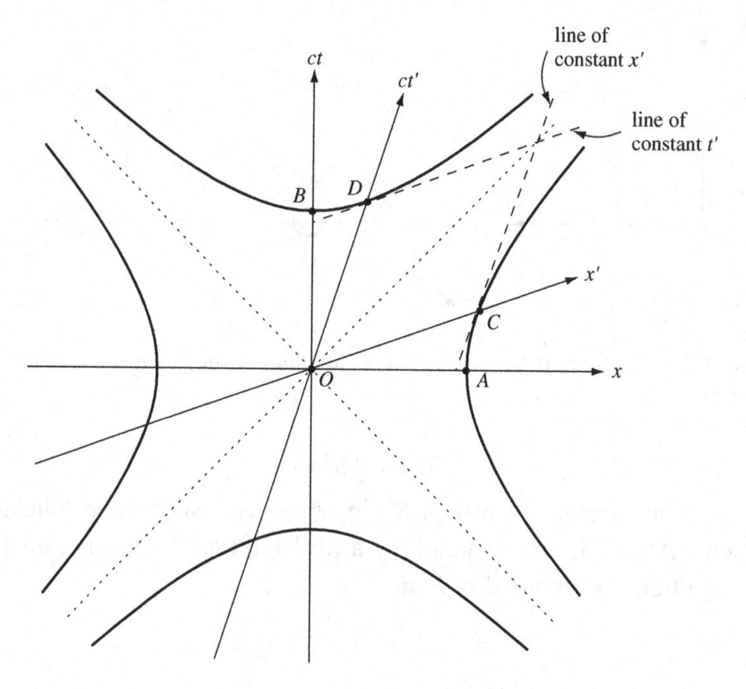

Figure 1.8 The invariant hyperbolae $c^2t^2 - x^2 = c^2t'^2 - x'^2 = \pm 1$.

distance along the x-axis. Now setting $ct' = 0$ we find that $x' = \pm 1$, so that OC is a unit distance along the x'-axis. Similarly, OB and OD are unit distances along the t- and t'- axes respectively. We also note that the tangents to the invariant hyperbolae at C and D are lines of constant x' and t' respectively.

The length contraction and time dilation effects can now be read off directly from the diagram. For example, the worldlines of the end-points of a unit rod OC in S', namely $x' = 0$ and $x' = 1$, cut the x-axis in less than unit distance. Similarly, worldlines $x = 0$ and $x = 1$ in S cut the x'-axis inside OC, illustrating the reciprocal nature of length contraction. Also, a clock at rest at the origin of S' will move along the t'-axis, reaching D with a reading of $t' = 1$. However, the event D has a t-coordinate that is greater than unity, thereby illustrating the time dilation effect.

1.9 The Minkowski spacetime line element

Let consider more closely the meaning of the interval between two events A and B in spacetime. Given that in a particular inertial frame S the coordinates of A and B are (t_A, x_A, y_A, z_A) and (t_B, x_B, y_B, z_B), we have so far taken the square of the interval between A and B to be

$$\Delta s^2 = c^2\Delta t^2 - \Delta x^2 - \Delta y^2 - \Delta z^2,$$

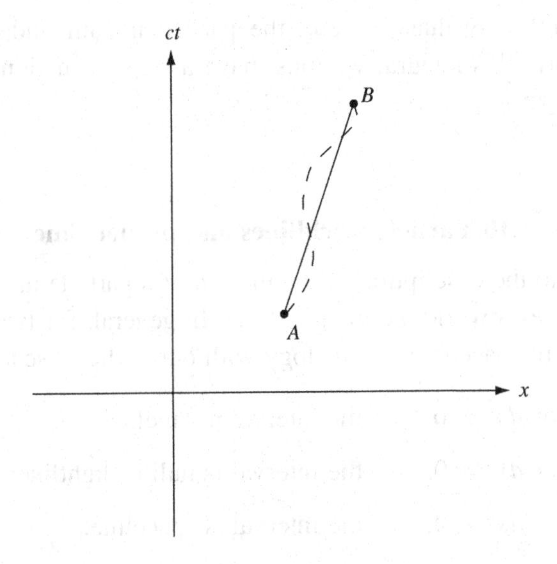

Figure 1.9 Two paths in spacetime connecting the events A and B.

where $\Delta t = t_B - t_A$ etc. This interval is invariant under Lorentz transformation and corresponds to the 'distance' in spacetime measured along the straight line in Figure 1.9 connecting A and B. This line may be interpreted as the worldline of a particle moving at constant velocity relative to S between events A and B. However, the question naturally arises of what interval is measured between A and B along some other path in spacetime, for example the 'wiggly' path shown in Figure 1.9.

To address this question, we must express the intrinsic geometry of the Minkowski spacetime in infinitesimal form. Clearly, if two infinitesimally separated events have coordinates (t, x, y, z) and $(t+dt, x+dx, y+dy, z+dz)$ in S then the square of the infinitesimal interval between them is given by[5]

$$\boxed{ds^2 = c^2dt^2 - dx^2 - dy^2 - dz^2,}$$

which is known as the *line element* of Minkowski spacetime, or the special-relativistic line element. From our earlier considerations, it is clear that ds^2 is invariant under any Lorentz transformation. The invariant interval between A and B along an arbitrary path in spacetime is then given by

$$\Delta s = \int_A^B ds,$$

[5] To avoid mathematical ambiguity, one should properly denote the squares of infinitesimal coordinate intervals by $(dt)^2$ etc., but this notation is not in common use in relativity textbooks. We will thus adopt the more usual form dt^2, but it should be remembered that this is *not* the differential of t^2.

where the integral is evaluated along the particular path under consideration. Clearly, to perform this integral we must have a set of equations describing the spacetime path.

1.10 Particle worldlines and proper time

Let us now turn to the description of the motion of a particle in spacetime terms. A particle describes a worldline in spacetime. In general, for two infinitesimally separated events in spacetime; by analogy with our earlier discussion we have:

for $ds^2 > 0$,	the interval is timelike;
for $ds^2 = 0$,	the interval is null or lightlike;
for $ds^2 < 0$,	the interval is spacelike.

However, relativistic mechanics prohibits the acceleration of a massive particle to speeds greater than or equal to c, which implies that its worldline must lie within the lightcone (Figure 1.3) at each event on it. In other words, the interval between any two infinitesimally separated events on the particle's worldline must be timelike (and future-pointing). For a massless particle such as a photon, any two events on its worldline are separated by a null interval. Figure 1.10 illustrates general worldlines for a massive particle and for a photon.

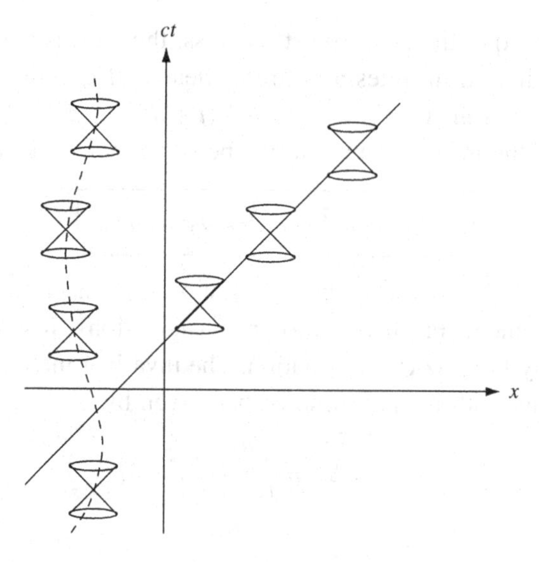

Figure 1.10 The worldlines of a photon (solid line) and a massive particle (broken line). The lightcones at seven events are shown.

A particle worldline may be described by giving x, y and z as functions of t in some inertial frame S. However, a more four-dimensional way of describing a worldline is to give the four coordinates (t, x, y, z) of the particle in S as functions of a parameter λ that varies monotonically along the worldline. Given the four functions $t(\lambda)$, $x(\lambda)$, $y(\lambda)$ and $z(\lambda)$, each value of λ determines a point along the curve. Any such parameter is possible, but a natural one to use for a massive particle is its *proper time*.

We define the proper time interval $d\tau$ between two infinitesimally separated events on the massive particle's worldline by

$$c^2 d\tau^2 = ds^2. \tag{1.6}$$

Thus, if the coordinate differences in S between the two events are dt, dx, dy, dz then we have

$$c^2 d\tau^2 = c^2 dt^2 - dx^2 - dy^2 - dz^2.$$

Hence the proper time interval between the events is given by

$$\boxed{d\tau = (1 - v^2/c^2)^{1/2} dt = dt/\gamma_v,}$$

where v is the speed of the particle with respect to S over this infinitesimal interval. If we integrate $d\tau$ between two points A and B on the worldline, we obtain the total elapsed proper time interval:

$$\Delta\tau = \int_A^B d\tau = \int_A^B \left[1 - \frac{v^2(t)}{c^2}\right]^{1/2} dt. \tag{1.7}$$

We see that if the particle is at rest in S then the proper time τ is just the coordinate time t measured by clocks at rest in S. If at any instant in the history of the particle we introduce an *instantaneous rest frame* S' such that the particle is momentarily at rest in S' then we see that *the proper time τ is simply the time recorded by a clock that moves along with the particle*. It is therefore an invariantly defined quantity, a fact that is clear from (1.6).

Thus the worldline of a massive particle can be described by giving the four coordinates (t, x, y, z) as functions of τ (see Figure 1.11). For example,

$$t = \tau(1 - v^2/c^2)^{-1/2},$$
$$x = v\tau(1 - v^2/c^2)^{-1/2},$$
$$y = z = 0$$

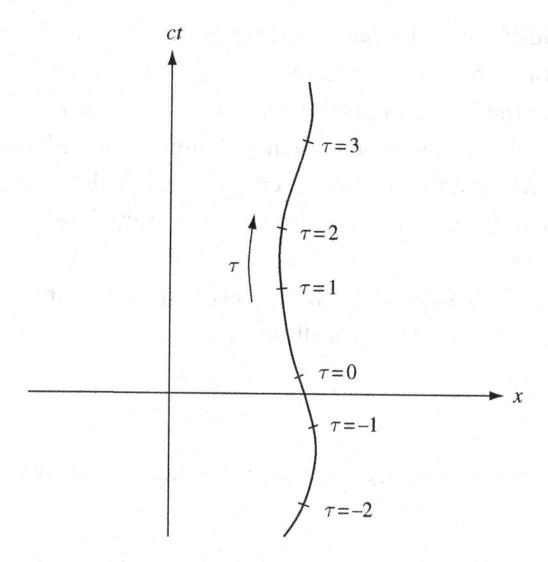

Figure 1.11 A path in the (t, x)-plane can be specified by giving one coordinate in terms of the other, for example $x = x(t)$, or alternatively by giving both coordinates as functions of a parameter λ along the curve: $t = t(\lambda)$, $x = x(\lambda)$. For massive particles the natural parameter to use is the proper time τ.

is the worldline of a particle, moving at constant speed v along the x-axis of S, which passes through the origin of S at $t = 0$.

1.11 The Doppler effect

A useful illustration of particle worldlines and the concept of proper time is provided by deriving the Doppler effect in a transparently four-dimensional manner. Let us consider an observer $\mathcal{O}$ at rest in some inertial frame S, and a radiation-emitting source $\mathcal{E}$ moving along the positive x-axis of S at a uniform speed v. Suppose that the source emits the first wavecrest of a photon at an event A, with coordinates (t_e, x_e) in S, and the next wavecrest at an event B with coordinates $(t_e + \Delta t_e, x_e + \Delta x_e)$. Let us assume that these two wavecrests reach the observer at the events C and D coordinates (t_o, x_o) and $(t_o + \Delta t_o, x_o)$ respectively. This situation is illustrated in Figure 1.12. From (1.7), the proper time interval experienced by $\mathcal{E}$ between the events A and B is

$$\Delta\tau_{AB} = \left(1 - v^2/c^2\right)^{1/2} \Delta t_e, \tag{1.8}$$

and the proper time interval experienced by $\mathcal{O}$ between the events C and D is

$$\Delta\tau_{CD} = \Delta t_o. \tag{1.9}$$

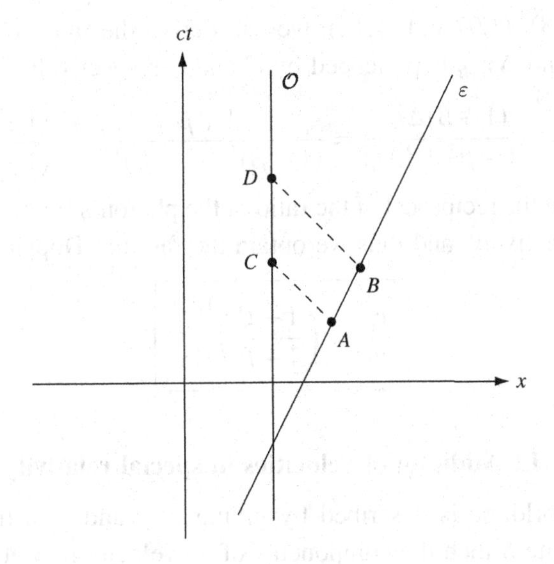

Figure 1.12 Spacetime diagram of the Doppler effect.

Along each of the worldlines representing the photon wavecrests,

$$ds^2 = c^2\,dt^2 - dx^2 - dy^2 - dz^2 = 0.$$

Thus, since we are assuming that $dy = dz = 0$, along the worldline connecting the events A and C we have

$$\int_{t_e}^{t_o} c\,dt = -\int_{x_e}^{x_o} dx, \tag{1.10}$$

where the minus sign on the right-hand side arises because the photon is travelling in the negative x-direction. From (1.10), we obtain the (obvious) result $c(t_o - t_e) = -(x_o - x_e)$. Similarly, along the worldline connecting B and D we have

$$\int_{t_e+\Delta t_e}^{t_o+\Delta t_o} c\,dt = -\int_{x_e+\Delta x_e}^{x_o} dx.$$

Rewriting the integrals on each side, we obtain

$$\left(\int_{t_e}^{t_o} + \int_{t_o}^{t_o+\Delta t_o} - \int_{t_e}^{t_e+\Delta t_e}\right) c\,dt = -\left(\int_{x_e}^{x_o} - \int_{x_e}^{x_e+\Delta x_e}\right) dx,$$

where the first integrals on each side of the equation cancel by virtue of (1.10). Thus we find that $c\Delta t_o - c\Delta t_e = \Delta x_e$, from which we obtain

$$\Delta t_o = \left(1 + \frac{1}{c}\frac{\Delta x_e}{\Delta t_e}\right)\Delta t_e = \left(1 + \frac{v}{c}\right)\Delta t_e. \tag{1.11}$$

Hence, using (1.8), (1.9) and (1.11), we can derive the ratio of the proper time intervals $\Delta\tau_{CD}$ and $\Delta\tau_{AB}$ experienced by $\mathcal{O}$ and $\mathcal{E}$ respectively:

$$\frac{\Delta\tau_{CD}}{\Delta\tau_{AB}} = \frac{(1+\beta)\Delta t_{\mathrm{e}}}{(1-\beta^2)^{1/2}\Delta t_{\mathrm{e}}} = \frac{1+\beta}{(1-\beta)^{1/2}(1+\beta)^{1/2}} = \frac{(1+\beta)^{1/2}}{(1-\beta)^{1/2}}.$$

This ratio must be the reciprocal of the ratio of the photon's frequency as measured by $\mathcal{E}$ and $\mathcal{O}$ respectively, and thus we obtain the familiar Doppler-effect formula

$$\boxed{\frac{\nu_{\mathcal{O}}}{\nu_{\mathcal{E}}} = \left(\frac{1-\beta}{1+\beta}\right)^{1/2}.} \tag{1.12}$$

1.12 Addition of velocities in special relativity

If a particle's worldline is described by giving x, y and z as functions of t in some inertial frame S then the components of its velocity in S at any point are

$$u_x = \frac{dx}{dt}, \qquad u_y = \frac{dy}{dt}, \qquad u_z = \frac{dz}{dt}.$$

The components of its velocity in some other inertial frame S' are usually obtained by taking differentials of the Lorentz transformation. For inertial frames S and S' related by a boost v in standard configuration, we have from (1.3)

$$dt' = \gamma_v(dt - v\,dx/c^2), \quad dx' = \gamma_v(dx - v\,dt), \quad dy' = dy, \quad dz' = dz,$$

where we have made explicit the dependence of γ on v. We immediately obtain

$$\boxed{\begin{aligned} u'_x &= \frac{dx'}{dt'} = \frac{u_x - v}{1 - u_x v/c^2}, \\ u'_y &= \frac{dy'}{dt'} = \frac{u_y}{\gamma_v(1 - u_x v/c^2)}, \\ u'_z &= \frac{dz'}{dt'} = \frac{u_z}{\gamma_v(1 - u_x v/c^2)}. \end{aligned}} \tag{1.13}$$

These replace the 'common sense' addition-of-velocities formulae of Newtonian mechanics. The inverse transformations are obtained by replacing v by $-v$.

The special-relativistic addition of velocities along the same direction is elegantly expressed using the rapidity parameter (Section 1.4). For example, consider three inertial frames S, S' and S''. Suppose that S' is related to S by a boost of speed v in the x-direction and that S'' is related to S' by a boost of speed u' in the x'-direction. Using (1.5), we quickly find that

$$ct'' = ct\cosh(\psi_v + \psi_{u'}) - x\sinh(\psi_v + \psi_{u'}),$$

$$x'' = -ct\sinh(\psi_v + \psi_{u'}) + x\cosh(\psi_v + \psi_{u'}),$$
$$y'' = y,$$
$$z'' = z,$$

where $\tanh\psi_v = v/c$ and $\tanh\psi_{u'} = u'/c$. This shows that S'' is connected to S by a boost in the x-direction with speed u, where $u/c = \tanh(\psi_v + \psi_{u'})$. Thus *we simply add the rapidities* (in a similar way to adding the angles of two spatial rotations about the same axis). This gives

$$u = c\tanh(\psi_v + \psi_{u'}) = c\,\frac{\tanh\psi_v + \tanh\psi_{u'}}{1 + \tanh\psi_v \tanh\psi_{u'}} = \frac{u' + v}{1 + u'v/c^2},$$

which is the special-relativistic formula for the addition of velocities in the same direction.

1.13 Acceleration in special relativity

The components of the acceleration of a particle in S are defined as

$$a_x = \frac{du_x}{dt}, \qquad a_y = \frac{du_y}{dt}, \qquad a_z = \frac{du_z}{dt},$$

and the corresponding quantities in S' are obtained from the differential forms of the expressions (1.13). For example,

$$du'_x = \frac{du_x}{\gamma_v^2(1 - u_x v/c^2)^2}.$$

Also, from the Lorentz transformation (1.3) we find that

$$dt' = \gamma_v(dt - v\,dx/c^2) = \gamma_v(1 - u_x v/c^2)\,dt.$$

So, for example, we have

$$a'_x = \frac{du'_x}{dt'} = \frac{1}{\gamma_v^3(1 - u_x v/c^2)^3}\,a_x. \tag{1.14}$$

Similarly, we obtain

$$a'_y = \frac{du'_y}{dt'} = \frac{1}{\gamma_v^2(1 - u_x v/c^2)^2}\,a_y + \frac{u_y v}{c^2\gamma_v^2(1 - u_x v/c^2)^3}\,a_x$$
$$a'_z = \frac{du'_z}{dt'} = \frac{1}{\gamma_v^2(1 - u_x v/c^2)^2}\,a_z + \frac{u_z v}{c^2\gamma_v^2(1 - u_x v/c^2)^3}\,a_x$$

We see from these transformation formulae that acceleration is not invariant in special relativity, unlike in Newtonian mechanics, as discussed in Section 1.2. However, it is clear that acceleration is an *absolute* quantity, that is, all observers agree upon whether a body is accelerating. If the acceleration is zero in one inertial frame, it is necessarily zero in any other frame.

Let us investigate the worldline of an accelerated particle. To make our illustration concrete, we consider a spaceship moving at a variable speed $u(t)$ relative to some inertial frame S and suppose that an observer B in the spaceship makes a continuous record of his accelerometer reading $f(\tau)$ as a function of his own proper time τ.

We begin by introducing an *instantaneous rest frame* (IRF) S', which, at each instant, is an inertial frame moving at the same speed v as the spaceship, i.e. $v = u$. Thus, at any instant, the velocity of the spaceship in the IRF S' is zero, i.e. $u' = 0$. Moreover, from the above discussion of proper time, it should be clear that at any instant an interval of proper time is equal to an interval of coordinate time in the IRF, i.e. $\delta\tau = \delta t'$. An accelerometer measures the rate of change of velocity, so that, during a small interval of proper time $\delta\tau$, B will record that his velocity has changed by an amount $f(\tau)\delta\tau$. Therefore, at any instant, in the IRF S' we have

$$\frac{du'}{dt'} = \frac{du'}{d\tau} = f(\tau).$$

From (1.14), we thus obtain

$$\frac{du}{dt} = \left(1 - \frac{u^2}{c^2}\right)^{3/2} f(\tau).$$

However, since $d\tau = (1 - u^2/c^2)^{1/2}\, dt$, we find that

$$\frac{du}{d\tau} = \left(1 - \frac{u^2}{c^2}\right) f(\tau),$$

which integrates easily to give

$$u(\tau) = c \tanh \psi(\tau),$$

where $c\psi(\tau) = \int_0^\tau f(\tau')\, d\tau'$ and we have taken $u(\tau = 0)$ to be zero. Thus we have an expression for the velocity of the spaceship in S as a function of B's proper time.

To parameterise the worldline of the spaceship in S, we note that

$$\frac{dt}{d\tau} = \left(1 - \frac{u^2}{c^2}\right)^{-1/2} = \cosh \psi(\tau),$$

$$\frac{dx}{d\tau} = u\left(1 - \frac{u^2}{c^2}\right)^{-1/2} = c \sinh \psi(\tau). \qquad (1.15)$$

Integration of these equations with respect to τ gives the functions $t(\tau)$ and $x(\tau)$.

1.14 Event horizons in special relativity

The presence of acceleration can produce surprising effects. Consider for simplicity the case of *uniform* acceleration. By this we do *not* mean that $du/dt = \text{constant}$, since this is inappropiate in special relativity because it would imply that $u \to \infty$ as $t \to \infty$, which is not permitted. Instead, uniform acceleration in special relativity means that the accelerometer reading $f(\tau)$ is constant. A spaceship whose engine is set at a constant emission rate would be uniformly accelerated in this sense.

Thus, if $f = \text{constant}$, we have $\psi = f\tau/c$. The equations (1.15) are then easily integrated to give

$$t = t_0 + \frac{c}{f}\sinh\frac{f\tau}{c},$$

$$x = x_0 + \frac{c^2}{f}\left(\cosh\frac{f\tau}{c} - 1\right),$$

where t_0 and x_0 are constants of integration. Setting $t_0 = x_0 = 0$ gives the path shown in Figure 1.13. The worldline takes the form of a hyperbola.

Imagine that an observer B has the resources to maintain an acceleration f indefinitely. Then there will be events that B will never be able to observe. The events in question lie on the future side of the asymptote to B's hyperbola; this asymptote (which is a null line) is the *event horizon* of B. Objects whose worldlines cross this horizon will disappear from B's view and will seem to take

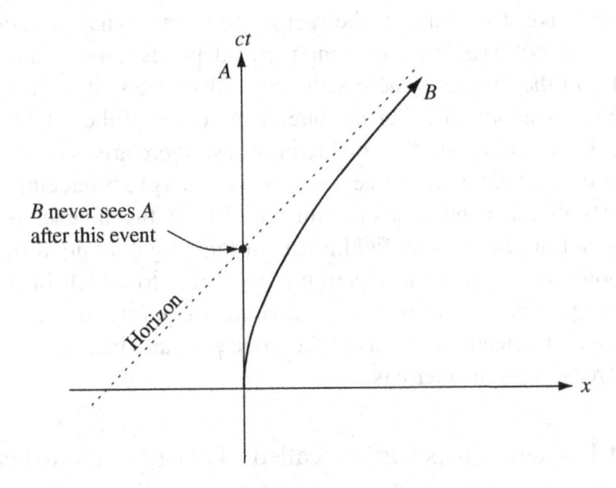

Figure 1.13 The worldline of a uniformly accelerated particle B starting from rest from the origin of S. If an observer A remains at $x = 0$, then the worldline of A is simply the t-axis. No message sent by A after $t = c/f$ will ever reach B.

for ever to do so. Nevertheless, the objects themselves cross the horizon in a finite proper time and still have an infinite lifetime ahead of them.

Appendix 1A: Einstein's route to special relativity

Most books on special relativity begin with some sort of description of the Michelson–Morley experiment and then introduce the Lorentz transformation. In fact, Einstein claimed that he was not influenced by this experiment. This is disputed by various historians of science and biographers of Einstein. One might think that these scholars are on strong ground, especially given that the experiment is referred to (albeit obliquely) in Einstein's papers. However, it may be worth taking Einstein's claim at face value.

Remember that Einstein was a *theorist* – one of the greatest theorists who has ever lived – and he had a *theorist's* way of looking at physics. A good theorist develops an intuition about how Nature works, which helps in the formulation of physical laws. For example, possible symmetries and conserved quantities are considered. We can get a strong clue about Einstein's thinking from the *title* of his famous 1905 paper on special relativity. The first paragraph is reproduced below.

ON THE ELECTRODYNAMICS OF MOVING BODIES
BY A. EINSTEIN

It is known that Maxwell's electrodynamics – as usually understood at the present time – when applied to moving bodies, leads to asymmetries which do not appear to be inherent in the phenomena. Take, for example, the reciprocal electrodynamic action of a magnet and a conductor. The observable phenomenon here depends only on the relative motion of the conductor and the magnet, whereas the customary view draws a sharp distinction between the two cases in which either the one or the other of these bodies is in motion. For if the magnet is in motion and the conductor at rest, there arises in the neighbourhood of the magnet an electric field with a certain definite energy, producing a current at the places where parts of the conductor are situated. But if the magnet is stationary and the conductor in motion, no electric field arises in the neighbourhood of the magnet. In the conductor, however, we find an electromotive force, to which in itself there is no corresponding energy, but which gives rise – assuming equality of relative motion in the two cases discussed – to electric currents of the same path and intensity as those produced by the electric forces in the former case.

You see that Einstein's paper is not called 'Transformations between inertial frames', or 'A theory in which the speed of light is assumed to be a universal constant'. Electrodynamics is at the heart of Einstein's thinking; Einstein realized that Maxwell's equations of electromagnetism *required* special relativity.

Maxwell's equations are

$$\vec{\nabla}\cdot\vec{D}=\rho, \qquad \vec{\nabla}\cdot\vec{B}=0,$$

$$\vec{\nabla}\times\vec{E}=-\frac{\partial\vec{B}}{\partial t}, \qquad \vec{\nabla}\times\vec{H}=\vec{j}+\frac{\partial\vec{D}}{\partial t},$$

where $\vec{D}=\epsilon_0\vec{E}+\vec{P}$ and $\vec{B}=\mu_0(\vec{H}+\vec{M})$, $\vec{P}$ and $\vec{M}$ being respectively the polarisation and the magnetisation of the medium in which the fields are present. In free space we can set $\vec{j}=\vec{0}$ and $\rho=0$, and we then get the more obviously symmetrical equations

$$\vec{\nabla}\cdot\vec{E}=0, \qquad \vec{\nabla}\cdot\vec{B}=0,$$

$$\vec{\nabla}\times\vec{E}=-\frac{\partial\vec{B}}{\partial t}, \qquad \vec{\nabla}\times\vec{B}=\mu_0\epsilon_0\frac{\partial\vec{E}}{\partial t}.$$

Taking the curl of the equation for $\vec{\nabla}\times\vec{E}$, applying the relation

$$\vec{\nabla}\times(\vec{\nabla}\times\vec{E})=\vec{\nabla}(\vec{\nabla}\cdot\vec{E})-\nabla^2\vec{E}$$

and performing a similar operation for $\vec{B}$ in the equation for $\vec{\nabla}\times\vec{B}$, we derive the equations for electromagnetic waves:

$$\nabla^2\vec{E}=\mu_0\epsilon_0\frac{\partial^2\vec{E}}{\partial t^2}, \qquad \nabla^2\vec{B}=\mu_0\epsilon_0\frac{\partial^2\vec{B}}{\partial t^2}.$$

These both have the form of a wave equation with a propagation speed $c=1/\sqrt{\mu_0\epsilon_0}$. Now, the constants μ_0 and ϵ_0 are properties of the 'vacuum':

μ_0, the permeability of a vacuum, equals $4\pi\times10^{-7}\,\mathrm{Hm}^{-1}$,

ϵ_0, the permittivity of a vacuum, equals $8.85\times10^{-12}\,\mathrm{Fm}^{-1}$.

This relation between the constants ϵ_0 and μ_0 and the speed of light was one of the most startling consequences of Maxwell's theory. But what do we mean by a 'vacuum'? Does it define an absolute frame of rest? If we deny the existence of an absolute frame of rest then how do we formulate a theory of electromagnetism? How do Maxwell's equations appear in frames moving with respect to each other? Do we need to change the value of c? If we do, what will happen to the values of ϵ_0 and μ_0?

Einstein solves all of these problems at a stroke by saying that Maxwell's equations take the same mathematical form in all inertial frames. The speed of light c is thus the same in all inertial frames. The theory of special relativity (including amazing conclusions such as $E=mc^2$) follows from a generalisation of this simple and theoretically compelling assumption. *Maxwell's equations therefore require special relativity.* You see that for a master theorist like Einstein, the

Michelson–Morley experiment might well have been a side issue. Einstein could 'see' special relativity lurking in Maxwell's equations.

Exercises

1.1 For two inertial frames S and S' in standard configuration, show that the coordinates of any given event in each frame are related by the Lorentz tranformations (1.3).

1.2 Two events A and B have coordinates (t_A, x_A, y_A, z_A) and (t_B, x_B, y_B, z_B) respectively. Show that both the time difference $\Delta t = t_B - t_A$ and the quantity

$$\Delta r^2 = \Delta x^2 + \Delta y^2 + \Delta z^2$$

are separately invariant under any Galilean transformation, whereas the quantity

$$\Delta s^2 = c^2 \Delta t^2 - \Delta x^2 - \Delta y^2 - \Delta z^2$$

is invariant under any Lorentz transformation.

1.3 In a given inertial frame two particles are shot out simultaneously from a given point, with equal speeds v in orthogonal directions. What is the speed of each particle relative to the other?

1.4 An inertial frame S' is related to S by a boost of speed v in the x-direction, and S'' is related to S' by a boost of speed u' in the x'-direction. Show that S'' is related to S by a boost in the x-direction with speed u, where

$$u = c \tanh(\psi_v + \psi_{u'});$$

$\tanh \psi_v = v/c$ and $\tanh \psi_{u'} = u'/c$.

1.5 An inertial frame S' is related to S by a boost $\vec{v}$ whose components in S are (v_x, v_y, v_z). Show that the coordinates (ct', x', y', z') and (ct, x, y, z) of an event are related by

$$\begin{pmatrix} ct' \\ x' \\ y' \\ z' \end{pmatrix} = \begin{pmatrix} \gamma & -\gamma\beta_x & -\gamma\beta_y & -\gamma\beta_z \\ -\gamma\beta_x & 1+\alpha\beta_x^2 & \alpha\beta_x\beta_y & \alpha\beta_x\beta_z \\ -\gamma\beta_y & \alpha\beta_y\beta_x & 1+\alpha\beta_y^2 & \alpha\beta_y\beta_z \\ -\gamma\beta_y & \alpha\beta_z\beta_x & \alpha\beta_z\beta_y & 1+\alpha\beta_z^2 \end{pmatrix} \begin{pmatrix} ct \\ x \\ y \\ z \end{pmatrix},$$

where $\vec{\beta} = \vec{v}/c$, $\gamma = (1 - |\vec{\beta}|^2)^{-1/2}$ and $\alpha = (\gamma - 1)/|\vec{\beta}|^2$. *Hint: The transformation must take the same form if both S and S′ undergo the same spatial rotation.*

1.6 An inertial frame S' is related to S by a boost of speed u in the positive x-direction. Similarly, S'' is related to S' by a boost of speed v in the y'-direction. Find the transformation relating the coordinates (ct, x, y, z) and (ct'', x'', y'', z'') and hence describe how S and S'' are physically related.

1.7 The frames S and S' are in standard configuration. A straight rod rotates at a uniform angular velocity ω' about its centre, which is fixed at the origin of S'. If the rod lies along the x'-axis at $t' = 0$, obtain an equation for the shape of the rod in S at $t = 0$.

1.8 Two events A and B have coordinates (t_A, x_A, y_A, z_A) and (t_B, x_B, y_B, z_B) respectively in some inertial frame S and are separated by a spacelike interval. Obtain an expression for the boost $\vec{v}$ required to transform to a new inertial frame S' in which the events A and B occur simultaneously.

1.9 Derive the Doppler effect (1.12) directly, using the Lorentz transformation formulae (1.3).

1.10 Two observers are moving along trajectories parallel to the y-axis in some inertial frame. Observer A emits a photon with frequency ν_A that travels in the positive x-direction and is received by observer B with frequency ν_B. Show that the Doppler shift ν_B/ν_A in the photon frequency is the same whether A and B travel in the same direction or opposite directions.

1.11 Astronauts in a spaceship travelling in a straight line past the Earth at speed $v = c/2$ wish to tune into Radio 4 on 198 kHz. To what frequency should they tune at the instant when the ship is closest to Earth?

1.12 Draw a spacetime diagram illustrating the coordinate transformation corresponding to two inertial frames S and S' in standard configuration (i.e. where S' moves at a speed v along the positive x-direction and the two frames coincide at $t = t' = 0$). Show that the angle between the x- and x'- axes is the same as that between the t- and t'- axes and has the value $\tan^{-1}(v/c)$.

1.13 Consider an event P separated by a timelike interval from the origin O of your diagram in Exercise 1.12. Show that the tangent to the invariant hyperbola passing through P is a line of simultaneity in the inertial frame whose time axis joins P to the origin. Hence, from your spacetime diagram, derive the formulae for length contraction and time dilation.

1.14 Alex and Bob are twins working on a space station located at a fixed position in deep space. Alex undertakes an extended return spaceflight to a distant star, while Bob stays on the station. Show that, on his return to the station, the proper time interval experienced by Alex must be less than that experienced by Bob, hence Bob is now the elder. How does Alex explain this age difference?

1.15 A spaceship travels at a variable speed $u(t)$ in some inertial frame S. An observer on the spaceship measures its acceleration to be $f(\tau)$, where τ is the proper time. If at $\tau = 0$ the spaceship has a speed u_0 in S show that

$$\frac{u(\tau) - u_0}{1 - u(\tau)u_0/c^2} = c \tanh \psi(\tau),$$

where $c\psi(\tau) = \int_0^\tau f(\tau')\, d\tau'$. Show that the velocity of the spaceship can never reach c.

1.16 If the spaceship in Exercise 1.15 left base at time $t = \tau = 0$ and travelled forever in a straight line with constant acceleration f, show that no signal sent by base later than time $t = c/f$ can ever reach the spaceship. By sketching an appropriate spacetime diagram show that light signals sent from the base appear increasingly redshifted to an observer on the spaceship. If the acceleration of the spaceship is g (for the comfort of its occupants), how long by the spaceship clock does it take to reach a star 10 light years from the base?

2
Manifolds and coordinates

Our discussion of special relativity has led us to model the physical world as a four-dimensional continuum, called spacetime, with a Minkowski geometry. This is an example of a *manifold*. As we shall see, the more complicated spacetime geometries of general relativity are also examples of manifolds. It is therefore worthwhile discussing manifolds in general. In the following we consider general properties of manifolds commonly encountered in physics, and we concentrate in particular on Riemannian manifolds, which will be central to our discussion of general relativity.

2.1 The concept of a manifold

In general, a manifold is any set that can be continuously parameterised. The number of independent parameters required to specify any point in the set uniquely is the *dimension* of the manifold, and the parameters themselves are the *coordinates* of the manifold. An abstract example is the set of all rigid rotations of Cartesian coordinate systems in three-dimensional Euclidean space, which can be parameterised by the Euler angles. So the set of rotations is a three-dimensional manifold: each point is a particular rotation, and the coordinates of the point are the three Euler angles. Similarly, the phase space of a particle in classical mechanics can be parameterised by three position coordinates (q_1, q_2, q_3) and three momentum coordinates (p_1, p_2, p_3), and thus the set of points in this phase space forms a six-dimensional manifold. In fact, one can regard 'manifold' as just a fancy word for 'space' in the general mathematical sense.

In its most primitive form a general manifold is simply an amorphous collection of points. Most manifolds used in physics, however, are 'differential manifolds', which are continuous and differentiable in the following way. A manifold is continuous if, in the neighbourhood of every point P, there are other points whose coordinates differ infinitesimally from those of P. A manifold is differentiable if it is possible to define a scalar field at each point of the manifold that can be differentiated everywhere. Both our examples above are differential manifolds.

The association of points with the values of their parameters can be thought of as a mapping of the points of a manifold into points of the Euclidean space of the same dimension. This means that 'locally' a manifold looks like the corresponding Euclidean space: it is 'smooth' and has a certain number of dimensions.

2.2 Coordinates

An N-dimensional manifold $\mathcal{M}$ of points is one for which N independent real coordinates $(x^1, x^2, \ldots, x^N)$ are required to specify any point completely.[1] These N coordinates are entirely general and are denoted collectively by x^a, where it is understood that $a = 1, 2, \ldots, N$.

As a technical point, we should mention that in general it may not be possible to cover the whole manifold with only one *non-degenerate* coordinate system, namely, one which ascribes a *unique* set of N coordinate values to each point, so that the correspondence between points and sets of coordinate values (labels) is one-to-one. Let us consider, for example, the points that constitute a plane. These points clearly form a two-dimensional manifold (called R^2). An example of a degenerate coordinate system on this manifold is the polar coordinates (r, ϕ) in the plane, which have a degeneracy at the origin because ϕ is indeterminate there. For this manifold, we could avoid the degeneracy at the origin by using, for example, Cartesian coordinates. For a general manifold, however, we might have no choice in the matter and might have to work with coordinate systems that cover only a portion of the manifold, called *coordinate patches*. For example, the set of points making up the surface of a sphere forms a two-dimensional manifold (called S^2). This manifold is usually 'parameterised' by the coordinates θ and ϕ, but ϕ is degenerate at the poles. In this case, however, it can be shown that there is no coordinate system that covers the whole of S^2 without degeneracy; the smallest number of patches needed is two. In general, a set of coordinate patches that covers the whole manifold is called an *atlas*.

Thus, in general, we do not require the whole of a manifold $\mathcal{M}$ to be covered by a single coordinate system. Instead, we may have a collection of coordinate systems, each covering some part of $\mathcal{M}$, and all these are on an equal footing. We do not regard any one coordinate system as in some way preferred.

2.3 Curves and surfaces

Given a manifold, we shall be concerned with points in it and with subsets of points that define *curves* and *surfaces*. We shall frequently define these curves

[1] The reason why the coordinates are written with superscripts rather than subscripts will become clear later.

and surfaces *parametrically*. Thus, since a curve has one degree of freedom, it depends on one parameter and so we define a curve in the manifold by the parametric equations

$$\boxed{x^a = x^a(u) \qquad (a = 1, 2, \ldots, N),}$$

where u is some parameter and $x^1(u), x^2(u), \ldots, x^N(u)$ denote N functions of u.

Similarly, since a *submanifold* or *surface* of M dimensions ($M < N$) has M degrees of freedom, it depends on M parameters and is given by the N parametric equations

$$\boxed{x^a = x^a(u^1, u^2, \ldots, u^M) \qquad (a = 1, 2, \ldots, N).} \tag{2.1}$$

If, in particular, $M = N - 1$ then the submanifold is called a *hypersurface*. In this case, the $N - 1$ parameters can be eliminated from these N equations to give one equation relating the coordinates, i.e.

$$f(x^1, x^2, \ldots, x^N) = 0.$$

From a different but equivalent point of view, a point in a manifold is characterised by N coordinates. If the point is restricted to lie in a particular hypersurface, i.e. an $(N - 1)$-dimensional subspace, then the point's coordinates must satisfy *one* constraint equation, namely

$$f(x^1, x^2, \ldots, x^N) = 0.$$

Similarly, points in an M-dimensional subspace ($M < N$) must satisfy $N - M$ constraints

$$\begin{aligned} f_1(x^1, x^2, \ldots, x^N) &= 0, \\ f_2(x^1, x^2, \ldots, x^N) &= 0, \\ &\vdots \\ f_{N-M}(x^1, x^2, \ldots, x^N) &= 0, \end{aligned}$$

which is an alternative to the parametric representation (2.1).

2.4 Coordinate transformations

To locate a point in a manifold we use a system of N coordinates, but the choice of these coordinates is *arbitrary*. The important idea is not the 'labels' but the points themselves and the geometrical and topological relationships between them.

We may relabel the points of a manifold by performing a *coordinate transformation* $x^a \to x'^a$ expressed by the N equations

$$\boxed{x'^a = x'^a(x^1, x^2, \ldots, x^N) \qquad (a = 1, 2, \ldots, N),} \tag{2.2}$$

giving each new coordinate as a function of the old coordinates. Hence we view a coordinate transformation *passively* as assigning the new primed coordinates $(x'^1, x'^2, \ldots, x'^N)$ to a point of the manifold whose old coordinates are $(x^1, x^2, \ldots, x^N)$.

We will assume that the functions involved in (2.2) are single-valued, continuous and differentiable over the valid ranges of their arguments. Thus by differentiating each equation in (2.2) with respect to each of the old coordinates x^b we obtain the $N \times N$ partial derivatives $\partial x'^a / \partial x^b$. These may be assembled into the $N \times N$ *transformation matrix*[2]

$$\left[\frac{\partial x'^a}{\partial x^b}\right] = \begin{pmatrix} \dfrac{\partial x'^1}{\partial x^1} & \dfrac{\partial x'^1}{\partial x^2} & \cdots & \dfrac{\partial x'^1}{\partial x^N} \\ \dfrac{\partial x'^2}{\partial x^1} & \dfrac{\partial x'^2}{\partial x^2} & \cdots & \dfrac{\partial x'^2}{\partial x^N} \\ \vdots & \vdots & & \vdots \\ \dfrac{\partial x'^N}{\partial x^1} & \dfrac{\partial x'^N}{\partial x^2} & \cdots & \dfrac{\partial x'^N}{\partial x^N} \end{pmatrix},$$

so that rows are labelled by the index in the numerator of the partial derivative and columns by the index in the denominator. The elements of the transformation matrix are functions of the coordinates, and so the numerical values of the matrix elements are in general different when evaluated at different points in the manifold. The determinant of the transformation matrix is called the *Jacobian* of the transformation and is denoted by

$$J = \det \left[\frac{\partial x'^a}{\partial x^b}\right].$$

Clearly, the numerical value of J also varies from point to point in the manifold.

If $J \neq 0$ for some range of the coordinates x^b then it follows that in this region we can (in principle) solve the equations (2.2) for the old coordinates x^b and obtain the *inverse transformation* equations

$$x^a = x^a(x'^1, x'^2, \ldots, x'^N) \qquad (a = 1, 2, \ldots, N).$$

[2] In general the notation [] denotes the matrix containing the elements within the square brackets.

In a similar manner to the above, we define the inverse transformation matrix $[\partial x^a/\partial x'^b]$ and the Jacobian of the inverse transformation $J' = \det[\partial x^a/\partial x'^b]$.

Using the chain rule, it is easy to show that the inverse transformation matrix is the inverse of the transformation matrix, since

$$\sum_{b=1}^{N} \frac{\partial x'^a}{\partial x^b}\frac{\partial x^b}{\partial x'^c} = \frac{\partial x'^a}{\partial x'^c} = \delta^a_c = \begin{cases} 1 & \text{if } a = c, \\ 0 & \text{if } a \neq c, \end{cases}$$

where we have defined the *Kronecker delta* δ^a_c and used the fact that

$$\frac{\partial x'^a}{\partial x'^c} = \frac{\partial x^a}{\partial x^c} = 0 \qquad \text{if } a \neq c,$$

because the coordinates in either the unprimed or the primed set are independent. Since the two transformation matrices are inverses of one another, it follows that $J' = 1/J$.

If we consider neighbouring points P and Q in the manifold, with coordinates x^a and $x^a + dx^a$ respectively, then in the new, primed, coordinate system the infinitesimal coordinate separation between P and Q is given by

$$dx'^a = \frac{\partial x'^a}{\partial x^1}\,dx^1 + \frac{\partial x'^a}{\partial x^2}\,dx^2 + \cdots + \frac{\partial x'^a}{\partial x^N}\,dx^N,$$

where it is understood that the partial derivatives on the right-hand side are evaluated at the point P. We can write this more economically as

$$dx'^a = \sum_{b=1}^{N} \frac{\partial x'^a}{\partial x^b}\,dx^b. \tag{2.3}$$

2.5 Summation convention

Our notation can be made more economical still by adopting Einstein's *summation convention*: whenever an index occurs twice in an expression, *once as a subscript and once as a superscript*, this is understood to imply a summation over the index from 1 to N, the dimension of the manifold.

Thus we can write (2.3) simply as

$$\boxed{dx'^a = \frac{\partial x'^a}{\partial x^b}\,dx^b,}$$

where, once again, it is understood that all the partial derivatives are evaluated at P. The index a appearing on each side of this equation is said to be a *free* index and may take on separately any value from 1 to N. We consider a superscript that appears in the denominator of a partial derivative as a subscript (and vice versa). Thus the index b on the right-hand side in effect appears once as a subscript and once as a superscript, and hence there is an implied summation from 1 to N. An index that is summed over in this way is called a *dummy* index, because it can be replaced by any other index not already in use. For example, we may write

$$\frac{\partial x'^a}{\partial x^b} dx^b = \frac{\partial x'^a}{\partial x^c} dx^c,$$

since c was not already in use in the expression.

Note that the proper use of the summation convention requires that, in any term, an index should not occur more than twice and that any repeated index must occur once as a subscript and once as a superscript.

2.6 Geometry of manifolds

So far, we have considered manifolds only in a very primitive form. We have assumed that the manifold is continuous and differentiable, but aside from these properties it remains an amorphous collection of points. We have not yet defined its geometry.

Consider two infinitesimally separated points P and Q in the manifold, with coordinates x^a and $x^a + dx^a$ respectively ($a = 1, 2, \ldots, N$). The *local geometry* of the manifold at the point P is determined by defining the invariant '*distance*' or '*interval*' ds between P and Q. In general, the distance between the points can be assigned to be any reasonably well-behaved function of the coordinates and their differentials, i.e.[3]

$$ds^2 = f(x^a, dx^a).$$

Clearly this function contains information on both the local geometry of the manifold at P *and* our chosen coordinate system. *It is the assignment at each point in the manifold of a distance between points with infinitesimally different values of the coordinates that determines the local geometry of the manifold.* To choose an example at random, a two-dimensional manifold, beloved of differential geometers for its richness, is the Finsler geometry, in which one may define coordinates ξ and ζ such that

$$ds^2 = (d\xi^4 + d\zeta^4)^{1/2}.$$

[3] It is conventional to give the expression for ds^2 rather than ds.

2.7 Riemannian geometry

For developing general relativity, we are not interested in the most general geometries and can confine our attention to manifolds in which the interval is given by an expression of the form[4] (assuming the summation convention)

$$\boxed{ds^2 = g_{ab}(x)\, dx^a dx^b.} \tag{2.4}$$

Thus, such an interval is *quadratic* in the coordinate differentials. We shall see below that the $g_{ab}(x)$ are the components of the *metric tensor* field in our chosen coordinate system. For the moment, however, we can consider them simply as a set of functions of the coordinates that determine the local geometry of the manifold at any point. Manifolds with a geometry expressible in the form (2.4) are called *Riemannian manifolds.* Strictly speaking, the manifold is only Riemannian if $ds^2 > 0$ always. If ds^2 can be positive or negative (or zero), as is the case in special relativity and general relativity, then the manifold should properly be called *pseudo-Riemannian* but is usually simply referred to as Riemannian.

The metric functions $g_{ab}(x)$ can be considered as the elements of a position-dependent $N \times N$ matrix. The metric functions can always be chosen so that $g_{ab}(x) = g_{ba}(x)$, i.e the matrix is symmetric. Suppose for argument's sake that the functions g_{ab} were not symmetric in a and b. Then we could always decompose the metric function into parts that are symmetric and antisymmetric respectively in a and b, i.e.

$$g_{ab}(x) = \tfrac{1}{2}[g_{ab}(x) + g_{ba}(x)] + \tfrac{1}{2}[g_{ab}(x) - g_{ba}(x)].$$

The contribution to ds^2 from the antisymmetric part would be $\frac{1}{2}[g_{ab}(x) - g_{ba}(x)]\, dx^a\, dx^b$, which vanishes identically, as is easily confirmed on swapping indices in one of the terms, so that any antisymmetric part of g_{ab} can safely be neglected. Thus in an N-dimensional Riemannian manifold there are $\frac{1}{2}N(N+1)$ independent metric functions $g_{ab}(x)$.

It is important to remember that the form of the metric functions can always be changed by making a change of coordinates. Since the interval between two points in the manifold is invariant under a coordinate transformation, using (2.4) and (2.3) we have

$$\begin{aligned} ds^2 &= g_{ab}(x)\, dx^a\, dx^b \\ &= g_{ab}(x) \frac{\partial x^a}{\partial x'^c} \frac{\partial x^b}{\partial x'^d}\, dx'^c\, dx'^d \\ &= g'_{cd}(x')\, dx'^c\, dx'^d, \end{aligned} \tag{2.5}$$

[4] As we shall see in Chapter 7, this is a consequence of the equivalence principle.

where the new metric functions $g'_{ab}(x')$ in the primed coordinate system are related to those in the unprimed coordinate system by

$$g'_{cd}(x') = g_{ab}(x(x'))\frac{\partial x^a}{\partial x'^c}\frac{\partial x^b}{\partial x'^d}.$$

Clearly, the metric functions $g'_{ab}(x')$ describe the same local geometry of the manifold as do the functions $g_{ab}(x)$.

Since there are N arbitrary coordinate transformations there are really only $\frac{1}{2}N(N+1) - N = \frac{1}{2}N(N-1)$ independent degrees of freedom associated with the $g_{ab}(x)$.

2.8 Intrinsic and extrinsic geometry

It is important to realise that the local geometry or *curvature* characterised by (2.4) is an *intrinsic* property of the manifold itself, i.e. it is *independent* of whether the manifold is *embedded* in some higher-dimensional space.

It is, of course, difficult (or impossible) to imagine higher-dimensional curved manifolds, so it is instructive to consider two-dimensional Riemannian manifolds, which can often be visualised as a surface embedded in a three-dimensional Euclidean space. It is important to make a distinction, however, between the *extrinsic* properties of the surface, which are dependent on how it is embedded into a higher-dimensional space, and properties that are intrinsic to the surface itself.

This distinction is traditionally made clear by considering the viewpoint of some two-dimensional being (called a 'bug') confined exclusively to the two-dimensional surface. Such a being would believe that it is able to look and measure in all directions, whereas it is in fact limited to making measurements of distance, angle etc. only within the surface. For example, it would receive light signals that had travelled within the two-dimensional surface. Properties of the geometry that are accessible to the bug are called *intrinsic*, whereas those that depend on the viewpoint of a higher-dimensional creature (who is able to see how the surface is shaped in the three-dimensional space) are called *extrinsic*.

The bug is able to define a coordinate system and measure distances in the surface (e.g. by counting how many steps it has to take) from one point to another. It can thus define a set of metric functions $g_{ab}(x)$ that characterise the intrinsic geometry of the surface (as expressed in the bug's chosen coordinate system).

Consider, for example, a two-dimensional plane surface, such as a flat sheet of paper, in our three-dimensional Euclidean space. The bug can label the entire sheet using rectangular Cartesian coordinates, so that the distance ds measured over the surface between any pair of points whose coordinate separations are dx and dy is given by

$$ds^2 = dx^2 + dy^2.$$

If this sheet is then rolled up into a cylinder, the bug would not be able to detect any differences in the geometrical properties of the surface (see Figure 2.1).

To the bug, the angles of a triangle still add up to 180°, the circumference of a circle is still $2\pi r$ etc. The proof of this fact is simple – the surface can simply be *un*rolled back to a flat surface without buckling, tearing or otherwise distorting it. A more mathematical approach is to note that if one parameterises the surface of the cylinder (of radius a) using cylindrical coordinates (z, ϕ), the distance ds measured over the surface between any two points whose coordinate separations are dz and $d\phi$ is given by

$$ds^2 = dz^2 + a^2\, d\phi^2.$$

By making the simple change of variables $x = z$ and $y = a\phi$ we recover the expression $ds^2 = dx^2 + dy^2$, which is valid over the whole surface, and so the intrinsic geometry is identical to that of a flat plane. Thus the surface of a cylinder is not intrinsically curved; its curvature is extrinsic and a result of the way it is embedded in three-dimensional space. Even if one were to crumple up the sheet of paper (without tearing it), so that its extrinsic geometry in three-dimensional space was very complicated, its intrinsic geometry would still be that of a plane.

The situation is somewhat different for a 2-sphere, i.e. a spherical surface, embedded in three-dimensional Euclidean space. Once again the surface is manifestly curved extrinsically on account of its embedding. Additionally, however, it cannot be formed from a flat sheet of paper without tearing or deformation. Its intrinsic geometry – based on measurements within the surface – differs from the intrinsic (Euclidean) geometry of the plane. This problem is well known to

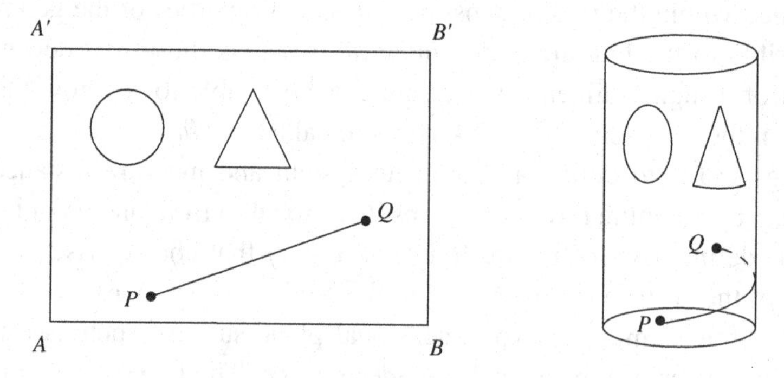

Figure 2.1 Rolling up a flat sheet of paper into a cylinder.

cartographers. Mathematically, if we parameterise a sphere (of radius a) by the usual angular coordinates (θ, ϕ) then

$$ds^2 = a^2(d\theta^2 + \sin^2\theta\, d\phi^2),$$

which cannot be transformed to the Euclidean form $ds^2 = dx^2 + dy^2$ *over the whole surface* by any coordinate transformation. Thus the surface of a sphere is intrinsically curved.

We note, however, that *locally* at any point A on the spherical surface we can define a set of Cartesian coordinates, so that $ds^2 = dx^2 + dy^2$ is valid in the neighbourhood of A. For example, the street layout of a town can be accurately represented by a flat map, whereas the entire globe can only be represented by performing projections that distort distance and/or angles. As an idea of what can happen to local Cartesian coordinate systems far from the point A where they are defined, consider Figure 2.2. If a bug starts at A and travels in the locally defined x-direction to B, it observes that C still lies in the y-direction. If instead the bug travels from A to C, it finds that B still lies in the x-direction. The non-Euclidean geometry of the spherical surface is also apparent from the fact that the angles of the triangle ABC sum to 270°.

We may take our discussion one step further, dispense with the three-dimensional space and embedding-related extrinsic geometry and consider the surfaces in isolation. Intrinsic geometry is all that remains with any meaning. For example, when we talk of the curvature of spacetime in general relativity, we must resist any temptation to think of spacetime as embedded in any 'higher' space. Any such embedding, whether or not it is physically realised, would be irrelevant to our discussion. Nevertheless, in developing our intuition for

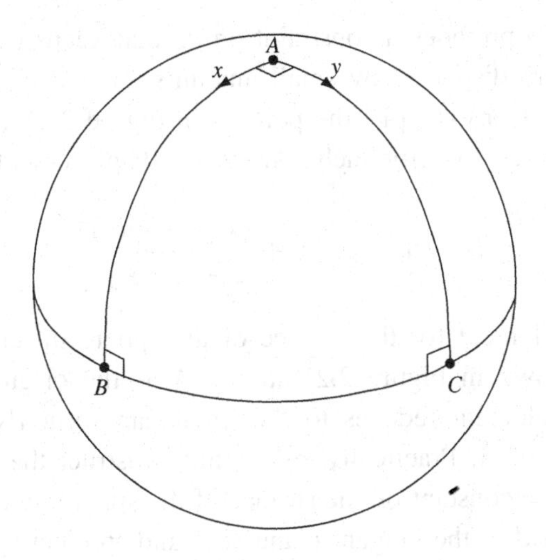

Figure 2.2 A two-dimensional spherical surface.

curved manifolds it oftens remains useful to imagine two-dimensional surfaces embedded in three-dimensional Euclidean space.

2.9 Examples of non-Euclidean geometry

Let us develop our intuition for non-Euclidean geometry by considering in more detail the surface of a sphere. We begin by imagining the usual Cartesian coordinate system (x, y, z) defining a Euclidean three-dimensional space with line element

$$ds^2 = dx^2 + dy^2 + dz^2. \tag{2.6}$$

Now, suppose that we have a sphere of radius a with its centre at the origin of our coordinate system. We will now ask the following question: what is the line element on the surface of the sphere?

The equation defining the sphere is

$$x^2 + y^2 + z^2 = a^2.$$

So, differentiating this equation, we obtain

$$2x\,dx + 2y\,dy + 2z\,dz = 0,$$

and we can write an equation for dz,

$$dz = -\frac{x\,dx + y\,dy}{z} = \frac{-(x\,dx + y\,dy)}{\left[a^2 - (x^2 + y^2)\right]^{1/2}}. \tag{2.7}$$

Thus, equation (2.7) provides a constraint on dz that keeps us on the surface of the sphere if we are displaced by small amounts dx and dy from an arbitrary point on the sphere (for example, the point A in Figure 2.2). Substituting for dz in (2.6) gives us the interval for such constrained displacements:

$$ds^2 = dx^2 + dy^2 + \frac{(x\,dx + y\,dy)^2}{a^2 - (x^2 + y^2)}, \tag{2.8}$$

which is the line element for the surface of the sphere in terms of our chosen coordinates (as shown in Figure 2.2), taking A as the origin of x and y. We see that this line element reduces to the Euclidean form $ds^2 = dx^2 + dy^2$ in the neighbourhood of A. Practically, one could construct the coordinate curves $x = \text{constant}$ and $y = \text{constant}$ on the surface of the sphere by creating a standard (x, y) coordinate grid in the tangent plane at A and 'projecting' vertically down onto the spherical surface.

We may obtain an alternative form for the line element by making the substitutions

$$x = \rho \cos\phi, \qquad y = \rho \sin\phi,$$

and after a little algebra we obtain[5]

$$\boxed{ds^2 = \frac{a^2 d\rho^2}{a^2 - \rho^2} + \rho^2 d\phi^2.} \tag{2.9}$$

As above, one can construct the ρ and ϕ coordinate curves on the sphere by creating a standard (ρ, ϕ) coordinate system in the tangent plane at A and projecting vertically down onto the surface. We also note that this line element contains a 'hidden symmetry', namely our freedom to choose an arbitrary point on the sphere as the origin $\rho = 0$.

The observant reader will have noticed that the line elements (2.8) and (2.9) have singularities at $\sqrt{x^2 + y^2} = a$, or, equivalently, $\rho = a$, corresponding to the equator of the sphere (relative to A). From our embedding picture, it is clear why the (x, y) and (ρ, ϕ) coordinates cover the surface of the sphere uniquely only up to this point. We note, however, that there is nothing pathological in the intrinsic geometry of the 2-sphere at the equator. What we have observed is only a *coordinate singularity*, which has resulted simply from choosing coordinates with a restricted domain of validity. Although the embedding picture we have adopted gives both the (x, y) and (ρ, ϕ) coordinate systems a clear geometrical meaning in our three-dimensional Euclidean space, it is important to realise that a bug confined to the two-dimensional surface of the sphere could, if it wished, have defined these coordinate systems to describe the intrinsic geometry without any reference to an embedding in higher dimensions.

We can make an analogous construction to find the metric for a *3-sphere* embedded in *four-dimensional* Euclidean space. The metric for the four-dimensional Euclidean space is

$$ds^2 = dx^2 + dy^2 + dz^2 + dw^2, \tag{2.10}$$

and, by analogy with the example above, the equation defining a 3-sphere is

$$x^2 + y^2 + z^2 + w^2 = a^2.$$

Differentiating as before gives

$$2x\,dx + 2y\,dy + 2z\,dz + 2w\,dw = 0,$$

[5] Note that the line elements (2.8) and (2.9) look different from the metric we would write down using standard spherical polars, $ds^2 = a^2 d\theta^2 + a^2 \sin^2\theta\, d\phi^2$. Nonetheless, both are valid line elements for the two-dimensional surface of a sphere.

and so substituting for dw in (2.10) gives the line element:

$$ds^2 = dx^2 + dy^2 + dz^2 + \frac{(x\,dx + y\,dy + z\,dz)^2}{a^2 - (x^2 + y^2 + z^2)}.$$

Transforming to spherical polar coordinates

$$x = r\sin\theta\cos\phi,$$
$$y = r\sin\theta\sin\phi,$$
$$z = r\cos\theta,$$

we obtain an alternative form for the line element:

$$\boxed{ds^2 = \frac{a^2}{a^2 - r^2}\,dr^2 + r^2 d\theta^2 + r^2\sin^2\theta\,d\phi^2.} \tag{2.11}$$

Notice that, in the limit $a \to \infty$, the metric tends to the form

$$ds^2 = dr^2 + r^2 d\theta^2 + r^2\sin^2\theta\,d\phi^2,$$

which is simply the metric of ordinary Euclidean three-dimensional space $ds^2 = dx^2 + dy^2 + dz^2$, rewritten in spherical polar coordinates. The line element (2.11) therefore describes a *non-Euclidean* three-dimensional space. We note that this line element also has a singularity, this time at $r = a$. As one might expect from our discussion above, this is once again just a coordinate singularity, although our existence as three-dimensional 'bugs' makes the geometric reason for this less straightforward to visualise!

2.10 Lengths, areas and volumes

For a given set of metric functions $g_{ab}(x)$, (2.4), it is useful to know how to compute the lengths of curves and the 'areas' and 'volumes' of subregions of the manifold.

The lengths of curves follow immediately from the line element. Suppose that the points A and B are joined by some path; then the length of this curve is given by

$$L_{AB} = \int_A^B ds = \int_A^B |g_{ab}\,dx^a\,dx^b|^{1/2},$$

where the integral is evaluated along the curve. As indicated, the absolute value of ds is taken before the square root is evaluated when considering pseudo-Riemannian manifolds. If the equation of the curve $x^a(u)$ is given in terms of some parameter u then

$$L_{AB} = \int_{u_A}^{u_B} \left| g_{ab} \frac{dx^a}{du} \frac{dx^b}{du} \right|^{1/2} du, \tag{2.12}$$

where u_A and u_B are the values of the parameter u at the endpoints of the curve.

For the calculation of areas and volumes, let us begin by considering the simple case where the metric is diagonal, i.e. $g_{ab}(x) = 0$ for $a \neq b$.[6] In this case the line element takes the form

$$ds^2 = g_{11}(dx^1)^2 + g_{22}(dx^2)^2 + \cdots + g_{NN}(dx^N)^2. \tag{2.13}$$

Such a system of coordinates is called *orthogonal* since, at all points in the manifold, any pair of coordinate curves cross at right angles, as is clear from (2.13). Thus, in orthogonal coordinate systems the ideas of area and volume can be built up simply. Consider, for example, an element of area in the (x^1, x^2)-surface defined by $x^a =$ constant for $a = 3, 4, \ldots, N$. Suppose that the area element is defined by the *coordinate* lengths dx^1 and dx^2 (see Figure 2.3). The *proper lengths* of the two line segments will be $\sqrt{g_{11}}\,dx^1$ and $\sqrt{g_{22}}\,dx^2$ respectively. Thus the element of area is[7]

$$dA = \sqrt{|g_{11}g_{22}|}\,dx^1\,dx^2. \tag{2.14}$$

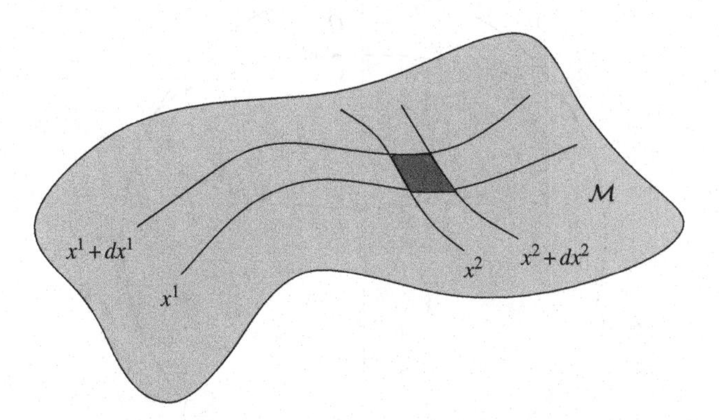

Figure 2.3 An element of area, on a manifold $\mathcal{M}$, defined by the coordinate intervals dx^1 and dx^2. The proper lengths dl^1 and dl^2 of these intervals are related to dx^1 and dx^2 by the metric functions. If the coordinate lines are orthogonal then the area of is $dl^1\,dl^2$.

[6] The general case is discussed in Section 2.14.

[7] We have implicitly assumed here that the manifold is strictly Riemannian. If the manifold is pseudo-Riemannian, some of the elements g_{ab} in (2.13) may be negative (see Section 2.13), and then we require the modulus signs.

Similarly, for 3-volumes in the (x^1, x^2, x^3)-surface defined by $x^a = \text{constant}$ for $a = 4, 5, \ldots, N$, we have

$$d^3V = \sqrt{|g_{11}g_{22}g_{33}|}\,dx^1\,dx^2\,dx^3. \tag{2.15}$$

We may, of course, define 3-volumes for any other three-dimensional subspace. We can define higher-dimensional 'volume' elements in a similar way until we reach the N-dimensional volume element

$$d^NV = \sqrt{|g_{11}g_{22}\cdots g_{NN}|}\,dx^1\,dx^2\cdots dx^N.$$

As examples of working with such metric functions, let us consider the non-Euclidean spaces discussed in Section 2.9. We begin with the line element (2.9),

$$ds^2 = \frac{a^2\,d\rho^2}{a^2 - \rho^2} + \rho^2\,d\phi^2, \tag{2.16}$$

which describes two-dimensional geometry on the surface of a sphere in terms of the coordinates (ρ, ϕ), the geometrical meanings of which are illustrated in Figure 2.4 assuming an embedding in three-dimensional Euclidean space. From (2.16) we see that this coordinate system is orthogonal, with $g_{\rho\rho} = a^2/(a^2 - \rho^2)$ and $g_{\phi\phi} = \rho^2$ (no sums on ρ or ϕ).[8] Let us consider a circle defined by $\rho = R$,

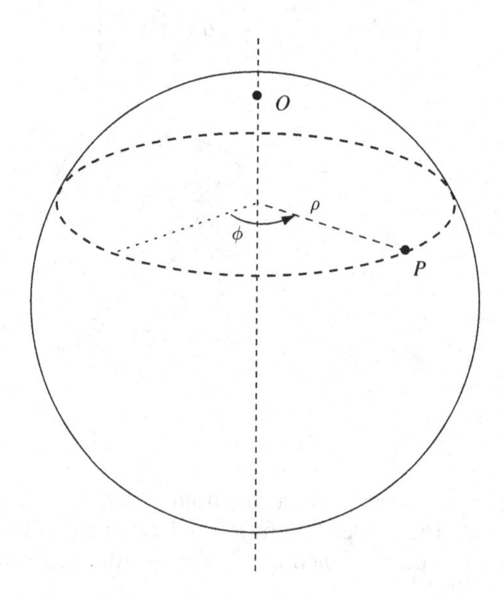

Figure 2.4 The surface of a sphere parameterised by the coordinates (ρ, ϕ) appearing in the line element (2.16).

[8] This form of notation is quite common, once a particular coordinate system has been chosen, and it is usually clear from the context that no summation is implied.

where R is some constant, and calculate its length, its area and the distance from its centre to the perimeter.

From (2.12) and (2.16), the distance *in the surface* from the centre to the perimeter along a line of constant ϕ is

$$D = \int_0^R \frac{a}{(a^2-\rho^2)^{1/2}}\,d\rho = a\sin^{-1}\left(\frac{R}{a}\right),$$

while the circumference of the circle is given by

$$C = \int_0^{2\pi} R\,d\phi = 2\pi R.$$

Similarly, from (2.14) we have, for the area of the spherical surface enclosed by C,

$$A = \int_0^{2\pi}\int_0^R \frac{a}{(a^2-\rho^2)^{1/2}}\rho\,d\rho\,d\phi = 2\pi a^2\left[1-\left(1-\frac{R^2}{a^2}\right)^{1/2}\right].$$

Note that if we rewrite the circumference C and area A in terms of the distance D then we obtain

$$C = 2\pi a\sin\left(\frac{D}{a}\right) \quad \text{and} \quad A = 2\pi a^2\left[1-\cos\left(\frac{D}{a}\right)\right]. \tag{2.17}$$

Thus, as D increases, the circumference of the circle increases *until the point when $D = \pi a/2$, after which C becomes smaller as D* increases.

In fact there is a slight subtlety here. As noted earlier, if we attempt to parameterise points beyond the equator of the sphere using the coordinates (ρ, ϕ), the system becomes degenerate, i.e. there is more than one point in the surface with the same coordinates. The degenerate nature of the (ρ, ϕ) coordinate system means that some care is required, for example, in calculating the total area of the surface. By symmetry this is given by

$$A_{\text{tot}} = 2\int_0^{2\pi}\int_0^a \frac{a}{(a^2-\rho^2)^{1/2}}\rho\,d\rho\,d\phi = 4\pi a^2.$$

Although we cannot easily visualise the geometry, we can perform similar calculations for the line element (2.11),

$$ds^2 = \frac{a^2}{a^2-r^2}dr^2 + r^2 d\theta^2 + r^2\sin^2\theta\,d\phi^2, \tag{2.18}$$

which describes a non-Euclidean three-dimensional space that tends to Euclidean three-dimensional space as $a \to \infty$. Let us consider a 2-sphere of coordinate radius $r = R$ and calculate the circumference around the equator, the area, the volume and the distance from its centre to the surface of the sphere.

From (2.12) and (2.18), the distance from the centre to the surface along a line $\theta = \text{constant}$, $\phi = \text{constant}$ is

$$D = \int_0^R \frac{a\,dr}{(a^2 - r^2)^{1/2}} = a \sin^{-1}\left(\frac{R}{a}\right).$$

Noting that the equator of the sphere is the curve $r = R$, $\theta = \pi/2$, its circumference is $C = \int_0^{2\pi} R\,d\phi = 2\pi R$, while the area of the surface $r = R$ and the volume it encloses are obtained from (2.14) and (2.15) and read

$$A = \int_0^{2\pi} \int_0^{\pi} R^2 \sin\theta\,d\theta\,d\phi = 4\pi R^2,$$

$$V = \int_0^{2\pi} \int_0^{\pi} \int_0^R \frac{ar^2 \sin\theta}{(a^2 - r^2)^{1/2}}\,dr\,d\theta\,d\phi$$

$$= 2\pi a^3 \left\{ \sin^{-1}\left(\frac{R}{a}\right) - \frac{R}{a}\left[1 - \left(\frac{R}{a}\right)^2\right]^{1/2} \right\}.$$

It is not difficult to see that the familiar results of three-dimensional Euclidean space are recovered when $R/a \ll 1$. Once again, we can rewrite our results in terms of D rather than R, and we find that C and A have maximum values at $D = \pi a/2$. By analogy with the above two-dimensional example, the total volume of this space is

$$V_{\text{tot}} = 2 \int_0^{2\pi} \int_0^{\pi} \int_0^a \frac{ar^2 \sin\theta}{(a^2 - r^2)^{1/2}}\,dr\,d\theta\,d\phi = 2\pi^2 a^3.$$

The three-dimensional non-Euclidean space described by the line element (2.18) thus has a *finite* volume. We can generate a line element for an *infinite* non-Euclidean three-dimensional space by making the substitution $a = ib$, i.e. choosing the 'radius' of the space to be pure imaginary. The line element (2.18) then becomes

$$ds^2 = \frac{b^2}{b^2 + r^2}\,dr^2 + r^2\,d\theta^2 + r^2 \sin^2\theta\,d\phi^2.$$

If we again consider the sphere defined by $r = R$, we find easily that in this space $C = 2\pi R$ and $A = 4\pi R^2$ as before but the distance from the centre of the sphere to its surface is now given by $D = b \sinh^{-1}(R/b)$. In this case, one finds that C, A and the volume V of the sphere are all monotonically increasing functions of D.

2.11 Local Cartesian coordinates

We now introduce a key property of Riemannian manifolds, to which we have alluded in earlier sections. For the moment we will confine our attention to manifolds that are strictly Riemannian, so that $ds^2 > 0$ always, but subsequently we will extend our discussion to pseudo-Riemannian spaces, in which ds^2 can be of either sign (or zero).

For a general Riemannian manifold, it is *not* possible to perform a coordinate transformation $x^a \to x'^a$ that will take the line element $ds^2 = g_{ab}(x)\,dx^a\,dx^b$ into the Euclidean form

$$ds^2 = (dx'^1)^2 + (dx'^2)^2 + \cdots + (dx'^N)^2 = \delta_{ab}\,dx'^a\,dx'^b,$$

at every point in the manifold. This is clear, since there are $N(N+1)/2$ independent metric functions $g_{ab}(x)$ but only N coordinate transformation functions $x'^a(x)$. As we shall now demonstrate, however, it is always possible to make a coordinate transformation such that *in the neighbourhood of some specified point* P the line element takes the Euclidean form. In other words, we can always find coordinates x'^a such that *at the point* P the new metric functions $g'_{ab}(x')$ satisfy

$$g'_{ab}(P) = \delta_{ab}, \tag{2.19}$$

$$\left.\frac{\partial g'_{ab}}{\partial x'^c}\right|_P = 0. \tag{2.20}$$

Thus, in the neighbourhood of P, we have

$$\boxed{g'_{ab}(x') = \delta_{ab} + \mathcal{O}[(x' - x'_P)^2].}$$

Such coordinates are called *local Cartesian coordinates* at P.

From (2.5), the general transformation rule for the metric functions is

$$g'_{ab} = \frac{\partial x^c}{\partial x'^a}\frac{\partial x^d}{\partial x'^b}g_{cd},$$

which we require to satisfy the conditions (2.19) and (2.20) at our chosen point P. If x^a is an arbitrary given coordinate system and x'^a is the desired system then there will be some relation $x^a(x')$ connecting the two sets of coordinates. Although we do not (as yet) know the required transformation, we can define it in terms of its Taylor expansion about P:

$$\begin{aligned} x^a(x') = x^a_P &+ \left(\frac{\partial x^a}{\partial x'^b}\right)_P (x'^b - x'^b_P) \\ &+ \frac{1}{2}\left(\frac{\partial^2 x^a}{\partial x'^b \partial x'^c}\right)_P (x'^b - x'^b_P)(x'^c - x'^c_P) \\ &+ \frac{1}{6}\left(\frac{\partial^3 x^a}{\partial x'^b \partial x'^c \partial x'^d}\right)_P (x'^b - x'^b_P)(x'^c - x'^c_P)(x'^d - x'^d_P) + \cdots. \end{aligned}$$

The numbers of free independent variables we have for this purpose are as follows:

$(\partial x^a/\partial x'^b)_P$ has N^2 independent values,

$(\partial^2 x^a/\partial x'^b \partial x'^c)_P$ has $\frac{1}{2}N^2(N+1)$ independent values,

$(\partial^3 x^a/\partial x'^b \partial x'^c \partial x'^d)_P$ has $\frac{1}{6}N^2(N+1)(N+2)$ independent values,

where we have made use of the fact that the second set of quantities is symmetric in b and c and the third set of quantities is totally symmetric in b, c and d. We may compare this with the number of independent parameters we may want to fix:

$g'_{ab}(P)$ has $\frac{1}{2}N(N+1)$ independent values,

$(\partial g'_{ab}/\partial x'^c)_P$ has $\frac{1}{2}N^2(N+1)$ independent values,

$(\partial^2 g'_{ab}/\partial x'^c \partial x'^d)_P$ has $\frac{1}{4}N^2(N+1)^2$ independent values.

The first question is whether we can satisfy the requirement (2.19). This condition consists of $N(N+1)/2$ independent equations, and to satisfy them we have N^2 free values in $(\partial x^a/\partial x'^b)_P$. Therefore, they can indeed be satisfied, leaving $N(N-1)/2$ numbers to spare! These spare degrees of freedom correspond exactly to the number of independent N-dimensional 'rotations' that leave δ_{ab} unchanged.

The next question is whether we can satisfy the requirement (2.20). This condition consists of $N^2(N+1)/2$ independent equations, and we can choose an equal number of free values $(\partial^2 x^a/\partial x'^b \partial x'^c)_P$ to satisfy them.

The final question is whether we can continue in this way to higher orders. In other words, can we find a set of coordinates x'^a such that $(\partial^2 g'_{ab}/\partial x'^c \partial x'^d)_P = 0$? This condition consists of $N^2(N+1)^2/4$ independent equations, but we have only $N^2(N+1)(N+2)/6$ free values in $(\partial^2 g'_{ab}/\partial x'^c \partial x'^d)_P$, so these equations cannot in general be satisfied. This means that there are $N^2(N^2-1)/12$ 'degrees of freedom' among the second derivatives $(\partial^2 g'_{ab}/\partial x'^c \partial x'^d)_P$, i.e. in general at least this number of second derivatives will not vanish.

Although we have shown, in principle, that it is always possible to define local Cartesian coordinates at any given point P, we have not shown explicitly how to find such coordinates. We will return to this point in Chapter 3.

2.12 Tangent spaces to manifolds

To aid our intuition of local Cartesian coordinates, it is useful to consider the simple example of a two-dimensional Riemannian manifold, which we can often

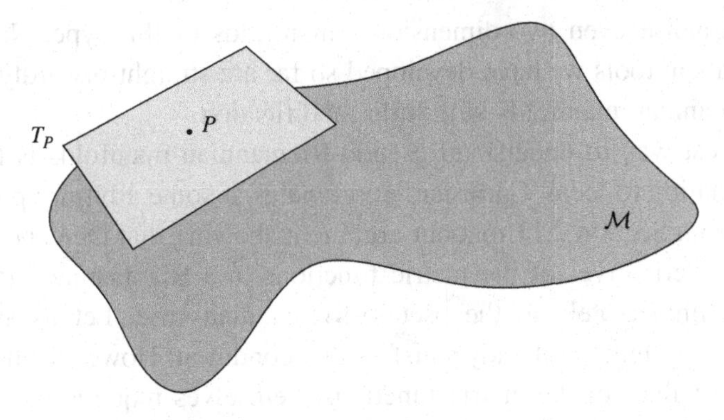

Figure 2.5 The tangent plane T_P to the curved surface $\mathcal{M}$ at the point P.

consider as a generally curved surface embedded in three-dimensional Euclidean space. A simple example is the surface of a sphere, shown in Figure 2.2. As we have shown, at any arbitrary point P we can find coordinates x and y (say) such that in the neighbourhood of P we have

$$ds^2 = dx^2 + dy^2.$$

It thus follows that a Euclidean two-dimensional space (a plane) will match the manifold locally at P. This Euclidean space is called the *tangent space* T_P to the manifold at P. In other words, in terms of our embedding picture a plane can always be drawn at any arbitrary point on a two-dimensional Riemannian surface in such a way that it is locally tangential to the surface (see Figure 2.5). Although the tangent plane to a surface at P gives a useful way of visualising the tangent space of a manifold at a point, this view can be misleading. As we stressed earlier, a manifold should be regarded as an entity in itself: there is no need for a higher-dimensional space in which it and its tangent spaces are embedded.

We may extend the idea of tangent spaces to higher dimensions. At an arbitrary point P in an N-dimensional Riemannian manifold we can find a coordinate system such that in the neighbourhood of P the line element is Euclidean. Thus, an N-dimensional Euclidean space matches the manifold locally at P. Just as each point P of an embedded two-dimensional surface has its tangent plane, making contact with the surface at P, so each point P of a manifold has a tangent space T_P attached to it.

2.13 Pseudo-Riemannian manifolds

Thus far we have confined our attention almost exclusively to strictly Riemannian manifolds, in which $ds^2 > 0$ always. In a pseudo-Riemannian manifold, however, ds^2 can be either positive, negative or zero and it is therefore much harder to visualise even two-dimensional manifolds of this type. Nevertheless, the mathematical tools we have developed so far are straightforwardly applied to pseudo-Riemannian manifolds with little modification.

The simplest way to understand pseudo-Riemannian manifolds is to consider the transformation to local 'Cartesian' coordinates at some arbitrary point P. You will notice from Section 2.11 that our argument showing that the condition (2.20) holds for the derivatives of the metric functions in a Riemannian manifold can be extended immediately to the pseudo-Riemannian case. Let us assume that the coordinate system x^a already satisfies this condition. However, the condition (2.19) on the values of the metric functions themselves requires further investigation. Let us now attempt to obtain a new coordinate system x'^a in which (2.19) is also satisfied. We note in passing that, in order for (2.20) to remain valid, the new coordinates x'^a must be related to the old ones x^a by a *linear* transformation, $x'^a = X^a{}_b x^b$, where the $X^a{}_b$ are *constants*.

In general, at a point P the metric functions in the new coordinate system are given in terms of the original metric functions by

$$g'_{ab}(P) = \left(\frac{\partial x^c}{\partial x'^a}\right)_P \left(\frac{\partial x^d}{\partial x'^b}\right)_P g_{cd}(P). \tag{2.21}$$

Let us define symmetric matrices G and G' having elements $g_{ab}(P)$ and $g'_{ab}(P)$ respectively. Similarly, we can define a matrix X having elements $(\partial x^a/\partial x'^b)_P$. Then, in matrix notation, (2.21) can be written as

$$\mathsf{G}' = \mathsf{X}^{\mathrm{T}}\mathsf{G}\mathsf{X}.$$

Since G is symmetric, it can be diagonalised by this similarity transformation, provided that we choose the columns of X to be the normalised eigenvectors of G. Then $\mathsf{G}' = \mathrm{diag}(\lambda_1, \lambda_2, \ldots, \lambda_N)$, where λ_a is the ath eigenvalue of G (the eigenvalues must all be real).

In a strictly Riemannian manifold, $ds^2 = g_{ab}\,dx^a dx^b$ is always positive at any point P. Thus the matrix $\mathsf{G} \equiv [g_{ab}]$ at any point must be positive definite, i.e. all its eigenvalues must be positive. At an arbitrary point in a pseudo-Riemannian manifold, however, ds^2 can be positive, negative or zero, depending on the direction in which one moves from P. Correspondingly, some of the eigenvalues of G are negative.

If we now scale our coordinates according to $x'^a \to x'^a/\sqrt{|\lambda_a|}$ (note that here there is no sum on a), we obtain at the point P

$$\mathsf{G}' = \mathrm{diag}(\pm 1, \pm 1, \ldots, \pm 1),$$

where the $+$ and $-$ signs depend on whether the corresponding eigenvalue is positive or negative. Thus, at any arbitrary point P in a pseudo-Riemannian manifold, it is always possible to find a coordinate system x'^a such that in the neighbourhood of P we have

$$\boxed{g'_{ab}(x') = \eta_{ab} + \mathcal{O}[(x' - x'_P)^2],}$$

where $[\eta_{ab}] = \text{diag}(\pm 1, \pm 1, \ldots, \pm 1)$. The number of positive entries minus the number of negative entries in $[\eta_{ab}]$ is called the *signature* of the manifold and is the same at all points.

It follows that, at any arbitrary point P in a pseudo-Riemannian manifold, an N-dimensional space with line element

$$ds^2 = \pm(dx^1)^2 \pm (dx^2)^2 \pm \cdots \pm (dx^N)^2$$

will match the manifold locally at P. Such a space is called *pseudo*-Euclidean and is the tangent space T_P to the pseudo-Riemannian manifold at P. An example of a pseudo-Euclidean space is the four-dimensional Minkowski spacetime of special relativity, which has the line element

$$ds^2 = d(ct)^2 - dx^2 - dy^2 - dz^2$$

when expressed in coordinates corresponding to a Cartesian inertial frame. Minkowski spacetime thus has a signature of -2.

2.14 Integration over general submanifolds

In Section 2.10, we restricted our calculation of 'volumes' to coordinate systems x^a that were orthogonal and to submanifolds that were obtained simply by allowing some of the coordinates to be constants. In fact neither of these simplifications is necessary, and we are now in a position to consider the general case.

Let us begin by calculating the full N-dimensional volume element $d^N V$ in an N-dimensional (pseudo-)Riemannian manifold. From Section 2.10, we know that if we are working in an orthogonal coordinate system then this volume element is given by

$$d^N V = \sqrt{|g_{11} g_{22} \cdots g_{NN}|}\, dx^1 dx^2 \cdots dx^N.$$

For such a coordinate system the matrix G is given by

$$\mathsf{G} \equiv [g_{ab}] = \text{diag}(g_{11}, g_{22}, \ldots, g_{NN}),$$

so that its determinant is simply the product of the diagonal elements,

$$\det \mathsf{G} = g_{11} g_{22} \cdots g_{NN}.$$

It is usual to denote $\det \mathsf{G}$ simply by the symbol g. Thus, we may rewrite the volume element as

$$\boxed{d^N V = \sqrt{|g|}\, dx^1 dx^2 \cdots dx^N.} \tag{2.22}$$

What we will now show is that this expression remains valid for an *arbitrary* coordinate system.

The key to proving the general result (2.22) for the volume element at some arbitrary point P in the manifold is to transform to local Cartesian coordinates x'^a at P. We know that a small N-dimensional region at P will have volume $d^N V = dx'^1 dx'^2 \cdots dx'^N$. In *any* other general coordinate system x^a it is a well-known result that

$$dx'^1 dx'^2 \cdots dx'^N = J\, dx^1 dx^2 \cdots dx^N, \tag{2.23}$$

where the Jacobian factor J is given by

$$J = \det \left[\frac{\partial x'^a}{\partial x^b} \right].$$

If, as in Section 2.13, we use X to denote the transformation matrix $[\partial x^a / \partial x'^b]$ then $J = \det(\mathsf{X}^{-1}) = (\det \mathsf{X})^{-1}$. Defining matrices G and G' as those having elements $g_{ab}(P)$ and $g'_{ab}(P)$ respectively, we have (see Section 2.13)

$$\mathsf{G}' = \mathsf{X}^{\mathrm{T}} \mathsf{G} \mathsf{X}. \tag{2.24}$$

Taking determinants of both sides of (2.24) and denoting $\det \mathsf{G}$ by g and $\det \mathsf{G}'$ by g' we obtain

$$g' = (\det \mathsf{X})^2 g = \frac{1}{J^2} g.$$

Since the x'^a are locally Cartesian coordinates, $\mathsf{G}' = \mathrm{diag}(\pm 1, \pm 1, \ldots, \pm 1)$, where the number of positive and negative signs depends on the signature of the manifold. Thus we have $g' = \pm 1$, so that $g = \pm J^2$. Hence, we obtain the required result:

$$d^N V = dx'^1 dx'^2 \cdots dx'^N = \sqrt{|g|}\, dx^1 dx^2 \cdots dx^N.$$

We now turn to the question how to integrate over submanifolds that are not defined simply by setting some of the coordinates x^a to be constant. Consider some M-dimensional subspace of an N-dimensional manifold. In general, the subspace can be defined by the N parametric equations

$$x^a = x^a(u^1, u^2, \ldots, u^M),$$

where the u^i $(i = 1, 2, \ldots, M)$ may be considered simply as a set of coordinates that parameterise the subspace. If we consider two neighbouring points in the subspace whose parameters differ by du^i then the coordinate separation between these points is simply

$$dx^a = \frac{\partial x^a}{\partial u^i}\, du^i.$$

Thus the distance ds between the points is given by

$$ds^2 = g_{ab} dx^a dx^b = g_{ab} \frac{\partial x^a}{\partial u^i} \frac{\partial x^b}{\partial u^j}\, du^i du^j,$$

which we may write as

$$ds^2 = h_{ij}\, du^i du^j,$$

where the h_{ij} are the *induced* metric functions on the subspace and are given by

$$h_{ij} = g_{ab} \frac{\partial x^a}{\partial u^i} \frac{\partial x^b}{\partial u^j}. \tag{2.25}$$

Thus we can now work simply in terms of this subspace and regard it as a manifold in itself. Thus the volume element for integrals over this subspace is given in terms of the parameters u^i by

$$\boxed{d^M V = \sqrt{|h|}\, du^1 du^2 \cdots du^M,}$$

where $h = \det[h_{ij}]$.

It is also worth noting here that the relation (2.25) is the key to determining whether one can embed a given manifold in another manifold of higher dimension. Suppose we begin with an M-dimensional manifold possessing the metric $h_{ij}(u)$ when labelled with the coordinates $u^i (i = 1, 2, \ldots, M)$. In order to embed this manifold in an N-dimensional manifold (where $N > M$) with metric $g_{ab}(x)$ in the coordinates $x^a (a = 1, 2, \ldots, N)$, then one must be able to satisfy the relation (2.25).

2.15 Topology of manifolds

In this chapter we have discussed only the local geometry of manifolds, which is defined at any point by the line element (2.4) giving the distance between points with infinitesimal coordinate separations. In addition to this local geometry a manifold also has a global geometry or *topology*. The topology of a manifold is defined by identifying certain sets of points, that is, regarding them as being coincident. For example, in Figure 2.1, we identified the line AA' with the line BB'. This property can be detected by a 'bug' on the surface, since by continuing in a straight line in a certain direction, it can get back to where it started. Thus a topology (in this case the fact that the space is periodic in one of the coordinates) is an intrinsic property of a manifold.

We shall see that general relativity is a 'local' theory, in which the local geometry (or curvature) of the four-dimensional spacetime manifold at any point is determined by the energy density of matter and/or radiation at that point. The field equations of general relativity do *not* constrain the global topology of the spacetime manifold.

Exercises

2.1 In three-dimensional Euclidean space R^3, write down expressions for the change of coordinates from Cartesian coordinates $[x^a] = (x, y, z)$ to spherical polar coordinates $[x'^a] = (r, \theta, \phi)$. Obtain expressions for the transformation and inverse transformation matrices in terms of the primed coordinates. By calculating the Jacobians J and J' for the transformation and its inverse, find where the transformation is non-invertible.

2.2 Write down the line element for three-dimensional Euclidean space in spherical polar coordinates x^a and cylindrical polar coordinates x'^a. Hence identify the metric functions in each coordinate system and show that they obey

$$g'_{cd}(x') = g_{ab}(x)\frac{\partial x^a}{\partial x'^c}\frac{\partial x^b}{\partial x'^d}.$$

2.3 In three-dimensional Euclidean space a coordinate system x'^a is related to the Cartesian coordinates x^a by

$$x^1 = x'^1 + x'^2, \qquad x^2 = x'^1 - x'^2, \qquad x^3 = 2x'^1 x'^2 + x'^3.$$

Describe the coordinate surfaces in the primed system. Obtain the metric functions g'_{ab} in the primed system and hence show that these coordinates are not orthogonal. Calculate the volume element dV in the primed coordinate system.

2.4 Consider the surface of a 2-surface embedded in three-dimensional Euclidean space. In a *stereographic projection*, one assigns coordinates (ρ, ϕ) to each point on the surface of the sphere. The ϕ-coordinate is the standard azimuthal polar angle. The ρ-coordinate of each point is obtained by drawing a straight line in three dimensions from the south pole of the sphere through the point in question and extending the line until it intersects the tangent plane to the north pole of the sphere; the ρ-coordinate is then the distance in the tangent plane from the north pole to the intersection point. Show that the line element for the surface of the sphere in these coordinates is

$$ds^2 = \left(1 + \frac{\rho^2}{4a^2}\right)^{-2} (d\rho^2 + \rho^2 d\phi^2).$$

At what point(s) on the sphere are these coordinates degenerate? If instead one works in terms of the Cartesian coordinates x and y in the tangent plane at the north pole, what is the corresponding form of the line element? At what point(s) on the sphere are these new coordinates degenerate?

2.5 Consider the surface of the Earth, which we assume for simplicity to be a 2-sphere of radius a. In terms of standard polar coordinates (θ, ϕ), the *longitude* of a point, in radians, rather than the usual degrees, is simply ϕ (measured eastwards from the Greenwich meridian), whereas its latitude $\lambda = \pi/2 - \theta$ radians. Show that the line element on the Earth's surface in these coordinates is

$$ds^2 = a^2(d\lambda^2 + \cos^2\lambda\, d\phi^2).$$

To make a map of the Earth's surface, we introduce the functions $x = x(\lambda, \phi)$ and $y = y(\lambda, \phi)$ and use them as Cartesian coordinates on a flat rectangular piece of paper. Each choice of the two functions corresponds to a different *map projection*. The *Mercator projection* is defined by

$$x = \frac{W\phi}{2\pi}, \qquad y = \frac{H}{2\pi}\ln\left[\tan\left(\frac{\pi}{4} + \frac{\lambda}{2}\right)\right],$$

where W and H are the width and height of the map respectively. Find the line element for this projection.

2.6 For the general map projection discussed in Exercise 2.5, show that the angle between two directions at some point on the Earth's surface will equal the angle between the corresponding directions on the map, provided that the functions x and y are chosen such that

$$\Omega(x, y)(dx^2 + dy^2) = a^2(d\lambda^2 + \cos^2\lambda\, d\phi^2),$$

for some function $\Omega(x, y)$. Show that the Mercator projection satisfies this condition. Write down the general requirement on x and y for an *equal-area* projection, in which the area of any region of the map is proportional to the corresponding area on the Earth's surface. Find such a projection. Is it possible to obtain a projection that simultaneously is equal-area and preserves angles?

2.7 A *conformal transformation* is not a change of coordinates but an actual change in the geometry of a manifold such that the metric tensor transforms as

$$\tilde{g}_{ab}(x) = \Omega^2(x) g_{ab}(x),$$

where $\Omega(x)$ is some non-vanishing scalar function of position. In a pseudo-Riemannian manifold, show that if $x^a(\lambda)$ is a null curve with respect to g_{ab} (i.e. $ds^2 = 0$ along the curve), then it is also a null curve with respect to $\tilde{g}_{ab}$. Is this true for timelike curves?

2.8 A curve on the surface of a 2-sphere of radius a is defined parametrically by $\theta = u$, $\phi = 2u - \pi$, where $0 \leq u \leq \pi$. Sketch the curve and show that its total length is

$$L = a\int_0^\pi \sqrt{1 + 4\sin^2 u}\, du.$$

Show that, in general, the length of a curve is independent of the parameter used to describe it.

2.9 Show that the line element of a 3-sphere of radius a embedded in four-dimensional Euclidean space can be written in the form

$$ds^2 = a^2[d\chi^2 + \sin^2\chi(d\theta^2 + \sin^2\theta\, d\phi^2)].$$

Hence, in this three-dimensional non-Euclidean space, calculate the area of the 2-sphere defined by $\chi = \chi_0$. Also find the total volume of the three-dimensional space.

2.10 Consider the three-dimensional space with line element

$$ds^2 = \frac{dr^2}{1 - 2\mu/r} + r^2(d\theta^2 + \sin^2\theta\, d\phi^2),$$

and calculate the following quantities:

(a) the area of a sphere of coordinate radius $r = R$;
(b) the 3-volume of a sphere of coordinate radius $r = R$;
(c) the radial distance between the sphere $r = 2\mu$ and the sphere $r = 3\mu$;
(d) the 3-volume contained between the two spheres in part (c).

Verify that your answers reduce to the usual Euclidean results in the limit $\mu \to 0$.

2.11 Prove the following results used in Section 2.11:

(a) $(\partial x^a/\partial x'^b)_P$ has N^2 independent values;
(b) $(\partial^2 x^a/\partial x'^b \partial x'^c)_P$ has $\frac{1}{2}N^2(N+1)$ independent values;
(c) $(\partial^3 x^a/\partial x'^b \partial x'^c \partial x'^d)_P$ has $\frac{1}{6}N^2(N+1)(N+2)$ independent values;
(d) $g'_{ab}(P)$ has $\frac{1}{2}N(N+1)$ independent values;
(e) $(\partial g'_{ab}/\partial x'^c)_P$ has $\frac{1}{2}N^2(N+1)$ independent values;
(f) $\left(\partial^2 g'_{ab}/\partial x'^c \partial x'^d\right)_P$ has $\frac{1}{4}N^2(N+1)^2$ independent values.

Hence show that, in a general Riemannian manifold, at least $N^2(N^2-1)/12$ of the second derivatives $\left(\partial^2 g'_{ab}/\partial x'^c \partial x'^d\right)_P$ will not vanish in any coordinate system.

2.12 Consider the two-dimensional space with line element

$$ds^2 = \frac{dr^2}{1 - 2\mu/r} + r^2\, d\phi^2.$$

Using the result (2.25), show that this geometry can be embedded in three-dimensional Euclidean space, and find the equations for the corresponding two-dimensional surface.

2.13 By identifying a suitable coordinate transformation, show that the line element

$$ds^2 = (c^2 - a^2 t^2)\, dt^2 - 2at\, dt\, dx - dx^2 - dy^2 - dz^2,$$

where a is a constant, can be reduced to the Minkowski line element.

3

Vector calculus on manifolds

The notion of a vector is extremely useful in describing physical processes and is employed in nearly all branches of mathematical physics. The reader should be familiar with vector calculus in two- and three-dimensional Euclidean spaces and with the description of vectors in terms of their components in simple coordinate systems such as Cartesian or spherical polar coordinates.

The concept of vectors is also very useful in both special and general relativity, and we now consider how to generalise our familiar Euclidean ideas in order to define vectors in a general (pseudo-)Riemannian manifold and in arbitrary coordinate systems. For illustration, however, we will often consider two-dimensional Riemannian manifolds that can be envisaged as surfaces embedded in three-dimensional Euclidean space. An example is the surface of a sphere, which we might take to be the surface of the Earth (remembering to consider ourselves as truly two-dimensional 'bugs'!).

3.1 Scalar fields on manifolds

Before considering vector fields on manifolds, let us briefly discuss scalar fields. A real (or complex) scalar field defined on (some region of) a manifold $\mathcal{M}$ assigns a real (or complex) value to each point P in (that region of) $\mathcal{M}$; an example is the air temperature on the surface of the Earth. If one labels the points in $\mathcal{M}$ using some coordinate system x^a then one can express the value at each point as a function of the coordinates $\phi(x^a)$. The value of the scalar field at any point P does not depend on the chosen coordinate system. Thus, under an arbitrary coordinate transformation $x^a \to x'^a$, the scalar field is described by a *different function* $\phi'(x'^a)$ of the new coordinates, such that

$$\boxed{\phi'(x'^a) = \phi(x^a).}$$

Indeed, this is the defining characteristic for a scalar field.

3.2 Vector fields on manifolds

A vector field defined on (some region of) a manifold $\mathcal{M}$ assigns a single vector to each point P in (that region of) $\mathcal{M}$. The vector at P is often drawn as an extended directed line segment with its base at P, but this convention requires careful interpretation on general manifolds. Once again it is convenient to illustrate our discussion by considering a two-dimensional manifold such as the spherical surface of the Earth. Let us consider, for example, the vector field defined by the wind velocity (at ground level). Wind velocity is measured at a given observation point and refers solely to that point, despite the visual convenience of showing it on a chart as an arrow apparently extending for a long distance. It is an example of a *local* vector. Other examples include momentum, current density and velocity in general. Such vectors are defined at a given point P. More accurately, they can be measured by an observer (bug) in a laboratory covering a *small* region of the manifold in the neighbourhood of P.

At an arbitrary point P in the manifold, any local vector $\boldsymbol{v}$ lies in the *tangent space* T_P to the manifold at P. Indeed, T_P consists of the set of all (local) vectors at the point P. This may be visualised simply for two-dimensional manifolds by embedding them as surfaces in three-dimensional Euclidean space (see Figure 3.1), but the idea is easily extended to higher dimensions and can be defined independently of any embedding. As we discussed in Chapter 2, the tangent space at any point of a (pseudo-)Riemannian manifold is a (pseudo-)Euclidean space of the *same* dimensionality. Moreover, at an arbitrary point P, local vectors obey all the usual rules of vector algebra in (pseudo-)Euclidean geometry.

It is important to realise, however, that local vectors defined at *different* points P and Q in the manifold lie in *different* tangent spaces. Thus there is no way of adding local vectors at different points. Other notions that must be abandoned are those of position vectors and displacement vectors, which clearly are not locally

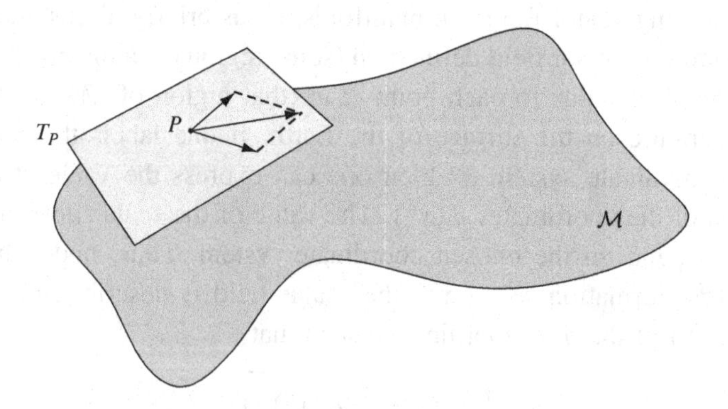

Figure 3.1 Local vectors defined at the point P lie in the tangent space T_P to the manifold at that point.

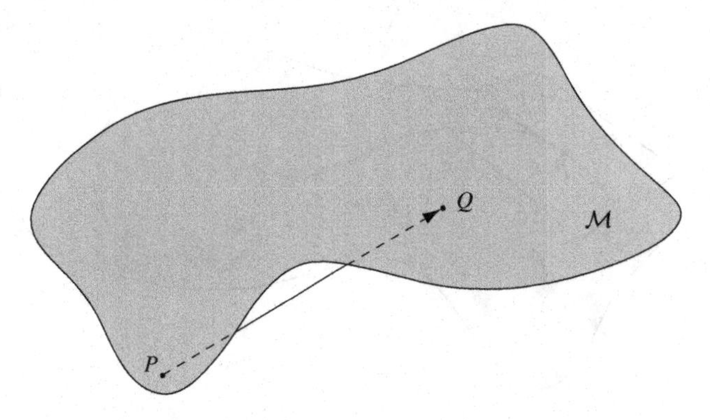

Figure 3.2 The displacement vector between two general points P and Q does not lie in the manifold $\mathcal{M}$, unless the manifold is itself Euclidean.

defined. Using an embedding picture of a two-dimensional manifold, this is may be visualised as shown in Figure 3.2. The 'displacement vector' connecting the points P and Q does not lie in the manifold and thus has no intrinsic geometrical meaning. Heuristically, however, we *can* define the displacement vector $\delta \boldsymbol{s}$ between two *nearby points* P and Q, since this is a local quantity. In the limit $Q \to P$, the vector $\delta \boldsymbol{s}$ lies in the tangent space at P.

Clearly, if the original manifold is itself (pseudo-)Euclidean then the tangent space at any point coincides with the manifold. Thus vectors defined at different points in the manifold do lie in the same space, and the notions of position and displacement vectors are valid. This reflects our common experience of vector algebra.

3.3 Tangent vector to a curve

The most obvious example of a vector field defined on (a subregion of) a manifold is the *tangent vector* to a curve $\mathcal{C}$, which is defined at each point along $\mathcal{C}$. The notion of a tangent vector to a curve is also central to our subsequent development of basis vectors, described below.

Consider a curve $\mathcal{C}$ in an N-dimensional manifold. This curve may be described by the N parametric equations $x^a(u)$, where u is some general parameter that varies along the curve. At any point P along $\mathcal{C}$, the tangent vector $\boldsymbol{t}$ to the curve, with respect to the parameter value u, is defined as

$$\boldsymbol{t} = \lim_{\delta u \to 0} \frac{\delta \boldsymbol{s}}{\delta u}, \tag{3.1}$$

where $\delta \boldsymbol{s}$ is the infinitesimal separation vector between the point P and some nearby point Q on the curve corresponding to the parameter value $u + \delta u$. Clearly $\boldsymbol{t}$ will lie in the tangent space T_P at the point P; this is illustrated in Figure 3.3.

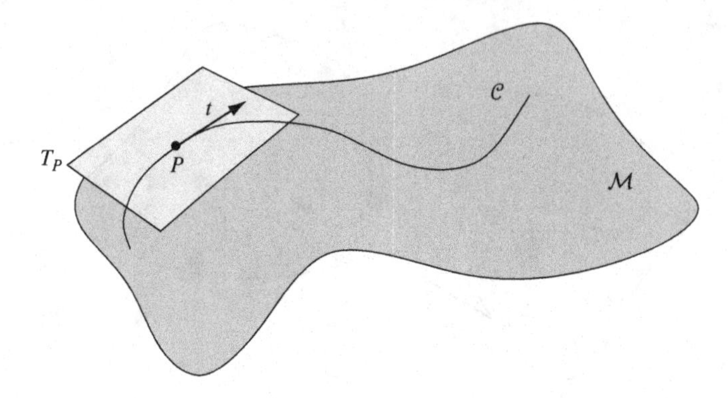

Figure 3.3 The tangent vector $\boldsymbol{t}$ to the curve $\mathcal{C}$ at a point P.

Although the heuristic approach we have adopted here is perfectly adequate for our purposes, in a general manifold the formal mathematical device for constructing the tangent vector to some curve at P is to identify $\boldsymbol{t}$ with the *directional derivative* operator along the curve at that point. This is discussed further in Appendix 3A and, in fact, enables one to give a precise mathematical meaning to the general notion of a vector in a non-Euclidean manifold.

3.4 Basis vectors

As we have seen, a vector field on a manifold is defined simply by giving, in a smooth manner, a prescription for a local vector $\boldsymbol{v}(x)$ at each point in the manifold. At each point P the vector lies in the tangent space T_P at that point. This vector is a geometrical entity, defined *independently* of any coordinate system with which we choose to label points in the manifold. Nevertheless, at each point P we can define a set of *basis vectors* $\boldsymbol{e}_a$ for the tangent space T_P, the number of such vectors being equal to the dimension of T_P and hence of $\mathcal{M}$ (how this may be achieved will be discussed shortly). Any vector at P can then be expressed as a linear combination of these basis vectors, provided that they are linearly independent, which we will assume is always the case. Thus, we can express the local vector field $\boldsymbol{v}(x)$ at each point in terms of basis vectors $\boldsymbol{e}_a(x)$ defined at each point:

$$\boxed{\boldsymbol{v}(x) = v^a(x)\, \boldsymbol{e}_a(x).}$$

The numbers $v^a(x)$ are known as the *contravariant components* of the vector field $\boldsymbol{v}(x)$ in the basis $\boldsymbol{e}_a(x)$.

For any set of basis vectors $\boldsymbol{e}_a(x)$, we can define a second set of vectors called the *dual basis vectors*. Instead of denoting the dual basis vectors by some other kernel letter, it is the convention to denote a member of this second basis set by

$\boldsymbol{e}^a(x)$. Although the positioning of the index may seem odd (not least because of the possible confusion with powers), it enables effective use of the summation convention that we shall adopt in due course. At any point P, the dual basis vectors are defined by the relation

$$\boxed{\boldsymbol{e}^a(x) \cdot \boldsymbol{e}_b(x) = \delta^a_b,} \tag{3.2}$$

so that $\boldsymbol{e}^a$ and $\boldsymbol{e}_a$ form *reciprocal* systems of vectors.

The dual basis vectors at P also lie in the tangent space T_P and form an alternative basis for it.[1] Thus, we can also express the local vector field $\boldsymbol{v}(x)$ at each point as a linear combination of the dual basis vectors $\boldsymbol{e}^a(x)$ defined at that point:

$$\boxed{\boldsymbol{v}(x) = v_a(x)\, \boldsymbol{e}^a(x).}$$

The numbers $v_a(x)$ are known as the *covariant components* of the vector $\boldsymbol{v}(x)$ in the basis $\boldsymbol{e}^a(x)$.

Using the relation (3.2) we can find simple expressions for the contravariant and covariant components of a vector $\boldsymbol{v}$. For example,[2]

$$\boldsymbol{v} \cdot \boldsymbol{e}^a = v^b \boldsymbol{e}_b \cdot \boldsymbol{e}^a = v^b \delta^a_b = v^a,$$

where we have used the fact that δ^b_a can be used to *replace one index with another*. Thus we may write $v^a = \boldsymbol{v} \cdot \boldsymbol{e}^a$. Similarly, we may show that $v_a = \boldsymbol{v} \cdot \boldsymbol{e}_a$. We now consider how a set of basis vectors (and their duals) may be constructed at each point P in the manifold.

Coordinate basis vectors

An obvious basis in which to describe local vectors is the *coordinate basis*. In any particular coordinate system x^a, we can define at every point P of the manifold a set of N *coordinate basis vectors*

$$\boxed{\boldsymbol{e}_a = \lim_{\delta x^a \to 0} \frac{\delta \boldsymbol{s}}{\delta x^a},} \tag{3.3}$$

where $\delta \boldsymbol{s}$ is the infinitesimal vector displacement between P and a nearby point Q whose coordinate separation from P is δx^a along the x^a coordinate curve. Thus $\boldsymbol{e}_a$ is the *tangent vector* to the x^a coordinate curve at the point P. This set of vectors provides a basis for the tangent space T_P at the point P (see Figure 3.4).

[1] More precisely, these vectors define the dual tangent space T^*_P at P, but this subtlety need not concern us here.

[2] From now on we will no longer make explicit the dependence of the basis vectors and components on the position x in the manifold, except where including this argument makes the explanation clearer.

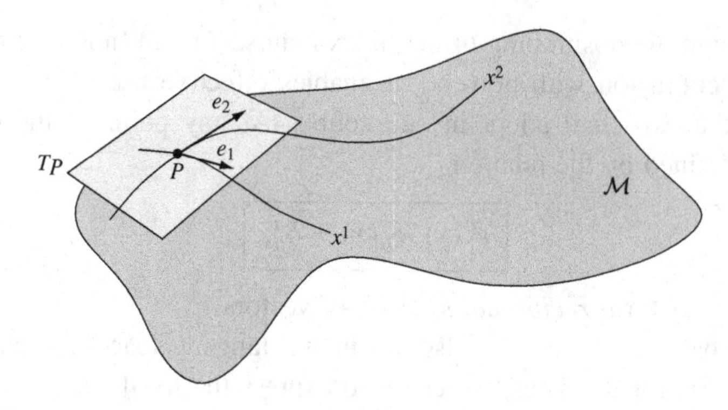

Figure 3.4 The coordinate basis vectors $\boldsymbol{e}_a$ at a point P in a manifold are the tangent vectors to the coordinate curves in the manifold and form a basis for the tangent space at P.

From the definition (3.3), we see that if two nearby points P and Q have coordinates x^a and $x^a + dx^a$ respectively, where now we allow dx^a to be non-zero for all a, then their infinitesimal vector separation is given by

$$d\boldsymbol{s} = \boldsymbol{e}_a(x)\, dx^a. \tag{3.4}$$

We can use this expression to relate the inner product of the coordinate basis vectors at some arbitrary point P to the value of the metric functions $g_{ab}(x)$ at that point. From (3.4), we have

$$ds^2 = d\boldsymbol{s} \cdot d\boldsymbol{s} = (dx^a \boldsymbol{e}_a) \cdot (dx^b \boldsymbol{e}_b) = (\boldsymbol{e}_a \cdot \boldsymbol{e}_b)\, dx^a\, dx^b.$$

Comparing this with the standard expression $ds^2 = g_{ab}(x)\, dx^a\, dx^b$, (2.4), for the line element, we find that

$$\boxed{\boldsymbol{e}_a(x) \cdot \boldsymbol{e}_b(x) = g_{ab}(x).} \tag{3.5}$$

Thus, quite generally, in a coordinate basis the scalar product of two vectors is given by

$$\boldsymbol{v} \cdot \boldsymbol{w} = (v^a \boldsymbol{e}_a) \cdot (w^b \boldsymbol{e}_b) = g_{ab} v^a w^b.$$

If the basis $\boldsymbol{e}^a(x)$ is dual to a coordinate basis $\boldsymbol{e}_a(x)$ then the a-coordinate distance between two nearby points separated by the displacement vector $d\boldsymbol{s}$ is given by

$$dx^a = \boldsymbol{e}^a \cdot d\boldsymbol{s}.$$

Moreover, in this case we may use the dual basis vectors to define the quantities

$$\boxed{g^{ab}(x) = \boldsymbol{e}^a(x) \cdot \boldsymbol{e}^b(x),} \tag{3.6}$$

which, as we will show, form the *contravariant components of the metric tensor* and are in general different from the quantites $g_{ab}(x)$; we will return to these later.

Orthonormal basis vectors at a point

At any given point P in a manifold, it is often useful to define a set of *orthonormal* basis vectors $\hat{\boldsymbol{e}}_a$ in T_P, which are chosen to be of unit length and orthogonal to one another. This is expressed mathematically by the requirement that at P

$$\boxed{\hat{\boldsymbol{e}}_a \cdot \hat{\boldsymbol{e}}_b = \eta_{ab},} \tag{3.7}$$

where $[\eta_{ab}] = \mathrm{diag}(\pm 1, \pm 1, \ldots, \pm 1)$ is the Cartesian line element of the tangent space T_P and depends on the signature of the (in general) pseudo-Riemannian manifold (see Section 2.13). These orthonormal basis vectors need not be related to any particular coordinate system that we are using to label the manifold, although they can always be defined by, for example, giving their components in a coordinate basis. Moreover, it is clear from (3.7) that the orthonormal basis vectors $\hat{\boldsymbol{e}}_a$ at P are in fact the *coordinate basis vectors* of a coordinate system for which $g_{ab}(P) = \eta_{ab}$ (or $g_{ab}(P) = \delta_{ab}$ for a strictly Riemannian manifold).

3.5 Raising and lowering vector indices

Unless otherwise stated, we will assume that we are working with a coordinate basis, as discussed above, and its dual. The contravariant and covariant components in these bases are equally good ways of specifying a vector. The link between them is found by considering the different ways in which one can write the scalar product $\boldsymbol{v} \cdot \boldsymbol{w}$ of two vectors. First, we can write

$$\boldsymbol{v} \cdot \boldsymbol{w} = (v^a \boldsymbol{e}_a) \cdot (w^b \boldsymbol{e}_b) = (\boldsymbol{e}_a \cdot \boldsymbol{e}_b) v^a w^b = g_{ab} v^a w^b,$$

where we have used the contravariant components of the two vectors. Similarly, using the covariant components, we can write the scalar product as

$$\boldsymbol{v} \cdot \boldsymbol{w} = (v_a \boldsymbol{e}^a) \cdot (w_b \boldsymbol{e}^b) = (\boldsymbol{e}^a \cdot \boldsymbol{e}^b) v_a w_b = g^{ab} v_a w_b.$$

Finally, we could express the scalar product in terms of the contravariant components of one vector and the covariant components of the other,

$$\boldsymbol{v} \cdot \boldsymbol{w} = (v^a \boldsymbol{e}_a) \cdot (w_b \boldsymbol{e}^b) = v^a w_b (\boldsymbol{e}_a \cdot \boldsymbol{e}^b) = v^a w_b \delta^b_a = v^a w_a;$$

similarly, we could write

$$\boldsymbol{v} \cdot \boldsymbol{w} = (v_a \boldsymbol{e}^a) \cdot (w^b \boldsymbol{e}_b) = v_a w^b (\boldsymbol{e}^a \cdot \boldsymbol{e}_b) = v_a w^b \delta^a_b = v_a w^a.$$

By comparing these four alternative expressions for the scalar product of two vectors, we can deduce one of the most useful properties of the quantities g_{ab} and g^{ab}. Since $g_{ab}v^a w^b = v^a w_a$ holds for any arbitrary vector $\boldsymbol{v}$, it follows that

$$\boxed{g_{ab}w^b = w_a,}$$

which illustrates the fact that the quantities g_{ab} can be used to *lower an index*. In other words, we can obtain the covariant components of a vector from its contravariant components. By a similar argument, we have

$$\boxed{g^{ab}w_b = w^a,}$$

so that the quantities g^{ab} can be used to perform the reverse process of *raising an index*. It is straightforward to show that the coordinate and dual basis vectors themselves are related in an analogous way by

$$\boxed{\boldsymbol{e}_a = g_{ab}\boldsymbol{e}^b \qquad \text{and} \qquad \boldsymbol{e}^a = g^{ab}\boldsymbol{e}_b.}$$

We will now prove the useful result that the matrix $[g^{ab}]$ containing the contravariant components of the metric tensor is the *inverse* of the matrix $[g_{ab}]$ that contains its covariant components. Using the index-lowering and index-raising action of g_{ab} and g^{ab} on the components of an arbitrary vector $\boldsymbol{v}$, we find that

$$\delta^a_c v^c = v^a = g^{ab}v_b = g^{ab}g_{bc}v^c,$$

but since $\boldsymbol{v}$ is arbitrary we must have

$$\boxed{g^{ab}g_{bc} = \delta^a_c.} \tag{3.8}$$

Denoting the matrix $[g_{ab}]$ by G and the matrix $[g^{ab}]$ by $\tilde{\mathsf{G}}$, this equation can be written in matrix form as $\mathsf{G}\tilde{\mathsf{G}} = \mathsf{I}$. Hence G and $\tilde{\mathsf{G}}$ are inverse matrices.

3.6 Basis vectors and coordinate transformations

Let us consider a coordinate transformation $x^a \to x'^a$ on a manifold. There is a simple relationship between the coordinate basis vectors $\boldsymbol{e}_a$ associated with the coordinate system x^a and the coordinate basis vectors $\boldsymbol{e}'_a$ associated with the new system of coordinates x'^a. It can be found by considering the infinitesimal displacement vector $d\boldsymbol{s}$ between two nearby points P and Q. Clearly, this displacement cannot depend on the coordinate system being used, so we must have

$$d\boldsymbol{s} = dx^a\boldsymbol{e}_a = dx'^a\boldsymbol{e}'_a.$$

Noting that $dx^a = (\partial x^a/\partial x'^b)\, dx'^b$, we find that at any point P the two sets of coordinate basis vectors are related by

$$\boxed{\boldsymbol{e}'_a = \frac{\partial x^b}{\partial x'^a}\, \boldsymbol{e}_b,} \tag{3.9}$$

where the partial derivative is evaluated at the point P. Repeating this calculation using the dual basis vectors, we find that

$$\boxed{\boldsymbol{e}'^a = \frac{\partial x'^a}{\partial x^b}\, \boldsymbol{e}^b.} \tag{3.10}$$

Using (3.9) and (3.10), we can now calculate how the components of any general vector $\boldsymbol{v}$ must transform under the coordinate transformation. Since a vector is a geometrical entity that is independent of the coordinate system, we have (for example)

$$\boldsymbol{v} = v^a \boldsymbol{e}_a = v'^a \boldsymbol{e}'_a.$$

So, the new contravariant components are given by

$$v'^a = \boldsymbol{e}'^a \cdot \boldsymbol{v} = \frac{\partial x'^a}{\partial x^b}\, \boldsymbol{e}^b \cdot \boldsymbol{v} \qquad \Rightarrow \qquad \boxed{v'^a = \frac{\partial x'^a}{\partial x^b} v^b.}$$

Similarly, the new covariant components are given by

$$v'_a = \boldsymbol{e}'_a \cdot \boldsymbol{v} = \frac{\partial x^b}{\partial x'^a}\, \boldsymbol{e}_b \cdot \boldsymbol{v} \qquad \Rightarrow \qquad \boxed{v'_a = \frac{\partial x^b}{\partial x'^a} v_b.}$$

3.7 Coordinate-independent properties of vectors

As we have seen, in a coordinate basis and its dual the scalar product $\boldsymbol{v}\cdot\boldsymbol{w}$ of two vectors at each point P of the manifold can be written in four ways:

$$\boxed{g_{ab}v^a w^b = g^{ab} v_a w_b = v^a w_a = v_a w^a.}$$

Using the transformation properties of the metric coefficients g_{ab} and those of the vector components, it is straightforward to show that these expressions yield the *same* result in any other coordinate system.

In a strictly Riemannian manifold the scalar product is *positive definite*, which means that $g_{ab}v^a v^b \geq 0$ for all vectors v^a, with $g_{ab}v^a v^b = 0$ only if $v^a = 0$. In a pseudo-Riemannian space, however, this condition is relaxed and leads to some rather odd properties, such as the possibility of non-zero vectors having zero length. We must therefore make definitions that allow us to deal with such properties in a way that extends and generalises familiar concepts in Euclidean space.

The *length* of a vector $|\boldsymbol{v}|$ is defined in terms of its components by

$$|g_{ab}v^a v^b|^{1/2} = |g^{ab}v_a v_b|^{1/2} = |v_a v^a|^{1/2}.$$

A *unit* vector has length unity. As remarked above, in a pseudo-Riemannian manifold we can have $|v_a v^a|^{1/2} = 0$ for $v^a \neq 0$, in which case the vector $\boldsymbol{v}$ is described as *null*.

The *angle* θ between two non-null vectors $\boldsymbol{v}$ and $\boldsymbol{w}$ is defined by

$$\cos\theta = \frac{v_a w^a}{|v_b v^b|^{1/2}|w_c w^c|^{1/2}}.$$

In a pseudo-Riemannian manifold, this formula can lead to $|\cos\theta| > 1$, resulting in a non-real value for θ.

Two vectors are *orthogonal* if their scalar product is zero. This definition makes sense even if one or both of the vectors is or are null. In fact, a null vector is a non-zero vector that is orthogonal to itself.

3.8 Derivatives of basis vectors and the affine connection

As we have said, local vectors at different points P and Q in a manifold lie in different tangent spaces, so there is no way of adding or subtracting them. In order to define the derivative of a vector field, however, one must compare vectors at different points, albeit in the limit where the distance between the points tends to zero. We will adopt here an intuitive approach that is sufficient for our purposes in developing vector calculus on curved manifolds and provides a simple geometrical picture. Specifically, on this occasion, we will assume the manifold to be embedded in a higher-dimensional (pseudo-)Euclidean space, which thus allows vectors at different points to be compared.[3]

In some arbitrary coordinate system x^a on the manifold, let us consider the basis vectors $\boldsymbol{e}_a$ at two nearby points P and Q with coordinates x^a and $x^a + \delta x^a$ respectively (see Figure 3.5). In general, the basis vectors at Q will differ infinitesimally from those at P, so that

$$\boldsymbol{e}_a(Q) = \boldsymbol{e}_a(P) + \delta\boldsymbol{e}_a.$$

[3] It is worth noting that one can embed any four-dimensional torsionless (pseudo-)Riemannian manifold in some (pseudo-)Euclidean space of sufficiently higher dimension; see, for example, J. Nash, The imbedding problem for Riemannian manifolds, *Annals of Mathematics* **63**, 20–63, 1956 and C. Clarke, On the global isometric embedding of pseudo-Riemannian manifolds, *Proceedings of the Royal Society* **A314**, 417–28, 1970. Indeed, recent theoretical work on *braneworld* models suggests that our spacetime may indeed be embedded in some higher-dimensional manifold! Alternatively, one can define the derivative of a vector field on a general manifold without using an embedding picture, but in a rather more formal manner; see, for example, R.M. Wald, *General Relativity*, University of Chicago Press, 1984.

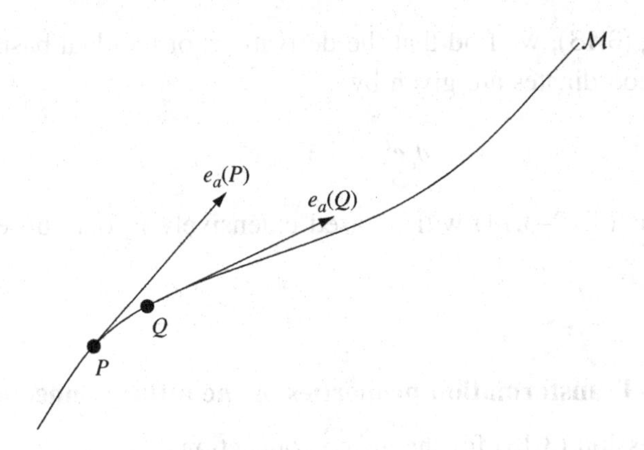

Figure 3.5 The basis vectors $\boldsymbol{e}_a(P)$ and $\boldsymbol{e}_a(Q)$ lie in the tangent spaces to the manifold $\mathcal{M}$ at the points P and Q respectively.

The standard partial derivative of the basis vector is given by $\delta \boldsymbol{e}_a/\delta x^c$ in the limit $\delta x^c \to 0$. In general, however, the resulting vector will not lie in the tangent space to the manifold at P. We thus define the derivative *in the manifold* of the coordinate basis vector by projecting into the tangent space at P,

$$\frac{\partial \boldsymbol{e}_a}{\partial x^c} \equiv \left(\lim_{\delta x^c \to 0} \frac{\delta \boldsymbol{e}_a}{\delta x^c} \right)_{\parallel T_P}. \tag{3.11}$$

Now we can expand this derivative vector in terms of the basis vectors $\boldsymbol{e}_a(P)$ at the point P, and write

$$\boxed{\frac{\partial \boldsymbol{e}_a}{\partial x^c} = \Gamma^b{}_{ac} \boldsymbol{e}_b,} \tag{3.12}$$

where the N^3 coefficients Γ^b_{ac} are known collectively as the *affine connection* or, in older textbooks, the *Christoffel symbol* (of the second kind) at the point P. From (3.11), it is also clear that the derivative operator obeys Leibnitz' theorem when acting on expressions containing only vectors in T_P. By taking the scalar product of (3.12) with the dual basis vector $\boldsymbol{e}^d$ and using the reciprocity relation (3.2), we can also write the affine connection as[4]

$$\Gamma^b{}_{ac} = \boldsymbol{e}^b \cdot \partial_c \boldsymbol{e}_a. \tag{3.13}$$

Furthermore, by differentiating the reciprocity relation $\boldsymbol{e}^a \cdot \boldsymbol{e}_b = \delta^b_a$ with respect to the coordinate x^c, we find that

$$\partial_c(\boldsymbol{e}^a \cdot \boldsymbol{e}_b) = (\partial_c \boldsymbol{e}^a) \cdot \boldsymbol{e}_b + \boldsymbol{e}^a \cdot (\partial_c \boldsymbol{e}_b) = 0.$$

[4] From now on, we shall often use the shorthand ∂_c to denote $\partial/\partial x^c$. We also note here that, in some textbooks, an even more terse notation is used, in which partial differentation is denoted by a comma. For example, the partial derivative $\partial_c v^a$ of the contravariant components of a vector would be written $v^a{}_{,c}$.

Then, on using (3.13), we find that the derivatives of the dual basis vectors with respect to the coordinates are given by

$$\partial_c \boldsymbol{e}^a = -\Gamma^a{}_{bc}\boldsymbol{e}^b. \tag{3.14}$$

The expressions (3.12–3.14) will be used extensively in our subsequent discussions.

3.9 Transformation properties of the affine connection

From the expression (3.13) for the affine connection,

$$\Gamma^a{}_{bc} = \boldsymbol{e}^a \cdot \frac{\partial \boldsymbol{e}_b}{\partial x^c}, \tag{3.15}$$

we see that, in some new coordinate system x'^a, it is given by

$$\Gamma'^a{}_{bc} = \boldsymbol{e}'^a \cdot \frac{\partial \boldsymbol{e}'_b}{\partial x'^c},$$

Substituting the expressions (3.9) and (3.10) for the new basis and dual basis vectors, we find

$$\begin{aligned}
\Gamma'^a{}_{bc} &= \frac{\partial x'^a}{\partial x^d}\,\boldsymbol{e}^d \cdot \frac{\partial}{\partial x'^c}\left(\frac{\partial x^f}{\partial x'^b}\,\boldsymbol{e}_f\right) \\
&= \frac{\partial x'^a}{\partial x^d}\,\boldsymbol{e}^d \cdot \left(\frac{\partial x^f}{\partial x'^b}\frac{\partial \boldsymbol{e}_f}{\partial x'^c} + \frac{\partial^2 x^f}{\partial x'^c \partial x'^b}\,\boldsymbol{e}_f\right) \\
&= \frac{\partial x'^a}{\partial x^d}\frac{\partial x^f}{\partial x'^b}\frac{\partial x^g}{\partial x'^c}\,\boldsymbol{e}^d \cdot \frac{\partial \boldsymbol{e}_f}{\partial x^g} + \frac{\partial x'^a}{\partial x^d}\frac{\partial^2 x^f}{\partial x'^c \partial x'^b}\,\boldsymbol{e}^d \cdot \boldsymbol{e}_f \\
&= \frac{\partial x'^a}{\partial x^d}\frac{\partial x^f}{\partial x'^b}\frac{\partial x^g}{\partial x'^c}\Gamma^d{}_{fg} + \frac{\partial x'^a}{\partial x^d}\frac{\partial^2 x^d}{\partial x'^c \partial x'^b},
\end{aligned} \tag{3.16}$$

where in the last line we have used the reciprocity relation (3.2) between the basis and dual basis vectors. We will see later that, because of the presence of the last term on the right-hand side of (3.16), the $\Gamma^a{}_{bc}$ do *not* transform as the components of a tensor.

By swapping derivatives with respect to x and x' in the last term on the right-hand side of (3.16), we arrive at an alternative (but equivalent) expression:

$$\boxed{\Gamma'^a{}_{bc} = \frac{\partial x'^a}{\partial x^d}\frac{\partial x^f}{\partial x'^b}\frac{\partial x^g}{\partial x'^c}\Gamma^d{}_{fg} - \frac{\partial x^d}{\partial x'^b}\frac{\partial x^f}{\partial x'^c}\frac{\partial^2 x'^a}{\partial x^d \partial x^f}.} \tag{3.17}$$

3.10 Relationship of the connection and the metric

The observant reader will have noticed that there was some arbitrariness in how we introduced the affine connection in (3.12). We could just as easily have written (3.12) with $\Gamma^b{}_{ac}$ replaced by $\Gamma^b{}_{ca}$, i.e. with the two subscripts interchanged. In a general Riemannian manifold, these two sets of quantities are *not* necessarily equal to one another. In fact, one can show that the quantities

$$T^b{}_{ac} = \Gamma^b{}_{ac} - \Gamma^b{}_{ca} \tag{3.18}$$

are the components of a third-rank tensor (see Chapter 4) called the *torsion tensor*. For our considerations of standard general relativity, however, we can assume that our manifolds are *torsionless*, so that $T^b{}_{ac} = 0$ in any coordinate system.[5] *Hence, from here onwards, we will assume (unless otherwise stated) that the affine connection is symmetric in its last two indices*, i.e.

$$\boxed{\Gamma^b{}_{ac} = \Gamma^b{}_{ca}.} \tag{3.19}$$

In a manifold that is torsionless, so that (3.19) is satisfied, there is a simple relationship between the affine connection $\Gamma^b{}_{ac}$ and the metric functions g_{ab}, which we now derive. From (3.5) we have $g_{ab} = \boldsymbol{e}_a \cdot \boldsymbol{e}_b$. Differentiating this expression with respect to x^c, we obtain

$$\begin{aligned} \partial_c g_{ab} &= (\partial_c \boldsymbol{e}_a) \cdot \boldsymbol{e}_b + \boldsymbol{e}_a \cdot (\partial_c \boldsymbol{e}_b) \\ &= \Gamma^d{}_{ac} \boldsymbol{e}_d \cdot \boldsymbol{e}_b + \boldsymbol{e}_a \cdot \Gamma^d{}_{bc} \boldsymbol{e}_d \\ &= \Gamma^d{}_{ac} g_{db} + \Gamma^d{}_{bc} g_{ad}. \end{aligned} \tag{3.20}$$

By cyclically permuting the indices a, b, c, we obtain two equivalent expressions,

$$\begin{aligned} \partial_b g_{ca} &= \Gamma^d{}_{cb} g_{da} + \Gamma^d{}_{ab} g_{cd}, \\ \partial_a g_{bc} &= \Gamma^d{}_{ba} g_{dc} + \Gamma^d{}_{ca} g_{bd}. \end{aligned}$$

Using these three expressions, we now form the combination

$$\begin{aligned} &\partial_c g_{ab} + \partial_b g_{ca} - \partial_a g_{bc} \\ &= \Gamma^d{}_{ac} g_{db} + \Gamma^d{}_{bc} g_{ad} + \Gamma^d{}_{cb} g_{da} + \Gamma^d{}_{ab} g_{cd} - \Gamma^d{}_{ba} g_{dc} - \Gamma^d{}_{ca} g_{bd} = 2\Gamma^d{}_{cb} g_{ad}, \end{aligned}$$

where, in obtaining the last line, we have used the assumed symmetry properties (3.19) of the affine connection and the symmetry of metric functions. Multiplying

[5] It is straightforward to show that any (pseudo-)Riemannian manifold that can be embedded in some (pseudo-) Euclidean space of higher dimension must be torsionless.

through by g^{ea}, recalling from (3.8) that $g^{ea}g_{ad} = \delta^e_a$ and relabelling indices, we finally obtain

$$\boxed{\Gamma^a{}_{bc} = \tfrac{1}{2}g^{ad}(\partial_b g_{dc} + \partial_c g_{bd} - \partial_d g_{bc}).} \tag{3.21}$$

In fact, the quantity defined by the right-hand side in (3.21) is properly called the *metric connection* and is often denoted by the symbol $\{{}^{\,a}_{bc}\}$. In a manifold with torsion, it will differ from the affine connection defined by (3.11) and (3.12). As we have shown, however, in a torsionless manifold the affine and metric connections are equivalent, and so $\Gamma^a{}_{bc}$ is usually referred to simply as the connection. Unless otherwise stated, we will follow this convention from now on.

Equation (3.21) is *very important*, because it tells us how to compute the connection at any point in a manifold. In other words, if one knows the metric g_{ab} in some coordinate system x^a then one can form the derivatives of g_{ab} appearing in (3.21) and hence calculate *all* the numbers $\Gamma^a{}_{bc}$ at *any* point.

We finish this section by establishing a few useful formulae involving the connection $\Gamma^a{}_{bc}$ and the related quantities

$$\boxed{\Gamma_{abc} \equiv g_{ad}\Gamma^d{}_{bc}.}$$

It is straightforward to show that $\Gamma^a{}_{bc} = g^{ad}\Gamma_{dbc}$. From (3.21), we find that

$$\Gamma_{abc} = \tfrac{1}{2}(\partial_b g_{ac} + \partial_c g_{ba} - \partial_a g_{bc}). \tag{3.22}$$

The quantity Γ_{abc} is traditionally known as a *Christoffel symbol of the first kind.* Adding Γ_{bac} to Γ_{abc} gives

$$\partial_c g_{ab} = \Gamma_{abc} + \Gamma_{bac}, \tag{3.23}$$

which allows us to express partial derivatives of the metric components in terms of the connection coefficients. If we denote the value of the determinant $\det[g_{ab}]$ by g then the cofactor of the element g_{ab} in this determinant is gg^{ab} (note that g is *not* a scalar: changing coordinates changes the value of g at any point). It follows that $\partial_c g = gg^{ab}(\partial_c g_{ab})$, so from (3.23) we have

$$\partial_c g = gg^{ab}(\Gamma_{abc} + \Gamma_{bac}) = g\left(\Gamma^b{}_{bc} + \Gamma^a{}_{ac}\right) = 2g\Gamma^a{}_{ac}. \tag{3.24}$$

The implied summation over a is an example of a *contraction* over a pair of indices (see Chapter 4); $\Gamma^a{}_{ac}$ means simply $\Gamma^1{}_{1c} + \Gamma^2{}_{2c} + \cdots + \Gamma^N{}_{Nc}$. Thus the contraction of the connection coefficients (3.21) is given by

$$\Gamma^a{}_{ab} = \tfrac{1}{2}g^{-1}\partial_b g = \tfrac{1}{2}\partial_b \ln|g|, \tag{3.25}$$

the modulus signs being needed if the manifold is *pseudo*-Riemannian. Alternatively, we can write

$$\Gamma^a{}_{ab} = \partial_b \ln \sqrt{|g|} = \frac{1}{\sqrt{|g|}} \partial_b \sqrt{|g|}. \qquad (3.26)$$

3.11 Local geodesic and Cartesian coordinates

In Chapter 2, we showed that at any point P in a pseudo-Riemannian manifold it is possible in principle to find *local Cartesian coordinates* x'^a such that

$$g'_{ab}(P) = \eta_{ab}, \qquad (3.27)$$

$$\left.\frac{\partial g'_{ab}}{\partial x'^c}\right|_P = 0, \qquad (3.28)$$

where $[\eta_{ab}] = \text{diag}(\pm 1, \pm 1, \ldots, \pm 1)$. The number of positive entries in $[\eta_{ab}]$ minus the number of negative entries is the signature of the manifold. Supposing that we start with some general system of coordinates x^a, we now show how to obtain local Cartesian coordinates in practice.

Let us begin by demanding that our new coordinate system x'^a satisfies the condition (3.28) but *not* necessarily the condition (3.27). From our expression (3.20) for the derivative of the metric in terms of the connection, we see that the condition (3.28) will be satisfied if the connection coefficients in the new coordinate system vanish at P, i.e.

$$\Gamma'^a{}_{bc}(P) = 0. \qquad (3.29)$$

Conversely, from (3.21) we see that the condition (3.28) implies (3.29). The condition (3.29) makes much simpler the mathematics of parallel transport, covariant differentiation and intrinsic differentiation (see later). Coordinates for which (3.29) holds are generally referred to as *geodesic coordinates* about P, but this is not always appropriate since they need not be based on geodesics (which we will also discuss later).

Suppose that we start with some arbitrary coordinate system x^a, the 'original' system, in which the point P has coordinates x^a_P. Let us now define a new system of coordinates x'^a by

$$\boxed{x'^a = x^a - x^a_P + \tfrac{1}{2}\Gamma^a{}_{bc}(P)\left(x^b - x^b_P\right)\left(x^c - x^c_P\right),} \qquad (3.30)$$

where the $\Gamma^a{}_{bc}(P)$ are the connection coefficients at P in the original coordinate system. Clearly, the origin of the new coordinate system is at P. Differentiation of (3.30) with respect to x^d yields

$$\frac{\partial x'^a}{\partial x^d} = \delta^a_d + \Gamma^a{}_{dc}(P)\,(x^c - x^c_P),$$

so that, at the point P, $\partial x'^a/\partial x^d = \delta^a_d$; its inverse is given by $\partial x^a/\partial x'^d = \delta^a_d$. Differentiating again we obtain

$$\frac{\partial^2 x'^a}{\partial x^e \partial x^d} = \Gamma^a{}_{dc}(P)\delta^c_e = \Gamma^a{}_{de}(P).$$

If we now substitute these results into the expression (3.17) for the transformation properties of the connection, we find that

$$\Gamma'^a{}_{bc}(P) = \delta^a_d \delta^f_b \delta^g_c \Gamma^d{}_{fg}(P) - \delta^d_b \delta^f_c \Gamma^a{}_{df}(P) = \Gamma^a{}_{bc}(P) - \Gamma^a{}_{bc}(P) = 0.$$

So in the new (primed) coordinate system the connection coefficients at P are zero, and from (3.29) we have a system of geodesic coordinates at P.

The metric functions $g'_{ab}(P)$ in the geodesic coordinates x'^a will not necessarily satisfy the condition (3.27). Nevertheless, we can obtain such a system of *local Cartesian coordinates* by making a second *linear* coordinate transformation

$$\boxed{x''^a = X^a{}_b x'^b,}$$

where the coefficients $X^a{}_b$ are *constants*. Thus we can bring the metric $g''_{ab}(P)$ in these coordinates into the form (3.27) *without* affecting its derivatives, so that (3.28) will still be satisfied. The required values of the coefficients $X^a{}_b$ were discussed in Section 2.13.

3.12 Covariant derivative of a vector

Suppose that a vector field $\boldsymbol{v}(x)$ is defined over some region of a manifold. We will consider the derivative of this vector field with respect to the coordinates labelling the points in the manifold. Let us begin by writing the vector in terms of its contravariant components $\boldsymbol{v} = v^a \boldsymbol{e}_a$. We thus obtain

$$\partial_b \boldsymbol{v} = (\partial_b v^a)\boldsymbol{e}_a + v^a(\partial_b \boldsymbol{e}_a), \tag{3.31}$$

where the second term arises because, in an arbitrary coordinate system, the coordinate basis vectors vary with the position in the manifold. If we defined locally Cartesian coordinates at some point P in the manifold then in the neighbourhood of this point the coordinate basis vectors are constant and so the second term would vanish at P (but not elsewhere, unless the manifold $\mathcal{M}$ is (pseudo-)Euclidean, so that the whole of $\mathcal{M}$ can be covered by a Cartesian coordinate system).

Using (3.13), we may write (3.31) as

$$\partial_b \boldsymbol{v} = (\partial_b v^a)\boldsymbol{e}_a + v^a \Gamma^c{}_{ab} \boldsymbol{e}_c.$$

Since a and c are dummy indices in the last term on the right-hand side, we may interchange them to obtain

$$\partial_b \boldsymbol{v} = (\partial_b v^a)\boldsymbol{e}_a + v^c \Gamma^a{}_{cb} \boldsymbol{e}_a = (\partial_b v^a + v^c \Gamma^a{}_{cb})\boldsymbol{e}_a.$$

The reason for interchanging the dummy indices is that we may then factor out $\boldsymbol{e}_a$. Thus, at any point P, we now have an expression for the derivative of a vector field with respect to the coordinates *in terms of the basis vectors of the coordinate system at P*. The quantity in brackets is called the *covariant derivative* of the vector components, and the standard notation for it is[6]

$$\boxed{\nabla_b v^a \equiv \partial_b v^a + \Gamma^a{}_{cb} v^c.} \tag{3.32}$$

Thus the derivative of the vector field $\boldsymbol{v}$ can be written in the compact notation

$$\partial_b \boldsymbol{v} = (\nabla_b v^a)\boldsymbol{e}_a.$$

We note that, in local geodesic coordinates about some point P, the second term in the covariant derivative (3.32) vanishes at P and thus reduces to the ordinary partial derivative.

So far we have considered only the covariant derivative of the contravariant components v^a of a vector. The corresponding result for the covariant components v_a may be found in a similar way, by considering the derivative of $\boldsymbol{v} = v_a \boldsymbol{e}^a$ and using (3.14) to obtain

$$\boxed{\nabla_b v_a = \partial_b v_a - \Gamma^c{}_{ab} v_c.} \tag{3.33}$$

Comparing the expressions (3.32) and (3.33) for the covariant derivatives of the contravariant and covariant components of a vector respectively, we see that there are some similarities and some differences. It may help to remember that the index with respect to which the covariant derivative is taken (b in this case) is also the last subscript on the connection; the remaining indices can then only be arranged in one way without raising or lowering them. Finally, the sign difference must be remembered: for a contravariant index (superscript) the sign is positive, whereas for a covariant index (subscript) the connection carries a minus sign.

We conclude this section by considering the covariant derivative of a scalar. The covariant derivative differs from the simple partial derivative only because the coordinate basis vectors change with position in the manifold. However, a

[6] In some textbooks, the covariant derivative is denoted by a semicolon, so that the covariant derivative $\nabla_b v^a$ would be written as $v^a{}_{;b}$.

scalar ϕ does not depend on the basis vectors at all, so its covariant derivative must be the same as its partial derivative, i.e.

$$\boxed{\nabla_b \phi = \partial_b \phi.} \tag{3.34}$$

3.13 Vector operators in component form

The equations of electromagnetism, fluid mechanics and many other areas of classical physics make use of vector calculus in three-dimensional Euclidean space, employing the gradient $\nabla\phi$ and the Laplacian $\nabla^2\phi$ of scalar fields, together with the divergence $\nabla \cdot \boldsymbol{v}$ and the curl $\nabla \times \boldsymbol{v}$ of a vector field. Explicit forms for these are given in many texts for useful coordinate systems such as Cartesian, cylindrical polar, spherical polar (typically the 11 coordinate systems in which Laplace's equation separates). The covariant derivative provides a unified picture of all these derivatives and a direct route to the explicit forms in an arbitrary coordinate system. Moreover, it allows for the generalisation of these operators to more general manifolds.

Gradient

The gradient of a scalar field ϕ is given simply by

$$\boxed{\nabla\phi = (\nabla_a \phi)\boldsymbol{e}^a = (\partial_a \phi)\boldsymbol{e}^a,} \tag{3.35}$$

since the covariant derivative of a scalar is the same as its partial derivative.

Divergence

Replacing the partial derivatives that occur in local Cartesian coordinates by covariant derivatives, which are valid in arbitrary coordinate systems, the divergence of a vector field is given by the scalar quantity

$$\nabla \cdot \boldsymbol{v} = \nabla_a v^a = \partial_a v^a + \Gamma^a{}_{ab} v^b.$$

Using the result (3.26) we can rewrite the divergence as

$$\boxed{\nabla \cdot \boldsymbol{v} \equiv \nabla_a v^a = \frac{1}{\sqrt{|g|}} \partial_a \left(\sqrt{|g|} v^a\right),} \tag{3.36}$$

where g is the determinant of the matrix $[g_{ab}]$.

Laplacian

If we replace $\boldsymbol{v}$ by $\nabla\phi$ in $\nabla\cdot\boldsymbol{v}$ then we obtain the Laplacian $\nabla^2\phi$. From (3.35), $\boldsymbol{v} = v_a\boldsymbol{e}^a = (\partial_a\phi)\boldsymbol{e}^a$, so the covariant components are $v_a = \partial_a\phi$. In (3.36), however, we require the contravariant components v^a. These may be obtained by raising the index with the metric to give

$$v^a = g^{ab}v_b = g^{ab}\partial_b\phi.$$

Substituting this into (3.36), we obtain

$$\boxed{\nabla^2\phi \equiv \nabla_a\nabla^a\phi = \frac{1}{\sqrt{|g|}}\partial_a\left(\sqrt{|g|}g^{ab}\partial_b\phi\right).}$$

It is worth noting that the symbol used for the Laplacian operator often depends on the dimensionality of the manifold being used. In particular, the triangular (three-sided) symbol ∇^2 that is commonly used in the three-dimensional (and N-dimensional cases) is replaced by the box-shaped (four-sided) symbol $\Box^2$ in four-dimensional spacetimes, in which case it is called the *d'Alembertian* operator.

Curl

The special form of the curl of a vector field, which is itself a vector, exists only in three dimensions. In its more general form, which is valid in higher dimensions, the curl is defined as a rank-2 antisymmetric tensor (see Chapter 4) with components

$$\boxed{(\text{curl}\,\boldsymbol{v})_{ab} = \nabla_a v_b - \nabla_b v_a.}$$

In fact this difference of covariant derivatives can be simplified, since

$$\nabla_a v_b - \nabla_b v_a = \partial_a v_b - \Gamma^c{}_{ba}v_c - \partial_b v_a + \Gamma^c{}_{ab}v_c = \partial_a v_b - \partial_b v_a,$$

where the connections have cancelled because of their symmetry properties.

3.14 Intrinsic derivative of a vector along a curve

Normally, we think of vector fields as functions of the coordinates x^a defined over some region of the manifold. However, we can also encounter vector fields that are defined only on some subspace of the manifold, and an extreme example occurs when the vector field $\boldsymbol{v}(u)$ is defined only along some curve $x^a(u)$ in the manifold; an example might be the spin 4-vector $\boldsymbol{s}(\tau)$ of a single particle along its worldline in spacetime. We now consider how to calculate the derivative of such a vector with respect to the parameter u along the curve.

Let us begin by writing the vector field at any point along the curve $\mathcal{C}$ as

$$\mathbf{v}(u) = v^a(u)\mathbf{e}_a(u),$$

where the $\mathbf{e}_a(u)$ are the coordinate basis vectors at the point on the curve corresponding to the parameter value u. Thus, the derivative of $\mathbf{v}$ along the curve $\mathcal{C}$ is given by

$$\frac{d\mathbf{v}}{du} = \frac{dv^a}{du}\mathbf{e}_a + v^a\frac{d\mathbf{e}_a}{du} = \frac{dv^a}{du}\mathbf{e}_a + v^a\frac{\partial \mathbf{e}_a}{\partial x^c}\frac{dx^c}{du},$$

where we have used the chain rule to rewrite the last term on the right-hand side; this is a valid procedure since the basis vectors $\mathbf{e}_a$ *are* also defined away from the curve $\mathcal{C}$. Using (3.13) to write the partial derivatives of the basis vectors in terms of the connection, we obtain

$$\frac{d\mathbf{v}}{du} = \frac{dv^a}{du}\mathbf{e}_a + \Gamma^b{}_{ac}v^a\frac{dx^c}{du}\mathbf{e}_b.$$

Interchanging the dummy indices a and b in the last term, we may factor out the basis vector, and we find that

$$\boxed{\frac{d\mathbf{v}}{du} = \left(\frac{dv^a}{du} + \Gamma^a{}_{bc}v^b\frac{dx^c}{du}\right)\mathbf{e}_a \equiv \frac{Dv^a}{Du}\mathbf{e}_a.} \tag{3.37}$$

The term in parentheses is called the *intrinsic* (or *absolute*) derivative of the components v^a along the curve $\mathcal{C}$ and is often denoted by Dv^a/Du as indicated. Similarly, the intrinsic derivative of the covariant components v_a of a vector is given by

$$\boxed{\frac{Dv_a}{Du} = \frac{dv_a}{du} - \Gamma^b{}_{ac}v_b\frac{dx^c}{du}.}$$

A convenient way to remember the form of the intrinsic derivative is to *pretend* that the vector $\mathbf{v}$ is in fact defined throughout (some region of) the manifold, i.e. not only along the curve $\mathcal{C}$. In some cases of interest, this may in fact be true anyway; for example, $\mathbf{v}$ might denote the 4-velocity of some distributed fluid. We can now differentiate the components v^a (say) with respect to the coordinates x^a. Thus we can write

$$\frac{dv^a}{du} = \frac{\partial v^a}{\partial x^c}\frac{dx^c}{du}.$$

Substituting this into (3.37), we can then factor out dx^c/du and recognise the other factor as the covariant derivative $\nabla_c v^a$. Thus we can write

$$\frac{Dv^a}{Du} = (\nabla_c v^a)\frac{dx^c}{du} \tag{3.38}$$

and similarly for the intrinsic derivatives of the covariant components. It must be remembered, however, that if $\boldsymbol{v}$ is only defined along the curve $\mathcal{C}$ then formally (3.38) is not defined and acts merely as an *aide-memoire*.

3.15 Parallel transport

Let us again consider some curve $\mathcal{C}$ in the manifold, given parameterically in some general coordinate system by $x^a(u)$. Moreover, let O be some initial point on the curve with parameter u_0 at which a vector $\boldsymbol{v}$ is defined. We can now think of 'transporting' $\boldsymbol{v}$ along $\mathcal{C}$ in such a way that

$$\frac{d\boldsymbol{v}}{du} = \mathbf{0} \tag{3.39}$$

is satisfied at each point along the curve. The result is a *'parallel' field of vectors at each point along* $\mathcal{C}$, generated by the *parallel transport* of $\boldsymbol{v}$.

In a (pseudo-)Euclidean manifold, the parallel transport of a vector has the simple geometrical interpretation that the vector $\boldsymbol{v}$ is transported without any change to its length or direction. This is illustrated in Figure 3.6 for a curve $\mathcal{C}$ in a two-dimensional Euclidean space (i.e. a plane). If the coordinates x^a are Cartesian, it is clear that the *components* v^a of the vector field satisfy

$$\frac{dv^a}{du} = 0. \tag{3.40}$$

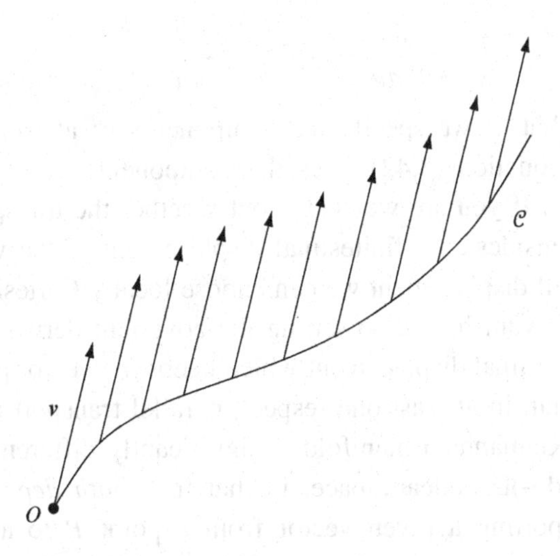

Figure 3.6 A parallel field of vectors $\boldsymbol{v}(u)$ generated by parallel transport along a curve $\mathcal{C}$ parameterised by u.

In an arbitrary coordinate system in the plane, however, (3.40) is no longer valid, and from (3.37) we see that it must be generalised to

$$\boxed{\frac{Dv^a}{Du} \equiv \frac{dv^a}{du} + \Gamma^a_{bc} v^b \frac{dx^c}{du} = 0.} \tag{3.41}$$

From the basic requirement (3.39), it is clear that (3.41) is equally valid for the parallel transport of a vector along a curve in any (pseudo-)Riemannian manifold in some arbitrary coordinate system x^a, although the geometrical interpretation is more subtle in this case. If one is willing to adopt a picture in which the (pseudo-) Riemannian manifold is embedded in a (pseudo-)Euclidean space of sufficiently higher dimension, then one can recover a simple geometrical interpretation of parallel transport. Consider some curve $\mathcal{C}$ in the (pseudo-)Riemannian manifold given in terms of some coordinate system in the manifold by $x^a(u)$. Let P and Q be two neighbouring points on the curve with affine parameter values u and $u + \delta u$ respectively. Starting with the vector $\mathbf{v}$ at P, which lies in the tangent space T_P, shift the vector to the neighbouring point Q while keeping it parallel to itself. In a Euclidean embedding space, this simply means transporting the vector without changing its length or direction. At the point Q the vector will not, in general, lie in the tangent space T_Q, on account of the curvature of the embedded manifold. Nevertheless, by considering only that part of the vector that is tangential to the embedded manifold at Q, we obtain a definite vector lying in T_Q. It is straightforward to show that this vector coincides with the parallel-transported vector at Q according to (3.41).

If we rewrite (3.41) as

$$\frac{dv^a}{du} = -\Gamma^a{}_{bc} v^b \frac{dx^c}{du}, \tag{3.42}$$

then we can see that, if we specify the components v^a at *some arbitrary point* along the curve, equation (3.42) fixes the components of v^a *along the entire length of the curve*. If you are worried about whether the transportation is really parallel, simply consider an infinitesimal displacement of the vector from some point P. For a small displacement we can choose locally Cartesian coordinates at P, in which the Γs vanish, and so setting the covariant derivative equal to zero describes an infinitesimal displacement which keeps the vector parallel ($dv^a = 0$).

We note here that, in at least one respect, parallel transport along curves in a general (pseudo-)Riemannian manifold is significantly different from that along curves in a (pseudo-)Euclidean space, in that it is *path dependent*: the vector obtained by transporting a given vector from a point P to a remote point Q depends on the route taken from P to Q. This path dependence is also apparent in transporting a vector around a closed loop, where on returning to the starting point the direction of the transported vector is (in general) different from the vector's initial direction. This path dependence can be demonstrated on a curved two-dimensional surface, and in general can be expressed mathematically in terms of the *curvature tensor* of the manifold. We will return to this topic in Chapter 7.

3.16 Null curves, non-null curves and affine parameters

So far, we have treated all curves in a manifold on a equal footing. In pseudo-Riemannian manifolds, however, it is important to distinguish between null curves and non-null curves. In the former, the interval ds between any two nearby points on the curve is zero, whereas in the latter case ds is non-zero. The distinction between these two types of curve may also be defined in terms of their tangent vectors, and this leads to the identification of a class of privileged parameters, called affine parameters, in terms of which the curves may be defined.

Consider some curve $x^a(u)$ in a general manifold. As discussed earlier, the tangent vector $\boldsymbol{t}$ to the curve at some point P, with respect to the parameter value u, is defined by (3.1). In a given coordinate system, we can write $\delta\boldsymbol{s} = \boldsymbol{e}_a \delta x^a$, where the $\boldsymbol{e}_a$ are coordinate basis vectors at P. We then obtain

$$\boxed{\boldsymbol{t} = \frac{dx^a}{du}\boldsymbol{e}_a.} \tag{3.43}$$

From this expression, we see that the *length* of the tangent vector $\boldsymbol{t}$ to the curve $x^a(u)$ at the point P is given by

$$\boxed{|\boldsymbol{t}| = |g_{ab}t^a t^b|^{1/2} = \left|g_{ab}\frac{dx^a}{du}\frac{dx^b}{du}\right|^{1/2} = \frac{|g_{ab}\,dx^a dx^b|^{1/2}}{du} = \left|\frac{ds}{du}\right|,}$$

where ds is the distance measured along the curve at P that corresponds to the parameter interval du along the curve.

A *non-null curve* is one for which the tangent vector at *every* point is not null, i.e. $|\boldsymbol{t}| \neq 0$. For such a curve, the length of the tangent vector at each point depends on the parameter u and, in general, can vary along the curve. However, we see that if the curve is parameterised in terms of a parameter u that is related to the distance s measured along the curve by $u = as + b$, where a and b are constants, with $a \neq 0$, then the length of the tangent vector will be constant along the curve. In this case u is called an *affine* parameter along the curve. Moreover, if we take $u = s$ then the tangent vector (with components dx^a/ds) is always of *unit* length.

A *null curve* is one for which the tangent vector is null, $|\boldsymbol{t}| = 0$, at *every* point along the curve; equivalently, the distance ds between any two points on a null curve is zero. Since s does not vary along the curve, we clearly cannot use it as a parameter. We are, however, free to use any other non-zero scalar parameter u that does vary along the curve. Moreover, even for null curves it is still possible to define a privileged family of *affine* parameters. The definition of an affine parameter for a null curve is best introduced through the study of geodesics.

3.17 Geodesics

A *geodesic* in Euclidean space is a straight line, which has two equivalent defining properties. First, its tangent vector always points in the same direction (along the line) and, second, it is the curve of shortest length between two points. We can use generalisations of either property to define geodesics in more general manifolds. The fixed direction of their tangent vectors can be used to define both non-null and null geodesics in a pseudo-Riemannian manifold, whereas clearly the extremal length can only be used to define non-null geodesics. In a manifold that is *torsionless* (so that (3.19) is satisfied) these two defining properties are equivalent, for non-null geodesics, and lead to the same curves.[7]

Let us begin by characterising a geodesic as a curve $x^a(u)$ described in terms of some general parameter u by the fixed direction of its tangent vector $\mathbf{t}(u)$. The equations satisfied by the functions $x^a(u)$ are thus determined by the requirement that, along the curve,

$$\frac{d\mathbf{t}}{du} = \lambda(u)\,\mathbf{t}, \tag{3.44}$$

where $\lambda(u)$ is some function of u. From (3.41), we see that the components t^a of the tangent vector in the coordinate basis must satisfy

$$\frac{Dt^a}{Du} = \frac{dt^a}{du} + \Gamma^a{}_{bc} t^b \frac{dx^c}{du} = \lambda(u) t^a.$$

Since the components of the tangent vector are $t^a = dx^a/du$ we find that the equations satisfied by a geodesic are

$$\frac{d^2x^a}{du^2} + \Gamma^a{}_{bc}\frac{dx^b}{du}\frac{dx^c}{du} = \lambda(u)\frac{dx^a}{du}. \tag{3.45}$$

Equation (3.45) is valid for both null and non-null geodesics parameterised in terms of some general parameter u. If the curve is parameterised in such a way that $\lambda(u)$ vanishes, however, then u is a privileged parameter called an *affine parameter*. From (3.44), we see that this corresponds to a parameterisation

[7] In a manifold *with* torsion, the two properties lead to different curves: a curve whose length is stationary with respect to small variations in the path is called a *metric geodesic*, whereas a curve whose tangent vector is constant along the path is an *affine geodesic*.

in which the tangent vector is the same at all points along the curve (i.e. it is parallel-transported), so that

$$\frac{d\mathbf{t}}{du} = \mathbf{0} \qquad \Rightarrow \qquad \frac{Dt^a}{Du} = 0. \tag{3.46}$$

The equations satisfied by an affinely parameterised geodesic are thus

$$\boxed{\frac{d^2x^a}{du^2} + \Gamma^a{}_{bc}\frac{dx^b}{du}\frac{dx^c}{du} = 0.} \tag{3.47}$$

Since one is always free to choose an affine parameter, we shall henceforth restrict ourselves to this simplified form. In particular, for non-null geodesics a convenient affine parameter is the distance s measured along the curve. The *geodesic equation* (3.47) is one of the most important results for our study of particle motion in general relativity.

Finally, we note how affine parameters are related to one another. If we change the parameterisation from an affine parameter u to some other parameter u' then the functions $x^a(u')$ describing $\mathcal{C}$ in terms of the new parameter will differ from the original functions $x^a(u)$. If, for some arbitrary new parameter u', we rewrite (3.47) in terms of derivatives with respect to u' then the geodesic equation does *not*, in general, retain the form (3.47) but instead becomes

$$\frac{d^2x^a}{du'^2} + \Gamma^a{}_{bc}\frac{dx^b}{du'}\frac{dx^c}{du'} = \left(\frac{d^2u/du'^2}{du/du'}\right)\frac{dx^a}{du'}. \tag{3.48}$$

It is clear from (3.48) that if u is an affine parameter then so too is any *linearly related* parameter $u' = au + b$, where a and b are constants (i.e. they do not depend on position along the curve) and $a \neq 0$.

3.18 Stationary property of non-null geodesics

Let us now consider non-null geodesics as curves of extremal length between two fixed points A and B in the manifold. Suppose that we describe the curve $x^a(u)$ in terms of some general (not necessarily affine) parameter u. The length along the curve is

$$L = \int_A^B ds = \int_A^B |g_{ab}\dot{x}^a\dot{x}^b|^{1/2}\, du,$$

where the overdot is a shorthand for d/du. Now consider the variation in path $x^a(u) \to x^a(u) + \delta x^a(u)$, where A and B are fixed. The requirement for $x^a(u)$ to be a geodesic is that $\delta L = 0$ with respect to the variation in the path. This is a calculus-of-variations problem, (3.66), in which the integrand $F = \dot{s} = |g_{ab}\dot{x}^a\dot{x}^b|^{1/2}$.

If we substitute this form for F directly into the Euler–Lagrange equations (3.67), i.e.

$$\frac{d}{du}\left(\frac{\partial F}{\partial \dot{x}^c}\right) - \frac{\partial F}{\partial x^c} = 0,$$

then we obtain

$$\frac{d}{du}\left(\frac{g_{ac}\dot{x}^a}{\dot{s}}\right) - \frac{1}{2\dot{s}}(\partial_c g_{ab})\dot{x}^a\dot{x}^b = 0. \tag{3.49}$$

Noting that $\dot{g}_{ac} = (\partial_b g_{ac})\dot{x}^b$, the u-derivative is given by

$$\frac{d}{du}\left(\frac{g_{ac}\dot{x}^a}{\dot{s}}\right) = \frac{1}{\dot{s}}\left[(\partial_b g_{ac})\dot{x}^a\dot{x}^b + g_{ac}\ddot{x}^a - \frac{\ddot{s}}{\dot{s}}g_{ac}\dot{x}^a\right].$$

Substituting this expression back into (3.49) and rearranging yields

$$g_{ac}\ddot{x}^a + (\partial_b g_{ac})\dot{x}^a\dot{x}^b - \tfrac{1}{2}(\partial_c g_{ab})\dot{x}^a\dot{x}^b = \left(\frac{\ddot{s}}{\dot{s}}\right) g_{ac}\dot{x}^a. \tag{3.50}$$

By interchanging dummy indices, we can write $(\partial_b g_{ac})\dot{x}^a\dot{x}^b = \frac{1}{2}(\partial_b g_{ac} + \partial_a g_{bc})\dot{x}^a\dot{x}^b$. Substituting this into (3.50), multiplying the whole equation by g^{dc} and remembering that $g^{dc}g_{ac} = \delta^d_c$, we find that

$$\ddot{x}^d + \tfrac{1}{2}g^{dc}(\partial_b g_{ac} + \partial_a g_{bc} - \partial_c g_{ab})\dot{x}^a\dot{x}^b = \left(\frac{\ddot{s}}{\dot{s}}\right)\dot{x}^d.$$

Finally, using the expression (3.21) for the connection in terms of the metric and relabelling indices, we obtain

$$\boxed{\ddot{x}^a + \Gamma^a{}_{bc}\dot{x}^b\dot{x}^c = \left(\frac{\ddot{s}}{\dot{s}}\right)\dot{x}^a.} \tag{3.51}$$

Comparing this equation with (3.48) we see that the two are equivalent. We also see that, for a non-null geodesic, an affine parameter u is related to the distance s measured along the curve by $u = as + b$, where a and b are constants ($a \neq 0$).

3.19 Lagrangian procedure for geodesics

In order to obtain the parametric equations $x^a = x^a(u)$ of an affinely parameterised geodesic, we must solve the system of differential equations (3.47). Bearing in mind that the equations (3.21), which define the $\Gamma^a{}_{bc}$, are already complicated, it would seem a formidable procedure to set up the geodesic equations, let alone solve them. Nevertheless, in the previous section we found that the equations for a non-null geodesic arise very naturally from a variational approach. Looking back at the derivation of (3.51), however, we note this requires that $\dot{s} \neq 0$. Thus the proof is not valid for null geodesics. Fortunately, it *is* possible to set up a variational procedure which generates the equations of an affinely parameterised geodesic and which remains valid for null geodesics. This very neat procedure also produces the connection coefficients $\Gamma^a{}_{bc}$ as a spin-off.

In standard classical mechanics, one can describe a system in terms of a set of generalised coordinates x^a that are functions of time t. These coordinates define a space with a line element

$$ds^2 = g_{ab}\, dx^a dx^b,$$

which, in classical mechanics, is called the *configuration space* of the system. One can form the *Lagrangian* for the system from the kinetic and potential energies,

$$L = T - V = \tfrac{1}{2} g_{ab} \dot{x}^a \dot{x}^b - V(x),$$

where $\dot{x}^a \equiv dx^a/dt$. By demanding that the *action*

$$S = \int_{t_i}^{t_f} L dt$$

is stationary with respect to small variations in the functions $x^a(t)$, the equations of motion of the system are then found as the Euler–Lagrange equations

$$\frac{d}{dt}\left(\frac{\partial L}{\partial \dot{x}^a}\right) - \frac{\partial L}{\partial x^a} = 0.$$

This should all be familiar to the reader (but is discussed in more detail in Chapter 19). Less familiar, perhaps, is how the equations of motion look if we write them out in full:

$$\ddot{x}^a + \Gamma^a{}_{bc} \dot{x}^b \dot{x}^c = -g^{ab} \partial_b V.$$

These are just the equations of an affinely parameterised geodesic with a force term on the right-hand side. In this case, the $\Gamma^a{}_{bc}$ are the metric connections of the configuration space. If the forces vanish then Lagrange's equations say that 'free' particles move along geodesics in the configuration space.

Thus, by analogy, in an arbitrary pseudo-Riemannian manifold we may obtain the equations for an affinely parameterised (null or non-null) geodesic $x^a(u)$ by considering the 'Lagrangian'

$$L = g_{ab} \dot{x}^a \dot{x}^b,$$

where $\dot{x}^a \equiv dx^a/du$ and we have omitted the irrelevant factor $\frac{1}{2}$. As can be shown directly, substituting this Lagrangian into the Euler–Lagrange equations

$$\frac{d}{du}\left(\frac{\partial L}{\partial \dot{x}^a}\right) - \frac{\partial L}{\partial x^a} = 0 \tag{3.52}$$

yields, as required,

$$\boxed{\ddot{x}^a + \Gamma^a{}_{bc}\dot{x}^b\dot{x}^c = 0.} \tag{3.53}$$

Performing this calculation, one finds that nowhere does it require $\dot{s} \neq 0$ and so is valid for both null and non-null geodesics. Thus the Euler–Lagrange equations provide a useful way of generating the geodesic equations, and the connection coefficients may be extracted from the latter.

We note that, in seeking solutions of the geodesic equations (3.53), it often helps to make use of the first integral of the equations. For *null* geodesics, the first integral is simply

$$\boxed{g_{ab}\dot{x}^a\dot{x}^b = 0,} \tag{3.54}$$

whereas, for *non-null* geodesics, if we choose the parameter $u = s$ then

$$\boxed{|g_{ab}\dot{x}^a\dot{x}^b| = 1.} \tag{3.55}$$

These results can prove extremely useful in solving the geodesics equations.

Demonstrating the equivalence of the geodesic and Euler–Lagrange equations allows us to make a useful observation. If the g_{ab} do not depend on some particular coordinate x^d (say) then (3.52) shows that

$$\frac{\partial L}{\partial \dot{x}^d} = g_{db}\dot{x}^b = \text{constant}.$$

However, $\dot{x}^b = t^b$, where $\boldsymbol{t}$ is the tangent vector to the geodesic, and so we find that

$$\boxed{t_d = \text{constant}.}$$

Thus, we have the important result that *if the metric coefficients g_{ab} do not depend on the coordinate x^d then the dth covariant component t_d of the tangent vector is a conserved quantity along an affinely parameterised geodesic.* We will use this result often in our discussion of particle motion in general relativity.

3.20 Alternative form of the geodesic equations

The most common form of the geodesic equations is that given in (3.53). It is sometimes useful, however, to recast the geodesic equations in different forms. Thus, we note here an alternative way of writing them that will be of particular practical use when we come to study particle motion in general relativity.

From (3.46), for a geodesic we have $d\mathbf{t}/du = 0$. In some coordinate system we may write this equation in terms of the intrinsic derivative of the *covariant* components of the tangent vector as

$$\frac{Dt_a}{Du} \equiv \frac{dt_a}{du} - \Gamma^b{}_{ac} t_b \frac{dx^c}{du} = 0.$$

Remembering that $t^c = \dot{x}^c = dx^c/du$, we thus have

$$\dot{t}_a = \Gamma^b{}_{ac} t_b t^c,$$

which, on rewriting the connection coefficients $\Gamma^b{}_{ac}$ using (3.21), becomes

$$\dot{t}_a = \tfrac{1}{2} g^{bd} (\partial_a g_{dc} + \partial_c g_{ad} - \partial_d g_{ac}) t_b t^c = \tfrac{1}{2} (\partial_a g_{dc} + \partial_c g_{ad} - \partial_d g_{ac}) t^d t^c.$$

Using the symmetry of the metric tensor, we see that the last two terms in the summation on d and c cancel. Thus, we obtain a useful alternative form of the geodesic equations,

$$\boxed{\dot{t}_a = \tfrac{1}{2} (\partial_a g_{cd}) t^c t^d.} \tag{3.56}$$

From this equation, we may immediately verify our earlier finding that if the metric g_{cd} does not depend on the coordinate x^a then $t_a =$ constant.

Appendix 3A: Vectors as directional derivatives

In an arbitrary manifold, the formal mathematical definition of a tangent vector to a curve at some point P is in terms of the *directional derivative* along the curve at that point. In particular, let us consider some curve $\mathcal{C}$ defined in terms of an arbitrary coordinate system by $x^a(u)$. In addition, suppose that some arbitrary scalar function $f(x^a)$ is defined on the manifold. At any point P on the curve, the directional derivative of f is defined simply as

$$\frac{df}{du} = \frac{\partial f}{\partial x^a} \frac{dx^a}{du}$$

at that point. However, $t^a \equiv dx^a/du$ gives the components of a tangent vector to the curve at P and, since f is arbitrary, we may write

$$\frac{d}{du} = t^a \frac{\partial}{\partial x^a}.$$

Thus, the components t^a define a unique directional derivative, which we may identify as the tangent vector $\boldsymbol{t}$. Moreover, it follows that the differential operators $\partial/\partial x^a$ are the coordinate basis vectors $\boldsymbol{e}_a$ at P, i.e. they are the tangent vectors to the coordinate curves at this point.

In fact, any set of vector components v^a defines a unique directional derivative

$$v^a \frac{\partial}{\partial x^a}, \tag{3.57}$$

and, conversely, this directional derivative defines a unique set of components v^a. We may thus identify (3.57) as the vector $\boldsymbol{v}$. Thus the definition of a vector as a directional derivative replaces the more familiar notion of a directed line segment, which cannot be generalised to non-Euclidean manifolds. It is straightforward to verify that all the usual rules of vector algebra and the behaviour of the components v^a under coordinate transformations follow immediately from (3.57).

Appendix 3B: Polar coordinates in a plane

As a simple example of the material presented in this chapter, let us consider the special case of a two-dimensional Euclidean plane. The most common way of labelling points in a plane is by using Cartesian coordinates (x, y), but it is sometimes convenient to use plane polar coordinates (ρ, ϕ). The two coordinate systems are related by the equations

$$\rho = (x^2 + y^2)^{1/2}, \quad \phi = \tan^{-1}(y/x),$$

and their inverses

$$x = \rho \cos\phi, \quad y = \rho \sin\phi.$$

The transformation matrices relating these two sets of coordinates are

$$\begin{pmatrix} \dfrac{\partial \rho}{\partial x} & \dfrac{\partial \rho}{\partial y} \\[2ex] \dfrac{\partial \phi}{\partial x} & \dfrac{\partial \phi}{\partial y} \end{pmatrix} = \begin{pmatrix} \cos\phi & \sin\phi \\[2ex] -\dfrac{1}{\rho}\sin\phi & \dfrac{1}{\rho}\cos\phi \end{pmatrix}$$

and

$$\begin{pmatrix} \dfrac{\partial x}{\partial \rho} & \dfrac{\partial x}{\partial \phi} \\[2ex] \dfrac{\partial y}{\partial \rho} & \dfrac{\partial y}{\partial \phi} \end{pmatrix} = \begin{pmatrix} \cos\phi & -\rho\sin\phi \\[2ex] \sin\phi & \rho\cos\phi \end{pmatrix},$$

which are easily shown to be inverses of one another. For convenience, in the following we will sometimes refer to the polar coordinates as the coordinate system $x^a (a = 1, 2)$.

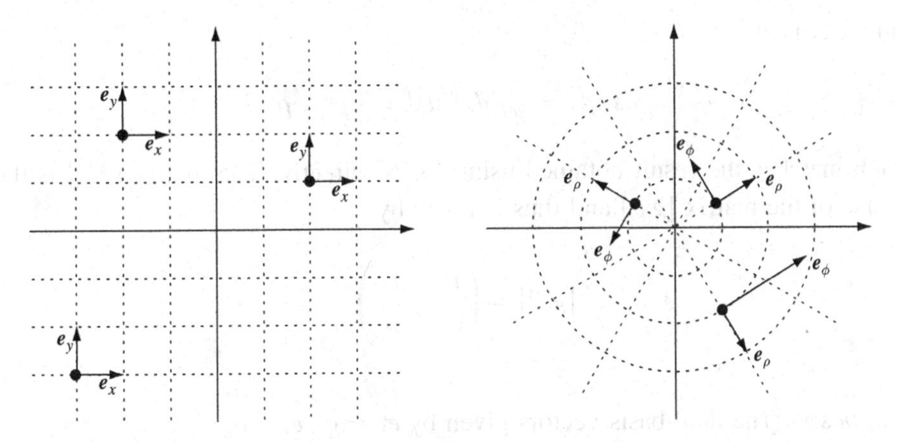

Figure 3.7 Labelling points in a plane with Cartesian coordinates and plane polar coordinates. Examples of basis vectors for the two systems are also shown.

Basis vectors Let us now consider the coordinate basis vectors in each system. The coordinate curves for each system are shown as dotted lines in Figure 3.7 and the basis vectors are tangents to these curves. For the Cartesian coordinates, $\boldsymbol{e}_x$ and $\boldsymbol{e}_y$ have the special property that they are the same at every point P in the plane. They are of unit length and point along the x- and y-directions respectively, and we can write

$$d\boldsymbol{s} = dx\,\boldsymbol{e}_x + dy\,\boldsymbol{e}_y.$$

In plane polar coordinates this becomes

$$d\boldsymbol{s} = (\cos\phi\, d\rho - \rho\sin\phi\, d\phi)\boldsymbol{e}_x + (\sin\phi\, d\rho + \rho\cos\phi\, d\phi)\boldsymbol{e}_y, \tag{3.58}$$

and so, using the definition (3.3) of the coordinate basis vectors, we obtain

$$\boldsymbol{e}_\rho = \cos\phi\,\boldsymbol{e}_x + \sin\phi\,\boldsymbol{e}_y, \tag{3.59}$$

$$\boldsymbol{e}_\phi = -\rho\sin\phi\,\boldsymbol{e}_x + \rho\cos\phi\,\boldsymbol{e}_y \tag{3.60}$$

Alternatively, we could have arrived at the same result using the transformation equations (3.9) for basis vectors. The basis vectors $\boldsymbol{e}_\rho$ and $\boldsymbol{e}_\phi$ are shown in Figure 3.7.

Metric components Substituting the expressions (3.59) and (3.60) into the result $g_{ab} = \boldsymbol{e}_a \cdot \boldsymbol{e}_b$, we find that in polar coordinates

$$[g_{ab}] = \begin{pmatrix} 1 & 0 \\ 0 & \rho^2 \end{pmatrix}.$$

Thus, we have

$$ds^2 = ds \cdot ds = g_{ab}\, dx^a\, dx^b = d\rho^2 + \rho^2\, d\phi^2,$$

which matches the result obtained using (3.58) directly. The matrix $[g^{ab}]$ is the inverse of the matrix $[g_{ab}]$ and thus is given by

$$[g^{ab}] = \begin{pmatrix} 1 & 0 \\ 0 & 1/\rho^2 \end{pmatrix}.$$

Dual basis The dual basis vectors given by $\boldsymbol{e}^a = g^{ab}\boldsymbol{e}_b$ are

$$\boldsymbol{e}^\rho = g^{\rho\rho}\boldsymbol{e}_\rho + g^{\rho\phi}\boldsymbol{e}_\phi = \boldsymbol{e}_\rho,$$

$$\boldsymbol{e}^\phi = g^{\phi\rho}\boldsymbol{e}_\rho + g^{\phi\phi}\boldsymbol{e}_\phi = \frac{1}{\rho^2}\boldsymbol{e}_\phi,$$

where no summation is implied over ρ or ϕ. These dual basis vectors are easily shown to obey the reciprocity relation $\boldsymbol{e}^a \cdot \boldsymbol{e}_b = \delta^a_b$.

Derivatives of basis vectors Since $\boldsymbol{e}_x$ and $\boldsymbol{e}_y$ are constant vector fields, the derivatives of the polar coordinate basis vectors are easily found as

$$\frac{\partial \boldsymbol{e}_\rho}{\partial \rho} = \frac{\partial}{\partial \rho}(\cos\phi\, \boldsymbol{e}_x + \sin\phi\, \boldsymbol{e}_y) = 0,$$

$$\frac{\partial \boldsymbol{e}_\rho}{\partial \phi} = \frac{\partial}{\partial \phi}(\cos\phi\, \boldsymbol{e}_x + \sin\phi\, \boldsymbol{e}_y) = -\sin\phi\, \boldsymbol{e}_x + \cos\phi\, \boldsymbol{e}_y = \frac{1}{\rho}\, \boldsymbol{e}_\phi.$$

These have a simple geometrical picture. At each of two nearby points P and Q the vector $\boldsymbol{e}_\rho$ must point away from the origin, and so in slightly different directions. The derivative of $\boldsymbol{e}_\rho$ with respect to ϕ is just the difference between $\boldsymbol{e}_\rho$ at P and Q divided by $\delta\phi$ (the angle between them). The difference in this case is clearly a vector parallel to $\boldsymbol{e}_\phi$, which makes the above results reasonable. Similarly,

$$\frac{\partial \boldsymbol{e}_\phi}{\partial \rho} = \frac{\partial}{\partial \rho}(-\rho\sin\phi\, \boldsymbol{e}_x + \rho\cos\phi\, \boldsymbol{e}_y) = -\sin\phi\, \boldsymbol{e}_x + \cos\phi\, \boldsymbol{e}_y = \frac{1}{\rho}\, \boldsymbol{e}_\phi,$$

$$\frac{\partial \boldsymbol{e}_\phi}{\partial \phi} = \frac{\partial}{\partial \phi}(-\rho\sin\phi\, \boldsymbol{e}_x + \rho\cos\phi\, \boldsymbol{e}_y) = -\rho\cos\phi\, \boldsymbol{e}_x - \rho\sin\phi\, \boldsymbol{e}_y = -\rho\, \boldsymbol{e}_\rho.$$

The student is encouraged to explain these formulae geometrically.

Connection coefficients Using the general formula $\partial_c \boldsymbol{e}_a = \Gamma^b{}_{ac} \boldsymbol{e}_b$, we can now read off the connection coefficients in plane polar coordinates:

$$\frac{\partial \boldsymbol{e}_\rho}{\partial \rho} = 0 \qquad \Rightarrow \qquad \Gamma^\rho{}_{\rho\rho} = 0, \qquad \Gamma^\phi{}_{\rho\rho} = 0$$

$$\frac{\partial \boldsymbol{e}_\rho}{\partial \phi} = \frac{1}{\rho}\,\boldsymbol{e}_\phi \qquad \Rightarrow \qquad \Gamma^\rho{}_{\rho\phi} = 0, \qquad \Gamma^\phi{}_{\rho\phi} = \frac{1}{\rho}$$

$$\frac{\partial \boldsymbol{e}_\phi}{\partial \rho} = \frac{1}{\rho}\,\boldsymbol{e}_\phi \qquad \Rightarrow \qquad \Gamma^\rho{}_{\phi\rho} = 0, \qquad \Gamma^\phi{}_{\phi\rho} = \frac{1}{\rho}$$

$$\frac{\partial \boldsymbol{e}_\phi}{\partial \phi} = -\rho\,\boldsymbol{e}_\rho \qquad \Rightarrow \qquad \Gamma^\rho{}_{\phi\phi} = -\rho, \quad \Gamma^\phi{}_{\phi\phi} = 0,$$

where no summation is assumed over repeated indices. Thus, although we computed the derivatives of $\boldsymbol{e}_\rho$ and $\boldsymbol{e}_\phi$ by using the constancy of $\boldsymbol{e}_x$ and $\boldsymbol{e}_y$, the Cartesian basis vectors do *not* appear in the above equations. The connection's importance is that it enables one to express these derivatives *without* using any other coordinates than polar. We can alternatively calculate the connection coefficients from the metric using the general result (3.21). For example,

$$\Gamma^\phi{}_{\rho\phi} = \tfrac{1}{2} g^{a\phi}(\partial_\phi g_{a\rho} + \partial_\rho g_{a\phi} - \partial_a g_{\rho\phi}),$$

where summation is implied only over the index a. Since $g^{\rho\phi} = 0$ and $g^{\phi\phi} = 1/\rho^2$, we have

$$\Gamma^\phi{}_{\rho\phi} = \frac{1}{2\rho^2}(\partial_\phi g_{\phi\rho} + \partial_\rho g_{\phi\phi} - \partial_\phi g_{\rho\phi}) = \frac{1}{2\rho^2}\partial_\rho g_{\phi\phi} = \frac{1}{2\rho^2}\partial_\rho(\rho^2) = \frac{1}{\rho}.$$

This is the same expression for $\Gamma^\phi{}_{\rho\phi}$ as that derived above. Indeed, this method of computing the connection is generally far more straightforward than calculating the derivatives of basis vectors.

Covariant derivative Given the connection coefficients, we can calculate the covariant derivative of a vector field in polar coordinates. As an example of its use, let us find an expression for the divergence $\nabla \cdot \boldsymbol{v}$ of a vector field. This is given by

$$\nabla \cdot \boldsymbol{v} = \nabla_a v^a = \partial_a v^a + \Gamma^a{}_{ba} v^b.$$

Now, the contracted connection coefficients are given by

$$\Gamma^a{}_{\rho a} = \Gamma^\rho{}_{\rho\rho} + \Gamma^\phi{}_{\rho\phi} = \frac{1}{\rho},$$

$$\Gamma^a{}_{\phi a} = \Gamma^\rho{}_{\phi\rho} + \Gamma^\phi{}_{\phi\phi} = 0,$$

so we have

$$\nabla \cdot \boldsymbol{v} = \frac{\partial v^\rho}{\partial \rho} + \frac{1}{\rho} v^\rho + \frac{\partial v^\phi}{\partial \phi} = \frac{1}{\rho}\frac{\partial}{\partial \rho}(\rho v^\rho) + \frac{\partial v^\phi}{\partial \phi}.$$

This formula may not be immediately familiar. The reason for this is that most often a vector $\boldsymbol{v}$ is expressed in terms of the *normalised* basis vectors $\hat{\boldsymbol{e}}_\rho = \boldsymbol{e}_\rho$ and $\hat{\boldsymbol{e}}_\phi = \boldsymbol{e}_\phi/\rho$. In this normalised basis the vector components are $\hat{v}^\rho = v^\rho$ and $\hat{v}^\phi = \rho v^\phi$, and the divergence takes its more usual form

$$\nabla \cdot \boldsymbol{v} = \frac{1}{\rho}\frac{\partial}{\partial \rho}(\rho \hat{v}^\rho) + \frac{1}{\rho}\frac{\partial \hat{v}^\phi}{\partial \phi}.$$

Geodesics Finally, let us consider a geodesic in a plane. We already know that the answer is a straight line, and this is trivially proven in Cartesian coordinates. For illustration, however, let us perform the calculation the hard way, i.e. in plane polar coordinates. There are two geodesic equations,

$$\frac{d^2 x^a}{ds^2} + \Gamma^a{}_{bc}\frac{dx^b}{ds}\frac{dx^c}{ds} = 0$$

for $a = \rho, \phi$, where we are using the arclength s as our parameter along the geodesic. The only non-zero connection coefficients are $\Gamma^\rho{}_{\phi\phi} = -\rho$ and $\Gamma^\phi{}_{\rho\phi} = \Gamma^\phi{}_{\phi\rho} = 1/\rho$. Thus, writing out the geodesic equations for $a = \rho$ and $a = \phi$, we have

$$\frac{d^2\rho}{ds^2} - \rho\left(\frac{d\phi}{ds}\right)^2 = 0, \tag{3.61}$$

$$\frac{d^2\phi}{ds^2} + \frac{2}{\rho}\frac{d\rho}{ds}\frac{d\phi}{ds} = 0. \tag{3.62}$$

Also, since in a Euclidean plane we can only have non-null geodesics, a first integral of these equations is provided by

$$g_{ab}\frac{dx^a}{ds}\frac{dx^b}{ds} = 1 \quad \Rightarrow \quad \left(\frac{d\rho}{ds}\right)^2 + \rho^2\left(\frac{d\phi}{ds}\right)^2 = 1. \tag{3.63}$$

Of course, this could have been obtained simply by dividing through $ds^2 = d\rho^2 + \rho^2\, d\phi^2$ by ds^2.

Equation (3.62) can be written as

$$\frac{1}{\rho^2}\frac{d}{ds}\left(\rho^2 \frac{d\phi}{ds}\right) = 0,$$

from which we obtain

$$\rho^2 \frac{d\phi}{ds} = k = \text{constant}. \tag{3.64}$$

Inserting this into (3.63), we find that

$$\frac{d\rho}{ds} = \left(1 - \frac{k^2}{\rho^2}\right)^{1/2}. \tag{3.65}$$

The shape of the geodesic is what really interests us, i.e. ρ as a function of ϕ or vice versa. Dividing (3.64) by (3.65), we obtain

$$\frac{d\phi}{d\rho} = \frac{k}{\rho^2}\left(1 - \frac{k^2}{\rho^2}\right)^{-1/2},$$

which can be integrated easily to give

$$\phi = \phi_0 + \cos^{-1}\left(\frac{k}{\rho}\right),$$

where ϕ_0 is the integration constant. The shape of the geodesic is given by

$$\rho\cos(\phi - \phi_0) = k,$$

which, on expanding the cosine and using $x = \rho\cos\phi$ and $y = \rho\sin\phi$, gives

$$x\cos\phi_0 + y\sin\phi_0 = k.$$

This is the general equation of a straight line. Thus we recover the familiar result in an unfamiliar coordinate system.

Appendix 3C: Calculus of variations

The calculus of variations provides a means of finding a function (or set of functions) that makes an integral dependent on the function(s) *stationary*, i.e. makes the value of the integral a local maximum or minimum. Let us consider the path integral

$$I = \int_A^B F(x^a, \dot{x}^a, u)\, du, \tag{3.66}$$

where A, B and the form of the integrand F are fixed, but the 'curve' or path $x^a(u)$ has to be chosen so as to make stationary the value of I. From (3.66), we see that we are considering quite a general case, in which the integrand F is a function of the $2N$ independent functions x^a and $\dot{x}^a \equiv dx^a/du$ and the parameter u.

Now consider making an arbitrary variation $x^a(u) \to x^a(u) + \delta x^a(u)$ in the path, keeping the endpoints A and B fixed. The corresponding first-order variation in the value of the integral is

$$\delta I = \int_A^B \delta F\, du = \int_A^B \left(\frac{\partial F}{\partial x^a}\delta x^a + \frac{\partial F}{\partial \dot{x}^a}\delta \dot{x}^a\right) du.$$

Integrating the last term by parts and requiring the variation δI to be zero, we obtain

$$\delta I = \left[\frac{\partial F}{\partial \dot{x}^a}\delta x^a\right]_A^B + \int_A^B \left[\frac{\partial F}{\partial x^a} - \frac{d}{du}\left(\frac{\partial F}{\partial \dot{x}^a}\right)\right]\delta x^a\, du = 0.$$

Since A and B are fixed, the first term vanishes. Then, since δx^a is arbitrary, our required extremal curve $x^a(u)$ must satisfy the N equations

$$\boxed{\frac{d}{du}\left(\frac{\partial F}{\partial \dot{x}^a}\right) - \frac{\partial F}{\partial x^a} = 0.} \tag{3.67}$$

These are the *Euler–Lagrange* equations for the problem.

Exercises

3.1 Show that, in general, $\boldsymbol{e}_a = g_{ab}\boldsymbol{e}^b$ and $\boldsymbol{e}^a = g^{ab}\boldsymbol{e}_b$. Show also that, under a coordinate transformation,

$$\boldsymbol{e}'_a = \frac{\partial x^b}{\partial x'^a}\boldsymbol{e}_b \qquad \text{and} \qquad \boldsymbol{e}'^a = \frac{\partial x'^a}{\partial x^b}\boldsymbol{e}^b.$$

3.2 Calculate the coordinate basis vectors $\boldsymbol{e}'_a$ of the coordinates system x'^a in Exercise 2.3 in terms of the coordinate basis vectors $\boldsymbol{e}_a$ of the Cartesian system. Hence verify that the metric functions g'_{ab} agree with those found earlier. Calculate the dual basis vectors $\boldsymbol{e}'^a$ in the primed system and hence the quantities g'^{ab}. Find the contravariant and covariant components of $\boldsymbol{e}_1$ in the primed basis. Hence verify that $\boldsymbol{e}_1$ is of unit length.

3.3 For any metric g_{ab} show that $g^{ab}g_{ab} = N$, where N is the dimension of the manifold.

3.4 Show that the affine connection can be written as $\Gamma^b{}_{ac} = \boldsymbol{e}^b \cdot \partial_c \boldsymbol{e}_a$. Show further that, in a torsionless manifold, $\partial_c \boldsymbol{e}_a = \partial_a \boldsymbol{e}_c$.

3.5 Show that, under a coordinate transformation, the affine connection transforms as

$$\Gamma'^a_{bc} = \frac{\partial x'^a}{\partial x^d}\frac{\partial x^f}{\partial x'^b}\frac{\partial x^g}{\partial x'^c}\Gamma^d{}_{fg} - \frac{\partial x^d}{\partial x'^b}\frac{\partial x^f}{\partial x'^c}\frac{\partial^2 x'^a}{\partial x^d \partial x^f}.$$

3.6 For a diagonal metric g_{ab}, show that the connection coefficients are given by (with $a \neq b \neq c$ and no summation over repeated indices)

$$\Gamma^a{}_{bc} = 0, \qquad \Gamma^b{}_{aa} = -\frac{1}{2g_{bb}}\partial_b g_{aa},$$

$$\Gamma^a{}_{ba} = \Gamma^a{}_{ab} = \partial_b\left(\ln\sqrt{|g_{aa}|}\right), \qquad \Gamma^a{}_{aa} = \partial_a\left(\ln\sqrt{|g_{aa}|}\right).$$

3.7 Let g be the determinant of the matrix $[g_{ab}]$. By considering the cofactor of the element g_{ab} in this determinant, or otherwise, show that $\partial_c g = g g^{ab}(\partial_c g_{ab})$.

3.8 In a manifold with *non-zero torsion*, show that the affine connection defined by (3.11) may be written as

$$\Gamma^a{}_{bc} = \begin{Bmatrix} a \\ bc \end{Bmatrix} - \frac{1}{2}(T^a{}_{cb} + T_c{}^a{}_b - T_{bc}{}^a),$$

where $\left\{ {a \atop bc} \right\}$ is the metric connection defined by the right-hand side of (3.21) and $T^a{}_{bc}$ is the torsion tensor defined in (3.18). Defining an index symmetrisation operation such that $\Gamma^a{}_{(bc)} \equiv \frac{1}{2}(\Gamma^a{}_{bc} + \Gamma^a{}_{cb})$, show further that

$$\Gamma^a{}_{(bc)} = \begin{Bmatrix} a \\ bc \end{Bmatrix} + T_{(bc)}{}^a.$$

3.9 In a manifold with *non-zero torsion*, show that the condition $\Gamma^a{}_{bc} = 0$ implies that $\partial_a g_{bc} = 0$ but *not* vice versa. Show further that, under a coordinate transformation of the form

$$x'^a = x^a - x^a_P + \frac{1}{2}\Gamma^a{}_{bc}(P)(x^b - x^b_P)(x^c - x^c_P),$$

the affine connection at the point P in the new coordinate system is given by

$$\Gamma'^a{}_{bc}(P) = \frac{1}{2}T'^a{}_{bc}(P)$$

and hence the transformation does not yield a set of geodesic coordinates. Is it still possible to define local Cartesian coordinates in a manifold with non-zero torsion?

3.10 Show that, for the covariant components v_a of a vector, the covariant derivative and the intrinsic derivative along a curve are given respectively by

$$\nabla_b v_a = \partial_b v_a - \Gamma^c{}_{ab} v_c \qquad \text{and} \qquad \frac{Dv_a}{Du} = \frac{dv_a}{du} - \Gamma^b{}_{ac} v_b \frac{dx^c}{du}.$$

3.11 Show that for a vector field with contravariant components v^b to have a vanishing covariant derivative $\nabla_a v^b$ everywhere in a manifold, it must satisfy the relation

$$(\partial_b \Gamma^d{}_{ac} - \partial_c \Gamma^d{}_{ab} + \Gamma^e{}_{ac}\Gamma^d{}_{eb} - \Gamma^e{}_{ab}\Gamma^d{}_{ec})v^a = 0.$$

Hint: Use the fact that partial derivatives commute.

3.12 If a vector field v^a vanishes on a hypersurface S that bounds a region V of an N-dimensional manifold, show that

$$\int_V (\nabla_a v^a)\sqrt{-g}\, d^N x = 0.$$

3.13 On the surface of a unit sphere, $ds^2 = d\theta^2 + \sin^2\theta\, d\phi^2$. Calculate the connection coefficients in the (θ, ϕ) coordinate system. A vector $\boldsymbol{v}$ of unit length is defined at the point $(\theta_0, 0)$ as parallel to the circle $\phi = 0$. Calculate the components of $\boldsymbol{v}$ after it has been parallel-transported around the circle $\theta = \theta_0$. Hence show that, in general, after parallel transport the direction of $\boldsymbol{v}$ is different but its length is unchanged.

3.14 If the two vectors with contravariant components v^a and w^a are each parallel-transported along a curve, show that $v^a w_a$ remains constant along the curve. Hence show that if a geodesic is timelike (or null or spacelike) at some point, it is timelike (or null or spacelike) at all points.

3.15 An affinely parameterised geodesic $x^a(u)$ satisfies

$$\frac{d^2x^a}{du^2} + \Gamma^a{}_{bc}\frac{dx^b}{du}\frac{dx^c}{du} = 0.$$

Show that the form of this equation remains unchanged by an arbitrary coordinate transformation $x^a \to x'^a$. Find the form of the geodesic equation for a geodesic described in terms of some general (non-affine) parameter λ. Hence show that all affine parameters are related by a linear transformation with constant coefficients.

3.16 If $x^\mu(\lambda)$ is an affinely parameterised geodesic, show that

$$\frac{Du^\mu}{D\lambda} = 0,$$

where $u^\mu = dx^\mu/d\lambda$. Hence show that the geodesic equations can be written as

$$\frac{du_\mu}{d\lambda} = \frac{1}{2}(\partial_\mu g_{\nu\sigma})u^\nu u^\sigma.$$

3.17 By substituting the 'Lagrangian' $L = g_{ab}\dot{x}^a\dot{x}^b$ into the Euler–Lagrange equations, show directly that

$$\ddot{x}^a + \Gamma^a{}_{bc}\dot{x}^b\dot{x}^c = 0,$$

where the dots denote differentiation with respect to an affine parameter.

3.18 By transforming from a local inertial coordinate system ξ^μ in which

$$ds^2 = c^2 d\tau^2 = \eta_{\mu\nu}\, d\xi^\mu\, d\xi^\nu,$$

to a general coordinate system x^μ, show that freely falling particles obey the geodesic equations of motion

$$\frac{d^2x^\lambda}{d\tau^2} + \Gamma^\lambda{}_{\mu\nu}\frac{dx^\mu}{d\tau}\frac{dx^\nu}{d\tau} = 0,$$

where

$$\Gamma^\lambda{}_{\mu\nu} = \frac{\partial x^\lambda}{\partial \xi^\alpha}\frac{\partial^2 \xi^\alpha}{\partial x^\mu \partial x^\nu}.$$

3.19 By considering the 'Lagrangian' $L = g_{ab}\dot{x}^a\dot{x}^b$, derive the equations for an affinely parameterised geodesic on the surface of a sphere in the coordinates (θ, ϕ). Hence show that, of all the circles of constant latitude on a sphere, only the equator is a geodesic. Use your geodesic equations to pick out the connection coefficients in this coordinate system.

3.20 In the 2-space with line element

$$ds^2 = \frac{dr^2 + r^2 d\theta^2}{r^2 - a^2} - \frac{r^2 dr^2}{(r^2 - a^2)^2},$$

where $r > a$, show that the differential equation for the geodesics may be written as

$$a^2 \left(\frac{dr}{d\theta} \right)^2 + a^2 r^2 = K r^4,$$

where K is a constant such that $K = 1$ if the geodesic is null. By setting $r\, d\theta/dr = \tan\phi$, show that the space is mapped onto a Euclidean plane in which (r, ϕ) are taken as polar coordinates and the geodesics are mapped to straight lines.

4
Tensor calculus on manifolds

The coordinates with which one labels points in a manifold are entirely arbitrary. For example, we could choose to parameterise the surface of a sphere in terms of the coordinates (θ, ϕ), taking *any* point as the north pole, or we could use any number of alternative coordinate systems. It is also clear, however, that our description of any physical processes occurring on the surface of the sphere should not depend on our chosen coordinate system. For example, at any point P on the surface one can say that, for example, the air temperature has a particular value or that the wind has a certain speed in a particular direction. These respectively scalar and vector physical quantities do not depend on which coordinates are used to label points in the surface. Thus in, order to describe these physical *fields* on the surface, we must formulate our equations in a way that is valid in *all coordinate systems*. We have already dealt with such a description for scalar and vector quantities on manifolds, but now we turn to the generalisation of these ideas to quantities that cannot be described as a scalar or a vector. This requires the introduction of the concept of *tensors*.

4.1 Tensor fields on manifolds

Let us begin by considering vector fields in a slightly different manner. Suppose we have some arbitrary vector field, defining a vector $\boldsymbol{t}$ at each point of a manifold. How can we obtain from $\boldsymbol{t}$ a scalar field? Clearly, the only way to do this is to take the scalar product of $\boldsymbol{t}$ with a vector $\boldsymbol{v}$ from another vector field. Thus, at each point P in the manifold, we can think of vector $\boldsymbol{t}$ in T_P as a *linear function* $\boldsymbol{t}(\cdot)$ that takes another vector in T_P as its argument and produces a real number. We can denote the number produced by the action of $\boldsymbol{t}$ on a particular vector $\boldsymbol{v}$ by

$$\boxed{\boldsymbol{t}(\boldsymbol{v}) \equiv \boldsymbol{t} \cdot \boldsymbol{v}.} \qquad (4.1)$$

It is now clear how we can generalise the notion of a vector: in the tangent space T_P, we can define a *tensor* $\boldsymbol{t}$ as a linear map from some number of vectors to the real numbers. The *rank* of the tensor is the number of vectors it has for its arguments. For example, we can write a third-rank tensor as $\boldsymbol{t}(\cdot,\cdot,\cdot)$. Once again, we denote the number that the tensor $\boldsymbol{t}$ produces from the vectors $\boldsymbol{u}$, $\boldsymbol{v}$ and $\boldsymbol{w}$ by

$$\boldsymbol{t}(\boldsymbol{u},\boldsymbol{v},\boldsymbol{w}).$$

The tensor is defined by the precise set of operations applied to the vectors $\boldsymbol{u}$, $\boldsymbol{v}$ and $\boldsymbol{w}$ to produce a scalar. Notice, however, that the definition of a tensor does *not* mention the components of the vectors; a tensor must give the same real number independently of the reference system in which the vector components are calculated. If at each point P in some region of the manifold we have a tensor defined then the result is a *tensor field* in this region.

In fact we have already encountered examples of tensors. Clearly, from our above discussion, any vector is a rank-1 tensor. Higher-rank tensors thus constitute a generalisation of the concept of a vector. For example, a particularly important second-rank tensor is the *metric tensor* $\boldsymbol{g}$, which we have already met. This defines a linear map of two vectors into the number that is their inner product, i.e.

$$\boxed{\boldsymbol{g}(\boldsymbol{u},\boldsymbol{v}) \equiv \boldsymbol{u}\cdot\boldsymbol{v}.}$$

We will investigate the properties of this special tensor shortly. Finally, we note also that a *scalar* function of position $\phi(x)$ is a real-valued function of no vectors at all, and is therefore classified as a *zero-rank tensor*.

The fact that a tensor is a *linear* map of the vectors into the reals is particularly useful. For simplicity, let us consider a rank-1 tensor. Linearity means that, for general vectors $\boldsymbol{u}$ and $\boldsymbol{v}$ and general scalars α and β,

$$\boxed{\boldsymbol{t}(\alpha\boldsymbol{u}+\beta\boldsymbol{v}) = \alpha\boldsymbol{t}(\boldsymbol{u})+\beta\boldsymbol{t}(\boldsymbol{v}).}$$

Similar expansions may be performed for tensors of higher rank. For a second-rank tensor, for example, we can write

$$\begin{aligned}\boldsymbol{t}(\alpha\boldsymbol{u}+\beta\boldsymbol{v},\gamma\boldsymbol{w}+\epsilon\boldsymbol{z}) &= \alpha\boldsymbol{t}(\boldsymbol{u},\gamma\boldsymbol{w}+\epsilon\boldsymbol{z})+\beta\boldsymbol{t}(\boldsymbol{v},\gamma\boldsymbol{w}+\epsilon\boldsymbol{z})\\ &= \alpha\gamma\boldsymbol{t}(\boldsymbol{u},\boldsymbol{w})+\alpha\epsilon\boldsymbol{t}(\boldsymbol{u},\boldsymbol{z})+\beta\gamma\boldsymbol{t}(\boldsymbol{v},\boldsymbol{w})+\beta\epsilon\boldsymbol{t}(\boldsymbol{v},\boldsymbol{z}).\end{aligned}$$

4.2 Components of tensors

When a tensor is evaluated with combinations of basis and dual basis vectors it yields its *components* in that particular basis. For example, the covariant and contravariant components of the rank-1 tensor (vector) in (4.1) in the basis $\boldsymbol{e}_a$ are given by

$$\boxed{\boldsymbol{t}(\boldsymbol{e}_a) = t_a \qquad \text{and} \qquad \boldsymbol{t}(\boldsymbol{e}^a) = t^a.}$$

Consider now a second-rank tensor $\boldsymbol{t}(\cdot, \cdot)$. Its covariant and contravariant components are given by

$$\boxed{\boldsymbol{t}(\boldsymbol{e}_a, \boldsymbol{e}_b) = t_{ab} \qquad \text{and} \qquad \boldsymbol{t}(\boldsymbol{e}^a, \boldsymbol{e}^b) = t^{ab}.}$$

For tensors of rank 2 and higher, however, we can also define sets of *mixed* components. For a rank-2 tensor there are two possible sets of mixed components,

$$\boxed{\boldsymbol{t}(\boldsymbol{e}^a, \boldsymbol{e}_b) = t^a{}_b \qquad \text{and} \qquad \boldsymbol{t}(\boldsymbol{e}_a, \boldsymbol{e}^b) = t_a{}^b.}$$

For a general rank-2 tensor these two sets of components need *not* be equal. The contravariant, covariant and mixed components of higher-rank tensors can be obtained in an analogous manner.

The components of a tensor in a particular basis set specify the action of the tensor on any other vectors in terms of *their* components. For example, using the linearity property, we find that

$$\boldsymbol{t}(\boldsymbol{u}, \boldsymbol{v}) = \boldsymbol{t}(u^a \boldsymbol{e}_a, v^b \boldsymbol{e}_b) = t_{ab} u^a v^b.$$

To obtain this result, we expressed $\boldsymbol{u}$ and $\boldsymbol{v}$ in terms of their contravariant components. We could have written either vector in terms of its contravariant or covariant components, however. Hence we find that there are numerous equivalent expressions for $\boldsymbol{t}(\boldsymbol{u}, \boldsymbol{v})$ in component notation:

$$\boxed{\boldsymbol{t}(\boldsymbol{u}, \boldsymbol{v}) = t_{ab} u^a v^b = t^{ab} u_a v_b = t_a{}^b u^a v_b = t^a{}_b u_a v^b.}$$

This illustrates the general rule that the *subscript and superscript positions of a dummy index can be swapped without affecting the result.*

4.3 Symmetries of tensors

A second-rank tensor $\boldsymbol{t}$ is called *symmetric* or *antisymmetric* if, for *all* pairs of vectors $\boldsymbol{u}$ and $\boldsymbol{v}$,

$$\boxed{\boldsymbol{t}(\boldsymbol{u}, \boldsymbol{v}) = \pm \boldsymbol{t}(\boldsymbol{v}, \boldsymbol{u}),}$$

with the plus sign for a symmetric tensor and the minus sign for an antisymmetric tensor. Setting $\boldsymbol{u} = \boldsymbol{e}_a$ and $\boldsymbol{v} = \boldsymbol{e}_b$, we see that the covariant components of a symmetric or antisymmetric tensor satisfy $t_{ab} = \pm t_{ba}$. By using different combinations of basis and dual basis vectors we also see that, for such a tensor, $t^{ab} = \pm t^{ba}$ and $t_a{}^b = \pm t^b{}_a$.

An arbitrary rank-2 tensor can always be split uniquely into the sum of its symmetric and antisymmetric parts. For illustration let us work with the covariant components t_{ab} of the tensor in some basis. We can always write

$$t_{ab} = \tfrac{1}{2}(t_{ab} + t_{ba}) + \tfrac{1}{2}(t_{ab} - t_{ba}),$$

which is clearly the sum of a symmetric and an antisymmetric part. A notation frequently used to denote the components of the symmetric and antisymmetric parts is

$$t_{(ab)} \equiv \tfrac{1}{2}(t_{ab} + t_{ba}) \qquad \text{and} \qquad t_{[ab]} \equiv \tfrac{1}{2}(t_{ab} - t_{ba}).$$

In an analogous manner, a general rank-N tensor $\boldsymbol{t}(\boldsymbol{u}, \boldsymbol{v}, \ldots, \boldsymbol{w})$ is symmetric or antisymmetric with respect to some permutation of its vector arguments if its value after permuting the arguments is equal to respectively plus or minus its original value. From an arbitrary rank-N tensor, however, we can always obtain a tensor that is symmetric with respect to *all* permutations of its vector arguments and one that is antisymmetric with respect to all permutations. In terms of the tensor's covariant components, these symmetric and antisymmetric parts are given by

$$t_{(ab\ldots c)} = \frac{1}{N!}(\text{sum over all permutations of the indices } a, b, \ldots, c),$$

$$t_{[ab\ldots c]} = \frac{1}{N!}(\text{alternating sum over all permutations of the indices } a, b, \ldots, c).$$

For example, the covariant components of the *totally antisymmetric part* of a third-rank tensor are given by

$$t_{[abc]} = \tfrac{1}{6}(t_{abc} - t_{acb} + t_{cab} - t_{cba} + t_{bca} - t_{bac}).$$

We may extend the notation still further in order to define tensors that are symmetric or antisymmetric to permutations of particular subsets of their indices.

To illustrate this, let us consider the covariant components t_{abcd} of a fourth-rank tensor. Typical expressions might include:

$$
\begin{aligned}
t_{(ab)cd} &= \tfrac{1}{2}(t_{abcd} + t_{bacd}), \\
t_{a[b|c|d]} &= \tfrac{1}{2}(t_{abcd} - t_{adcb}), \\
t_{(a|b|cd)} &= \tfrac{1}{6}(t_{abcd} + t_{abdc} + t_{dbac} + t_{dbca} + t_{cbda} + t_{cbad}), \\
t_{[ab](cd)} &= \tfrac{1}{2}\left[t_{ab(cd)} - t_{ba(cd)}\right] \\
&= \tfrac{1}{2}\left[\tfrac{1}{2}(t_{abcd} + t_{abdc}) - \tfrac{1}{2}(t_{bacd} + t_{badc})\right] \\
&= \tfrac{1}{4}(t_{abcd} + t_{abdc} - t_{bacd} - t_{badc}).
\end{aligned}
$$

The symbols || are used to exclude unwanted indices from the (anti-) symmetrisation implied by ( ) and [ ].

4.4 The metric tensor

The most important tensor that one can define on a manifold is the *metric tensor* $\boldsymbol{g}$. This defines a linear map of two vectors into the number that is their inner product, i.e.

$$
\boxed{\boldsymbol{g}(\boldsymbol{u}, \boldsymbol{v}) \equiv \boldsymbol{u} \cdot \boldsymbol{v}.} \qquad (4.2)
$$

From this definition, it is clear that $\boldsymbol{g}$ is a symmetric second-rank tensor. Its covariant and contravariant components are given by

$$
\boxed{g_{ab} = \boldsymbol{g}(\boldsymbol{e}_a, \boldsymbol{e}_b) = \boldsymbol{e}_a \cdot \boldsymbol{e}_b \quad \text{and} \quad g^{ab} = \boldsymbol{g}(\boldsymbol{e}^a, \boldsymbol{e}^b) = \boldsymbol{e}^a \cdot \boldsymbol{e}^b,}
$$

which, from (4.2), clearly match our earlier definitions. As we showed in Chapter 3, the matrix $[g^{ab}]$ containing the contravariant components of the metric tensor is the *inverse* of the matrix $[g_{ab}]$ that contains its covariant components. The mixed components of $\boldsymbol{g}$ are given by

$$
\boxed{\boldsymbol{g}(\boldsymbol{e}^b, \boldsymbol{e}_a) = \boldsymbol{g}(\boldsymbol{e}_a, \boldsymbol{e}^b) = \delta^b_a,}
$$

where the last equality is a result of the reciprocity relation between basis vectors and their duals.

4.5 Raising and lowering tensor indices

The contravariant and covariant components of the metric tensor can be used for raising and lowering general tensor indices, just as they are used for vector indices. As we have seen, when a tensor acts on different combinations of basis and dual basis vectors it yields different components. Consider, for example, a third-rank tensor $\boldsymbol{t}$. Its covariant components are given by

$$\boldsymbol{t}(\boldsymbol{e}_a, \boldsymbol{e}_b, \boldsymbol{e}_c) = t_{abc}, \tag{4.3}$$

whereas one possible set of *mixed* components of the tensor is given by

$$\boldsymbol{t}(\boldsymbol{e}_a, \boldsymbol{e}_b, \boldsymbol{e}^c) = t_{ab}{}^c.$$

As we stated earlier, in general these two sets of components will differ, since the basis and dual basis vectors are related by the metric: $\boldsymbol{e}_c = g_{cd}\boldsymbol{e}^d$. Thus, for example,

$$t_{abc} = g_{cd}t_{ab}{}^d.$$

In a similar way we can raise or lower more than one index at a time. For example,

$$t^a{}_{bc} = g^{ad}g_{ce}t_{db}{}^e.$$

4.6 Mapping tensors into tensors

Tensors can be thought of not just as maps between vectors and real numbers but also as maps between tensors and other tensors. Consider, for example, a third-rank tensor $\boldsymbol{t}$, but let us *not* 'fill' all of its argument 'slots' with vectors. If, for instance, we fill just its last slot with some fixed vector $\boldsymbol{u}$, we have the object

$$\boldsymbol{t}(\cdot, \cdot, \boldsymbol{u}). \tag{4.4}$$

What sort of object is this? Well, it is clear that, if we supply two further vectors to this object, we will obtain a real number. Thus the object (4.4) is itself a second-rank tensor, which we could denote by $\boldsymbol{s}$ (say). Thus the third-rank tensor $\boldsymbol{t}$ has 'mapped' the vector $\boldsymbol{u}$ into the second-rank tensor $\boldsymbol{s}$. The covariant components (say) of $\boldsymbol{s}$ are given by

$$s_{ab} \equiv \boldsymbol{s}(\boldsymbol{e}_a, \boldsymbol{e}_b) = \boldsymbol{t}(\boldsymbol{e}_a, \boldsymbol{e}_b, \boldsymbol{u}) = t_{abc}u^c,$$

where, in the last slot, we have expressed $\boldsymbol{u}$ as $u^c\boldsymbol{e}_c$. By expressing this vector as $u_c\boldsymbol{e}^c$ instead, we obtain the equivalent expression $s_{ab} = t_{ab}{}^c u_c$.

As a further example of mapping between tensors, let us fill both the first and last slots of $\boldsymbol{t}$ with fixed vectors $\boldsymbol{v}$ and $\boldsymbol{u}$ respectively to obtain the object

$$\boldsymbol{t}(\boldsymbol{v}, \cdot, \boldsymbol{u}).$$

Clearly, this object is a first-rank tensor (or vector), which we denote by $\boldsymbol{w}$. Thus the third-rank tensor $\boldsymbol{t}$ has mapped the two vectors $\boldsymbol{v}$ and $\boldsymbol{u}$ into the vector $\boldsymbol{w}$. The covariant components (say) of $\boldsymbol{w}$ are

$$w_b = \boldsymbol{w}(\boldsymbol{e}_b) = \boldsymbol{t}(\boldsymbol{v}, \boldsymbol{e}_b, \boldsymbol{u}),$$

which can be expressed in several equivalent ways, i.e.

$$w_b = t_{abc}v^a u^c = t^a{}_{bc}v_a u^c = t_{ab}{}^c v^a u_c = t^a{}_b{}^c v_a u_c.$$

The number of free indices in such expressions is the rank of the resulting tensor.

4.7 Elementary operations with tensors

Tensor calculus is concerned with *tensorial operations*, that is, operations on tensors which result in quantities that are still tensors. We now consider some elementary tensorial operations.

Addition (and subtraction)

It is clear from the definition of a tensor that the sum and difference of two tensors of rank N are both themselves tensors of rank N. For example, the covariant components (say) of the sum $\boldsymbol{s}$ and difference $\boldsymbol{d}$ of two rank-2 tensors are given straightforwardly by

$$\boxed{\begin{aligned} s_{ab} &= \boldsymbol{s}(\boldsymbol{e}_a, \boldsymbol{e}_b) = \boldsymbol{t}(\boldsymbol{e}_a, \boldsymbol{e}_b) + \boldsymbol{r}(\boldsymbol{e}_a, \boldsymbol{e}_b) = t_{ab} + r_{ab}, \\ d_{ab} &= \boldsymbol{d}(\boldsymbol{e}_a, \boldsymbol{e}_b) = \boldsymbol{t}(\boldsymbol{e}_a, \boldsymbol{e}_b) - \boldsymbol{r}(\boldsymbol{e}_a, \boldsymbol{e}_b) = t_{ab} - r_{ab}. \end{aligned}} \tag{4.5}$$

Multiplication by a scalar

If $\boldsymbol{t}$ is a rank-N tensor then so too is $\alpha\boldsymbol{t}$, where α is some arbitrary real constant. Clearly, its components are all multiplied by α.

Outer product

The *outer* or *tensor product* of two tensors produces a tensor of higher rank. The simplest example of an outer product is that of two vectors. This is defined as the rank-2 tensor, denoted by $\boldsymbol{u} \otimes \boldsymbol{v}$, such that

$$\boxed{(\boldsymbol{u} \otimes \boldsymbol{v})(\boldsymbol{p}, \boldsymbol{q}) \equiv \boldsymbol{u}(\boldsymbol{p})\boldsymbol{v}(\boldsymbol{q}),}$$

where $\boldsymbol{p}$ and $\boldsymbol{q}$ are arbitrary vector arguments (this notation is not to be confused with the vector product $\boldsymbol{u} \times \boldsymbol{v}$ of two vectors, which is itself a vector). Note that the outer product is *not*, in general, commutative, so that $\boldsymbol{u} \otimes \boldsymbol{v}$ and $\boldsymbol{v} \otimes \boldsymbol{u}$ are different rank-2 tensors. The covariant components (say) of $\boldsymbol{u}\otimes\boldsymbol{v}$ in some basis are given by

$$(\boldsymbol{u}\otimes\boldsymbol{v})(\boldsymbol{e}_a,\boldsymbol{e}_b)\equiv\boldsymbol{u}(\boldsymbol{e}_a)\boldsymbol{v}(\boldsymbol{e}_b)=u_a v_b.$$

The outer product of higher-rank tensors is a simple generalisation of the outer product of two vectors. For example, the outer product of a rank-2 tensor $\boldsymbol{t}$ with a rank-1 tensor $\boldsymbol{s}$ is defined by

$$(\boldsymbol{t}\otimes\boldsymbol{s})(\boldsymbol{p},\boldsymbol{q},\boldsymbol{r})\equiv\boldsymbol{t}(\boldsymbol{p},\boldsymbol{q})\boldsymbol{s}(\boldsymbol{r}).$$

This is a rank-3 tensor, which we could call $\boldsymbol{h}$. The mixed components, for instance, of this tensor are given by

$$h^a{}_{bc}=\boldsymbol{t}(\boldsymbol{e}^a,\boldsymbol{e}_b)\boldsymbol{s}(\boldsymbol{e}_c)=t^a{}_b s_c. \tag{4.6}$$

In general, the outer product of an Nth-rank tensor with an Mth-rank tensor will produce an $(N+M)$th-rank tensor.

Contraction (and inner product)

The contraction of a tensor is performed by summing over the basis and dual basis vectors in two of its vector arguments, and it results in a tensor of *lower rank*. Let us take as an example a rank-3 tensor $\boldsymbol{h}$ and consider the quantity

$$\boxed{\boldsymbol{q}(\cdot)=\boldsymbol{h}(\boldsymbol{e}^a,\cdot,\boldsymbol{e}_a).}$$

This is clearly a rank-1 tensor with covariant components (say) given by

$$q_b=\boldsymbol{h}(\boldsymbol{e}^a,\boldsymbol{e}_b,\boldsymbol{e}_a)=h^a{}_{ba}. \tag{4.7}$$

Thus in terms of tensor components, contraction amounts to setting a subscript equal to a superscript and summing, as the summation convention requires. In general, performing a single contraction on an Nth-rank tensor will produce a tensor of rank $N-2$.

Contraction may be combined with tensor multiplication to obtain the *inner product* of two tensors. For example, if $h^a{}_{bc}$ were in fact given by (4.6), then (4.7) could be written as

$$q_b=\boldsymbol{t}(\boldsymbol{e}^a,\boldsymbol{e}_b)\boldsymbol{s}(\boldsymbol{e}_a)=t^a{}_b s_a,$$

which is the inner product of the tensors $\boldsymbol{t}$ and $\boldsymbol{s}$. Alternatively, one could view the q_b as a contraction of the rank-3 tensor having components $t^a{}_b s_c$, which is the outer product $\boldsymbol{t}\otimes\boldsymbol{s}$.

If two tensors $\boldsymbol{t}$ and $\boldsymbol{s}$ are rank 2 or lower then we can denote their inner product unambiguously by $\boldsymbol{t}\cdot\boldsymbol{s}$. Note, however, that in general such an inner product is *not* commutative. For example, if $\boldsymbol{t}$ is a rank-2 tensor and $\boldsymbol{s}$ is rank 1 then the contravariant components (say) of the vectors $\boldsymbol{t}\cdot\boldsymbol{s}$ and $\boldsymbol{s}\cdot\boldsymbol{t}$ are respectively

$$t^{ab}s_b \quad \text{and} \quad t^{ab}s_a.$$

Clearly, the 'dot' notation for the inner product becomes ambiguous if either tensor is rank 3 or higher, since there is then a choice concerning which indices to contract.

4.8 Tensors as geometrical objects

We have seen that a rank-1 tensor $\boldsymbol{t}(\cdot)$ can be identified as a vector. The covariant and contravariant components of this vector in some basis are given by

$$\boldsymbol{t}(\boldsymbol{e}_a) = t_a \qquad \text{and} \qquad \boldsymbol{t}(\boldsymbol{e}^a) = t^a.$$

We are used to thinking of a vector $\boldsymbol{t}$ as a geometrical object which can be made up from a linear combination of the basis vectors,

$$\boldsymbol{t} = t^a\boldsymbol{e}_a = t_a\boldsymbol{e}^a. \tag{4.8}$$

Tensors of higher rank are generalisations of the concept of a vector and can also be regarded as geometrical entities. In a particular basis, a general tensor can expressed as a linear combination of a *tensor basis* made up from the basis vectors and their duals.

Let us consider the outer product $\boldsymbol{e}_a\otimes\boldsymbol{e}_b$ of two basis vectors of some coordinate system. The contravariant components of this rank-2 tensor *in this basis* are very simple,

$$(\boldsymbol{e}_a\otimes\boldsymbol{e}_b)(\boldsymbol{e}^c,\boldsymbol{e}^d) = \boldsymbol{e}_a(\boldsymbol{e}^c)\boldsymbol{e}_b(\boldsymbol{e}^d) = \delta^c_a\delta^d_b.$$

Now suppose that we have some general rank-2 tensor $\boldsymbol{t}$, whose contravariant components in our basis are t^{ab}. Let us consider the quantity $t^{ab}(\boldsymbol{e}_a\otimes\boldsymbol{e}_b)$. This is a sum of rank-2 tensors, which must therefore also be a rank-2 tensor (see above). If we consider its action on two basis vectors, we find

$$t^{ab}(\boldsymbol{e}_a\otimes\boldsymbol{e}_b)(\boldsymbol{e}^c,\boldsymbol{e}^d) = t^{ab}\delta^c_a\delta^d_b = t^{cd};$$

the t^{cd} are simply the contravariant components of $\boldsymbol{t}$. Thus, in an analogous way to the vector in (4.8), we may express the rank-2 tensor $\boldsymbol{t}$ as a linear combination basis tensors,

$$\boxed{\boldsymbol{t} = t^{ab}(\boldsymbol{e}_a\otimes\boldsymbol{e}_b).}$$

By considering different tensor bases, constructed from other combinations of the basis and dual basis vectors, we can also write $\boldsymbol{t}$ in several different ways:

$$\boldsymbol{t} = t_{ab}(\boldsymbol{e}^a \otimes \boldsymbol{e}^b) = t_a{}^b(\boldsymbol{e}^a \otimes \boldsymbol{e}_b) = t^a{}_b(\boldsymbol{e}_a \otimes \boldsymbol{e}^b).$$

This idea is extended straightforwardly to higher-rank tensors.

4.9 Tensors and coordinate transformations

The description of tensors as a geometrical objects lends itself naturally to a discussion of the behaviour of tensor components under a coordinate transformation $x^a \to x'^a$ on the manifold. As shown in Chapter 3, there is a simple relationship between the coordinate basis vectors $\boldsymbol{e}_a$ associated with the coordinate system x^a and the coordinate basis vectors $\boldsymbol{e}'_a$ associated with a new system of coordinates x'^a. We found that at any point P the two sets of coordinate basis vectors are related by

$$\boldsymbol{e}'_a = \frac{\partial x^b}{\partial x'^a}\boldsymbol{e}_b, \tag{4.9}$$

where the partial derivative is evaluated at the point P. A similar relationship holds between the two sets of dual basis vectors:

$$\boldsymbol{e}'^a = \frac{\partial x'^a}{\partial x^b}\boldsymbol{e}^b. \tag{4.10}$$

Using (4.9) and (4.10), we can now calculate how the components of any general tensor must transform under the coordinate transformation.

As shown in Chapter 3, the contravariant components of a vector $\boldsymbol{t}$ in the new coordinate basis are given by

$$t'^a = \boldsymbol{t}(\boldsymbol{e}'^a) = \frac{\partial x'^a}{\partial x^b}\boldsymbol{t}(\boldsymbol{e}^b) = \frac{\partial x'^a}{\partial x^b}t^b.$$

Similarly, the covariant components of $\boldsymbol{t}$ are given by

$$t'_a = \boldsymbol{t}(\boldsymbol{e}'_a) = \frac{\partial x^b}{\partial x'^a}\boldsymbol{t}(\boldsymbol{e}_b) = \frac{\partial x^b}{\partial x'^a}t_b.$$

It is important to remember that the unprimed and primed components describe the *same* vector $\boldsymbol{t}$ in terms of *different* basis vectors, i.e. $\boldsymbol{t} = t^a\boldsymbol{e}_a = t'^a\boldsymbol{e}'_a$. The vector $\boldsymbol{t}$ is a geometric entity that does not depend on the choice of coordinate system.

The transformation properties of the components of higher-rank tensors may be found in a similar way. For example, if $\boldsymbol{t}$ is a second-rank tensor then

$$\boxed{\begin{aligned} t'_{ab} &= \frac{\partial x^c}{\partial x'^a}\frac{\partial x^d}{\partial x'^b}t_{cd}, \\ t'^{ab} &= \frac{\partial x'^a}{\partial x^c}\frac{\partial x'^b}{\partial x^d}t^{cd}, \\ t'^{\ b}_{a} &= \frac{\partial x^c}{\partial x'^a}\frac{\partial x'^b}{\partial x^d}t_c^{\ d}. \end{aligned}} \tag{4.11}$$

Once again, these components describe the *same* tensor (which is a geometric entity) in terms of different bases. For example,

$$\boldsymbol{t} = t^{ab}(\boldsymbol{e}_a \otimes \boldsymbol{e}_b) = t'^{ab}(\boldsymbol{e}'_a \otimes \boldsymbol{e}'_b).$$

In general, when transforming the components of a tensor of arbitrary rank, each superscript inherits a transformation 'matrix' $\partial x'^a/\partial x^c$ and each subscript a transformation matrix $\partial x^c/\partial x'^a$. Thus, for example,

$$\boxed{t'_{ab}{}^{c} = \frac{\partial x^d}{\partial x'^a}\frac{\partial x^e}{\partial x'^b}\frac{\partial x'^c}{\partial x^f}t_{de}{}^{f}.} \tag{4.12}$$

Indeed, *the basic requirement for a set of quantities to be the components of a tensor is that they transform in such a way under a change of coordinates.* We shall return to this point later.

4.10 Tensor equations

Given a coordinate system (and hence a coordinate basis and its dual), it is convenient to work in terms of the components of a tensor $\boldsymbol{t}$ in this system rather than with the geometrical entity $\boldsymbol{t}$ itself. Therefore, from here onwards we shall adopt a much-used convention, which is *to confuse a tensor with its components.* This allows us to refer simply to *the tensor* $t_{ab}{}^{c}$, rather than *the tensor with components* $t_{ab}{}^{c}$.

We now come to the reason why tensors are important in mathematical physics. Let us illustrate this by way of an example. Suppose we find that in one particular coordinate system two tensors are equal, for example,

$$t_{ab} = s_{ab}. \tag{4.13}$$

Let us multiply both sides by $\partial x^a/\partial x'^c$ and $\partial x^b/\partial x'^d$ and take the implied summations to obtain

$$\frac{\partial x^a}{\partial x'^c}\frac{\partial x^b}{\partial x'^d}t_{ab} = \frac{\partial x^a}{\partial x'^c}\frac{\partial x^b}{\partial x'^d}s_{ab}.$$

Since t_{ab} and s_{ab} are both covariant components of tensors of rank 2, this equation can be restated as $t'_{ab} = s'_{ab}$. In other words, the equation (4.13) holds *in any other coordinate system*. In short, a tensor equation which holds in one coordinate system necessarily holds in *all* coordinate systems. Put another way, tensor equations are coordinate independent, which is in fact obvious from the geometric approach we have adopted since the outset. One particularly useful fact that emerges clearly from this discussion, and the transformation law (4.12), is that *if all the components of a tensor are zero in one coordinate system then they vanish in all coordinate systems*. This is useful in proving many tensor relations.

4.11 The quotient theorem

Not all objects with indices are the components of a tensor. An important example is provided by the connection coefficients $\Gamma^a{}_{bc}$, which vanish in a locally Cartesian coordinate system but not in other coordinate systems. Moreover, in Chapter 3 we derived the transformation properies of $\Gamma^a{}_{bc}$ and found that these were not of the form (4.12).

As mentioned above, the *fundamental* requirement that a set of quantities form the components of a tensor is that they obey a transformation law of the kind (4.12) under a change of coordinates. The *quotient theorem* provides a means of establishing this requirement in a particular case without having to demonstrate explicitly that the transformation law holds. It states that *if a set of quantities when contracted with a tensor produces another tensor then the original set of quantities is also a tensor*. Rather than give a general statement of the theorem and its proof, which tend to become obscured by a mass of indices, we shall give an example that illustrates the gist of the theorem.

In an N-dimensional manifold, suppose that with each system of coordinates about a point P there are associated N^3 numbers $t^a{}_{bc}$ and it is known that, for *arbitrary* contravariant vector components v^a, the N^2 numbers $t^a{}_{bc}v^c$ transform as the components of a rank-2 tensor at P under a change of coordinates. This means that

$$t'^a{}_{bc}v'^c = \frac{\partial x'^a}{\partial x^d}\frac{\partial x^e}{\partial x'^b}t^d{}_{ef}v^f, \tag{4.14}$$

where the $t'^a{}_{bc}$ are the corresponding N^3 numbers associated with the primed coordinate system. Then we may deduce that *the $t^a{}_{bc}$ are the components of a* *rank-3 tensor*, as follows. Since $v^f = (\partial x^f/\partial x'^c)v'^c$, equation (4.14) yields

$$t'^a{}_{bc}v'^c = \frac{\partial x'^a}{\partial x^d}\frac{\partial x^e}{\partial x'^b}t^d{}_{ef}\frac{\partial x^f}{\partial x'^c}v'^c$$

which, on rearrangement gives

$$\left(t'^a{}_{bc} - \frac{\partial x'^a}{\partial x^d}\frac{\partial x^e}{\partial x'^b}\frac{\partial x^f}{\partial x'^c}t^d{}_{ef}\right)v'^c = 0.$$

This holds for arbitrary vector components v'^c, so the expression in parentheses must vanish identically. Thus

$$t'^a{}_{bc} = \frac{\partial x'^a}{\partial x^d}\frac{\partial x^e}{\partial x'^b}\frac{\partial x^f}{\partial x'^c}t^d{}_{ef},$$

and therefore the $t^a{}_{bc}$ must be the components of a third-rank tensor.

Thus the gist of the quotient theorem is that if a set of numbers displays tensor characteristics when some of their indices are 'killed off' by summation with the components of an *arbitrary* tensor then the original numbers are the components of a tensor.

4.12 Covariant derivative of a tensor

It is straightforward to show that in an arbitrary coordinate system (unlike in local Cartesian coordinates) the differentiation of the components of a general tensor, other than a scalar, with respect to the coordinates does *not* in general result in the components of another tensor. For example, consider the derivative of the contravariant components v^a of a vector. Under a change of coordinates we have

$$\begin{aligned}\frac{\partial v'^a}{\partial x'^b} &= \frac{\partial x^c}{\partial x'^b}\frac{\partial v'^a}{\partial x^c}\\ &= \frac{\partial x^c}{\partial x'^b}\frac{\partial}{\partial x^c}\left(\frac{\partial x'^a}{\partial x^d}v^d\right)\\ &= \frac{\partial x^c}{\partial x'^b}\frac{\partial x'^a}{\partial x^d}\frac{\partial v^d}{\partial x^c} + \frac{\partial x^c}{\partial x'^b}\frac{\partial^2 x'^a}{\partial x^c \partial x^d}v^d. \end{aligned} \quad (4.15)$$

The presence of the second term on the right-hand side of (4.15) shows that the derivatives $\partial v^a/\partial x^b$ do not form the components of a second-order tensor. This term arises because the 'transformation matrix' $[\partial x'^a/\partial x^b]$ changes with position in the manifold (this is not true in local Cartesian coordinates, for which the second term vanishes).

To avoid this difficulty, in Chapter 3 we introduced the covariant derivative of a vector,

$$\nabla_b v^a = \partial_b v^a + \Gamma^a{}_{cb}v^c,$$

in terms of which we may write $\partial_b \mathbf{v} = (\nabla_b v^a)\mathbf{e}_a$. Using the transformation properties of the connection, derived in Chapter 3, it is straightforward to show that the $\nabla_b v^a$ are the (mixed) components of a rank-2 tensor, which is in fact clear from their definition. We denote this rank-2 tensor by $\nabla \mathbf{v}$, which is formally the outer product of the vector differential operator ∇ with the vector $\mathbf{v}$, although it is usual to omit the symbol $\otimes$ in outer products containing ∇. In a given basis we have $\nabla = \mathbf{e}^a \partial_a$, so we may write, for example,

$$\boxed{\nabla \mathbf{v} = \mathbf{e}^a \partial_a \otimes v^b \mathbf{e}_b = \mathbf{e}^a \otimes \partial_a (v^b \mathbf{e}_b) = (\nabla_a v^b)\mathbf{e}^a \otimes \mathbf{e}_b.}$$

Similarly, the $\nabla_b v_a$ form the covariant components of this tensor, i.e. $\nabla \mathbf{v} = (\nabla_a v_b)\mathbf{e}^a \otimes \mathbf{e}^b$. Indeed, it is easy to check that $\nabla_b v^a$ and $\nabla_b v_a$ satisfy the required transformation laws for being the components of a tensor.

We can extend the idea of the covariant derivative to higher-rank tensors. For example, let us consider an arbitrary rank-2 tensor $\mathbf{t}$ and derive the form of the covariant derivative $\nabla_c t^{ab}$ of its contravariant components. Expressing $\mathbf{t}$ in terms of its contravariant components, we have

$$\partial_c \mathbf{t} = \partial_c (t^{ab} \mathbf{e}_a \otimes \mathbf{e}_b) = (\partial_c t^{ab})\, \mathbf{e}_a \otimes \mathbf{e}_b + t^{ab} (\partial_c \mathbf{e}_a) \otimes \mathbf{e}_b + t^{ab} \mathbf{e}_a \otimes (\partial_c \mathbf{e}_b).$$

We can rewrite the derivatives of the basis vectors in terms of connection coefficients to obtain

$$\partial_c \mathbf{t} = (\partial_c t^{ab}) \mathbf{e}_a \otimes \mathbf{e}_b + t^{ab} \Gamma^d{}_{ac} \mathbf{e}_d \otimes \mathbf{e}_b + t^{ab} \mathbf{e}_a \otimes \Gamma^d{}_{bc} \mathbf{e}_d.$$

Interchanging the dummy indices a and d in the second term on the right-hand side and b and d in the third term, this becomes

$$\partial_c \mathbf{t} = \left(\partial_c t^{ab} + \Gamma^a{}_{dc} t^{db} + \Gamma^b{}_{dc} t^{ad}\right) \mathbf{e}_a \otimes \mathbf{e}_b,$$

where the expression in parentheses is the required covariant derivative,

$$\nabla_c t^{ab} = \partial_c t^{ab} + \Gamma^a{}_{dc} t^{db} + \Gamma^b{}_{dc} t^{ad}. \tag{4.16}$$

Using (4.16), the derivative of the tensor $\mathbf{t}$ with respect to x^c can be written in terms of its contravariant components as

$$\partial_c \mathbf{t} = (\nabla_c t^{ab}) \mathbf{e}_a \otimes \mathbf{e}_b.$$

Similar results may be obtained for the the covariant derivatives of the mixed and covariant components of the second-order tensor $\boldsymbol{t}$. Collecting these results together, we have

$$\boxed{\begin{aligned}\nabla_c t^{ab} &= \partial_c t^{ab} + \Gamma^a{}_{dc} t^{db} + \Gamma^b{}_{dc} t^{ad},\\ \nabla_c t^a{}_b &= \partial_c t^a{}_b + \Gamma^a{}_{dc} t^d{}_b - \Gamma^d{}_{bc} t^a{}_d,\\ \nabla_c t_{ab} &= \partial_c t_{ab} - \Gamma^d_{ac} t_{db} - \Gamma^d_{bc} t_{ad}.\end{aligned}} \tag{4.17}$$

The positions of the indices in these expressions are once again very systematic. The last index on each connection coefficient matches that on the covariant derivative, and the remaining indices can only be logically arranged in one way. For each contravariant index (superscript) on the left-hand side we add a term on the right-hand side containing a Christoffel symbol with a plus sign, and for every covariant index (subscript) we add a corresponding term with a minus sign. This is extended straightforwardly to tensors with an arbitrary number of contravariant and covariant indices. We note that the quantities $\nabla_c t^{ab}$, $\nabla_c t^a{}_b$ and $\nabla_c t_{ab}$ are the components of the *same* third-order tensor $\nabla\boldsymbol{t}$ with respect to different tensor bases, i.e.

$$\nabla\boldsymbol{t} = (\nabla_c t^{ab})\boldsymbol{e}^c \otimes \boldsymbol{e}_a \otimes \boldsymbol{e}_b = (\nabla_c t^a{}_b)\boldsymbol{e}^c \otimes \boldsymbol{e}_a \otimes \boldsymbol{e}^b = (\nabla_c t_{ab})\boldsymbol{e}^c \otimes \boldsymbol{e}^a \otimes \boldsymbol{e}^b.$$

One particularly important result is that the covariant derivative of the metric tensor $\boldsymbol{g}$ is identically zero at all points in a manifold, i.e.

$$\boxed{\nabla\boldsymbol{g} = \boldsymbol{0}.}$$

Alternatively, we can write this in terms of the components in *any* basis as

$$\boxed{\nabla_c g_{ab} = 0 \qquad \text{and} \qquad \nabla_c g^{ab} = 0.} \tag{4.18}$$

This result follows immediately from comparing, for example, the third result in (4.17) with our expression (3.20), derived in Chapter 3, for the partial derivative of the metric in terms of the affine connection. We note, in particular, that the expression (3.20) holds even in a manifold with non-zero torsion, and therefore so too must the result (4.18).[1]

The result (4.18) has an important consequence, which considerably simplifies tensor manipulations. This is that *we can interchange the order of raising or*

[1] In fact, for a general manifold with non-zero torsion, it is *not* necessary that (4.18) holds since one can, in principle, define the affine connection and the metric *independently*. In arriving at our earlier expression (3.20), we had in fact already assumed implicitly that the affine connection was *metric-compatible*, in which case (4.18) holds automatically. This topic is, however, beyond the scope of our discussion.

lowering an index and performing covariant differentiation without affecting the result. For example, consider the contravariant components t^{ab} of some rank-2 tensor. Using (4.18), we can write, for example,

$$\nabla_c t^{ab} = \nabla_c(g^{bd} t^a{}_d) = (\nabla_c g^{bd}) t^a{}_d + g^{bd}(\nabla_c t^a{}_d) = g^{bd}(\nabla_c t^a{}_d).$$

We also note that the covariant derivative obeys the standard rule for the differentiation of a product.

4.13 Intrinsic derivative of a tensor along a curve

In Chapter 3 we encountered vector fields that are defined only on some subspace of the manifold, an extreme example being when the vector field $\boldsymbol{v}(u)$ is defined only along some curve $x^a(u)$ in the manifold (as for the spin $\boldsymbol{s}(\tau)$ of a single particle along its worldline in spacetime). In a similar way a tensor field $\boldsymbol{t}(u)$ could be defined only along some curve $\mathcal{C}$. We now consider how to calculate the derivative of such a tensor with respect to the parameter u along the curve.

Let us begin by expressing the tensor at any point along the curve in terms of its contravariant components (say),

$$\boldsymbol{t}(u) = t^{ab}(u)\, \boldsymbol{e}_a(u) \otimes \boldsymbol{e}_b(u),$$

where the $\boldsymbol{e}_a(u)$ are the coordinate basis vectors at the point on the curve corresponding to the parameter value u. Thus, the derivative of $\boldsymbol{t}$ along the curve $\mathcal{C}$ is given by

$$\frac{d\boldsymbol{t}}{du} = \frac{dt^{ab}}{du}\boldsymbol{e}_a \otimes \boldsymbol{e}_b + t^{ab}\frac{d\boldsymbol{e}_a}{du} \otimes \boldsymbol{e}_b + t^{ab}\boldsymbol{e}_a \otimes \frac{d\boldsymbol{e}_b}{du}.$$

Using the chain rule to rewrite the derivatives of the basis vectors, we obtain

$$\frac{d\boldsymbol{t}}{du} = \frac{dt^{ab}}{du}\boldsymbol{e}_a \otimes \boldsymbol{e}_b + t^{ab}\frac{dx^c}{du}\frac{\partial \boldsymbol{e}_a}{\partial x^c} \otimes \boldsymbol{e}_b + t^{ab}\boldsymbol{e}_a \otimes \frac{dx^c}{du}\frac{\partial \boldsymbol{e}_b}{\partial x^c}.$$

Finally, by writing the partial derivatives of the basis vectors in terms of the connection and relabelling indices, we find that

$$\frac{d\boldsymbol{t}}{du} = \left(\frac{dt^{ab}}{du} + \Gamma^a_{dc} t^{db}\frac{dx^c}{du} + \Gamma^b{}_{dc} t^{ad}\frac{dx^c}{du}\right)\boldsymbol{e}_a \otimes \boldsymbol{e}_b. \tag{4.19}$$

The term in brackets is called the *intrinsic* (or *absolute*) derivative of the components t^{ab} along the curve $\mathcal{C}$ and is denoted

$$\boxed{\frac{Dt^{ab}}{Du} = \frac{dt^{ab}}{du} + \Gamma^a{}_{dc} t^{db}\frac{dx^c}{du} + \Gamma^b_{dc} t^{ad}\frac{dx^c}{du}.}$$

Similar results may be obtained for the covariant and mixed components of the tensor $\boldsymbol{t}$. For example, the derivative of $\boldsymbol{t}$ along the curve may be written

$$\frac{d\boldsymbol{t}}{du} = \frac{Dt^{ab}}{Du}\boldsymbol{e}_a \otimes \boldsymbol{e}_b = \frac{Dt_{ab}}{Du}\boldsymbol{e}^a \otimes \boldsymbol{e}^b = \frac{Dt^a_{\ b}}{Du}\boldsymbol{e}_a \otimes \boldsymbol{e}^b.$$

Clearly, the method can be extended easily to higher-rank tensors.

In a similar way to vectors, a tensor $\boldsymbol{t}$ is said to be *parallel-transported* along a curve $\mathcal{C}$ if $d\boldsymbol{t}/du = \boldsymbol{0}$ or, equivalently, in terms of its components, if for example $Dt^{ab}/Du = 0$.

Following our discussion of the intrinsic derivative of a vector in Chapter 3, a convenient way to remember the form of the intrinsic derivative is to *pretend* that the tensor $\boldsymbol{t}$ is in fact defined throughout (some region of) the manifold, i.e. not only along the curve $\mathcal{C}$. If this were the case then we could differentiate $\boldsymbol{t}$ with respect to the coordinates x^a. Thus we could write

$$\frac{dt^{ab}}{du} = \frac{\partial t^{ab}}{\partial x^c}\frac{dx^c}{du}.$$

Substituting this into (4.19), we could then factor out dx^c/du and recognise the other factor as the covariant derivative $\nabla_c t^{ab}$. Thus we could write

$$\frac{Dt^{ab}}{Du} = \nabla_c t^{ab}\frac{dx^c}{du}, \qquad (4.20)$$

with similar expressions for the intrinsic derivatives of its other components. It must be remembered, however, that if $\boldsymbol{t}$ is only defined along the curve $\mathcal{C}$ then formally (4.20) is not defined and acts merely as an *aide-memoire*.

Exercises

4.1 If $\boldsymbol{t}$ is a rank-2 tensor, show that

$$\boldsymbol{t}(\boldsymbol{u}+\boldsymbol{v}, \boldsymbol{w}+\boldsymbol{z}) = t_{ab}(u^a + v^a)(w^b + z^b).$$

4.2 If $s_{ab} = s_{ba}$ and $t_{ab} = -t_{ba}$ are the component of a symmetric and an antisymmetric tensor respectively, show that $s_{ab}t^{ab} = 0$.

4.3 If t_{ab} are the components of an antisymmetric tensor and v_a the components of a vector, show that

$$v_{[a}t_{bc]} = \tfrac{1}{3}(v_a t_{bc} + v_c t_{ab} + v_b t_{ca}).$$

4.4 If t_{ab} are the components of a symmetric tensor and v_a the components of a vector, show that if

$$v_a t_{bc} + v_c t_{ab} + v_b t_{ca} = 0$$

then either $t_{ab} = 0$ or $v_a = 0$.

4.5 If the tensor t_{abcd} satisfies $t_{abcd}v^a w^b v^c w^d = 0$ for arbitrary vectors v^a and w^a, show that

$$t_{abcd} + t_{cdab} + t_{cbad} + t_{adcb} = 0.$$

4.6 Consider the infinitesimal coordinate transformation

$$x'^a = x^a + \epsilon v^a(x),$$

where $v^a(x)$ is a vector field and ϵ is a small scalar quantity. Show that, to first order in ϵ,

$$g'_{ab}(x') = g_{ab}(x) - \epsilon(g_{ac}\partial_b v^c + g_{cb}\partial_a v^c).$$

4.7 By investigating their transformation properties, show that $\nabla_b v^a$ are the mixed components of a rank-2 tensor.

4.8 If v_a are the covariant components of a vector and A_{ab} are the components of an antisymmetric rank-2 tensor, show that

$$\nabla_a v_b - \nabla_b v_a = \partial_a v_b - \partial_b v_a,$$

$$\nabla_a A_{bc} + \nabla_c A_{ab} + \nabla_b A_{ca} = \partial_a A_{bc} + \partial_c A_{ab} + \partial_b A_{ca}.$$

Determine the symmetry properties of the rank-3 tensor

$$B_{abc} = \partial_a A_{bc} + \partial_c A_{ab} + \partial_b A_{ca}.$$

4.9 Show that covariant differentiation obeys the usual product rule, e.g.

$$\nabla_a(A_{bc}B^{cd}) = (\nabla_a A_{bc})B^{cd} + A_{bc}(\nabla_a B^{cd}).$$

Hint: Use local Cartesian coordinates.

4.10 For a general rank-2 tensor T^{ab}, show that the covariant divergence is given by

$$\nabla_a T^{ab} = \frac{1}{\sqrt{|g|}}\partial_a(\sqrt{|g|}\,T^{ab}) + \Gamma^b{}_{ca}T^{ac}.$$

Show further that if $A^{ab} = -A^{ba}$ are the components of an antisymmetric rank-2 tensor then

$$\nabla_a A^{ab} = \frac{1}{\sqrt{|g|}}\partial_a(\sqrt{|g|}\,A^{ab}).$$

Hence show that if the antisymmetric tensor field A^{ab} vanishes on a hypersurface S that bounds a region V of an N-dimensional manifold then

$$\int_V (\nabla_a A^{ab})\sqrt{-g}\,d^N x = 0.$$

4.11 Any coordinate transformation $x^a \to x'^a$ under which the metric is *form invariant*, i.e. such that

$$g'_{ab}(x) = g_{ab}(x)$$

is called an *isometry* (note that the argument is the same on both sides of the above equation). Show that the infinitesimal coordinate transformation in Exercise 4.6 is an isometry, to first order in ϵ, provided that v^a satisfies

$$g_{ac}\partial_b v^c + g_{cb}\partial_a v^c + v^c \partial_c g_{ab} = 0.$$

Show further that this expression can be written as

$$\nabla_a v_b + \nabla_b v_a = 0.$$

This is *Killing's equation* and any vector satisfying it is known as a *Killing vector* of the metric g_{ab}. Show that if v^a and w^a are both Killing vectors then so too is any linear combination $\lambda v^a + \mu w^a$, where λ and μ are constants.

5

Special relativity revisited

Now that we have the machinery of tensor calculus in place, let us return to special relativity and consider how to express this theory in a more formal manner.

5.1 Minkowski spacetime in Cartesian coordinates

In the language of Chapter 2, the Minkowski spacetime of special relativity is a fixed four-dimensional pseudo-Euclidean manifold. As such, there exists a privileged class of Cartesian coordinate systems (t, x, y, z) covering the whole manifold, so that at *every* point (or event) the squared line element takes the form

$$ds^2 = c^2\,d\tau^2 = c^2\,dt^2 - dx^2 - dy^2 - dz^2,$$

where we have taken the opportunity to define the proper time interval $d\tau^2 = ds^2/c^2$. It is convenient to introduce the indexed coordinates $x^\mu\,(\mu = 0, 1, 2, 3)$,[1] so that

$$x^0 \equiv ct, \qquad x^1 \equiv x, \qquad x^2 \equiv y, \qquad x^3 \equiv z,$$

and to write the line element as

$$\boxed{ds^2 = \eta_{\mu\nu}\,dx^\mu\,dx^\nu,}$$

[1] It is conventional to use Greek indices when discussing four-dimensional spacetimes rather than the Latin indices a, b, c etc. from the start of the alphabet, which are used for abstract N-dimensional manifolds. Moreover, in relativity theory, it is more common for a Greek index to run from 0 to 3 than from 1 to 4 (although the latter usage is found in some textbooks). Also, it is conventional to use Latin letters from the middle of the alphabet, such as i, j, k etc., for indices that run from 1 to 3.

where the $\eta_{\mu\nu}$ are the covariant components of the metric tensor and are given by

$$[\eta_{\mu\nu}] = \begin{pmatrix} 1 & 0 & 0 & 0 \\ 0 & -1 & 0 & 0 \\ 0 & 0 & -1 & 0 \\ 0 & 0 & 0 & -1 \end{pmatrix}. \tag{5.1}$$

From now on we will often use the shorthand notation $[\eta_{\mu\nu}] = \mathrm{diag}(1, -1, -1, -1)$. It is clear that the contravariant components of the metric are identical, i.e. $[\eta^{\mu\nu}] = \mathrm{diag}(1, -1, -1, -1)$. With this definition of the metric, Minkowski spacetime has a signature of -2.[2] We also note that, since the metric coefficients are constant, the connection $\Gamma^{\mu}{}_{\nu\sigma}$ vanishes everywhere in this coordinate system.

5.2 Lorentz transformations

Cartesian coordinates, which we are using in the context of special relativity, have a direct physical interpretation and correspond to distances and times measured by an observer at rest in some inertial frame S that is labelled using three-dimensional Cartesian coordinates[3] (remember that, in Chapter 1, we defined an inertial frame as one in which a free particle moves in a straight line with fixed speed). Transforming to a different Cartesian inertial frame corresponds to performing a coordinate transformation on the Minkowski spacetime to a new system x'^{μ}. Since we require that the new coordinate system x'^{μ} also corresponds to a Cartesian inertial frame, the (squared) line element ds^2 must take the *same form* in these primed coordinates as it did in the unprimed coordinates, i.e.

$$ds^2 = \eta_{\mu\nu}\, dx^{\mu}\, dx^{\nu} = \eta_{\mu\nu}\, dx'^{\mu}\, dx'^{\nu}.$$

In other words the metric in the new coordinates must also be given by (5.1). From the transformation properties of a second-rank tensor, this means that the transformation $x^{\mu} \to x'^{\mu}$ must satisfy

$$\boxed{\eta_{\mu\nu} = \frac{\partial x'^{\rho}}{\partial x^{\mu}} \frac{\partial x'^{\sigma}}{\partial x^{\nu}} \eta_{\rho\sigma},} \tag{5.2}$$

which is the necessary and sufficient condition that a transformation $x^{\mu} \to x'^{\mu}$ is a *Lorentz transformation* between two Cartesian inertial coordinate systems. From (5.2), we see that the elements of the transformation matrix must be

[2] Note that some relativists use an alternative, but equivalent, definition $[\eta_{\mu\nu}] = \mathrm{diag}(-1, 1, 1, 1)$ in which the signature is $+2$.

[3] We shall prove this shortly.

constants. Thus the transformation between two inertial coordinate systems *must be linear*, i.e.

$$\boxed{x'^{\mu} = \Lambda^{\mu}{}_{\nu} x^{\nu} + a^{\mu}} \qquad (5.3)$$

where the $\Lambda^{\mu}{}_{\nu}$ and a^{ν} are constants. This has the form of a general inhomogeneous Lorentz transformation (or Poincaré transformation). We will generally take the (unimportant) constants a^{μ} to be zero, in which case (5.3) reduces to a normal, homogeneous, Lorentz transformation. As discussed in Chapter 1, the constants $\Lambda^{\mu}{}_{\nu}$ in the transformation matrix depend upon the relative speed and orientation of the two inertial frames. If the unprimed and primed coordinates correspond to inertial frames S and S' in *standard configuration*, with S' moving at a speed v relative to S, then the transformation matrix can be written in two equivalent forms:

$$[\Lambda^{\mu}{}_{\nu}] = \left[\frac{\partial x'^{\mu}}{\partial x^{\nu}}\right] = \begin{pmatrix} \gamma & -\beta\gamma & 0 & 0 \\ -\beta\gamma & \gamma & 0 & 0 \\ 0 & 0 & 1 & 0 \\ 0 & 0 & 0 & 1 \end{pmatrix} = \begin{pmatrix} \cosh\psi & -\sinh\psi & 0 & 0 \\ -\sinh\psi & \cosh\psi & 0 & 0 \\ 0 & 0 & 1 & 0 \\ 0 & 0 & 0 & 1 \end{pmatrix}, \qquad (5.4)$$

where $\beta = v/c$, $\gamma = (1-\beta^2)^{-1/2}$ and the rapidity is defined by $\psi = \tanh^{-1}\beta$. Clearly, if the axes of S' and S are rotated with respect to one another then the transformation is more complicated.

The transformation inverse to (5.4) is clearly obtained by putting $v \to -v$ (or equivalently $\psi \to -\psi$). In general, the inverse transformation matrix is denoted by

$$[\Lambda_{\mu}{}^{\nu}] = \left[\frac{\partial x^{\nu}}{\partial x'^{\mu}}\right]$$

and may be calculated from the forward transform using the index-raising and index-lowering properties of the metric, i.e.

$$\boxed{\Lambda_{\mu}{}^{\nu} = \eta_{\mu\rho}\eta^{\nu\sigma}\Lambda^{\rho}{}_{\sigma}.}$$

That this is indeed the required inverse may be shown using the condition (5.2), which gives

$$\Lambda_{\mu}{}^{\nu}\Lambda^{\mu}{}_{\tau} = \eta_{\mu\rho}\eta^{\nu\sigma}\Lambda^{\rho}{}_{\sigma}\Lambda^{\mu}{}_{\tau} = \eta_{\tau\sigma}\eta^{\nu\sigma} = \delta^{\nu}_{\tau}.$$

5.3 Cartesian basis vectors

Figure 5.1 shows the coordinate curves for two systems of coordinates x^a and x'^a, corresponding to Cartesian inertial frames S and S' in standard configuration (with the 2- and 3- directions suppressed). In any coordinate system the coordinate

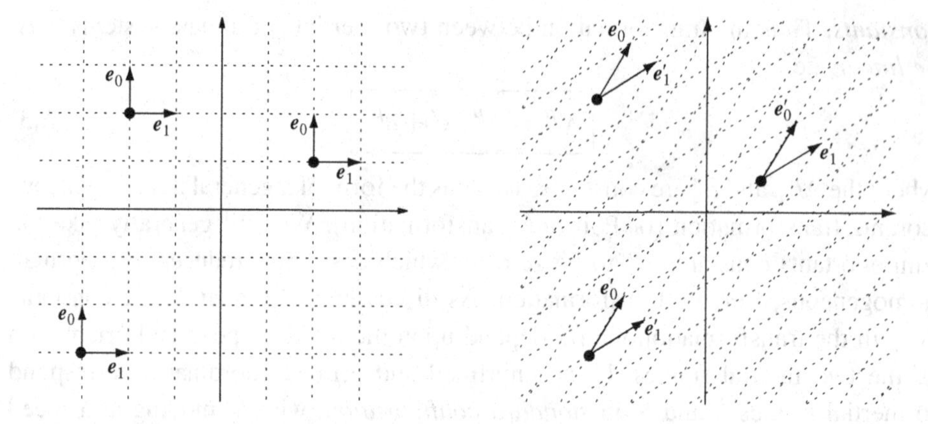

Figure 5.1 The coordinate curves (dotted lines) for two systems of coordinates x^a and x'^a, corresponding to Cartesian inertial frames S and S' in standard configuration. The coordinate basis vectors for each system are also shown. The 2- and 3-directions are suppressed, and null vectors would lie at 45 degrees to the vertical axis.

basis vectors are tangents to the coordinate curves; these are shown for S and S' in Figure 5.1 (in this diagram, null vectors would lie at 45 degrees to the vertical axis). In general, the two sets of basis vectors are related by

$$\boxed{\boldsymbol{e}'_\mu = \Lambda_\mu{}^\nu \boldsymbol{e}_\nu \qquad \text{and} \qquad \boldsymbol{e}_\mu = \Lambda^\nu{}_\mu \boldsymbol{e}'_\nu,}$$

which tells us how to draw one set of basis vectors in terms of the other set.

The two sets of basis vectors satisfy

$$\boxed{\boldsymbol{e}_\mu \cdot \boldsymbol{e}_\nu = \boldsymbol{e}'_\mu \cdot \boldsymbol{e}'_\nu = \eta_{\mu\nu},}$$

and so *both* sets form an orthonormal basis at each point in the pseudo-Euclidean Minkowski spacetime. As drawn in Figure 5.1 the vectors $\boldsymbol{e}_\mu$ appear mutually perpendicular, but the $\boldsymbol{e}'_\mu$ do not. This is an artefact of representing a pseudo-Euclidean space on a Euclidean piece of paper. As we shall see, the notion of an orthonormal set of basis vectors at any point in the spacetime is of fundamental importance for our description of observers.

We can also define dual basis vectors for each system as

$$\boldsymbol{e}^\mu = \eta^{\mu\nu} \boldsymbol{e}_\nu \qquad \text{and} \qquad \boldsymbol{e}'^\mu = \eta^{\mu\nu} \boldsymbol{e}'_\nu.$$

These vectors also form orthonormal sets, since

$$\boldsymbol{e}^\mu \cdot \boldsymbol{e}^\nu = \boldsymbol{e}'^\mu \cdot \boldsymbol{e}'^\nu = \eta^{\mu\nu},$$

and the components $\eta^{\mu\nu}$ are identical to the components $\eta_{\mu\nu}$.

5.4 Four-vectors and the lightcone

As in any manifold, we can define vectors at any point P in Minkowski spacetime (and thus vector fields).[4] In relativity, vectors defined on a four-dimensional spacetime manifold are called *4-vectors*. These 4-vectors are *geometrical entities* in spacetime, which can be defined without any reference to a basis (or coordinate system). Nevertheless, in a particular coordinate system, we can write a general 4-vector $\boldsymbol{v}$ at P in terms of the coordinate basis vectors at P:

$$\boxed{\boldsymbol{v} = v^{\mu}\boldsymbol{e}_{\mu}.}$$

Let us assume for the moment that we are using Cartesian coordinates x^{μ} corresponding to some inertial frame S. At each point P in spacetime we have a constant set of orthonormal basis vectors $\boldsymbol{e}_{\mu}$. The square of the length of a vector $\boldsymbol{v}$ at a point P (which is a coordinate-independent quantity) is then given by

$$\boldsymbol{v} \cdot \boldsymbol{v} = v_{\mu}v^{\mu} = \eta_{\mu\nu}v^{\mu}v^{\nu}.$$

We have that

$$\text{for } \eta_{\mu\nu}v^{\mu}v^{\nu} > 0 \text{ the vector is } \textit{timelike}; \tag{5.5}$$

$$\text{for } \eta_{\mu\nu}v^{\mu}v^{\nu} = 0 \text{ the vector is } \textit{null}; \tag{5.6}$$

$$\text{for } \eta_{\mu\nu}v^{\mu}v^{\nu} < 0 \text{ the vector is } \textit{spacelike}. \tag{5.7}$$

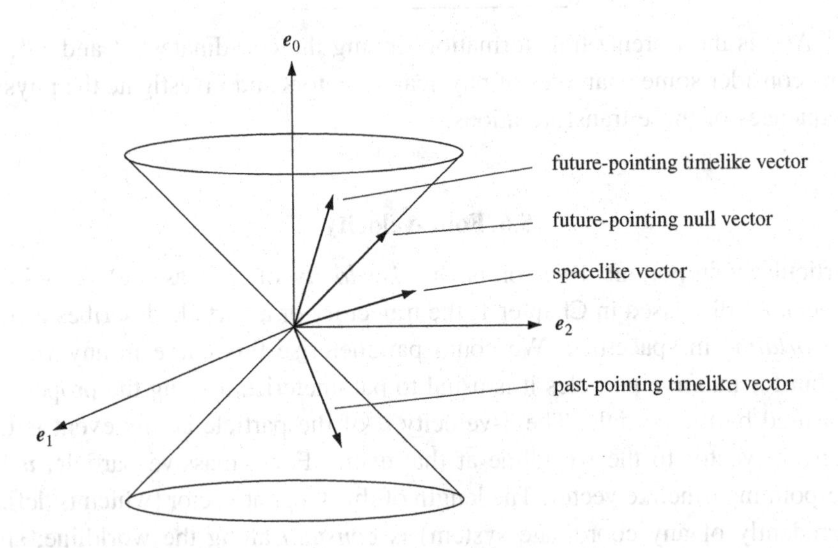

Figure 5.2 The lightcone at some point P in Minkowski spacetime (with one spatial dimension suppressed).

[4] In fact, since Minkowski spacetime is pseudo-Euclidean, the tangent space T_P at any point P coincides with the manifold itelf. Thus, in this special case, we are not restricted to local vectors and can reinstate the notions of position vector and of the displacement vector between arbitrary points in the manifold.

Thus, as we would expect, the coordinate basis vector $\boldsymbol{e}_0$, which has components (1, 0, 0, 0), is timelike. Similarly, the basis vectors $\boldsymbol{e}_i (i = 1, 2, 3)$ are spacelike. Moreover, for any timelike or null vector $\boldsymbol{v}$, if $\boldsymbol{v} \cdot \boldsymbol{e}_0 > 0$ then $\boldsymbol{v}$ is called *future-pointing* whereas if $\boldsymbol{v} \cdot \boldsymbol{e}_0 < 0$ then $\boldsymbol{v}$ is *past-pointing*.

At any point P in the Minkowski spacetime, the set of all null vectors at P forms the *lightcone* or *null-cone*. The structure of the lightcone is illustrated in Figure 5.2, with one spatial dimension suppressed.

5.5 Four-vectors and Lorentz transformations

Suppose that the Cartesian coordinates x^μ and x'^μ correspond to inertial frames S and S'. Thus, at each point P in the Minkowski spacetime we have two sets of (constant) basis vectors $\boldsymbol{e}_\mu$ and $\boldsymbol{e}'_\mu$, and a general 4-vector $\boldsymbol{v}$ defined at P can be expressed in terms of *either* set:

$$\boldsymbol{v} = v^\mu \boldsymbol{e}_\mu = v'^\mu \boldsymbol{e}'_\mu.$$

Thus, the components in the two bases are related by

$$\boxed{\begin{aligned} v'^\mu &= \boldsymbol{v} \cdot \boldsymbol{e}'^\mu = \Lambda^\mu{}_\nu v^\nu \\ v^\mu &= \boldsymbol{v} \cdot \boldsymbol{e}^\mu = \Lambda_\nu{}^\mu v'^\nu, \end{aligned}} \tag{5.8}$$

where $\Lambda^\mu{}_\nu$ is the Lorentz transformation linking the coordinates x^μ and x'^μ. Let us now consider some examples of physical 4-vectors and investigate the physical consequences of these transformations.

5.6 Four-velocity

A particularly important 4-vector is the *4-velocity* of a (massive) particle (or observer). As discussed in Chapter 1, the trajectory of a particle describes a curve $\mathcal{C}$ or *worldline* in spacetime. We could parameterise this curve in any way we wish, but for massive particles it is usual to parameterise it using the proper time τ measured by the particle. The 4-velocity $\boldsymbol{u}$ of the particle at any event is then the tangent vector to the worldline at that event. For a massive particle, $\boldsymbol{u}$ is a future-pointing timelike vector. The length of this tangent vector (which is defined independently of any coordinate system) is *constant* along the worldline, since (as shown in Chapter 3)

$$\boxed{\boldsymbol{u} \cdot \boldsymbol{u} = \left(\frac{ds}{d\tau}\right)^2 = c^2.} \tag{5.9}$$

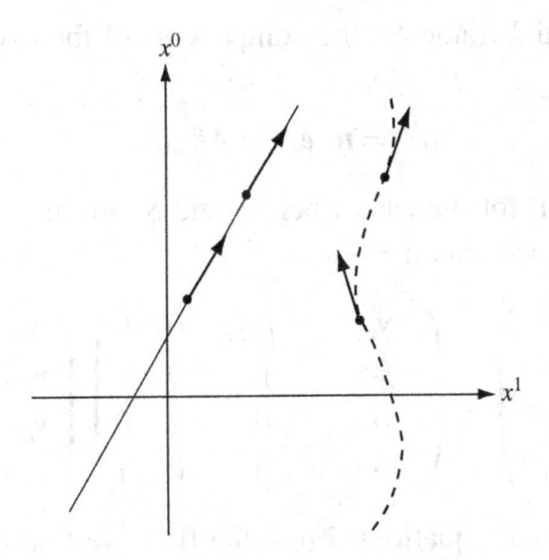

Figure 5.3 The 4-velocity at events along the worldlines of a particle travelling at uniform speed in S (solid line) and a particle accelerating with respect to S (broken line).

Since τ is proportional to the interval s along the worldline, it is an *affine* parameter (see Chapter 3).

Suppose that we label spacetime with some Cartesian coordinate system corresponding to an inertial frame S. We can then write the worldline of a particle in this coordinate system as $x^\mu = x^\mu(\tau)$. Figure 5.3 shows the 4-velocity at two events on the worldline of a particle moving at uniform velocity in the frame S. In this case the *direction* of the 4-velocity is also constant along the worldline. The figure also shows the 4-velocity at two events on the worldline of a particle that is accelerating (back and forth) with respect to the frame S. Clearly, in this case, the direction of the 4-velocity changes along the worldline.

The (contravariant) components of the 4-velocity in the frame S are given by

$$u^\mu = \boldsymbol{u} \cdot \boldsymbol{e}^\mu = \frac{dx^\mu}{d\tau}. \tag{5.10}$$

Setting $x^0 = ct$ for the moment, and noting that $d\tau = dt/\gamma_u$, where $\gamma_u = (1 - |\vec{u}|^2/c^2)^{-1/2}$, we can write these components as

$$[u^\mu] = \gamma_u \left(c, \frac{dx^1}{dt}, \frac{dx^2}{dt}, \frac{dx^3}{dt} \right) = \gamma_u(c, \vec{u}), \tag{5.11}$$

where in the last line (with a slight abuse of notation) we have introduced the *relative 3-vector* $\vec{u} = (\vec{u}^1, \vec{u}^2, \vec{u}^3)$, which is the familiar (three-dimensional) velocity vector of the particle as measured by an observer at rest in S, and $u^2 = \vec{u} \cdot \vec{u}$.

In some other inertial frame S', the components of the 4-velocity of the particle are

$$u'^{\mu} = \boldsymbol{u}\cdot\boldsymbol{e}'^{\mu} = \Lambda^{\mu}{}_{\nu}u^{\nu}.$$

Writing this out in full for the case where S and S' are in standard configuration with relative speed v, we obtain

$$\begin{pmatrix}\gamma_{u'}c\\ \gamma_{u'}\vec{u}'^1\\ \gamma_{u'}\vec{u}'^2\\ \gamma_{u'}\vec{u}'^3\end{pmatrix} = \begin{pmatrix}\gamma_v & -\beta\gamma_v & 0 & 0\\ -\beta\gamma_v & \gamma_v & 0 & 0\\ 0 & 0 & 1 & 0\\ 0 & 0 & 0 & 1\end{pmatrix}\begin{pmatrix}\gamma_u c\\ \gamma_u\vec{u}^1\\ \gamma_u\vec{u}^2\\ \gamma_u\vec{u}^3\end{pmatrix}.$$

This is equivalent to four equations. From the first, we find that

$$\frac{\gamma_u}{\gamma_{u'}} = \frac{1}{\gamma_v}\frac{1}{1-\vec{u}^1 v/c^2},$$

and from the others we obtain the 3-velocity addition law in special relativity,

$$\vec{u}'^1 = \frac{\vec{u}^1 - v}{(1-\vec{u}^1 v/c^2)},$$

$$\vec{u}'^2 = \frac{\vec{u}^2}{\gamma_v(1-\vec{u}^1 v/c^2)},$$

$$\vec{u}'^3 = \frac{\vec{u}^3}{\gamma_v(1-\vec{u}^1 v/c^2)}.$$

Note that this approach has allowed us to derive the 3-velocity addition law in an almost trivial way.

5.7 Four-momentum of a massive particle

The 4-momentum of a massive particle of rest mass m_0 is defined in terms of its four-velocity $\boldsymbol{u}$ by

$$\boxed{\boldsymbol{p} = m_0\boldsymbol{u}.}$$

At any point P along the particle's worldline the square of the length of this vector is

$$\boldsymbol{p}\cdot\boldsymbol{p} = m_0\boldsymbol{u}\cdot m_0\boldsymbol{u} = m_0^2c^2. \tag{5.12}$$

In Cartesian coordinates x^{μ} corresponding to some inertial frame S, the components of the 4-momentum are simply $p^{\mu} = \boldsymbol{p}\cdot\boldsymbol{e}^{\mu}$. According to convention we write

$$[p^{\mu}] = (E/c, p^1, p^2, p^3) = (E/c, \vec{p}), \tag{5.13}$$

where E is the *energy* of the particle as measured in the frame S and $\vec{p}$ is its *3-momentum* measured in S. Comparing (5.13) with (5.11) we see that, in special relativity,

$$E = \gamma_u m_0 c^2, \tag{5.14}$$

$$\vec{p} = \gamma_u m_0 \vec{u}. \tag{5.15}$$

In the frame S, the squared length of the 4-momentum is given by $p^\mu p_\mu$. Thus, from (5.13) and (5.12), we find that

$$E^2 - p^2 c^2 = m_0^2 c^4,$$

where $p^2 = \vec{p} \cdot \vec{p}$. This is the well-known *energy–momentum invariant*.

5.8 Four-momentum of a photon

The above discussion concerned particles of non-zero rest mass, which move at speeds less than c. We now consider particles such as photons and perhaps neutrinos, which move at the speed of light. The worldline of a massless particle is a null curve, along which $d\tau = 0$. Thus, we cannot parameterise such a worldline using the proper time τ. Nevertheless, there are many other parameters that we can use. For example, in an inertial frame, a photon travelling in the positive x-direction will describe the path $x = ct$. This could be written parametrically as

$$x^\mu = u^\mu \sigma, \tag{5.16}$$

where σ is the parameter and $[u^\mu] = (1, 1, 0, 0)$. Using (3.43), the tangent vector to the worldline is then

$$\boldsymbol{u} = \frac{dx^\mu}{d\sigma} \boldsymbol{e}_\mu = u^\mu \boldsymbol{e}_\mu.$$

Since the worldline is a null curve, we have

$$\boldsymbol{u} \cdot \boldsymbol{u} = 0, \tag{5.17}$$

in contrast with (5.9). Moreover, with this choice of parameter σ we see that

$$\boxed{\frac{d\boldsymbol{u}}{d\sigma} = \boldsymbol{0},} \tag{5.18}$$

which is the *equation of motion* for a photon. We note that although this has been derived using the fact that the Cartesian basis vectors $\boldsymbol{e}_\mu$ do not change with position, it is a vector equation and therefore will hold in any basis (i.e. any coordinate system).

Our choice of parameterisation in (5.16) may appear somewhat arbitrary. Indeed, it is true that there exists an unlimited number of parameterisations that could be used. For example, suppose that we replaced σ by α^2 (say). As the new parameter α varies between $-\infty$ and ∞, the *same* worldline $x = ct$ would be described in the spacetime. Since this is a null curve, the condition (5.17) would continue to be true (as may be verified explicitly). In the new parameterisation, however, the equation of motion (5.18) would *not* still hold. The special class of parameters for which the equation of motion has the simple form (5.18) is the class of *affine* parameters (as discussed in Section 3.16). Since one is always free to choose such a parameter, we will assume from here on that equation (5.18) is satisfied.

So far, we have not mentioned the frequency (or energy) of the photon. Clearly, the tangent vector $\boldsymbol{u}$ can be multipled by any scalar constant and will still satisfy the equations (5.17) and (5.18). The 4-momentum of a photon is therefore defined as

$$\boxed{\boldsymbol{p} = \alpha \boldsymbol{u},}$$

for a constant α chosen such that, in an arbitrary inertial frame S, the components of $\boldsymbol{p}$ are

$$[p^\mu] = (E/c, \vec{p}),$$

where E is the energy of the photon as measured in S and $\vec{p}$ is its 3-momentum in S. From (5.17) we thus have $E = pc$.

For photons, it is also common to introduce the *4-wavevector* $\boldsymbol{k}$, which is related to the four-momentum by $\boldsymbol{p} = \hbar \boldsymbol{k}$. Thus, in the frame S, the 4-wavevector has components given by

$$[k^\mu] = (2\pi/\lambda, \vec{k}),$$

where λ is the wavelength of the photon as measured in S and $\vec{k} = (2\pi/\lambda)\vec{n}$ and $\vec{n}$ is a unit 3-vector in the direction of propagation.

5.9 The Doppler effect and relativistic aberration

An example of the usefulness of the 4-vector approach (and particularly the photon 4-wavevector) is provided by the Doppler effect. Suppose that an observer $\mathcal{O}$ is at rest in some Cartesian inertial frame S defined by the coordinates x^μ in spacetime. Let us also suppose that a source of radiation is moving relative to S with a speed v in the positive x^1-direction and that at some event P the observer receives a photon of wavelength λ in a direction that makes an angle θ with the positive

x^1-direction. Thus, at the event P the components $k^\mu = \boldsymbol{k} \cdot \boldsymbol{e}^\mu$ of the photon's 4-wavevector in this coordinate system are

$$[k^\mu] = \frac{2\pi}{\lambda}(1, \cos\theta, \sin\theta, 0).$$

The photon observed at the event P must have been emitted by the source at some other event Q (say). However, the equation of motion of a photon implies that its 4-momentum $\boldsymbol{p}$, and hence its 4-wavevector $\boldsymbol{k}$, is *constant* along its worldline. Thus the photon's 4-wavevector $\boldsymbol{k}$ at the event Q is the same as that at the event P.

Let us denote the Cartesian inertial frame in which the radiation source is at rest by S' (whose spatial axes are assumed not to be rotated with respect to those of S); this frame is represented by the coordinates x'^μ in spacetime. Thus, at the event Q the components in S' of the photon's 4-wavevector are given by $k'^\mu = \boldsymbol{k} \cdot \boldsymbol{e}'^\mu$ and read

$$k'^\mu = \Lambda^\mu{}_\nu k^\nu, \tag{5.19}$$

where $[\Lambda^\mu{}_\nu]$ is given by (5.4).

We denote these components in S' by

$$[k'^\mu] = \frac{2\pi}{\lambda'}(1, \cos\theta', \sin\theta', 0).$$

The zeroth component of (5.19) yields the ratio of the proper wavelength and the observed wavelength:

$$\boxed{\frac{\lambda}{\lambda'} = \gamma(1 - \beta\cos\theta).}$$

This equation contains all the familiar *Doppler effect* results as special cases. If $\theta = 0$, the source must be approaching the observer along the negative x^1-axis. If $\theta = \pi$, the source is receding from the observer along the positive x^1-axis. Finally, if $\theta = \pm\pi/2$ we obtain the transverse Doppler effect. Similarly, from the 2- and 3- components of (5.19) we obtain immediately

$$\boxed{\tan\theta' = \frac{\tan\theta}{\gamma[1 - (v/c)\sec\theta]},}$$

which is a version of the *relativistic aberration formula.*

5.10 Relativistic mechanics

In relativistic mechanics, the equation of motion of a massive particle is given by

$$\boxed{\frac{d\boldsymbol{p}}{d\tau}=\boldsymbol{f},}$$

where $\boldsymbol{f}$ is the *4-force*. In some Cartesian inertial frame S (for which the basis vectors are constant throughout the spacetime) the components f^μ of the 4-force are given by the familiar expression

$$f^\mu = \boldsymbol{e}^\mu \cdot \frac{d\boldsymbol{p}}{d\tau} = \boldsymbol{e}^\mu \cdot \frac{d}{d\tau}(p^\nu \boldsymbol{e}_\nu) = \frac{dp^\nu}{d\tau}\delta^\mu_\nu = \frac{dp^\mu}{d\tau},$$

where we have used the fact that $\boldsymbol{e}^\mu$ and $\boldsymbol{e}_\mu$ are reciprocal sets of vectors. Noting that $d\tau = dt/\gamma_u$, we may write

$$[f^\mu] = \gamma_u \frac{d}{dt}\left(\frac{E}{c}, \vec{p}\right) = \gamma_u\left(\frac{\vec{f}\cdot\vec{u}}{c}, \vec{f}\right),$$

where in the last equality we have introduced the familiar *3-force* $\vec{f}$ as measured in the frame S, and $\vec{u}$ is the 3-velocity in this frame. Writing the components in this way, the time and space parts of the equation of motion in S are (as required)

$$\frac{1}{\gamma_u}\frac{dE}{d\tau} = \frac{dE}{dt} = \vec{f}\cdot\vec{u}, \tag{5.20}$$

$$\frac{1}{\gamma_u}\frac{d\vec{p}}{d\tau} = \frac{d\vec{p}}{dt} = \vec{f}, \tag{5.21}$$

where E and $\vec{p}$ are given by (5.14) and (5.15) respectively.

There is, however, a certain rarely discussed subtlety in relativistic mechanics. Let us consider the scalar product $\boldsymbol{u}\cdot\boldsymbol{f}$, which is of course invariant under coordinate transformations. This is given by

$$\begin{aligned}\boldsymbol{u}\cdot\boldsymbol{f} = \boldsymbol{u}\cdot\frac{d\boldsymbol{p}}{d\tau} &= \boldsymbol{u}\cdot\left(\frac{dm_0}{d\tau}\boldsymbol{u} + m_0\frac{d\boldsymbol{u}}{d\tau}\right)\\ &= c^2\frac{dm_0}{d\tau} + m_0\boldsymbol{u}\cdot\frac{d\boldsymbol{u}}{d\tau}\\ &= c^2\frac{dm_0}{d\tau},\end{aligned}$$

where we have (twice) used the fact that $\boldsymbol{u}\cdot\boldsymbol{u} = c^2$. Thus, we see that in special relativity the action of a force can alter the rest mass of a particle! A force that preserves the rest mass is called a *pure* force and must satisfy $\boldsymbol{u}\cdot\boldsymbol{f} = 0$.

If so desired, one can also introduce the *4-acceleration* of a particle, $\boldsymbol{a} = d\boldsymbol{u}/d\tau$, in terms of which a *pure* 4-force takes the familiar form $\boldsymbol{f} = m_0\boldsymbol{a}$. In some Cartesian inertial frame S, the components of the 4-acceleration are

$$[a^\mu] = \left[\frac{du^\mu}{d\tau}\right] = \gamma_u \frac{d}{dt}(\gamma_u c, \gamma_u \vec{u}) = \gamma_u \left(c\frac{d\gamma_u}{dt}, \gamma_u \frac{d\vec{u}}{dt} + \vec{u}\frac{d\gamma_u}{dt}\right),$$

$$= \gamma_u \left(c\frac{d\gamma_u}{dt}, \gamma_u \vec{a} + \vec{u}\frac{d\gamma_u}{dt}\right),$$

where $\vec{a} = d\vec{u}/dt$ is the 3-acceleration in the frame S.

5.11 Free particles

We now come to a very important observation concerning relativistic mechanics. In the absence of any forces, the equation of motion of a massive particle is

$$\boxed{\frac{d\boldsymbol{p}}{d\tau} = \boldsymbol{0},} \tag{5.22}$$

where the proper time τ is an affine parameter along the particle's worldline. Similarly, the equation of motion of a photon is

$$\boxed{\frac{d\boldsymbol{p}}{d\sigma} = \boldsymbol{0},} \tag{5.23}$$

where σ is some affine parameter along the photon's worldline. However, in each case the 4-momentum $\boldsymbol{p}$ at some point on the worldline is simply a fixed multiple of the tangent vector to the worldline at that point. Thus, equations (5.22) and (5.23) say that tangent vectors to the worldlines of free particles and of photons form a parallel field of vectors along the worldline. From Chapter 2 we know that this is the definition of an affinely parameterised geodesic. Thus, *in special relativity the worldlines of free particles and photons are respectively non-null and null geodesics in Minkowski spacetime.*

5.12 Relativistic collisions and Compton scattering

We note from (5.22) and (5.23) that the conservation of energy and momentum for a free particle or photon is represented by the single equation $\boldsymbol{p} = \text{constant}$. We can, of course, add the 4-momenta of different particles. Thus for a system of n interacting particles $i = 1, 2, \ldots, n$ with no external forces, we have $\sum_{i=1}^{n} \boldsymbol{p}_i =$ constant, which is very useful in relativistic-collision calculations.

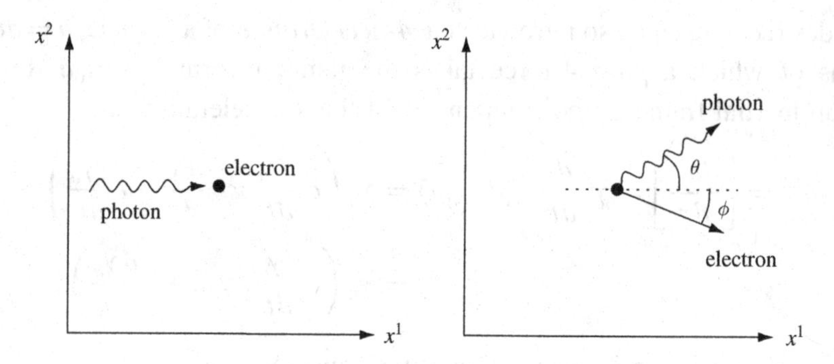

Figure 5.4 The Compton effect.

An important example of a relativistic collision is Compton scattering, in which a photon of 4-momentum $\boldsymbol{p}$ collides with an electron of 4-momentum $\boldsymbol{q}$. It is easiest to consider the collision in the inertial frame S in which the electron is at rest and the photon is travelling along the positive x^1-axis (see Figure 5.4). Thus the components of $\boldsymbol{p}$ and $\boldsymbol{q}$ in S are

$$[p^\mu] = (h\nu/c, h\nu/c, 0, 0),$$
$$[q^\mu] = (m_e c, 0, 0, 0),$$

where ν is the frequency of the photon as measured by a stationary observer in S, and m_e is the rest mass of the electron. Let us assume that, after the collision, the electron and photon have 4-momenta $\bar{\boldsymbol{p}}$ and $\bar{\boldsymbol{q}}$ such that they move off in the plane $x^3 = 0$, making angles θ and ϕ respectively with the x^1-axis. Thus

$$[\bar{p}^\mu] = (h\bar{\nu}/c, (h\bar{\nu}/c)\cos\theta, (h\bar{\nu}/c)\sin\theta, 0),$$
$$[\bar{q}^\mu] = (\gamma_u m_e c, \gamma_u m_e u\cos\phi, -\gamma_u m_e \sin\phi, 0),$$

where u is the electron's speed and $\bar{\nu}$ is the photon frequency as measured by a stationary observer in S after the collision. Conservation of total 4-momentum means that

$$p^\mu + q^\mu = \bar{p}^\mu + \bar{q}^\mu,$$

which gives

$$h\nu/c + m_e c = h\bar{\nu}/c + \gamma_u m_e c, \tag{5.24}$$
$$h\nu/c = (h\bar{\nu}/c)\cos\theta + \gamma_u m_e u\cos\phi, \tag{5.25}$$
$$0 = (h\bar{\nu}/c)\sin\theta - \gamma_u m_e u\sin\phi. \tag{5.26}$$

Eliminating u and ϕ from these equations leads to the formula for Compton scattering, which gives the frequency of the photon in S after the collision:

$$\boxed{\bar{\nu} = \nu\left[1 + \frac{h\nu}{m_e c^2}(1 - \cos\theta)\right]^{-1}.}$$

The components of the 4-momentum $\bar{\boldsymbol{p}}$ (or $\bar{\boldsymbol{q}}$) in any other inertial frame S' can be found easily by using $\bar{p}'^{\mu} = \Lambda^{\mu}{}_{\nu}\bar{p}^{\nu}$, where $\Lambda^{\mu}{}_{\nu}$ are the elements of the Lorentz transformation matrix connecting the frames S and S'.

5.13 Accelerating observers

So far we have only considered inertial observers, who move at uniform speeds with respect to one another. Let us now consider a general observer $\mathcal{O}$, who may be *accelerating* with respect to some inertial frame S. If the observer has a 4-velocity $\boldsymbol{u}(\tau)$, where τ is the proper time measured along the worldline, then his 4-acceleration is given by

$$\boldsymbol{a}(\tau) = \frac{d\boldsymbol{u}}{d\tau}.$$

It is worth noting that, at any given event P, the 4-acceleration $\boldsymbol{a}$ is always *orthogonal* to the corresponding 4-velocity $\boldsymbol{u}$, since

$$\boldsymbol{a}\cdot\boldsymbol{u} = \frac{d}{d\tau}\left(\tfrac{1}{2}\boldsymbol{u}\cdot\boldsymbol{u}\right) = \frac{d}{d\tau}\left(\tfrac{1}{2}c^2\right) = 0. \tag{5.27}$$

An accelerating observer has no inertial frame in which he or she is always at rest. Nevertheless, at any event P along the worldline we can define an *instantaneous rest frame* S', in which the observer $\mathcal{O}$ is *momentarily* at rest. Since the observer is at rest in S', the timelike basis vector $\boldsymbol{e}'_0$ of this frame must be parallel to the 4-velocity $\boldsymbol{u}$ of the observer. The remaining spacelike basis vectors $\boldsymbol{e}'_i$ $(i = 1, 2, 3)$ of S' are all orthogonal to $\boldsymbol{e}'_0$ and to one another and will depend on the relative velocity of S and S' and the relative orientation of their spatial axes. Observations made by $\mathcal{O}$ *at the event* P thus correspond to measurements made in the instantaneous rest frame (IRF) S' at P. This is illustrated in Figure 5.5.

Thus, the notion of a localised laboratory can be idealised as follows. An observer (whether accelerating or not) carries along four orthogonal unit vectors $\boldsymbol{e}'_{\mu}(\tau)$ (or *tetrad*), which vary along his worldline but always satisfy

$$\boxed{\boldsymbol{e}'_{\mu}(\tau)\cdot\boldsymbol{e}'_{\nu}(\tau) = \eta_{\mu\nu}.} \tag{5.28}$$

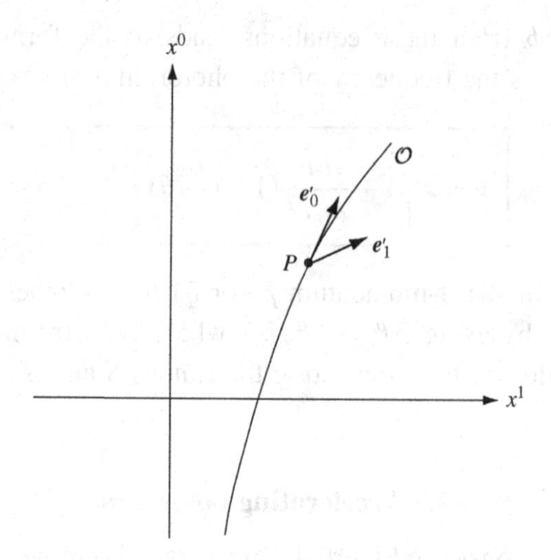

Figure 5.5 The basis vectors $\boldsymbol{e}'_0, \boldsymbol{e}'_1$ at the event P in the instantaneous rest frame S' of an observer $\mathcal{O}$ who is accelerating with respect to the inertial frame S.

In particular, the timelike unit vector is given by

$$\boxed{\boldsymbol{e}'_0(\tau) = \hat{\boldsymbol{u}}(\tau),} \tag{5.29}$$

where $\hat{\boldsymbol{u}}(\tau)$ is the *normalised* 4-velocity of the observer and is simply $\boldsymbol{u}(\tau)/c$. At any event P along the observer's worldline, the tetrad comprises the basis vectors of the Cartesian IRF at the event P and defines a time direction and three space directions to which the observer will refer all measurements. *Thus, the results of any measurement made by the observer at the event P are given by projections of physical quantities (i.e. vectors and tensors) onto these tetrad vectors.*

An important example occurs when the worldline of the observer intersects the worldline of some particle at the event P (at which we take the observer's proper time to be τ). If $\boldsymbol{p}$ is the 4-momentum of the particle at this event then the energy E' of the particle as measured by the observer is given by

$$\frac{E'}{c} = \boldsymbol{p} \cdot \boldsymbol{e}'_0(\tau) \quad \Rightarrow \quad E' = \boldsymbol{p} \cdot \boldsymbol{u}(\tau).$$

Similarly, the covariant components p'_i of the spatial momentum of the particle as measured by the observer are given by

$$p'_i = \boldsymbol{p} \cdot \boldsymbol{e}'_i(\tau).$$

Another example is provided by the 4-acceleration $\boldsymbol{a}$. Since at any event P on the worldline we have $\boldsymbol{e}'_0 = \hat{\boldsymbol{u}}$, the orthogonality condition (5.27) and the fact that in the IRF $[u'^{\mu}] = (c, \vec{0})$ imply that the components of the 4-acceleration in the IRF are $[a'^{\mu}] = (0, \vec{a}')$. Thus the magnitude of the 3-acceleration in the IRF can be computed as the simple invariant $-\boldsymbol{a}\cdot\boldsymbol{a}$.

It is interesting to consider how the tetrad of basis vectors changes along the worldline of an observer whose acceleration varies arbitrarily with time. As it is transported along the observer's worldline, the tetrad *must* satisfy the two requirements (5.28) and (5.29). Clearly, given $\boldsymbol{u}(\tau)$ the condition (5.29) determines the timelike basis vector $\boldsymbol{e}'_0(\tau)$ uniquely. Unfortunately, condition (5.28) is obviously insufficient to determine uniquely the evolution of the spacelike basis vectors $\boldsymbol{e}'_i(\tau) (i = 1, 2, 3)$, which reflect the different ways in which the observer's local laboratory might be spinning and tumbling. An important special case, however, is when the tetrad is 'non-rotating'.

This last requirement requires some clarification. Clearly, the basis vectors of the tetrad at any proper time τ are related to the basis vectors $\boldsymbol{e}_\mu$ of some given inertial frame by the Lorentz transformation

$$\boldsymbol{e}'_\mu(\tau) = \Lambda_\mu{}^\nu(\tau)\boldsymbol{e}_\nu.$$

Thus the tetrad basis vectors at two successive instants must also be related to each other by a Lorentz transformation, which can be thought of as a 'rotation' in spacetime. A 'non-rotating' tetrad is one where the basis vectors $\boldsymbol{e}'_\mu(\tau)$ change from instant to instant by precisely the amount implied by the rate of change of $\boldsymbol{u}$ but with *no additional rotation*. In other words, we accept the inevitable rotation in the timelike plane defined by $\boldsymbol{u}$ and $\boldsymbol{a}$ but *rule out any ordinary rotation of the 3-space vectors*.

Since we wish to treat the time and space directions on an equal footing, we must seek a general expression for the rate of change $d\boldsymbol{e}'_\mu/d\tau$ of a basis vector along the worldline such that: (i) it generates the appropriate Lorentz transformation if $\boldsymbol{e}'_\mu$ lies in the timelike plane defined by $\boldsymbol{u}$ and $\boldsymbol{a}$, and (ii) it excludes any rotation if $\boldsymbol{e}'_\mu$ lies in any other plane, in particular any spacelike plane. A little reflection shows that the unique answer to these requirements is

$$\boxed{\frac{d\boldsymbol{e}'_\mu}{d\tau} = \frac{1}{c^2}\left[(\boldsymbol{u}\cdot\boldsymbol{e}'_\mu)\boldsymbol{a} - (\boldsymbol{a}\cdot\boldsymbol{e}'_\mu)\boldsymbol{u}\right].} \tag{5.30}$$

Any vector that undergoes the above transformation is said to be *Fermi–Walker transported* along the worldline. From (5.30), we find that if $\boldsymbol{e}'_\mu$ is orthogonal to both $\boldsymbol{u}$ and $\boldsymbol{a}$ then $d\boldsymbol{e}'_\mu/d\tau = 0$ as required. Moreover, we see that $d\boldsymbol{e}'_0/d\tau = \boldsymbol{a}/c$, again as required.

A physical example of a 3-space vector that does not rotate along the worldline is the spin (i.e. the angular momentum vector) of a gyroscope that the observer accelerates with himself by means of forces applied to its centre of mass (so that there are no torques). Indeed, a careful observer could set up a non-rotating tetrad by aligning his three spatial axes using such gyroscopes.

5.14 Minkowski spacetime in arbitrary coordinates

There is no need to label events in Minkowski spacetime with the Cartesian inertial coordinates we have used thus far. The advantage of Cartesian coordinates X^μ, which put the line element into the form[5]

$$ds^2 = \eta_{\mu\nu}\, dX^\mu dX^\nu \tag{5.31}$$

(even just at a particular event P), is that they have a clear physical meaning, i.e. they correspond to time and distances measured by an observer at P who is at rest in some inertial frame S labelled using three-dimensional Cartesian coordinates (we will prove this below). Nevertheless, we are free to label events in spacetime using any arbitrary system of coordinates x^μ although, in general, the coordinates in such an arbitrary system may not have simple physical meanings.

Since the path of a free massive particle is a geodesic in Minkowski spacetime, its worldline $x^\mu(\tau)$ in some arbitrary coordinate system is given by the geodesic equations

$$\frac{d^2x^\mu}{d\tau^2} + \Gamma^\mu{}_{\nu\sigma}\frac{dx^\nu}{d\tau}\frac{dx^\sigma}{d\tau} = 0. \tag{5.32}$$

An inertial frame S is defined as one in which a free particle moves in a straight line with fixed speed. Thus from (5.31) it is clear that coordinates X^μ, such that (5.31) holds, define an inertial frame. In this case, the connection $\Gamma^\mu{}_{\nu\sigma}$ vanishes, and so the worldline of a particle is given by

$$\frac{d^2X^\mu}{d\tau^2} = 0. \tag{5.33}$$

Setting $[X^\mu] = (cT, X, Y, Z)$ for the moment, the $\mu = 0$ equation (5.33) shows that $dT/d\tau = \text{constant}$. Thus the $\mu = 1, 2, 3$ equations read

$$\frac{d^2X}{dT^2} = \frac{d^2Y}{dT^2} = \frac{d^2Z}{dT^2} = 0,$$

from which we see immediately that a free particle moves in a straight line with constant speed in S.

We could label the inertial frame S using three-dimensional spatial coordinates that are *not* Cartesian, however. For example, we could use spherical polar

[5] In the interest of clarity, in this section we will denote Cartesian inertial coordinates by X^μ and an arbitrary coordinate system by x^μ.

coordinates. This would correspond to making a change of variables in Minkowski spacetime to the new system $[x^\mu] = (ct, r, \theta, \phi)$, where

$$T = t, \qquad X = r\sin\theta\cos\phi, \qquad Y = r\sin\theta\sin\phi, \qquad Z = r\cos\theta.$$

In this case, the line element becomes

$$ds^2 = c\,dt^2 - dr^2 - r^2 d\theta^2 - r^2\sin^2\theta\; d\phi^2,$$

so the metric is $[g_{\mu\nu}] = \text{diag}(1, -1, -r^2, -r^2\sin^2\theta)$. From the metric we can show that the non-vanishing components of the connection in this coordinate system are (with $c = 1$)

$$\Gamma^1{}_{22} = -r, \qquad \Gamma^1{}_{33} = r\sin^2\theta,$$

$$\Gamma^2{}_{12} = 1/r, \qquad \Gamma^2{}_{33} = -\sin\theta\cos\theta,$$

$$\Gamma^3{}_{13} = 1/r, \qquad \Gamma^1{}_{22} = \cot\theta.$$

Thus, from (5.32), the geodesic equations for the worldline $x^\mu(\tau)$ of a free particle are *very complicated* in these coordinates (exercise), in spite of the fact that, to an observer with fixed (r, θ, ϕ) coordinates (i.e. at rest in S), a free particle still moves in a straight line with fixed speed.

Alternatively, we could use three-dimensional Cartesian coordinates to label points in a *non-inertial* frame S' that is accelerating with respect to S. As an example, consider transforming from $[X^\mu] = (cT, X, Y, Z)$ to a new system of coordinates $[x^\mu] = (ct, x, y, z)$, where t, x, y, z are defined by the equations[6]

$$T = t, \qquad X = x\cos\omega t - y\sin\omega t, \qquad Y = x\sin\omega t + y\cos\omega t, \qquad Z = z.$$

Thus points with constant x, y, z values (i.e. the values are fixed in S') rotate with angular speed ω about the Z-axis of S (see Figure 5.6). Substituting these definitions into (5.31), the line element becomes

$$ds^2 = [c^2 - \omega^2(x^2 + y^2)]dt^2 + 2\omega y dt dx - 2\omega x dt dy - dx^2 - dy^2 - dz^2,$$

and the geodesic equations (5.32) are (exercise)

$$\ddot{t} = 0,$$

$$\ddot{x} - \omega^2 x\dot{t}^2 - 2\omega\dot{y}\dot{t} = 0,$$

$$\ddot{y} - \omega^2 y\dot{t}^2 + 2\omega\dot{x}\dot{t} = 0,$$

$$\ddot{z} = 0,$$

[6] For a full discussion, see for example J. Foster & J. D. Nightingale, *A Short Course in General Relativity*, Springer-Verlag, 1995.

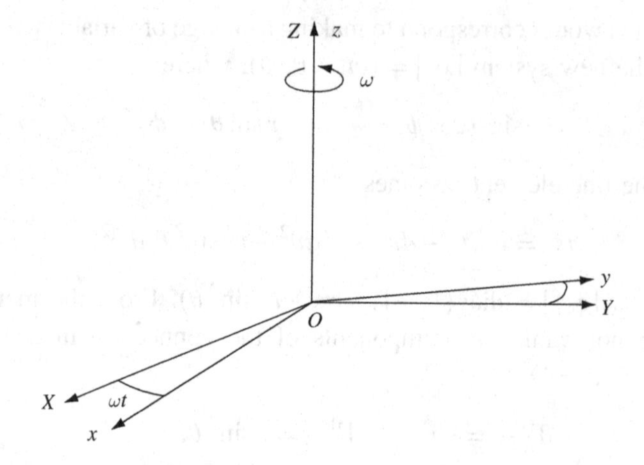

Figure 5.6 The coordinate system (x, y, z) rotating relative to the inertial coordinate system (X, Y, Z).

where the dots denote differentiation with respect to proper time τ. These equations give the worldline $x^\mu(\tau)$ of a free particle in this coordinate system. Once again, the first equation implies that $dt/d\tau = \text{constant}$, so that we can replace the dots in the remaining three equations with derivatives with respect to t. Multiplying through by the rest mass m of the particle and rearranging, these equations become

$$m\frac{d^2x}{dt^2} = m\omega^2 x + 2m\omega\frac{dy}{dt},$$
$$m\frac{d^2y}{dt^2} = m\omega^2 y - 2m\omega\frac{dx}{dt},$$
$$m\frac{d^2z}{dt^2} = 0,$$

or, in 3-vector notation,

$$m\frac{d^2\vec{x}}{dt^2} = -m\vec{\omega}\times(\vec{\omega}\times\vec{x}) - 2m\vec{\omega}\times\frac{d\vec{x}}{dt}, \tag{5.34}$$

where $\vec{x} = (x, y, z)$ and $\vec{\omega} = (0, 0, \omega)$. Thus we recover the equation of motion for a free particle in a rotating frame of reference. We note, however, that the coordinate t is the time measured by clocks at rest in the non-rotating system S, since we have set $t = T$. It is possible to rewrite the equation of motion in terms of the proper time measured by an observer at some some fixed position in S', but to do so would involve replacing (5.34) by a more complicated equation that tends to conceal the Coriolis and centrifugal forces. Note that t is exactly the proper time for an observer situated at the common origin O of the two systems, so observers close to O who are at rest in S' would accept (5.34) as (approximately) valid.

From these examples, we see that in general the geodesic equations can be rather complicated both for non-inertial frames *and* for inertial frames labelled by non-Cartesian spatial coordinates. Thus, when describing physical effects in an inertial frame, it is conventional to use Cartesian spatial coordinates to label points in the frame and so to work in a coordinate system X^μ for which (5.31) is valid. It is then much easier to disentangle the physical effects from artefacts of the coordinate system.

Exercises

5.1 Show that the transformation matrix for a Lorentz transformation from S to S' in standard configuration is given by (5.4).

5.2 Show that, under a Lorentz transformation, the covariant components of a vector transform as $v'_\mu = \Lambda_\mu{}^\nu v_\nu$. Hence show explicitly in component form that, for two 4-vectors $\boldsymbol{v}$ and $\boldsymbol{w}$, the scalar product $\boldsymbol{v}\cdot\boldsymbol{w}$ is invariant under a Lorentz transformation.

5.3 Prove that, for any timelike vector $\boldsymbol{v}$ in Minkowski space, there exists an inertial frame in which the spatial components are zero.

5.4 Prove (a) that the sum of any two spacelike vectors is spacelike; and (b) that a timelike vector and a null vector cannot be orthogonal.

5.5 For the spaceship discussed in Section 1.14, which maintains a uniform acceleration a in the x-direction of some inertial frame S, the worldline is given by

$$t(\tau) = \frac{c}{a}\sinh\frac{a\tau}{c}, \quad x(\tau) = \frac{c^2}{a}\left(\cosh\frac{a\tau}{c} - 1\right), \quad y(\tau) = 0, \quad z(\tau) = 0,$$

where τ is the proper time of an astronaut on the spaceship. Show that the 4-velocity of the rocket in the coordinate system (ct, x, y, z) is given by

$$[u^\mu] = \left(c\cosh\frac{a\tau}{c}, c\sinh\frac{a\tau}{c}, 0, 0\right).$$

Hence show explicitly that $u^\mu u_\mu = c^2$ and that the spaceship's 3-velocity is

$$\vec{u} = \left(c\tanh\frac{a\tau}{c}, 0, 0\right).$$

5.6 Show that the 4-acceleration of the spaceship in Exercise 5.5 is given by

$$[a^\mu] = \left(a\sinh\frac{a\tau}{c}, a\cosh\frac{a\tau}{c}, 0, 0\right).$$

Hence show that $a^\mu a_\mu = -a^2$ and that the magnitude of the spaceship's 3-acceleration in its own instantaneous rest frame is also a.

5.7 A spaceship has constant acceleration g in the x-direction in its locally comoving frame, i.e. the IRF. Show that, in an inertial frame, the spaceship's 4-velocity $[u^\mu] =$

$(u^0, u^1, 0, 0)$ and 4-acceleration $[a^\mu] = (a^0, a^1, 0, 0)$ satisfy $a^1 = gu^0/c$ and $a^0 = gu^1/c$. Show also that

$$\frac{d^2u^\mu}{d\tau^2} = \frac{g^2u^\mu}{c^2},$$

where τ is the proper time as measured by an occupant of the spaceship. A spaceship accelerates at a constant rate $g = 9.5\,\text{m s}^{-2}$ in its own locally comoving frame. It starts out towards the centre of the Galaxy 10 kpc distant. After going 5 kpc it decelerates at the same rate to come to rest again at the Galactic centre. The outward journey is then repeated in reverse to come back home. Show that, in the spaceship's frame, the elapsed travel time is 41.5 years. What is the elapsed time for the waiting observer (or descendants) on Earth?

5.8 Show that in its own instantaneous rest frame (IRF), a particle's 4-acceleration is given by $[a^\mu] = (0, \vec{a})$, where $\vec{a}$ is the 3-acceleration of the particle in the IRF.

5.9 Show that, in an inertial frame in which a particle's 3-acceleration $\vec{a}$ is orthogonal to its 3-velocity $\vec{u}$, the particle's 4-acceleration is given by $[a^\mu] = \gamma_u^2(0, \vec{a})$.

5.10 Show that when an electron and a positron annihilate, more than one photon must be produced.

5.11 Show that if a photon is reflected from a mirror moving parallel to its plane, then the angle of incidence of the photon is equal to the angle of reflection.

5.12 An inertial frame S' moves with constant velocity u along the x-axis with respect to frame S. A photon in frame S' is fired at an angle θ' to the forward direction of motion. Show that the angle θ measured in frame S is

$$\tan\theta = \frac{\tan\theta'(1-\beta^2)^{1/2}}{1+\beta\sec\theta'},$$

where $\beta = u/c$.

5.13 A photon with energy E collides with a stationary electron whose rest mass is m_0. As a result of the collision the direction of the photon's motion is deflected through an angle θ and its energy is reduced to E'. Show that

$$m_0c^2\left(\frac{1}{E'} - \frac{1}{E}\right) = 1 - \cos\theta.$$

Deduce that the wavelength of the photon is increased by

$$\Delta\lambda = \frac{2h}{m_0c}\sin^2\left(\frac{\theta}{2}\right),$$

where h is Planck's constant. At what angle to the initial photon direction does the electron move? Show that, if the photon is deflected through a right angle, and the photon energy satisfies $E \ll m_0c^2$, then after the interaction the angle of the electron's motion to the direction of the photon's initial motion is $\alpha = -\pi/4$.

5.14 Inverse Compton scattering occurs whenever a photon scatters off a particle moving with a speed very nearly equal to that of light. Suppose that a particle of rest mass

m_0 and total energy E collides head on with a photon of energy E_γ. Show that the scattered photon has energy

$$E\left(1+\frac{m_0^2c^4}{4EE_\gamma}\right)^{-1}.$$

Ultra-high-energy cosmic rays have energies up to 10^{20} eV. How much energy can a cosmic ray proton transfer to a microwave background photon?

5.15 For a pure 4-force $\boldsymbol{f}$ acting on a particle of rest mass m_0, show that the corresponding 3-force $\vec{f}$ satisfies

$$\vec{f} = \gamma_u m_0 \vec{a} + \frac{\vec{f}\cdot\vec{u}}{c^2}\vec{u}.$$

Hence show that $\vec{a}$ is only parallel to $\vec{f}$ when $\vec{f}$ is either parallel or orthogonal to $\vec{u}$. Show further that, in these two cases, one has $\vec{f} = \gamma_u^3 m_0 \vec{a}$ and $\vec{f} = \gamma_u m_0 \vec{a}$ respectively.

5.16 For a pure 4-force $\boldsymbol{f}$ acting on a particle of rest mass m_0, show that

$$m_0\frac{d\vec{u}}{d\tau} = \gamma_u \vec{f}.$$

5.17 In Minkowski spacetime, consider an emitter $\mathcal{E}$ moving at speed v along the positive x^1-axis of the frame S in which a receiver $\mathcal{R}$ is at rest. Prove the Doppler shift formula

$$\frac{\lambda_{\mathcal{R}}}{\lambda_{\mathcal{E}}} = \gamma_v\left(1-\frac{v}{c}\cos\theta\right),$$

where θ is the angle made by the photon trajectory with the x^1-axis of S. Show that this expression can be written in the manifestly covariant way

$$\frac{\lambda_{\mathcal{R}}}{\lambda_{\mathcal{E}}} = \frac{u^\mu_{\mathcal{E}} k_\mu}{u^\nu_{\mathcal{R}} k_\nu},$$

where $\boldsymbol{k}$ is the photon 4-wavevector and $\boldsymbol{u}_{\mathcal{E}}$ and $\boldsymbol{u}_{\mathcal{R}}$ are the 4-velocities of $\mathcal{E}$ and $\mathcal{R}$ respectively.

5.18 An astronaut on the space rocket in Exercise 5.5 refers all his measurements to an orthonormal tetrad $\{\boldsymbol{e}'_\mu(\tau)\}$ that comprises the basis vectors of a Cartesian instantaneous rest frame S' at proper time τ. Suppose that at $\tau = 0$ the tetrad coincides with the fixed basis vectors $\{\boldsymbol{e}_\mu\}$ of the (ct, x, y, z) coordinate system in the inertial frame S and that the rocket is not rotating in any way. Show that, in the (ct, x, y, z) coordinate system, the components of the astronaut's orthonormal tetrad at some later proper time τ are

$$\boldsymbol{e}'_0(\tau) = \left(\cosh\frac{a\tau}{c}, \sinh\frac{a\tau}{c}, 0, 0\right),$$

$$\boldsymbol{e}'_1(\tau) = \left(\sinh\frac{a\tau}{c}, \cosh\frac{a\tau}{c}, 0, 0\right),$$

$$\boldsymbol{e}'_2(\tau) = (0, 0, 1, 0),$$

$$\boldsymbol{e}'_3(\tau) = (0, 0, 0, 1).$$

The astronaut observes photons that were emitted with frequency ν_0 from a star that is stationary at the origin of S. Show that the frequency of the photons as measured by the astronaut at proper time τ is given by

$$\nu(\tau) = \nu_0 \exp(-a\tau/c).$$

5.19 At some event P in Minkowski spacetime, the worldline of a particle (either massive or massless) and an observer cross. If, at this event, the particle has 4-momentum $\boldsymbol{p}$ and the observer has 4-velocity $\boldsymbol{u}$ then show that the observer measures the magnitude of the spatial momentum of the particle to be

$$|\vec{p}| = \left[\frac{(\boldsymbol{p}\cdot\boldsymbol{u})^2}{c^2} - \boldsymbol{p}\cdot\boldsymbol{p}\right]^{1/2}.$$

5.20 Repeat Exercise 1.10 using 4-vectors.

5.21 In Minkowski spacetime, the coordinates (cT, X, Y, Z) correspond to a Cartesian inertial frame. The coordinates (ct, r, θ, ϕ) are related to them by the equations

$$X = r\sin\theta\cos\phi, \qquad Y = r\sin\theta\sin\phi, \qquad Z = r\cos\theta.$$

Obtain the special-relativistic equations of motion of a free particle in the (ct, r, θ, ϕ) coordinate system, and interpret these equations physically.

5.22 Repeat Exercise 5.21 for the coordinates (ct, ρ, ϕ, z), that are related to the Cartesian inertial coordinates (cT, X, Y, Z) by

$$T = t,$$

$$X = \rho\cos\phi\cos\omega t - \rho\sin\phi\sin\omega t,$$

$$Y = \rho\cos\phi\sin\omega t + \rho\sin\phi\cos\omega t,$$

$$Z = z,$$

where ω is a constant.

6
Electromagnetism

At the time special relativity was devised only two forces were known, electromagnetism and gravity. As mentioned in Chapter 1, it was electromagnetism that actually led to the development of special relativity. Therefore, we now discuss electromagnetism in some detail; in particular its relativistic formulation. This will introduce a number of ideas that we will use later in developing and applying a relativistic formulation of gravity, namely general relativity. Our guiding principle here is to derive tensorial equations in Minkowski spacetime. This makes it possible to express the theory in a form that is independent of the coordinate system used. We will see that *a consistent theory of electromagnetism follows from saying that there exists a pure 4-force that depends linearly on 4-velocity and also on a certain property of a particle, namely its charge q*. Even if one has no prior knowledge of electromagnetism, one can derive the complete theory in a few lines using this basic assumption and occasional appeals to simplicity.

6.1 The electromagnetic force on a moving charge

In some inertial frame S, the 3-force on a particle of charge q moving in an electromagnetic field is

$$\vec{f} = q(\vec{E} + \vec{u} \times \vec{B}),$$

where $\vec{u}$ is the particle's 3-velocity in S. The 3-vector fields $\vec{E}$ and $\vec{B}$ are the electric and magnetic fields as measured in S. This equation suggests that for the proper relativistic formulation we should write down a tensor equation in four-dimensional spacetime in which the electromagnetic 4-force $\boldsymbol{f}$ depends linearly on the particle's 4-velocity $\boldsymbol{u}$. Thus we are led to an equation of the form

$$\boxed{\boldsymbol{f} = q\boldsymbol{F} \cdot \boldsymbol{u},} \tag{6.1}$$

where $\boldsymbol{F}$ must be a rank-2 tensor in order to make a 4-force from a 4-velocity. We call $\boldsymbol{F}$ the *electromagnetic field tensor*. The scalar q is some property of the particle that determines the strength of the electromagnetic force upon it (i.e. its charge).

We could develop the theory entirely in terms of coordinate-independent 4-vectors and 4-tensors. Nevertheless, if we label points in spacetime with some arbitrary coordinate system x^μ, we may express (6.1) in component form as

$$\boxed{f_\mu = qF_{\mu\nu}u^\nu,}$$

where the $F_{\mu\nu}$ are the covariant components of $\boldsymbol{F}$ in our chosen coordinate system. In order that the rest mass of a particle is not altered by the action of the electromagnetic force we require the latter to be a *pure* force, so that *for any 4-velocity* $\boldsymbol{u}$ we have $\boldsymbol{u}\cdot\boldsymbol{f} = 0$. In component form this reads

$$f_\mu u^\mu = qF_{\mu\nu}u^\mu u^\nu = 0,$$

which implies that the electromagnetic field tensor must be *antisymmetric*, i.e.

$$\boxed{F_{\mu\nu} = -F_{\nu\mu}.}$$

The contravariant components of $\boldsymbol{F}$ are given by

$$F^{\mu\nu} = g^{\mu\sigma}g^{\nu\rho}F_{\sigma\rho},$$

where the $g^{\mu\nu}$ are the contravariant components of the metric tensor in our coordinate system. Since $g^{\mu\nu}$ is symmetric, it is clear that $F^{\mu\nu} = -F^{\nu\mu}$ also.

6.2 The 4-current density

So far we have found only the relativistic form of the electromagnetic force on an idealised point particle with charge q and 4-velocity $\boldsymbol{u}$, in terms of some as yet undetermined rank-2 antisymmetric tensor $\boldsymbol{F}$. In order to develop the theory further, we must now construct the *field equations* of the theory, which determine the electromagnetic field tensor $\boldsymbol{F}(x)$ at any point in spacetime in terms of charges and currents. To construct these field equations, we must first find a properly relativistic (or *covariant*) way of expressing the *source term*. In other words, we need to identify the 4-tensor, defined at each event in spacetime, that acts as the source of the electromagnetic field.

Let us consider some general time-dependent charge distribution. At each event P in spacetime we can characterise the distribution completely by giving the charge density ρ and 3-velocity $\vec{u}$ as measured in some inertial frame. For simplicity, let us consider the fluid in the frame S in which $\vec{u} = \vec{0}$ at P. In this

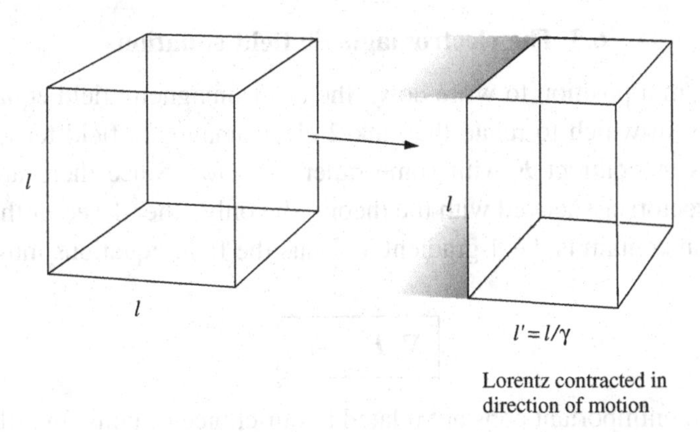

Figure 6.1 The Lorentz contraction of a fluid element in the direction of motion.

frame, the (proper) charge density is given by $\rho_0 = qn_0$, where q is the charge on each particle and n_0 is the number of particles in a unit volume. In some other frame S', moving with speed v relative to S, the volume containing a fixed number of particles will be Lorentz contracted along the direction of motion (see Figure 6.1). Hence in S' the number density of particles is $n' = \gamma_v n_0$, from which we obtain

$$\boxed{\rho' = \gamma_v \rho_0.}$$

Thus we see that the charge density is *not* a 4-scalar but *does* transform as the 0-component of a 4-vector. This suggests that the source term in the electromagnetic field equations should be a 4-vector. At each point in spacetime, the obvious choice is

$$\boxed{\boldsymbol{j}(x) = \rho_0(x)\boldsymbol{u}(x),}$$

where $\rho_0(x)$ is the proper charge density of the fluid (i.e. that measured by an observer comoving with the local flow) and $\boldsymbol{u}(x)$ is its 4-velocity. The squared length of this *4-current density* $\boldsymbol{j}$ at any event is

$$\boldsymbol{j} \cdot \boldsymbol{j} = \rho_0^2 c^2.$$

In an inertial frame S the components of the *4-current density* $\boldsymbol{j}$ are

$$[\, j^\mu] = \rho_0 \gamma_u (c, \vec{u}) = (c\rho, \vec{j}),$$

where ρ is the charge density as measured in S and $\vec{j}$ is the relativistic 3-current density in S. Thus, we see that $c^2\rho^2 - j^2$ is a Lorentz invariant, where $j^2 = \vec{j} \cdot \vec{j}$.

6.3 The electromagnetic field equations

We are now in a position to write down the electromagnetic field equations. The *simplest* way in which to relate the rank-2 electromagnetic field tensor $\boldsymbol{F}$ to the 4-vector $\boldsymbol{j}$ is to contract $\boldsymbol{F}$ with some other 4-vector. Since there are no more physical 4-vectors associated with the theory, the only other 4-vector that the field equations can contain is the 4-gradient ∇. Thus the field equations must be of the form

$$\boxed{\nabla \cdot \boldsymbol{F} = k\boldsymbol{j},} \tag{6.2}$$

where k is an unimportant constant related to our choice of units. In order to make our final results more familiar, let us work in Cartesian inertial coordinates x^{μ} corresponding to some inertial frame S. In such a system, the covariant derivative reduces simply to the partial derivative, and so we can write (6.2) in component form as

$$\boxed{\partial_{\mu} F^{\mu\nu} = k j^{\nu}.} \tag{6.3}$$

We can use this field equation to obtain the law for the conservation of charge. If we take the partial derivative ∂_{ν} of (6.3), we obtain

$$\partial_{\nu}\partial_{\mu} F^{\mu\nu} = k\partial_{\nu} j^{\nu}. \tag{6.4}$$

However, since $F^{\mu\nu}$ is antisymmetric, we can write the scalar on the left-hand side as

$$\partial_{\nu}\partial_{\mu} F^{\mu\nu} = -\partial_{\nu}\partial_{\mu} F^{\nu\mu} = -\partial_{\mu}\partial_{\nu} F^{\mu\nu} = -\partial_{\nu}\partial_{\mu} F^{\mu\nu},$$

from which we deduce that $\partial_{\nu}\partial_{\mu} F^{\mu\nu} = 0$. Thus the right-hand side of (6.4) must also be zero, so that

$$\boxed{\partial_{\mu} j^{\mu} = 0.}$$

Using 3-vector notation in the frame S, we may write this in a more familiar way:

$$\frac{\partial \rho}{\partial t} + \vec{\nabla} \cdot \vec{j} = 0,$$

which expresses the conservation of charge. This equation has the same form as the non-relativistic equation of charge continuity, but the *relativistic* expressions for ρ and $\vec{j}$ must be used in it.

It is clear, however, that we do not yet have a viable theory. The field equations of the theory are given by (6.3), but there are *six* independent components in $F^{\mu\nu}$ and only *four* field equations. Evidently our theory is under-determined as it stands. This suggests that $\boldsymbol{F}$ could be constructed from a 4-vector 'potential' $\boldsymbol{A}$. Again working in Cartesian inertial coordinates x^μ, let us write

$$\boxed{F_{\mu\nu} = \partial_\mu A_\nu - \partial_\nu A_\mu.} \tag{6.5}$$

Thus $F_{\mu\nu}$ is antisymmetric by construction and contains only four independent fields A_μ. Using the field equation (6.3), we can write

$$kj_\lambda = k\eta_{\lambda\nu}j^\nu = \partial_\mu F^\mu{}_\lambda = \eta^{\mu\sigma}\partial_\mu F_{\sigma\lambda},$$

where we have used the fact that the metric coefficients $\eta_{\mu\nu}$ in Cartesian inertial coordinates x^μ are constants.[1] Hence, by substituting into the expression (6.5), we obtain the electromagnetic field equations in terms of the 4-vector potential $\boldsymbol{A}$ as

$$\boxed{\eta^{\mu\sigma}(\partial_\mu\partial_\sigma A_\lambda - \partial_\mu\partial_\lambda A_\sigma) = kj_\lambda.} \tag{6.6}$$

Alternatively, we can express electromagnetism entirely in terms of the electromagnetic field tensor $F^{\mu\nu}$. In this case, we require the *two* field equations

$$\boxed{\begin{aligned} \partial_\mu F^{\mu\nu} &= kj^\nu, \\ \partial_\sigma F_{\mu\nu} + \partial_\nu F_{\sigma\mu} + \partial_\mu F_{\nu\sigma} &= 0, \end{aligned}} \tag{6.7}$$

where the second of these is straightforwardly derived from (6.5). Using the antisymmetrisation operation described in Section 4.3, the second equation can also be written very succinctly as $\partial_{[\sigma}F_{\mu\nu]} = 0$. The constant k may be found by demanding consistency with the standard Maxwell equations (see Section 6.5). In SI units we have $k = \mu_0$, where $\epsilon_0\mu_0 = 1/c^2$.

6.4 Electromagnetism in the Lorenz gauge

Suppose that we add an arbitrary 4-vector $\boldsymbol{Q}$ to the 4-potential $\boldsymbol{A}$. Thus, in component form (in Cartesian inertial coordinates, x^μ, for example) we have

$$A_\mu^{(\text{new})} = A_\mu + Q_\mu. \tag{6.8}$$

Note that this is *not* a coordinate transformation. We are still working in the *same* set of coordinates x^μ but have defined a new vector $\boldsymbol{A}^{(\text{new})}$, whose components

[1] In fact, such an operation is valid in *any* coordinate system. As we showed in Chapter 4, the covariant derivative of the metric tensor is identically zero, which means that we can interchange the order of index raising or lowering and covariant differentiation *without* affecting the result.

in this basis are given by (6.8). The new electromagnetic field tensor is then given by

$$F^{(\text{new})}_{\mu\nu} = \partial_\mu A^{(\text{new})}_\nu - \partial_\nu A^{(\text{new})}_\mu = \partial_\mu A_\nu - \partial_\nu A_\mu + \partial_\mu Q_\nu - \partial_\nu Q_\mu.$$

Clearly, we will recover the *original* electromagnetic field tensor provided that

$$\partial_\mu Q_\nu = \partial_\nu Q_\mu.$$

This equation can be satisfied if $\boldsymbol{Q}$ is the gradient of some scalar field ψ (say), so that $Q_\mu = \partial_\mu \psi$. Thus we have uncovered a *gauge freedom* in the theory: we are free to add the gradient of any scalar field ψ to the 4-vector potential $\boldsymbol{A}$, giving

$$\boxed{A^{(\text{new})}_\mu = A_\mu + \partial_\mu \psi,} \tag{6.9}$$

and still recover the same electromagnetic field tensor and hence the same electromagnetic field equations. The transformation (6.9) is an example of a *gauge transformation* and, as stated above, is distinct from a coordinate transformation.

In the field equations

$$\eta^{\mu\sigma}(\partial_\mu \partial_\sigma A_\lambda - \partial_\mu \partial_\lambda A_\sigma) = \mu_0 j_\lambda,$$

the second term on the left-hand side can be written as $\partial_\lambda \partial_\mu A^\mu$. Thus, we can make this term zero by choosing a scalar field ψ such that

$$\boxed{\partial_\mu A^\mu = 0.} \tag{6.10}$$

This condition is called the *Lorenz gauge*. It is worth noting that the condition (6.10) is preserved by any further gauge transformation $A_\mu \to A_\mu + \partial_\mu \psi$ if and only if $\partial_\mu \partial^\mu \psi = 0$.

Adopting the Lorenz gauge allows the electromagnetic field equations to be written very simply as

$$\eta^{\mu\sigma} \partial_\mu \partial_\sigma A_\lambda = \partial_\mu \partial^\mu A_\lambda = \mu_0 j_\lambda.$$

It is usual to write the four-dimensional Laplacian $\partial_\mu \partial^\mu$ using the notation $\Box^2 = \partial^\mu \partial_\mu = \partial_\mu \partial^\mu$, where $\Box^2$ is the *d'Alembertian* operator.[2] In Cartesian inertial coordinates (ct, x, y, z),

$$\Box^2 = \frac{1}{c^2}\frac{\partial^2}{\partial t^2} - \frac{\partial^2}{\partial x^2} - \frac{\partial^2}{\partial y^2} - \frac{\partial^2}{\partial z^2}.$$

[2] This operator should properly be written $\boldsymbol{\nabla}^2$, which is the inner product $\boldsymbol{\nabla} \cdot \boldsymbol{\nabla}$ of the 4-gradient with itself. However, the notation we have adopted is quite common, since it makes clearer the distinction between the four-dimensional Laplacian and the three-dimensional Laplacian $\vec{\nabla}^2 = \vec{\nabla} \cdot \vec{\nabla}$.

Then the electromagnetic field equations in the Lorenz gauge take the especially simple form

$$\boxed{\Box^2 A_\mu = \mu_0 j_\mu,}$$

together with the attendant gauge condition (6.10). Moreover, in the absence of charges and currents, the right-hand side becomes zero and so A_μ has wave solutions travelling at the speed of light, as do the components of $F_{\mu\nu}$ since in this case we also have $\Box^2 F_{\mu\nu} = 0$.

6.5 Electric and magnetic fields in inertial frames

We have not yet identified the components of $\boldsymbol{F}$ (or $\boldsymbol{A}$) with the familiar electric and magnetic 3-vector fields $\vec{E}$ and $\vec{B}$ as observed in some Cartesian inertial frame S. This is simply a matter of convention; we just have to name the components of $\boldsymbol{A}$ (say) in a way which results in 3-vector equations in S that describe the physics correctly in terms of the traditionally defined 3-vectors $\vec{E}$ and $\vec{B}$. Thus, in some Cartesian inertial frame S, the components of $\boldsymbol{A}$ are taken to be as follows:

$$[A^\mu] = \left(\frac{\phi}{c}, \vec{A}\right),$$

where ϕ is the electrostatic potential and $\vec{A}$ is the traditional three-dimensional vector potential. In terms of ϕ and $\vec{A}$, the Lorenz gauge condition becomes

$$\vec{\nabla} \cdot \vec{A} + \frac{1}{c^2}\frac{\partial \phi}{\partial t} = 0,$$

and, in this gauge, the field equations take the form

$$\Box^2 \vec{A} = \mu_0 \vec{j} \qquad \text{and} \qquad \Box^2 \phi = \frac{\rho}{\epsilon_0}.$$

In terms of ϕ and $\boldsymbol{A}$, the electric and magnetic fields in S are given by

$$\vec{B} = \vec{\nabla} \times \vec{A} \qquad \text{and} \qquad \vec{E} = -\vec{\nabla}\phi - \frac{\partial \vec{A}}{\partial t}. \tag{6.11}$$

It is straightforward to show that these equations lead to the Maxwell equations in their familiar form,

$$\boxed{\begin{aligned} &\vec{\nabla} \cdot \vec{E} = \frac{\rho}{\epsilon_0}, \qquad &&\vec{\nabla} \times \vec{E} = -\frac{\partial \vec{B}}{\partial t}, \\ &\vec{\nabla} \cdot \vec{B} = 0, \qquad &&\vec{\nabla} \times \vec{B} = \mu_0 \vec{j} + \mu_0 \epsilon_0 \frac{\partial \vec{E}}{\partial t}. \end{aligned}}$$

From the expressions (6.11) and (6.5) we have

$$E^i = -\delta^{ij}\partial_j\phi - c\partial_0 A^i = -c\delta^{ij}(\partial_j A^0 - \partial_0 A_j) = -c\delta^{ij}F_{j0},$$

where we have used the fact that $A^0 = \eta^{0\nu}A_\nu = A_0$. Also, we have

$$B^1 = \partial_2 A^3 - \partial_3 A^2 = \partial_3 A_2 - \partial_2 A_3 = F_{32},$$

where we have used the fact that $A^i = \eta^{i\nu}A_\nu = -A_i$. Similar results hold for B^2 and B^3. Thus we find that the covariant components of $\boldsymbol{F}$ in S are given by

$$[F_{\mu\nu}] = \begin{pmatrix} 0 & E^1/c & E^2/c & E^3/c \\ -E^1/c & 0 & -B^3 & B^2 \\ -E^2/c & B^3 & 0 & -B^1 \\ -E^3/c & -B^2 & B^1 & 0 \end{pmatrix}.$$

The corresponding electric and magnetic fields $\vec{E}'$ and $\vec{B}'$ in some other Cartesian inertial frame S' are most easily obtained by calculating the components of the electromagnetic field tensor $\boldsymbol{F}$ or the 4-potential $\boldsymbol{A}$ in this frame. For example, if S' is moving at speed v relative to S in standard configuration then the components in S' are given by

$$A'^\mu = \Lambda^\mu{}_\nu A^\nu \quad \text{and} \quad F'^{\mu\nu} = \Lambda^\mu{}_\sigma \Lambda^\nu{}_\rho F^{\sigma\rho},$$

where the matrix $[\Lambda^\mu{}_\nu]$ is given in Chapter 5.

6.6 Electromagnetism in arbitrary coordinates

So far we have developed electromagnetic theory in Cartesian inertial coordinates. In general, however, we are free to label points in the Minkowski spacetime using any *arbitrary* coordinate system x^μ. We could have developed the entire theory in such an arbitrary system, or even in a coordinate-independent way by using the 4-tensors themselves rather than their components in some coordinate system. Nevertheless, having expressed the theory in Cartesian inertial coordinates, it is now trivial to re-express it in a form valid in arbitrary coordinates.

As shown in (6.7), the electromagnetic field equations in Cartesian inertial coordinates, when expressed in terms of $\boldsymbol{F}$, are given by

$$\partial_\mu F^{\mu\nu} = \mu_0 j^\nu,$$

$$\partial_\sigma F_{\mu\nu} + \partial_\nu F_{\sigma\mu} + \partial_\mu F_{\nu\sigma} = 0.$$

In such a coordinate system, the partial derivative ∂_μ is identical to the covariant derivative ∇_μ, so we can rewrite these equations as

$$\boxed{\begin{aligned} \nabla_\mu F^{\mu\nu} &= \mu_0 j^\nu, \\ \nabla_\sigma F_{\mu\nu} + \nabla_\nu F_{\sigma\mu} + \nabla_\mu F_{\nu\sigma} &= 0. \end{aligned}} \tag{6.12}$$

These new equations are now fully covariant tensor equations, however, so that *if they are valid in one system of coordinates then they are valid in all coordinate systems*. Thus, (6.12) gives the electromagnetic field equations in an arbitrary coordinate system! Once again, using the antisymmetrisation operation discussed in Section 4.3, one can write the second equation simply as $\nabla_{[\sigma}F_{\mu\nu]} = 0$.

A similar procedure can be performed for the electromagnetic field equations when expressed in terms of the 4-vector potential A. From (6.6), in Cartesian inertial coordinates we have

$$\eta^{\mu\sigma}(\partial_\mu \partial_\sigma A_\lambda - \partial_\mu \partial_\lambda A_\sigma) = \mu_0 j_\lambda.$$

Once again, we can replace ∂_μ by ∇_μ, but in this case we must also replace $\eta^{\mu\sigma}$ by $g^{\mu\sigma}$, to obtain

$$\boxed{g^{\mu\sigma}(\nabla_\mu \nabla_\sigma A_\lambda - \nabla_\mu \nabla_\lambda A_\sigma) = \mu_0 j_\lambda.}$$

Again we have a fully covariant tensor equation, which must therefore be valid in any arbitrary coordinate system, the metric coefficients of which are $g^{\mu\sigma}$.

In arbitrary coordinates, the electromagnetic field equations still permit the gauge transformation

$$A_\mu^{(\text{new})} = A_\mu + \nabla_\mu \psi = A_\mu + \partial_\mu \psi,$$

where the last equality holds because the covariant derivative of the scalar field ψ is simply its partial derivative. We can again choose a scalar field ψ, so that

$$\boxed{\nabla_\mu A^\mu = 0,}$$

which is the Lorenz gauge condition in arbitrary coordinates. In this case the electromagnetic field equations can again be written in the form

$$\boxed{\Box^2 A_\mu = \mu_0 j_\mu,}$$

but now the d'Alembertian operator is given by $\Box^2 = g^{\mu\nu}\nabla_\mu \nabla_\nu = \nabla^\mu \nabla_\mu$. *In vacuo*, we may again write $\Box^2 A_\mu = 0$ and $\Box^2 F_{\mu\nu} = 0$. Also, charge conservation is given in arbitrary coordinates by

$$\boxed{\nabla_\mu j^\mu = 0.}$$

Finally, we note that the components of $\boldsymbol{F}$ and $\boldsymbol{A}$ in two different arbitrary coordinate systems x^μ and x'^μ are related by

$$A'^\mu = \frac{\partial x'^\mu}{\partial x^\nu} A^\nu \qquad \text{and} \qquad F'^{\mu\nu} = \frac{\partial x'^\mu}{\partial x^\sigma}\frac{\partial x'^\nu}{\partial x^\rho} F^{\sigma\rho}.$$

6.7 Equation of motion for a charged particle

From our original considerations in Section 6.1, we see that the coordinate-invariant manner of writing the equation of motion of a charged particle in an electromagnetic field is

$$\boxed{\frac{d\boldsymbol{p}}{d\tau} = m_0 \frac{d\boldsymbol{u}}{d\tau} = q\boldsymbol{F} \cdot \boldsymbol{u},}$$

where m_0 is the rest mass of the particle, $\boldsymbol{p}$ is its 4-momentum, $\boldsymbol{u}$ is its 4-velocity and τ is the proper time measured along its worldline. Note that the first equality holds because the electromagnetic force is a pure force.

In Cartesian inertial coordinates, this becomes

$$m_0 \frac{du^\mu}{d\tau} = qF^\mu{}_\nu u^\nu.$$

In a general coordinate system, however, the left-hand side is no longer valid since the ordinary derivative of the components of the 4-velocity along the particle's worldline must be replaced by the intrinsic derivative along the worldline. Using the expression for the intrinsic derivative given in Chapter 3, we find that in an arbitary coordinate system the equation of motion of a particle in an electromagnetic field is

$$m_0 \frac{Du^\mu}{D\tau} = m_0 \left(\frac{du^\mu}{d\tau} + \Gamma^\mu{}_{\nu\sigma} u^\nu u^\sigma \right) = qF^\mu{}_\nu u^\nu,$$

where we have written $dx^\sigma/d\tau$ as u^σ since the 4-velocity is the tangent to the particle's worldline $x^\mu(\tau)$.

The equation for the particle's worldline in arbitrary coordinates is thus given by

$$\boxed{\frac{d^2 x^\mu}{d\tau^2} + \Gamma^\mu{}_{\nu\sigma} \frac{dx^\nu}{d\tau}\frac{dx^\sigma}{d\tau} = \frac{q}{m_0} F^\mu{}_\nu \frac{dx^\nu}{d\tau}.} \tag{6.13}$$

In the absence of an electromagnetic field (or for an uncharged particle), the right-hand side is zero and we can recognise the result as the equation of a geodesic.

In summary, the general procedure for converting an equation valid in Cartesian inertial coordinates into one that is valid in an arbitrary coordinate system is as follows:

- *replace partial derivatives with covariant derivatives;*
- *replace ordinary derivatives along curves with intrinsic derivatives;*
- *replace $\eta_{\mu\nu}$ by $g_{\mu\nu}$.*

Exercises

6.1 Show that the second Maxwell equation in (6.7) can be written as $\partial_{[\sigma}F_{\mu\nu]} = 0$.

6.2 Show that the Maxwell equation (6.6) is unchanged under the gauge transformation (6.9).

6.3 In some Cartesian inertial frame S, the contravariant components of the electric and magnetic fields are E^i and B^i respectively. Show that the corresponding electromagnetic field-strength tensor has the contravariant components

$$[F^{\mu\nu}] = \begin{pmatrix} 0 & -E^1/c & -E^2/c & -E^3/c \\ E^1/c & 0 & -B^3 & B^2 \\ E^2/c & B^3 & 0 & -B^1 \\ E^3/c & -B^2 & B^1 & 0 \end{pmatrix}.$$

6.4 In a Cartesian inertial coordinate system in Minkowski spacetime the field equations of electromagnetism can be written

$$\partial_\mu F^{\mu\nu} = \mu_0 j^\nu,$$

$$\partial_\sigma F_{\mu\nu} + \partial_\nu F_{\sigma\mu} + \partial_\mu F_{\nu\sigma} = 0.$$

Show that these equations are equivalent to the standard form of Maxwell's equations *in vacuo*.

6.5 Two Cartesian inertial frames S and S' are in standard configuration. Show that the components of electric and magnetic fields in the two frames are related as follows:

$$E'^1 = E^1, \qquad B'^1 = B^1,$$

$$E'^2 = \gamma(E^2 - vB^3), \qquad B'^2 = \gamma\left(B^2 + \frac{v}{c^2}E^3\right),$$

$$E'^3 = \gamma(E^3 + vB^2), \qquad B'^3 = \gamma\left(B^3 - \frac{v}{c^2}E^2\right).$$

Show further that $c^2\vec{B}^2 - \vec{E}^2$ is Lorentz invariant.

6.6 Show that the transformation equations derived in Exercise 6.5 can be written as

$$\vec{E}'_{\parallel} = \vec{E}_{\parallel}, \qquad \vec{E}'_{\perp} = \gamma(\vec{E}_{\perp} + \vec{v} \times \vec{B}_{\perp}),$$
$$\vec{B}'_{\parallel} = \vec{B}_{\parallel}, \qquad \vec{B}'_{\perp} = \gamma\left(\vec{B}_{\perp} - \frac{1}{c^2}\vec{v} \times \vec{E}_{\perp}\right),$$

where $\vec{v} = (v, 0, 0)$, and $\vec{E}_{\parallel}$ and $\vec{E}_{\perp}$ denote the projections of $\vec{E}$ parallel and orthogonal to $\vec{v}$ respectively (and similarly for $\vec{B}$). Explain why these equations must hold for a Lorentz boost $\vec{v}$ in an arbitrary direction with respect to the axes of S.

6.7 Show that one may eliminate the explicit reference to the projections of $\vec{E}$ and $\vec{B}$ in Exercise 6.6 and write the transformations as

$$\vec{E}' = \gamma(\vec{E} + \vec{v} \times \vec{B}) + \frac{1-\gamma}{v^2}(\vec{v} \cdot \vec{E})\vec{v},$$
$$\vec{B}' = \gamma\left(\vec{B} - \frac{1}{c^2}\vec{v} \times \vec{E}\right) + \frac{1-\gamma}{v^2}(\vec{v} \cdot \vec{B})\vec{v}.$$

6.8 Show that $\vec{E} \cdot \vec{B}$ is a Lorentz invariant.

6.9 In an arbitrary coordinate system, the second Maxwell equation reads

$$\nabla_\sigma F_{\mu\nu} + \nabla_\nu F_{\sigma\mu} + \nabla_\mu F_{\nu\sigma} = 0.$$

Show that this can be written as

$$\partial_\sigma F_{\mu\nu} + \partial_\nu F_{\sigma\mu} + \partial_\mu F_{\nu\sigma} = 0,$$

and hence show that $\nabla_{[\sigma} F_{\mu\nu]} = 0$.

6.10 In Cartesian inertial coordinates, the equation of motion for a charged particle in an electromagnetic field is

$$m_0 \frac{du^\mu}{d\tau} = q F^\mu{}_\nu u^\nu.$$

Show that

$$\frac{d\vec{p}}{dt} = q(\vec{E} + \vec{u} \times \vec{B}) \qquad \text{and} \qquad \frac{d\mathcal{E}}{dt} = q\vec{E} \cdot \vec{u},$$

where $\vec{p}$ and $\mathcal{E}$ are the 3-momentum and the energy respectively of the particle in S. Interpret these results physically.

6.11 In some inertial frame S, show that the 3-acceleration of a charged particle in an electromagnetic field is

$$\vec{a} = \frac{d\vec{u}}{dt} = \frac{q}{\gamma m_0}\left[\vec{E} + \vec{u} \times \vec{B} - \frac{1}{c^2}(\vec{u} \cdot \vec{E})\vec{u}\right].$$

7

The equivalence principle and spacetime curvature

We are now in a position to use the experience gained in deriving a relativistic formulation of electromagnetism (together with some flashes of inspiration from Einstein!) to begin our formulation of a relativistic theory of gravity, namely general relativity.

7.1 Newtonian gravity

In our development of electromagnetism, we began by considering the electromagnetic 3-force on a charged particle. Let us therefore start our discussion of gravity by considering the description of the gravitational force in the classical, non-relativistic, theory of Newton. In the Newtonian theory, the gravitational force $\vec{f}$ on a (test) particle of *gravitational* mass m_G at some position is

$$\vec{f} = m_G\vec{g} = -m_G\vec{\nabla}\Phi,$$

where $\vec{g}$ is the gravitational field derived from the *gravitational potential* Φ at that position. In turn, the gravitational potential is determined by Poisson's equation:

$$\boxed{\vec{\nabla}^2\Phi = 4\pi G\rho,} \tag{7.1}$$

where ρ is the *gravitational matter density* and G is Newton's gravitational constant. This is the field equation of Newtonian gravity.

It is clear from (7.1) that Newtonian gravity is *not* consistent with special relativity. There is no explicit time dependence, which means that the potential Φ (and hence the gravitational force on a particle) responds *instantaneously* to a disturbance in the matter density ρ; this violates the special-relativistic requirement that signals cannot propagate faster than c. We might try to remedy this by noting that the Laplacian operator $\vec{\nabla}^2$ in (7.1) is equivalent to minus the d'Alembertian operator $\Box^2$ in the limit $c \to \infty$, and thus postulate the modified field equation

$$\Box^2 \Phi = -4\pi G\rho.$$

However, this equation does *not* yield a consistent relativistic theory. It is still not Lorentz covariant, since the matter density ρ does not transform as a Lorentz scalar. We shall discuss the transformation properties of the matter density later.

In addition to the incompatibility of Newtonian gravity with special relativity, there is a second fundamental difference between the electromagnetic and gravitational forces. The equation of motion of a particle of *inertial* mass m_{I} in a gravitational field is given by

$$\boxed{\frac{d^2\vec{x}}{dt^2} = -\frac{m_{\mathrm{G}}}{m_{\mathrm{I}}}\vec{\nabla}\Phi.} \tag{7.2}$$

It is a well-established experimental fact, however, that the ratio $m_{\mathrm{G}}/m_{\mathrm{I}}$ appearing in the equation of motion is *the same for all particles*. By an appropriate choice of units one may thus arrange for this ratio to equal unity. In contrast, the ratio q/m_{I} occurring in the equation of motion of a charged particle in an electromagnetic field is *not* the same for all particles. From (7.2), we thus see that the trajectory through space of a particle in a gravitational field is *independent of the nature of the particle*.

This equivalence of the gravitational and inertial masses (which allows us to refer simply to 'the mass'), is a *truly remarkable* coincidence in the Newtonian theory. In this theory, there is no a-priori reason why the quantity that determines the magnitude of the gravitational force on the particle should equal the quantity that determines the particle's 'resistance' to an applied force in general. It appears as an isolated experimental result, which has since been verified to an accuracy of at least one part in 10^{11} (by Dicke and co-workers).

7.2 The equivalence principle

The equality of the gravitational and inertial masses of a particle led Einstein to his classic 'elevator' thought experiment. Consider an observer in a freely falling elevator (i.e. after the lift cable has been cut). Objects released from rest relative to the elevator cabin remain floating 'weightless' in the cabin. A projectile shot from one side of the elevator to the other appears to move in a straight line at constant velocity, rather than in the usual curved trajectory. All this follows from the fact that the acceleration of any particle relative to the elevator is zero: the particle and the elevator cabin have the *same* acceleration relative to the Earth as a result of the equivalence of gravitational and inertial mass.

All these observations would hold *exactly* if the gravitational field of the Earth were truly *uniform*. Of course, the gravitational field of the Earth is not uniform but acts radially inwards towards its centre of mass, with a strength proportional to $1/r^2$. Thus, if the elevator were left to free-fall for a long time or if it were very large (i.e. a significant fraction of the Earth's radius), two particles released from rest near the walls of the elevator would gradually drift inwards, since they would both be falling along radial lines towards the centre of the Earth (see Figure 7.1). Furthermore, as a result of the varying strength of the gravitational field, particles released from rest near the floor of the elevator would gradually drift downwards whereas those near the ceiling would drift upwards. What the observer in the elevator would be experiencing would be the *tidal forces* resulting from the residual inhomogeneity in the strength and direction of the gravitational field once the main acceleration has been subtracted. It should always be remembered that these tidal forces can never be completely abolished in an elevator (laboratory) of finite, i.e. non-zero, size.

Nevertheless, provided that we consider the elevator cabin over a *short time period* and that it is *spatially small*, then a freely falling elevator (which may have (x, y, z) coordinates marked on its walls and an elevator clock measuring time t) resembles a Cartesian *inertial frame of reference*, and therefore *the laws of special relativity hold inside the elevator.*[1] These observations lead to

The equivalence principle: In a freely falling (non-rotating) laboratory occupying a small region of spacetime, the laws of physics are those of special relativity.[2]

7.3 Gravity as spacetime curvature

These observations led Einstein to make a profound proposal that simultaneously provides for a relativistic description of gravity *and* incorporates in a natural way the equivalence principle (and consequently the equivalence of gravitational and inertial mass). Einstein's proposal was that *gravity should no longer be regarded as a force in the conventional sense but rather as a manifestation of the curvature of the spacetime, this curvature being induced by the presence of matter*. This is the central idea underpinning the theory of general relativity.

[1] The elevator cabin must not only occupy a small region of spacetime but also be non-rotating with respect to distant matter in the universe. This statement is related to *Mach's principle*.

[2] This is in fact a statement of the *strong* equivalence principle, since it refers to all the Laws of physics. The more modest *weak* equivalence principle refers only to the trajectories of freely falling particles.

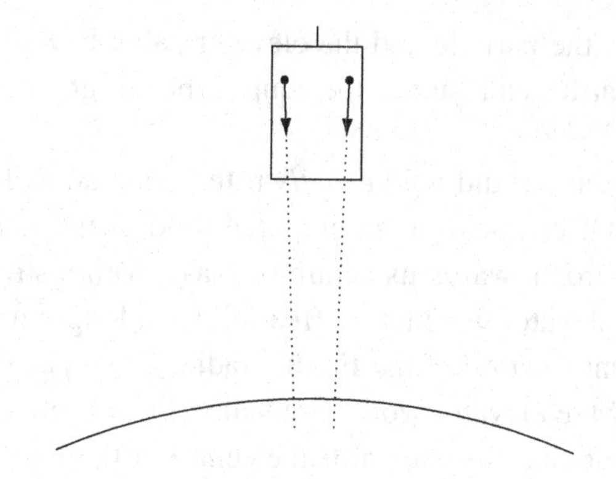

Figure 7.1 An elevator in free-fall towards the Earth.

If gravity is regarded a manifestation of the curvature of spacetime itself, and *not* as the action of some 4-force $\boldsymbol{f}$ defined on the manifold then the equation of motion of a particle moving only under the influence of gravity must be that of a 'free' particle in the curved spacetime, i.e.

$$\boxed{\frac{d\boldsymbol{p}}{d\tau} = \boldsymbol{0},}$$

where $\boldsymbol{p}$ is the particle's 4-momentum and τ is the proper time measured along the particle's worldline. Thus, *the worldline of a particle freely falling under gravity is a geodesic in the curved spacetime.*

The equivalence principle restricts the possible geometry of the curved spacetime to *pseudo-Riemannian*, as follows. The mathematical meaning of the equivalence principle is that it requires that *at any event P* in the spacetime manifold we must be able to define a coordinate system X^μ such that, *in the local neighbourhood* of P, the line element of spacetime takes the form

$$ds^2 \approx \eta_{\mu\nu}\, dX^\mu\, dX^\nu,$$

where exact equality holds at the event P. From the geodesic equation (as shown in Chapter 5), in such a coordinate system the path of a 'free' particle, i.e. one moving only under the influence of gravity, in the vicinity of the event P is given by

$$\frac{d^2X^i}{dT^2} \approx 0,$$

where $i = 1, 2, 3$ and we have denoted X^0 by cT (once again the equality in the above equations holds exactly at P). Thus, in the vicinity of P the coordinates X^μ define a *local* Cartesian inertial frame (like our small elevator considered over a short time interval), in which the laws of special relativity hold locally. In order that we can construct such a system, spacetime must be a pseudo-Riemannian manifold (which is curved and four-dimensional). For such a manifold, in some arbitrary coordinate system x^μ the line element takes the general form

$$\boxed{ds^2 = g_{\mu\nu}\, dx^\mu\, dx^\nu.}$$

7.4 Local inertial coordinates

The curvature of spacetime means that it is *not* possible to find coordinates in which the metric $g_{\mu\nu} = \eta_{\mu\nu}$ at all points in the manifold. Thus, it is not possible to define *global* Cartesian inertial frames as we could in the pseudo-Euclidean Minkowski spacetime. Instead, we are forced to use arbitrary coordinate systems x^μ to label events in spacetime, and these coordinates often do not have simple physical meanings. It is often the case that x^0 is a timelike coordinate and the $x^i (i = 1, 2, 3)$ are spacelike (i.e. the tangent vector to the x^0 coordinate curve is timelike at all points, and similarly the tangent vectors to the x^i coordinate curves are always spacelike). This allocation of coordinates is not necessary, however, and it is sometimes useful to define null coordinates. In any case, the arbitrary coordinates x^μ need not have any direct physical interpretation.

Nevertheless, as demanded by the equivalence principle, problems of physical meaning can always be overcome by transforming, at any event P in the curved spacetime, to a *local inertial coordinate system* X^μ, which, in a limited region of spacetime about P, corresponds to a freely falling, non-rotating, Cartesian frame over a short time interval. Mathematically, this corresponds to constructing about the event P a coordinate system X^μ such that

$$\boxed{g_{\mu\nu}(P) = \eta_{\mu\nu} \qquad \text{and} \qquad (\partial_\sigma g_{\mu\nu})_P = 0.} \tag{7.3}$$

This also means that $\Gamma^\mu{}_{\nu\sigma}(P) = 0$ and that the coordinate basis vectors at the event P form an orthonormal set, i.e.

$$\boxed{\boldsymbol{e}_\mu(P) \cdot \boldsymbol{e}_\nu(P) = \eta_{\mu\nu}.} \tag{7.4}$$

There are in fact an infinite number of local inertial coordinate systems at P, all of which are related to one another by Lorentz transformations. In other words, if a coordinate system X^μ satisfies the conditions (7.3), and hence the condition (7.4), then so too will the coordinate system

$$X'^\mu = \Lambda^\mu{}_\nu X^\nu,$$

where $\Lambda^\mu{}_\nu$ defines a Lorentz transformation. Thus, local Cartesian freely falling (non-rotating) frames at an event P are related to one another by boosts, spatial rotations or combinations of the two. For any one of these coordinate systems, the timelike basis vector $\boldsymbol{e}_0(P)$ is simply the normalised 4-velocity vector $\hat{\boldsymbol{u}}(P)$ of the origin of that frame at the event P, and the three mutually orthogonal spacelike vectors $\boldsymbol{e}_i(P)(i = 1, 2, 3)$ define the orientation of the spatial axes of the frame.

For points near to P, the metric in a local inertial coordinate system X^μ (whose origin is at P) is given by

$$g_{\mu\nu} = \eta_{\mu\nu} + \tfrac{1}{2}(\partial_\sigma \partial_\rho g_{\mu\nu})_P X^\sigma X^\rho + \cdots .$$

The sizes of the second derivatives $(\partial_\sigma \partial_\rho g_{\mu\nu})_P$ thus determine the region over which the approximation $g_{\mu\nu} \approx \eta_{\mu\nu}$ remains valid. We shall see the significance of these second derivatives shortly.

7.5 Observers in a curved spacetime

We discussed the subject of observers in Minkowski spacetime in Chapter 5, but let us now consider the subject in its full generality, in a curved spacetime. An observer will trace out some general (timelike) worldline $x^\mu(\tau)$ through spacetime, as expressed in some arbitrary coordinate system, where τ is the observer's proper time. An idealisation of his local laboratory is a frame of four orthonormal vectors $\hat{\boldsymbol{e}}_\alpha(\tau)$ (or tetrad) satisfying

$$\boxed{\hat{\boldsymbol{e}}_\alpha(\tau) \cdot \hat{\boldsymbol{e}}_\beta(\tau) = \eta_{\alpha\beta},}$$

which are carried with him along his worldline (these vectors may, in general, be totally unrelated to the basis vectors $\boldsymbol{e}_\mu$ of the coordinate system that we are using to label points in spacetime, although we can always express one set of vectors in terms of the other). In particular, at any point along his worldline the timelike vector $\hat{\boldsymbol{e}}_0(\tau)$ coincides with the normalised 4-velocity $\hat{\boldsymbol{u}}(\tau) = \boldsymbol{u}(\tau)/c$ of the observer. Similarly, the evolution of the spacelike vectors $\hat{\boldsymbol{e}}_i(\tau)$ along the worldline reflect the different ways in which his local laboratory may be spinning or tumbling. *Quantities measured in this laboratory correspond to projections of the relevant physical 4-vectors and 4-tensors onto this orthonormal frame.*

As shown in Chapter 5, if the observer has a 4-acceleration $\boldsymbol{a}(\tau) = d\boldsymbol{u}/d\tau$ but is not rotating, the tetrad basis vectors are Fermi–Walker-transported along the observer's worldline:

$$\boxed{\frac{d\hat{\boldsymbol{e}}_\mu}{d\tau} = \frac{1}{c^2}\left[(\boldsymbol{u} \cdot \hat{\boldsymbol{e}}_\mu)\boldsymbol{a} - (\boldsymbol{a} \cdot \hat{\boldsymbol{e}}_\mu)\boldsymbol{u}\right].} \tag{7.5}$$

This expression holds equally well in a curved spacetime. An important special case is that of a non-rotating, *freely falling* observer, i.e one who is moving only under the influence of gravity. The vectors $\hat{e}_\alpha(\tau)$ then define what is called a *freely falling frame* (FFF). Free from any external forces, the observer's worldline traces out a geodesic in the curved spacetime. Thus the timelike vector $\hat{\boldsymbol{e}}_0$ changes with proper time along the worldline according to

$$\frac{d\hat{\boldsymbol{e}}_0}{d\tau} = \boldsymbol{0}.$$

In other words, $\hat{\boldsymbol{e}}_0$ is parallel-transported along the worldline, and the observer's 4-acceleration $\boldsymbol{a}$ is zero. In this case we see from (7.5) that Fermi–Walker transport reduces to parallel transport. Thus the spacelike frame vectors $\hat{\boldsymbol{e}}_i$ $(i = 1, 2, 3)$ are *also* parallel-transported along the geodesic, so that

$$\frac{d\hat{\boldsymbol{e}}_i}{d\tau} = \boldsymbol{0}.$$

Hence, in an arbitrary coordinate system x^μ, the components $(\hat{e}_\alpha)^\mu(\tau) = \hat{\boldsymbol{e}}_\alpha(\tau) \cdot \boldsymbol{e}^\mu$ of any frame vector evolve as follows:

$$\frac{D(\hat{e}_\alpha)^\mu}{D\tau} = \frac{d(\hat{e}_\alpha)^\mu}{d\tau} + \Gamma^\mu{}_{\nu\sigma}(\hat{e}_\alpha)^\nu u^\sigma = 0.$$

This equation is extremely useful for determining what a freely falling observer would measure at a given event in spacetime. It is also clear that the frame vectors $\hat{\boldsymbol{e}}_\alpha$ at any event P along the observer's worldline are the basis vectors of a local Cartesian inertial coordinate system at P.

7.6 Weak gravitational fields and the Newtonian limit

It is clear that, by construction, our description of gravity in terms of spacetime curvature reduces to special relativity in local inertial frames. It is important to check, however, that such a description also reduces to Newtonian gravity in the appropriate limits.

In the absence of gravity, spacetime has a Minkowski geometry. Therefore a weak gravitational field corresponds to a region of spacetime that is only 'slightly' curved. In other words, in such a region there exist coordinates x^μ in which the metric takes the form

$$\boxed{g_{\mu\nu} = \eta_{\mu\nu} + h_{\mu\nu}, \qquad \text{where } |h_{\mu\nu}| \ll 1.} \qquad (7.6)$$

Note that it is important to say 'there exist coordinates' since (7.6) does not hold for all coordinates; as we saw in Chapter 5, one can find coordinates even in Minkowski space in which $g_{\mu\nu}$ is not close to the simple form $\eta_{\mu\nu}$. Let us assume that in the coordinate system (7.6) the metric is *stationary*, which means that all the derivatives $\partial_0 g_{\mu\nu}$ are zero. An example of such a coordinate system might be a fixed Cartesian frame at some point on the surface of the (non-rotating) Earth.

The worldline of a particle freely falling under gravity is given in general by the geodesic equation

$$\frac{d^2x^\mu}{d\tau^2} + \Gamma^\mu{}_{\nu\sigma}\frac{dx^\nu}{d\tau}\frac{dx^\sigma}{d\tau} = 0.$$

We shall assume, however, that the particle is slow-moving, so that the components of its 3-velocity satisfy $dx^i/dt \ll c (i = 1, 2, 3)$, where t is defined by $x^0 \equiv ct$. This is equivalent to demanding that, for $i = 1, 2, 3$,

$$\frac{dx^i}{d\tau} \ll \frac{dx^0}{d\tau}.$$

Thus we can ignore the 3-velocity terms in the geodesic equation to obtain

$$\frac{d^2x^\mu}{d\tau^2} + \Gamma^\mu{}_{00}c^2\left(\frac{dt}{d\tau}\right)^2 = 0. \tag{7.7}$$

Now, recalling the expression (3.21) giving the connection in terms of the metric and using the form (7.6) for $g_{\mu\nu}$, we find that the connection coefficients $\Gamma^\mu{}_{00}$ are given by

$$\Gamma^\mu{}_{00} = \tfrac{1}{2}g^{\kappa\mu}(\partial_0 g_{0\kappa} + \partial_0 g_{0\kappa} - \partial_\kappa g_{00}) = -\tfrac{1}{2}g^{\kappa\mu}\partial_\kappa g_{00} = -\tfrac{1}{2}\eta^{\kappa\mu}\partial_\kappa h_{00},$$

where the last equality is valid to first order in $h_{\mu\nu}$. Since we have assumed that the metric is stationary, we have

$$\Gamma^0{}_{00} = 0 \qquad \text{and} \qquad \Gamma^i{}_{00} = \tfrac{1}{2}\delta^{ij}\partial_j h_{00},$$

where the Latin index runs over $i = 1, 2, 3$. Inserting these coefficients into (7.7) gives

$$\frac{d^2t}{d\tau^2} = 0 \qquad \text{and} \qquad \frac{d^2\vec{x}}{d\tau^2} = -\tfrac{1}{2}c^2\left(\frac{dt}{d\tau}\right)^2\vec{\nabla}h_{00}.$$

The first equation implies that $dt/d\tau = \text{constant}$, and so we can combine the two equations to yield the following equation of motion for the particle:

$$\frac{d^2\vec{x}}{dt^2} = -\tfrac{1}{2}c^2\vec{\nabla}h_{00}.$$

If we compare this equation with the usual *Newtonian* equation of motion for a particle in a gravitational field (7.2), we see that the two are *identical* if we make the identification $h_{00} = 2\Phi/c^2$. Hence for a slowly moving particle our description of gravity as spacetime curvature tends to the Newtonian theory if the metric is such that, in the limit of a weak gravitational field,

$$\boxed{g_{00} = \left(1 + \frac{2\Phi}{c^2}\right).} \tag{7.8}$$

How big is the correction to the Minkowski metric? Some values of Φ/c^2 for various systems are as follows:

$$\frac{\Phi}{c^2} = -\frac{GM}{c^2 r} = \begin{cases} -10^{-9} & \text{at the surface of the Earth} \\ -10^{-6} & \text{at the surface of the Sun} \\ -10^{-4} & \text{at the surface of a white dwarf star.} \end{cases}$$

Thus, we see that even at the surface of a dense object like a white dwarf, the size of Φ/c^2 is much smaller than unity and hence the weak-field limit will be an excellent approximation.

From (7.8), the observant reader will have noticed that the description of gravity in terms of spacetime curvature has another immediate consequence, namely that the time coordinate t does not, in general, measure proper time. If we consider a clock at rest at some point in our coordinate system (i.e. $dx^i/dt = 0$), the proper time interval $d\tau$ between two 'clicks' of the clock is given by

$$c^2 d\tau^2 = g_{\mu\nu}\, dx^\mu\, dx^\nu = g_{00} c^2\, dt^2,$$

from which we find that

$$\boxed{d\tau = \left(1 + \frac{2\Phi}{c^2}\right)^{1/2} dt.}$$

This gives the interval of proper $d\tau$ corresponding to an interval dt of coordinate time for a stationary observer near a massive object, in a region where the gravitational potential is Φ. Since Φ is negative, this proper time interval is shorter than the corresponding interval for a stationary observer at a large distance from the object, where $\Phi \to 0$ and so $d\tau = dt$. Thus, as a bonus, our analysis has also yielded the formula for *time dilation* in a weak gravitational field.

7.7 Electromagnetism in a curved spacetime

Before going on to discuss the mathematics of curvature in detail, let us look back at our development of electromagnetism in Chapter 6. It is clear that our derivation of the electromagnetic field equations in arbitrary coordinates did *not* depend on the intrinsic geometry of the manifold on which the electromagnetic field tensor $\boldsymbol{F}$ and the 4-current $\boldsymbol{j}$ are defined. In other words, one can arrive at these equations without assuming the spacetime to have a Minkowski geometry. Thus, in the presence of gravitating matter, spacetime becomes curved but the field equations of electromagnetism in an arbitrary coordinate system are still given by

$$\boxed{\begin{aligned} \nabla_\mu F^{\mu\nu} &= \mu_0 j^\nu, \\ \nabla_\sigma F_{\mu\nu} + \nabla_\nu F_{\sigma\mu} + \nabla_\mu F_{\nu\sigma} &= 0. \end{aligned}} \tag{7.9}$$

The effects of gravitation are automatically included in these field equations through the covariant derivatives, which depend on the metric $g_{\mu\nu}$ describing the spacetime geometry. Moreover, if we construct a local Cartesian coordinate system about some point P in the manifold then (as discussed above) these coordinates correspond to a local inertial frame in the neighbourhood of P. In these coordinates, the equations of electromagnetism then take their familiar special relativistic forms.

An electromagnetic field tensor $\boldsymbol{F}$ defined on a curved spacetime gives rise (as in Minkowski space) to a 4-force $\boldsymbol{f} = q\boldsymbol{F} \cdot \boldsymbol{u}$, which acts on a particle of charge q with 4-velocity $\boldsymbol{u}$. Thus the equation of motion of a charged particle moving under the influence of an electromagnetic field in a curved spacetime has the same form as that in Minkowski spacetime, i.e.

$$\boxed{m_0 \frac{d\boldsymbol{u}}{d\tau} = q\boldsymbol{F} \cdot \boldsymbol{u},}$$

where m_0 is the rest mass of the particle. In this case, however, because of the curvature of spacetime the particle is moving under the influence of both electromagnetic forces *and* gravity. In some arbitrary coordinate system, the particle's worldline is again given by

$$\boxed{\frac{d^2 x^\mu}{d\tau^2} + \Gamma^\mu{}_{\nu\sigma} \frac{dx^\nu}{d\tau}\frac{dx^\sigma}{d\tau} = \frac{q}{m_0} F^\mu{}_\nu \frac{dx^\nu}{d\tau}.}$$

Obviously, in the absence of an electromagnetic field (or for an uncharged particle), the right-hand side is zero and we recover the equation of a geodesic.

We must remember, however, that the energy and momentum of the electromagnetic field will *itself* induce a curvature of spacetime, so the metric in this case is determined not only by the matter distribution but also by the radiation.

7.8 Intrinsic curvature of a manifold

Since the notion of curvature is central to general relativity, we must now investigate how to quantify the intrinsic curvature of a manifold at any given point P.[3] A manifold (or region of a manifold) is *flat* if there exist coordinates X^μ such that, throughout the region, the line element can be written

$$ds^2 = \epsilon_1(dX^1)^2 + \epsilon_2(dX^2)^2 + \cdots + \epsilon_N(dX^N)^2, \tag{7.10}$$

where $\epsilon_a = \pm 1$ (in other words 'flat' is a shorthand for pseudo-Euclidean). If, however, points in the manifold are labelled with some arbitrary coordinate system x^a then in general the line element ds^2 will not be of the above form. Thus, if for some manifold the line element is given by

$$ds^2 = g_{ab}(x)\, dx^a\, dx^b,$$

how can we tell whether the intrinsic geometry of the manifold in some region is flat or curved in some way?

Consider, for example, the following line element for a three-dimensional space:

$$ds^2 = dr^2 + r^2 d\theta^2 + r^2 \sin^2\theta\, d\phi^2.$$

Of course, we recognise this as the line element of ordinary three-dimensional Euclidean space written in spherical polar coordinates. In other words, the transformation

$$x = r\sin\theta\cos\phi, \qquad y = r\sin\theta\sin\phi, \qquad z = r\cos\theta$$

will turn the above line element into the form

$$ds^2 = dx^2 + dy^2 + dz^2. \tag{7.11}$$

But what about other line elements? For example, recall from Chapter 2 the three-dimensional space described by the line element (2.11):

$$ds^2 = \frac{a^2}{a^2 - r^2}\, dr^2 + r^2\, d\theta^2 + r^2 \sin^2\theta\, d\phi^2.$$

How can we tell whether this metric, or a more complicated metric, corresponds to flat space but merely looks complicated because of a weird choice of coordinates? It would be immensely tedious to try to discover whether there exists a coordinate transformation that reduces a metric to the form (7.11). We therefore need some means of telling whether a manifold is flat directly from the metric g_{ab}, *independently of the coordinate system being used.*

[3] Since the material presented here is applicable to *any* N-dimensional pseudo-Riemannian manifold, we will use indices a, b etc. that have a range 1 to N, rather than μ, ν etc., with a range 0 to 3. Of course, the final application to general relativity will govern the scope of our results.

The *physical* significance of this to general relativity is as follows. If, throughout some region of a four-dimensional spacetime, we can reduce the line element

$$ds^2 = g_{\mu\nu}\, dx^\mu\, dx^\nu$$

to Minkowski form then there can be no gravitational field in this region. The equivalence of a general line element to that of Minkowski spacetime therefore guarantees that the gravitational field will vanish. The solution to our mathematical problem of finding a coordinate-independent way of defining the curvature of spacetime will lead us to the field equations of gravity.

7.9 The curvature tensor

We can find a solution to the problem of measuring the curvature of a manifold at any point by considering changing the order of covariant differentiation. Covariant differentiation is clearly a generalisation of partial differentiation. There is one important respect in which it differs, however: it matters in which order covariant differentiation is performed, and changing the order (in general) changes the result.

Since for a scalar field the covariant derivative is simply the partial derivative, the order of differentiation does not matter. However, let us consider some arbitrary vector field defined on a manifold, with covariant components v_a. The covariant derivative of the v_a is given by

$$\nabla_b v_a = \partial_b v_a - \Gamma^d{}_{ab} v_d.$$

A second covariant differentiation then yields

$$\begin{aligned}
\nabla_c \nabla_b v_a &= \partial_c(\nabla_b v_a) - \Gamma^e{}_{ac} \nabla_b v_e - \Gamma^e{}_{bc} \nabla_e v_a \\
&= \partial_c \partial_b v_a - (\partial_c \Gamma^d{}_{ab}) v_d - \Gamma^d{}_{ab} \partial_c v_d \\
&\quad - \Gamma^e{}_{ac}(\partial_b v_e - \Gamma^d{}_{eb} v_d) - \Gamma^e{}_{bc}(\partial_e v_a - \Gamma^d{}_{ae} v_d),
\end{aligned}$$

which follows since $\nabla_b v_a$ is itself a rank-2 tensor. Swapping the indices b and c to obtain a corresponding expression for $\nabla_b \nabla_c v_a$ and then subtracting gives

$$\nabla_c \nabla_b v_a - \nabla_b \nabla_c v_a = R^d{}_{abc} v_d, \tag{7.12}$$

where

$$\boxed{R^d{}_{abc} \equiv \partial_b \Gamma^d{}_{ac} - \partial_c \Gamma^d{}_{ab} + \Gamma^e{}_{ac} \Gamma^d{}_{eb} - \Gamma^e{}_{ab} \Gamma^d{}_{ec}.} \tag{7.13}$$

To determine directly whether the N^4 quantities $R^d{}_{abc}$ transform as the components of a tensor under a coordinate transformation would be an arduous algebraic task. Fortunately the quotient theorem (Section 4.11) provides a much shorter route. The left-hand side of (7.12) is a tensor, for arbitrary vectors v_a, so the contraction of $R^d{}_{abc}$ with v_d is also a tensor. Since $R^d{}_{abc}$ does not depend on v_a, we conclude from the quotient theorem that the $R^d{}_{abc}$ are indeed the components of some rank-4 tensor $\boldsymbol{R}$. This tensor is called the *curvature tensor* (or *Riemann tensor*), and equation (7.13) shows that it is defined in terms of the metric tensor g_{ab} and its first and second derivatives.

We must now establish how the tensor (7.13) is related to the curvature of the manifold. In a *flat* region of a manifold, we may choose coordinates such that the line element takes the form (7.10) throughout the region. In these coordinates $\Gamma^a{}_{bc}$ *and* its derivatives are zero, and hence

$$\boxed{R^d{}_{abc} = 0}$$

at every point in the region. This is a tensor relation, however, and so it must hold in any coordinate system. Conversely, if $R^d{}_{abc} = 0$ at every point in some region of a manifold, then it may be shown that it is possible to introduce a coordinate system in which the line element takes the form (7.10), and hence this region is *flat*.[4] Thus the vanishing of the curvature tensor is a necessary and sufficient condition for a region of a manifold to be flat.

7.10 Properties of the curvature tensor

The curvature tensor (7.13) possesses a number of symmetries and satisfies certain identities, which we now discuss. The symmetries of the curvature tensor are most easily derived in terms of its covariant components

$$R_{abcd} = g_{ae}R^e{}_{bcd}.$$

For completeness, we note that in an arbitrary coordinate system an explicit form for these components is found, after considerable algebra, to be

$$R_{abcd} = \tfrac{1}{2}(\partial_d\partial_a g_{bc} - \partial_d\partial_b g_{ac} + \partial_c\partial_b g_{ad} - \partial_c\partial_a g_{bd}) - g^{ef}(\Gamma_{eac}\Gamma_{fbd} - \Gamma_{ead}\Gamma_{fbc}).$$

One could use this expression straightforwardly to derive the symmetry properties of the curvature tensor, but we take the opportunity here to illustrate a general mathematical device that is often useful in reducing the algebraic burden of tensor manipulations.

Let us choose some arbitrary point P in the manifold and construct a geodesic coordinate system about this point (see Section 3.11), in which the connection vanishes, $\Gamma^a{}_{bc}(P) = 0$, although in general its derivatives will not. In this

[4] For a proof of this result, see (for example) P. A. M. Dirac, *General Theory of Relativity*, Princeton Landmarks in Physics Series, Princeton University Press, 1996.

coordinate system, one may easily show directly from (7.13) that the covariant components of the curvature tensor at P are given by

$$(R_{abcd})_P = \tfrac{1}{2}(\partial_d \partial_a g_{bc} - \partial_d \partial_b g_{ac} + \partial_c \partial_b g_{ad} - \partial_c \partial_a g_{bd})_P.$$

From this expression one may immediately establish the following symmetry properties at P:

$$R_{abcd} = -R_{bacd}, \tag{7.14}$$

$$R_{abcd} = -R_{abdc}, \tag{7.15}$$

$$R_{abcd} = R_{cdab}. \tag{7.16}$$

The first two properties show that the curvature tensor is antisymmetric with respect to swapping the order of either the first two indices or the second two indices. The third property shows that it is symmetric with respect to swapping the first *pair* of indices with the second *pair* of indices. Moreover, we may also easily deduce the *cyclic identity*

$$R_{abcd} + R_{acdb} + R_{adbc} = 0, \tag{7.17}$$

which on using (7.15) may be written more succinctly as $R_{a[bcd]} = 0$. Although the results (7.14–7.17) have been derived in a special coordinate system, each condition is a tensor relation and so if it is valid in one coordinate system then it is valid in all. Moreover, since the point P is arbitrary, the results hold everywhere.

Although first appearances might suggest that the curvature tensor has N^4 components, the conditions (7.14–7.17) reduce the number of independent components to $N^2(N^2-1)/12$. Recall from Section 2.11 that this is also the number of degrees of freedom among the second derivatives $\partial_d \partial_c g_{ab}$. This is not surprising since, at any point P in a manifold, we can perform a transformation to local Cartesian coordinates in which $g_{ab}(P) = \eta_{\mu\nu}$ and $(\partial_c g_{ab})_P = 0$. Thus, a general metric at any point P is characterised by the $N^2(N^2-1)/12$ second derivatives that cannot be made to vanish there.

For manifolds of different dimensions we have the following results:

No. of dimensions	2	3	4
No. of independent components of R_{abcd}	1	6	20

You can see from this table that in four dimensions the number of independent components is reduced from a possible 256 to 20. You will also see that in one dimension the curvature tensor is always equal to zero: $R_{1111} = 0$. How can this be? Can a line not be curved? Think about this – the curvature measures the 'inner' properties of the space. When we say that a line is curved we refer to a particular embedding in a higher-dimensional space, but this does not tell us about the inner properties of the space. In one dimension, it is evident that we can *always* find a coordinate transformation that will reduce an arbitrary metric to the form (7.10). As a two-dimensional example, in Appendix 7A we calculate the single independent component of the curvature tensor for the surface of a sphere. The *Gaussian curvature* K of a two-dimensional surface is given by

$$K = \frac{R_{1212}}{g},$$

where $g = \det[g_{ab}]$ is the determinant of the metric tensor.

The curvature tensor also satisfies a differential identity, which may be derived as follows. Let us once again adopt a geodesic coordinate system about some arbitrary point P. In this coordinate system, differentiating and then evaluating the result at P gives

$$(\nabla_e R_{abcd})_P = (\partial_e R_{abcd})_P = (\partial_e \partial_c \Gamma_{abd} - \partial_e \partial_d \Gamma_{abc})_P.$$

Cyclically permuting c, d and e to obtain two further analogous relations and adding, one finds that at P

$$\boxed{\nabla_e R_{abcd} + \nabla_c R_{abde} + \nabla_d R_{abec} = 0.} \qquad (7.18)$$

This is, however, a tensor relation and thus holds in all coordinate systems; moreover, since P is arbitrary the relationship holds everywhere. This result is known as the *Bianchi identity* and, using the antisymmetry relation (7.14), it may be written more succinctly as

$$\nabla_{[e} R_{ab]cd} = 0.$$

7.11 The Ricci tensor and curvature scalar

It follows from the symmetry properties (7.14–7.16) of the curvature tensor that it possesses only two independent contractions. We may find these by contracting either on the first two indices or on the first and last indices respectively. From (7.14), raising the index a and then contracting on the first two indices gives

$$R^a{}_{acd} = 0.$$

Contracting on the first and last indices, however, gives in general a non-zero result and this leads to a new tensor, the *Ricci tensor*. It is traditional to use the same kernel letter for the Ricci tensor as for the curvature tensor, so we denote its components by

$$\boxed{R_{ab} \equiv R^{c}{}_{abc}.}$$

By raising the index a in the cyclic identity (7.17) and contracting with d, one may easily show that the Ricci tensor is symmetric. Thus we have $R^{a}{}_{b} = R_{a}{}^{b}$ and we can denote both by R^{b}_{a}.

A further contraction gives the *curvature scalar* (or *Ricci scalar*)

$$\boxed{R \equiv g^{ab} R_{ab} = R^{a}_{a},}$$

where again the same kernel letter is used. This is a scalar quantity defined at each point of the manifold.

The covariant derivatives of the Ricci tensor and the curvature scalar obey a particularly important relation, which will be central to our development of the field equations of general relativity. Raising a in the Bianchi identity (7.18) and contracting with d gives

$$\nabla_e R_{bc} + \nabla_c R^{a}{}_{bae} + \nabla_a R^{a}{}_{bec} = 0,$$

which, on using the antisymmetry property (7.15) in the second term, gives

$$\nabla_e R_{bc} - \nabla_c R_{be} + \nabla_a R^{a}{}_{bec} = 0.$$

If we now raise b and contract with e, we find

$$\nabla_b R^{b}_{c} - \nabla_c R + \nabla_a R^{ab}{}_{bc} = 0. \tag{7.19}$$

Using the symmetry property (7.16) we may write the third term as

$$\nabla_a R^{ab}{}_{bc} = \nabla_a R^{ba}{}_{cb} = \nabla_a R^{a}_{c} = \nabla_b R^{b}_{c},$$

so the first and last terms in (7.19) are identical and we obtain

$$2\nabla_b R^{b}_{c} - \nabla_c R = \nabla_b \left(2R^{b}_{c} - \delta^{b}_{c} R\right) = 0.$$

Finally, raising the index c, we obtain the important result

$$\boxed{\nabla_b \left(R^{bc} - \tfrac{1}{2} g^{bc} R\right) = 0.}$$

The term in parentheses is called the *Einstein tensor*

$$\boxed{G^{ab} \equiv R^{ab} - \tfrac{1}{2} g^{ab} R.}$$

It is clearly symmetric and thus possesses only one independent divergence $\nabla_a G^{ab}$, which vanishes (by construction). As we will see, it is this tensor that describes the curvature of spacetime in the field equations of general relativity.

7.12 Curvature and parallel transport

In Chapter 3, we remarked that parallel transport in a curved manifold was path dependent. We now have a more formal description of curvature. If a region of manifold is flat then the curvature tensor vanishes throughout the region; otherwise, it is curved. Thus there must be some link between the curvature tensor and parallel transport.

Let us consider the parallel transport of a vector $\boldsymbol{v}$ around a closed curve $\mathcal{C}$ in a manifold. We can define an arbitrary surface $\mathcal{A}$ bounding the curve $\mathcal{C}$ and break this surface up into a lot of small areas each bounded by closed curves $\mathcal{C}_N$, as indicated in Figure 7.2. The change in the components v^a on being parallel-transported around the closed curve $\mathcal{C}$ is then

$$\Delta v^a = \sum_N (\Delta v^a)_N,$$

where $(\Delta v^a)_N$ is the change in v^a around the small closed curve $\mathcal{C}_N$. This follows because the changes in Δv^a around any of the interior closed curves cancel, leaving just the contributions around the outer edges that bound the curve $\mathcal{C}$.

Let us now calculate $(\Delta v^a)_N$ around the small closed curve $\mathcal{C}_N$ defined by the parametric equations $x^a(u)$. The equation for parallel transport is given by (3.42):

$$\frac{dv^a}{du} = -\Gamma^a{}_{bc} v^b \frac{dx^c}{du}.$$

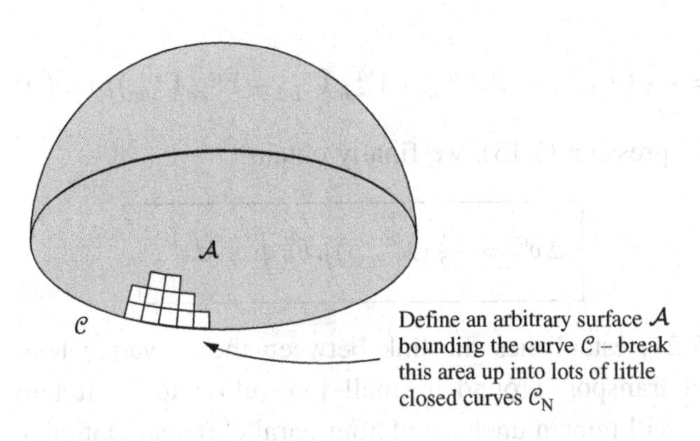

Figure 7.2 An arbitrary surface $\mathcal{A}$ bounding a closed curve $\mathcal{C}$.

Thus, if v^a is parallel-transported along the small closed curve $\mathcal{C}_N$ from some initial point P then at some other point along this curve we have

$$v^a(u) = v^a_P - \int_{u_P}^{u} \Gamma^a{}_{bc} v^b \frac{dx^c}{du}\, du. \tag{7.20}$$

However, since the closed curve is small we can expand the factors in the integrand about P to first order in $x^a - x^a_P$:

$$\Gamma^a{}_{bc}(u) = (\Gamma^a{}_{bc})_P + (\partial_d \Gamma^a{}_{bc})_P \left[x^d(u) - x^d_P\right] + \cdots,$$
$$v^a(u) = v^a_P - (\Gamma^a{}_{bc})_P\, v^b_P \left[x^c(u) - x^c_P\right] + \cdots.$$

Substituting these expressions into (7.20) and retaining terms only up to first order in $x^a - x^a_P$, we obtain

$$\begin{aligned} v^a(u) = v^a_P &- (\Gamma^a{}_{bc})_P\, v^b_P \int_{u_P}^{u} \frac{dx^c}{du}\, du \\ &- (\partial_d \Gamma^a{}_{bc} - \Gamma^a{}_{ec}\Gamma^e{}_{bd})_P\, v^b_P \int_{u_P}^{u} (x^d - x^d_P) \frac{dx^c}{du}\, du. \end{aligned}$$

If we integrate the coordinate differentials around a closed loop we have $\oint dx^c = 0$, and so we find that

$$\Delta v^a = -(\partial_d \Gamma^a{}_{bc} - \Gamma^a{}_{ec}\Gamma^e{}_{bd})_P\, v^b_P \oint_{u_P}^{u} x^d\, dx^c.$$

We may obtain an analogous result by interchanging the dummy indices c and d. Now using the result

$$\oint d(x^c x^d) = \oint (x^c\, dx^d + x^d\, dx^c) = 0,$$

we find that

$$\Delta v^a = -\tfrac{1}{2}(\partial_c \Gamma^a{}_{bd} - \partial_d \Gamma^a{}_{bc} + \Gamma^a{}_{ec}\Gamma^e{}_{bd} - \Gamma^a{}_{ed}\Gamma^e{}_{bc})_P\, v^b_P \oint x^c\, dx^d.$$

On using the expression (7.13), we finally obtain

$$\boxed{\Delta v^a = -\tfrac{1}{2}(R^a{}_{bcd})_P\, v^b_P \oint x^c\, dx^d.} \tag{7.21}$$

Equation (7.21) establishes the link between the curvature tensor at a point P and parallel transport around a small loop close to P. It tells us that the components v^a will remain unchanged after parallel transportation around a small closed loop near P *if and only if* the curvature tensor vanishes at P. So, returning

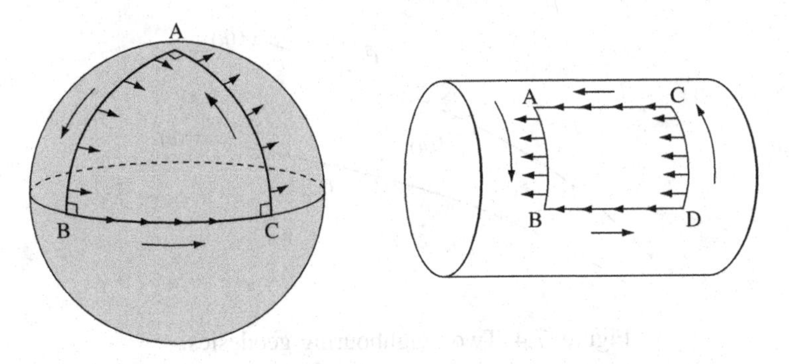

Figure 7.3 Parallel transport around a closed curve on the surface of a sphere and the surface of a cylinder.

to our construction of $(\Delta v^a)_N$, the vector components v^a will not change on parallel transportation around the entire closed curve $\mathcal{C}$ if the curvature tensor $R^a{}_{bcd}$ vanishes over the entire area $\mathcal{A}$ bounding the curve.

As an example, consider the parallel transportation of a vector around the closed triangle ABC on the surface of a sphere (see Figure 7.3). As shown in Appendix 7A, the curvature tensor is nowhere zero, and it is evident that the vector changes direction after parallel transportation around the triangle. However, as also mentioned in Appendix 7A, the curvature tensor vanishes everywhere on the surface of a cylinder and hence the components of a vector will remain unchanged if the vector is parallel-transported around any closed curve (see Figure 7.3).

7.13 Curvature and geodesic deviation

Another important consequence of curvature is that two nearby geodesics that are initially parallel either converge or diverge, depending on the local curvature. This is embodied in the *equation of geodesic deviation*, which we now derive.

Consider two neighbouring geodesics, $\mathcal{C}$ given by $x^a(u)$ and $\bar{\mathcal{C}}$ given by $\bar{x}^a(u)$, where u is an affine parameter, and let $\xi^a(u)$ be the small 'vector' connecting points on the two geodesics with the same parameter value (see Figure 7.4), i.e.

$$\bar{x}^a(u) = x^a(u) + \xi^a(u).$$

In particular, let us suppose that for some arbitrary value of u the vector $\xi^a(u)$ connects the point P on $\mathcal{C}$ to the point Q on $\bar{\mathcal{C}}$.

Once again our derivation is simplified considerably by constructing local geodesic coordinates about the point P, in which the connection coefficients

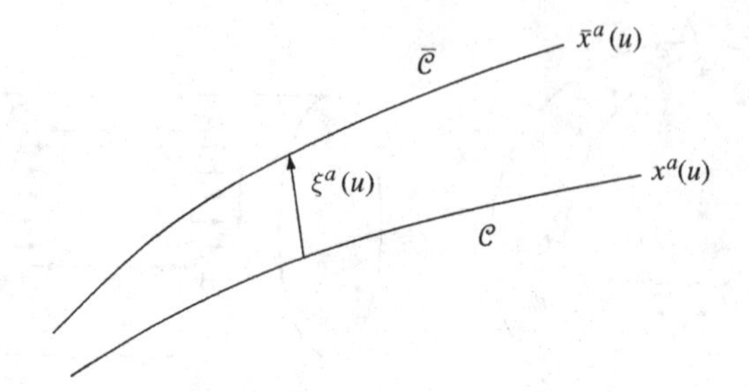

Figure 7.4 Two neighbouring geodesics.

vanish at P but their derivatives are in general non-zero there. In this coordinate system, since $\mathcal{C}$ and $\bar{\mathcal{C}}$ are geodesics we have

$$\left(\frac{d^2x^a}{du^2}\right)_P = 0, \tag{7.22}$$

$$\left(\frac{d^2\bar{x}^a}{du^2} + \Gamma^a{}_{bc}\frac{d\bar{x}^b}{du}\frac{d\bar{x}^c}{du}\right)_Q = 0, \tag{7.23}$$

at the points P and Q respectively. However, to first order in ξ^a,

$$\Gamma^a{}_{bc}(Q) = \Gamma^a{}_{bc}(P) + (\partial_d\Gamma^a{}_{bc})_P\,\xi^d = (\partial_d\Gamma^a{}_{bc})_P\,\xi^d.$$

Thus, subtracting (7.22) from (7.23) gives, to first order, at P

$$\ddot{\xi}^a + (\partial_d\Gamma^a{}_{bc})\,\dot{x}^b\dot{x}^c\xi^d = 0,$$

where the dots denote d/du. However, in our geodesic coordinates the second-order intrinsic derivative of ξ^a at P is given by

$$\frac{D^2\xi^a}{Du^2} = \frac{d}{du}\left(\dot{\xi}^a + \Gamma^a{}_{bc}\xi^b\dot{x}^c\right) = \ddot{\xi}^a + (\partial_d\Gamma^a{}_{bc})\,\xi^b\dot{x}^c\dot{x}^d,$$

where we have used the fact that $\Gamma^a{}_{bc}(P) = 0$; we note that nevertheless the derivatives of $\Gamma^a{}_{bc}$ at P may not vanish. Thus, combining the last two equations and relabelling dummy indices, we find that at P

$$\frac{D^2\xi^a}{Du^2} + (\partial_b\Gamma^a{}_{cd} - \partial_d\Gamma^a{}_{bc})\,\xi^b\dot{x}^c\dot{x}^d = 0.$$

We may now identify the terms in parentheses on the left-hand side as components $R^a{}_{cbd}$ of the Riemann tensor when expressed in local geodesic coordinates about P. Thus we may write the above result as

$$\boxed{\frac{D^2\xi^a}{Du^2} + R^a{}_{cbd}\xi^b\dot{x}^c\dot{x}^d = 0,} \tag{7.24}$$

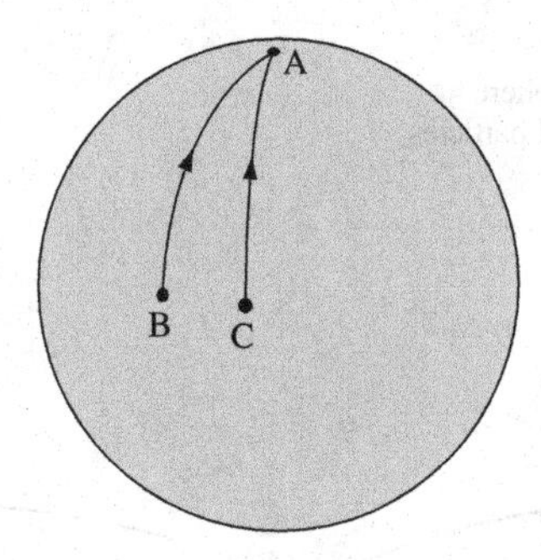

Figure 7.5 Converging geodesics on the surface of a sphere.

which is clearly a tensor relation and is hence valid in *any* coordinate system. Moreover, since P is an arbitrary point on $\mathcal{C}$, this relation is valid everywhere along the curve. The result (7.24) is the *equation of geodesic deviation.*

The geometric meaning of (7.24) is straightforward. In a flat region of a manifold, $R^a{}_{bcd} = 0$ and we may adopts Cartesian coordinates throughout. In this case, $D/Du = d/du$ and the equation of geodesic deviation reduces to $d^2\xi^a/du^2 = 0$, which implies that $\xi^a(u) = A^a u + B^a$ where A^a and B^a are constants. So in a flat region the separation vector $\xi^a(u)$ connecting the two geodesics (which are simple straight lines in this case) in general increases *linearly* with u. In the special case where the two lines are initially parallel then they will remain so and hence never intersect. In a curved region of a manifold, however $R^a{}_{bcd} \neq 0$ and so neighbouring geodesics either converge or diverge. For example, the two neighbouring geodesics AB and AC on the surface of a sphere (see Figure 7.5) converge as we approach the point A at the pole because the surface is positively curved. Equation (7.24) allows us to compute the rates of convergence or divergence of neighbouring geodesics for Riemannian spaces of arbitrary complexity. All one needs to do is to compute the curvature tensor (7.13) at each point using the metric.

7.14 Tidal forces in a curved spacetime

Now that we have derived the equation of geodesic variation (7.24), we can give a more quantitative account of the *gravitational tidal forces* mentioned in our discussion of the equivalence principle in Section 7.2. Let us begin by working in Newtonian gravity and consider an initially spherical distribution of non-interacting particles freely falling towards the Earth (see Figure 7.6). Each particle moves on a straight line through the centre of the Earth, but those nearer the Earth fall faster because the gravitational attraction is stronger. Thus the sphere no longer remains a sphere but is distorted into an ellipsoid of the same

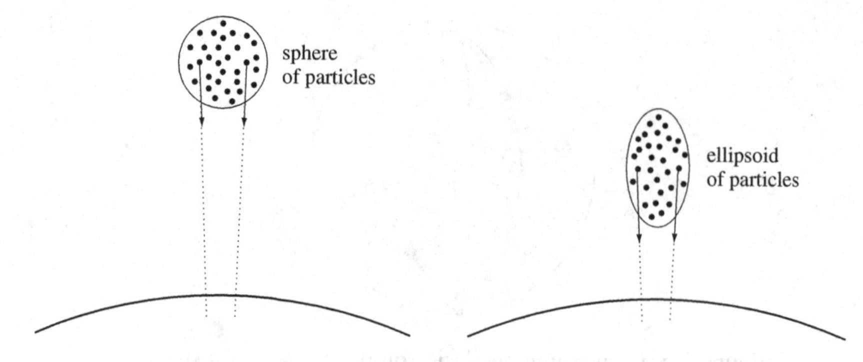

Figure 7.6 Tidal force on a collection of non-interacting particles.

volume (to first order in the particle separation): gravity has produced a tidal force in the sphere of particles that results in an elongation of the distribution in the direction of motion and a compression of the distribution in the transverse directions. Indeed, it is straightforward to show that, for two nearby particles with trajectories $x^i(t)$ and $\bar{x}^i(t) (i = 1, 2, 3)$ respectively in Cartesian coordinates, that the components of the separation vector $\zeta^i = x^i - \bar{x}^i$ evolve as

$$\frac{d^2\zeta^i}{dt^2} = -\left(\frac{\partial^2\Phi}{\partial x^i \partial x^j}\right)\zeta^j,$$

where Φ is the Newtonian gravitational potential (see Exercise 7.21).

A similar tidal effect occurs in general relativity and can be understood in terms of the curvature of the spacetime. In particular, we can gain some idea of the general-relativistic tidal forces by considering the equation of geodesic deviation (7.24). Consider any pair of our non-interacting particles. Each one is in free fall and so they must move along the timelike geodesics $x^\mu(\tau)$ and $\bar{x}^\mu(\tau)$ respectively, where τ is the proper time experienced by the first particle (say). If we define a small separation vector between the two particle worldlines by $\xi^\mu(\tau) = \bar{x}^\mu(\tau) - x^\mu(\tau)$, then (7.24) shows that it evolves according to the equation

$$\boxed{\frac{D^2\xi^\mu}{D\tau^2} = S^\mu{}_\nu \xi^\nu,} \tag{7.25}$$

where we have defined the *tidal stress tensor*

$$\boxed{S^\mu{}_\nu \equiv R^\mu{}_{\sigma\rho\nu} u^\sigma u^\rho,} \tag{7.26}$$

in which $u^\sigma \equiv du^\sigma/d\tau$ is the 4-velocity of the first particle. Note that in defining $S^\mu{}_\nu$ we have made use of the fact that the curvature tensor is antisymmetric in its last two indices. The result (7.25) is a fully covariant tensor equation and therefore holds in any coordinate system.

To understand the physical consequences of the geodesic deviation effect, it is helpful to consider how some observer will view the relative spatial acceleration of the two particles. Suppose that our observer is sitting on the first particle, the worldline $x^\mu(\tau)$ of which passes through some event P. In order to calculate the relative spatial acceleration measured by our observer, we may erect a set of orthonormal basis vectors $\hat{\boldsymbol{e}}_\alpha$ at P that define the instantaneous rest frame (IRF) of the first particle (and the observer) at this event. The timelike basis vector is given simply by $\hat{\boldsymbol{e}}_0 = \hat{\boldsymbol{u}}$, where $\boldsymbol{u}$ is the 4-velocity at P of the first particle, and we may choose the spacelike basis vectors $\hat{\boldsymbol{e}}_i$ in any way, provided that the full set satisfies

$$\hat{\boldsymbol{e}}_\alpha \cdot \hat{\boldsymbol{e}}_\beta = \eta_{\alpha\beta}.$$

In this way, the duals of these basis vectors, which are given by $\hat{\boldsymbol{e}}^\alpha = \eta^{\alpha\beta}\hat{\boldsymbol{e}}_\beta$, also form an orthonormal set. The general situation is illustrated schematically in Figure 7.7.

The components of the separation vector ξ with respect to our new frame are

$$\xi^{\hat{\alpha}} \equiv \hat{\boldsymbol{e}}^\alpha \cdot \xi = (\hat{\boldsymbol{e}}^\alpha)_\mu \xi^\mu;$$

these components give the temporal and spatial separations of the events P and Q on the two particle worldlines, as measured by our observer. Since the $\hat{\boldsymbol{e}}_\alpha (\alpha = 0, 1, 2, 3)$ are the basis vectors of an inertial Cartesian coordinate system at P, the intrinsic derivative in this coordinate system is simply equal to the ordinary

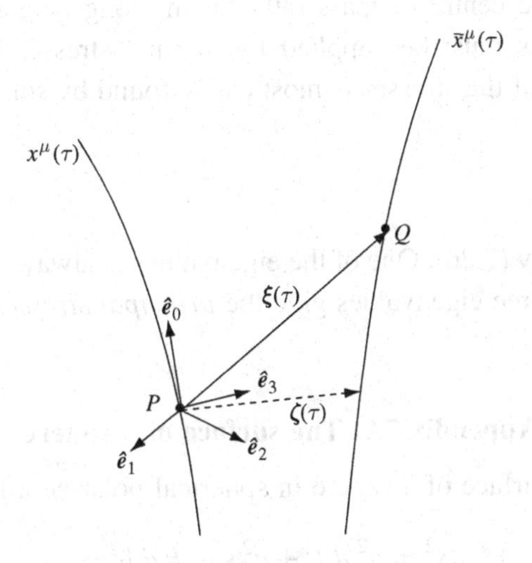

Figure 7.7 Schematic illustration of the basis vectors of the instantaneous rest frame at P. A general connecting vector $\boldsymbol{\xi}$ and the orthogonal connecting vector $\boldsymbol{\zeta}$ are also shown.

derivative. Moreover, with respect to the IRF, the 4-velocity of the first particle is simply $[u^{\hat{\alpha}}] = (c, \vec{0})$. Thus from (7.25) we have

$$\boxed{\frac{d^2\xi^{\hat{\alpha}}}{d\tau^2} = c^2 R^{\hat{\alpha}}{}_{\hat{0}\hat{0}\hat{\gamma}}\xi^{\hat{\gamma}},} \qquad (7.27)$$

where the components of the curvature tensor in the Cartesian inertial frame at P may be written as

$$R^{\hat{\alpha}}{}_{\hat{\beta}\hat{\gamma}\hat{\delta}} \equiv R^{\mu}{}_{\sigma\nu\rho}(\hat{e}^{\alpha})_{\mu}(\hat{e}_{\beta})^{\sigma}(\hat{e}_{\gamma})^{\nu}(\hat{e}_{\delta})^{\rho}. \qquad (7.28)$$

Equation (7.27) in fact holds for any orthonormal freely falling frame $\hat{e}_{\alpha}$.

Clearly, the general separation vector $\boldsymbol{\xi}$ is inappropriate for our discussion of the evolution of the *spatial* separation seen by our observer at P, since typically $\boldsymbol{\xi}$ will have some temporal component in the observer's frame. Thus, we must work instead with the *orthogonal connecting vector* $\boldsymbol{\zeta}(\tau)$ shown in Figure 7.7, which has a zero component in the $\hat{e}_0$-direction, i.e. $\zeta^{\hat{0}} = 0$. Since (7.27) is valid for any small connecting vector it must also hold for the orthogonal connecting vector $\boldsymbol{\zeta}$, but we must remember that $\zeta^{\hat{0}}(\tau) = 0$ for all τ.

A useful alternative interpretation of (7.25) or (7.27) is that it gives the force per unit mass required to keep two particles moving along parallel curves; this force must be supplied by some mechanical means. For example, the worldline of the centre of mass of a rigid body in free fall is a timelike geodesic, but this is not true of the other parts of the object, which are constrained to move along curves parallel to the centre of mass rather than along neighbouring geodesics. The necessary forces must be supplied by internal stresses in the object. The physical magnitude of the stresses is most easily found by solving the eigenvalue problem

$$S^{\mu}{}_{\nu}v^{\nu} = \lambda v^{\mu}, \qquad (7.29)$$

where $S^{\mu}{}_{\nu}$ is given by (7.26). One of the eigenvalues is always zero (for $v^{\mu} = u^{\mu}$), and the remaining three eigenvalues give the *principal stresses* in the object.

Appendix 7A: The surface of a sphere

The metric[5] of the surface of a sphere in spherical polar coordinates is

$$ds^2 = a^2\, d\theta^2 + a^2 \sin^2\theta\, d\phi^2.$$

[5] Note that this term is often applied, as here, to the line element itself.

To get used to handling problems involving curved spaces you should calculate the components of the affine connection, starting from this metric. The definition of the affine connection is

$$\Gamma^a{}_{bc} = \tfrac{1}{2} g^{ad}(\partial_b g_{dc} + \partial_c g_{bd} - \partial_d g_{bc}),$$

as given in (3.21), and in two dimensions there are six independent connection coefficients,

$$\Gamma^1{}_{11},\quad \Gamma^1{}_{12},\quad \Gamma^1{}_{22},\quad \Gamma^2{}_{11},\quad \Gamma^2{}_{12},\quad \Gamma^2{}_{22}.$$

These coefficients are given by (exercise):

$$\begin{aligned}
\Gamma^1{}_{11} &= \tfrac{1}{2} g^{11}(\partial_1 g_{11} + \partial_1 g_{11} - \partial_1 g_{11}) = 0,\\
\Gamma^1{}_{12} &= \tfrac{1}{2} g^{11}(\partial_2 g_{11} + \partial_2 g_{21} - \partial_1 g_{12}) = 0,\\
\Gamma^1{}_{22} &= \tfrac{1}{2} g^{11}(\partial_2 g_{21} + \partial_2 g_{21} - \partial_1 g_{22}) = -\tfrac{1}{2} g^{11}\partial_1 g_{22},\\
\Gamma^2{}_{11} &= \tfrac{1}{2} g^{22}(\partial_2 g_{12} + \partial_2 g_{12} - \partial_2 g_{11}) = 0,\\
\Gamma^2{}_{12} &= \tfrac{1}{2} g^{22}(\partial_1 g_{22} + \partial_2 g_{12} - \partial_2 g_{11}) = \tfrac{1}{2} g^{22}\partial_1 g_{22},\\
\Gamma^2{}_{22} &= \tfrac{1}{2} g^{22}(\partial_2 g_{22} + \partial_2 g_{22} - \partial_2 g_{22}) = 0.
\end{aligned}$$

So, the only two non-zero coefficients are

$$\Gamma^1{}_{22} = -\frac{1}{2a^2} 2a^2 \sin\theta\cos\theta = -\sin\theta\cos\theta,$$

$$\Gamma^2{}_{12} = \frac{1}{2a^2\sin^2\theta} 2a^2 \sin\theta\cos\theta = \frac{\cos\theta}{\sin\theta}.$$

The curvature tensor is

$$R_{abcd} = \tfrac{1}{2}(\partial_d\partial_a g_{bc} - \partial_d\partial_b g_{ac} + \partial_c\partial_b g_{ad} - \partial_c\partial_a g_{bd}) - g_{ef}(\Gamma^e{}_{ac}\Gamma^f{}_{bd} - \Gamma^e{}_{ad}\Gamma^f{}_{bc})$$

and in two dimensions the symmetry properties of this tensor mean that there is only *one* independent component. We can take this to be R_{1212}, so fortunately we only have to calculate this single component:

$$\begin{aligned}
R_{1212} &= \tfrac{1}{2}(\partial_2\partial_1 g_{21} - \partial_2^2 g_{11} + \partial_1\partial_2 g_{12} - \partial_1^2 g_{22}) - g_{ef}(\Gamma^e{}_{11}\Gamma^f{}_{22} - \Gamma^e{}_{12}\Gamma^f{}_{21})\\
&= -\tfrac{1}{2}\partial_1^2 g_{22} - g_{11}(\Gamma^1{}_{11}\Gamma^1{}_{22} - \Gamma^1{}_{12}\Gamma^1{}_{12}) - g_{22}(\Gamma^2{}_{11}\Gamma^2{}_{22} - \Gamma^2{}_{21}\Gamma^2{}_{21})\\
&= a^2\sin^2\theta.
\end{aligned}$$

Thus the Gaussian curvature K of a spherical surface is given by

$$K = \frac{R_{1212}}{g} = \frac{a^2\sin^2\theta}{a^4\sin^2\theta} = \frac{1}{a^2}.$$

Instead of a spherical surface, we could instead consider the surface of a cylinder of radius a. The metric of the surface in cylindrical polar coordinates is

$$ds^2 = a^2\, d\theta^2 + dz^2,$$

and it is obvious that this two-dimensional space is spatially flat because we can transform the metric into the form

$$ds^2 = dx^2 + dz^2$$

by the coordinate transformation $x = a\theta$. It therefore follows that the curvature of a cylindrical surface vanishes.

Exercises

7.1 From Poisson's equation $\nabla^2\Phi = 4\pi G\rho$ show that the gravitational potential outside a spherical object of mass M at a radial distance r from its centre is given by $\Phi(r) = -GM/r$. What is the form of $\Phi(r)$ inside a uniform spherical body?

7.2 A charged object held stationary in a laboratory on the surface of the Earth does not emit electromagnetic radiation. If the object is then dropped so that it is in free fall, it will begin to radiate. Reconcile these observations with the principle of equivalence. *Hint: Consider the spatial extent of the electric field of the charge.*

7.3 If X^μ is a local Cartesian coordinate system at some event P, show that so too is the coordinate system $X'^\mu = \Lambda^\mu{}_\nu X^\nu$, where $\Lambda^\mu{}_\nu$ defines a Lorentz transformation.

7.4 If two vectors $\boldsymbol{v}$ and $\boldsymbol{w}$ are Fermi–Walker-transported along some observer's worldline, show that their scalar product $\boldsymbol{v}\cdot\boldsymbol{w}$ is preserved at all points along the line.

7.5 Photons of frequency ν_E are emitted from the surface of the Sun and observed by an astronaut with fixed spatial coordinates at a large distance away. Obtain an expression for the frequency ν_O of the photons as measured by the astronaut. Hence estimate the observed redshift of the photon.

7.6 An experimenter A drops a pebble of rest mass m in a uniform gravitational field g. At a distance h below A, experimenter B converts the pebble (with no energy loss) into a photon of frequency ν_B. The photon passes by A, who observes it to have frequency ν_A. Use simple physical arguments to show that to a first approximation

$$\frac{\nu_B}{\nu_A} = 1 + \frac{gh}{c^2}.$$

Use this result to argue that for two stationary observers A and B in a weak gravitational field with potential Φ, the ratio of the rates at which their laboratory clocks run is $1 + \Delta\Phi/c^2$, where $\Delta\Phi$ is the potential difference between A and B.

7.7 A satellite is in circular polar orbit of radius r around the Earth (radius R, mass M). A standard clock C on the satellite is compared with an identical clock C_0 at the south pole on Earth. Show that the ratio of the rate of the orbiting clock to that of the clock on Earth is approximately

$$1+\frac{GM}{Rc^2}-\frac{3GM}{2rc^2}.$$

Note that the orbiting clock is faster only if $r>\frac{3}{2}R$, i.e. if $r-R>3184$ km.

7.8 Consider the limit of a weak gravitational field in a coordinate system in which $g_{\mu\nu}=\eta_{\mu\nu}+h_{\mu\nu}$, with $|h_{\mu\nu}|\ll 1$, and $\partial_0 g_{\mu\nu}=0$. Keeping only terms that are first order in v/c, show that the equation of motion for a slowly moving test particle takes the form

$$\frac{d^2x^i}{dt^2}\approx -\tfrac{1}{2}c^2\delta^{ij}\partial_j h_{00}+c\delta^{ik}(\partial_j h_{0k}-\partial_k h_{0j})v^j.$$

Give a physical interpretation of the second term on the right-hand side.

7.9 Show that in a two-dimensional Riemannian manifold all the components of R_{abcd} are equal either to zero or to $\pm R_{1212}$.

7.10 Show that the line element $ds^2=y^2\,dx^2+x^2\,dy^2$ represents the Euclidean plane, but the line element $ds^2=y\,dx^2+x\,dy^2$ represents a curved two-dimensional manifold.

7.11 For a two-dimensional manifold with line element $ds^2=dr^2+f^2(r)\,d\theta^2$, show that the Gaussian curvature is given by $K=-f''/f$, where a prime denotes d/dr.

7.12 By calculating the components of the curvature tensor $R^d{}_{abc}$ in each case, show that the line element

$$ds^2=\frac{a^2}{a^2-r^2}dr^2+r^2\,d\theta^2+r^2\sin^2\theta\,d\phi^2$$

represents a curved three-dimensional manifold. Show that the manifold is flat in the limit $a\to 0$.

7.13 A spacetime has the metric

$$ds^2=c^2\,dt^2-a^2(t)(dx^2+dy^2+dz^2).$$

Show that the only non-zero connection coefficients are are

$$\Gamma^0{}_{11}=\Gamma^0{}_{22}=\Gamma^0{}_{33}=a\dot{a}\qquad\text{and}\qquad\Gamma^1{}_{10}=\Gamma^2{}_{20}=\Gamma^3{}_{30}=\dot{a}/a.$$

Deduce that particles may be at rest in such a spacetime and that for such particles the coordinate t is their proper time. Show further that the non-zero components of the Ricci tensor are

$$R_{00}=3\ddot{a}/a\qquad\text{and}\qquad R_{11}=R_{22}=R_{33}=-a\ddot{a}-2\dot{a}^2.$$

Hence show that the 00-component of the Einstein tensor is $G_{00}=-3\dot{a}^2/a^2$.

7.14 Show that the covariant components of the curvature tensor are given by

$$R_{abcd}=\tfrac{1}{2}(\partial_d\partial_a g_{bc}-\partial_d\partial_b g_{ac}+\partial_c\partial_b g_{ad}-\partial_c\partial_a g_{bd})-g^{ef}(\Gamma_{eac}\Gamma_{fbd}-\Gamma_{ead}\Gamma_{fbc}),$$

and hence verify its symmetries. Show further that, for an N-dimensional manifold, the number of independent components is $N^2(N^2-1)/12$.

7.15 Show that for any two-dimensional manifold the covariant curvature tensor has the form

$$R_{abcd} = K(g_{ac}g_{bd} - g_{ad}g_{bc}),$$

where the scalar K may be a function of the coordinates. Why does this result not generalise to arbitrary manifolds of higher dimension?

7.16 If v^a are the contravariant components of a vector and T^{ab} are the contravariant components of a rank-2 tensor, prove the results

$$\nabla_c\nabla_b v^a - \nabla_b\nabla_c v^a = -R^a{}_{dbc}v^d,$$

$$\nabla_d\nabla_c T^{ab} - \nabla_c\nabla_d T^{ab} = -R^a{}_{ecd}T^{eb} - R^b{}_{ecd}T^{ae}.$$

Can you guess the corresponding result for the mixed components $T^{ab}{}_c$ of a rank-3 tensor?

7.17 Show that any Killing vector v^a, as defined in Exercise 4.11, satisfies the relations

$$\nabla_c\nabla_b v^a = R^a{}_{bcd}v^d,$$

$$v^a\nabla_a R = 0.$$

7.18 Calculate explicit forms for the Ricci tensor R_{ab} and the Ricci scalar R in terms of the metric, the connection and its partial derivatives.

7.19 Prove that the Ricci tensor R_{ab} is symmetric.

7.20 A *conformal transformation*, such as that in Exercise 2.7, is not a change of coordinates but an actual change in the geometry of a manifold such that the metric tensor transforms as

$$\tilde{g}_{ab}(x) = \Omega^2(x)g_{ab}(x),$$

where $\Omega(x)$ is some non-vanishing scalar function of position. Show that, under such a transformation, the metric connection transforms as

$$\widetilde{\Gamma}^a{}_{bc} = \Gamma^a{}_{bc} + \frac{1}{\Omega}\left(\delta^a_c\partial_b\Omega + \delta^a_b\partial_c\Omega - g_{bc}g^{ad}\partial_d\Omega\right).$$

Hence show that the curvature tensor, the Ricci tensor and the Ricci scalar transform respectively as

$$\widetilde{R}^a{}_{bcd} = R^a{}_{bcd} - 2\left(\delta^a_{[c}\delta^e_{d]}\delta^f_b - g_{b[c}\delta^e_{d]}g^{af}\right)\frac{\nabla_e\nabla_f\Omega}{\Omega}$$
$$+2\left(2\delta^a_{[c}\delta^e_{d]}\delta^f_b - 2g_{b[c}\delta^e_{d]}g^{af} + g_{b[c}\delta^a_{d]}g^{ef}\right)\frac{(\nabla_e\Omega)(\nabla_f\Omega)}{\Omega^2},$$

$$\widetilde{R}_{bc} = R_{bc} + \left[(N-2)\delta^f_b\delta^e_c + g_{bc}g^{ef}\right]\frac{\nabla_e\nabla_f\Omega}{\Omega}$$
$$-\left[2(N-2)\delta^f_b\delta^e_c - (N-3)g_{bc}g^{ef}\right]\frac{(\nabla_e\Omega)(\nabla_f\Omega)}{\Omega^2},$$

$$\widetilde{R} = \frac{R}{\Omega^2} + 2(N-1)g^{ef}\frac{\nabla_e\nabla_f\Omega}{\Omega^3} + (N-1)(N-4)g^{ef}\frac{(\nabla_e\Omega)(\nabla_f\Omega)}{\Omega^4},$$

where N is the dimension of the manifold.

7.21 Show that parallel transportation of a vector around the closed triangle ABC on the surface of a sphere, as shown in Figure 7.3, results in a vector that is orthogonal to its original direction.

7.22 On the surface of a sphere, show that, along the geodesic $\phi = \text{constant}$, the geodesic deviation vector ξ^i satisfies

$$\frac{D^2 \xi^\theta}{Ds^2} = 0, \qquad \frac{D^2 \xi^\phi}{Ds^2} = -\xi^\phi \left(\frac{d\theta}{ds}\right)^2.$$

Choose a geodesic $\phi = \phi_0$ with path length $s = \theta$ measured from $\theta = 0$, and a neighbouring geodesic $\phi = \phi_0 + \delta\phi_0$, also with $s = \theta$, and define $\xi^i(\theta)$ as the vector between $s = \theta$ on one and $s = \theta$ on the other. Show that $\xi^i(\theta) = (0, \xi^\phi)$ for all θ. Show in addition that if $\xi^\phi = 0$ when $\theta = 0$ then

$$\xi^\phi \xi_\phi = l^2 \sin^2\theta,$$

where l^2 is a constant, and that the two geodesics pass through $\theta = \pi$.

7.23 In Newtonian gravity, consider two nearby particles with trajectories $x^i(t)$ and $\bar{x}^i(t)(i = 1, 2, 3)$ respectively in Cartesian coordinates. Show that the components of the separation vector $\zeta^i = x^i - \bar{x}^i$ evolve as

$$\frac{d^2 \zeta^i}{dt^2} = -\left(\frac{\partial^2 \Phi}{\partial x^i \partial x^j}\right) \zeta^j,$$

where Φ is the Newtonian gravitational potential.

7.24 In the weak-field, Newtonian, limit of general relativity, we may choose coordinates such that $g_{\mu\nu} = \eta_{\mu\nu} + h_{\mu\nu}$, where $|h_{\mu\nu}| \ll 1$, and we assume that all particle velocities are small compared with c. By considering the equation of geodesic deviation, show that the general-relativistic tidal force reduces to the Newtonian limit given in Exercise 7.23.

8

The gravitational field equations

Let us now follow Einstein's suggestion that gravity is a manifestation of spacetime curvature induced by the presence of matter. We must therefore obtain a set of equations that describe quantitatively how the curvature of spacetime at any event is related to the matter distribution at that event. These will be the *gravitational field equations*, or *Einstein equations*, in the same way that the Maxwell equations are the field equations of electromagnetism.

Maxwell's equations relate the electromagnetic field $\boldsymbol{F}$ at any event to its source, the 4-current density $\boldsymbol{j}$ at that event. Similarly, Einstein's equations relate spacetime curvature to *its* source, the energy–momentum of matter. As we shall see, the analogy goes further. In any given coordinate system, Maxwell's equations are second-order partial differential equations for the components $F_{\mu\nu}$ of the electromagnetic field tensor (or equivalently for the components A_μ of the electromagnetic potential). We shall find that Einstein's equations are also a set of second-order partial differential equations, but instead for the metric coefficients $g_{\mu\nu}$ of spacetime.

8.1 The energy–momentum tensor

To construct the gravitational field equations, we must first find a properly relativistic (or *covariant*) way of expressing the *source term*. In other words, we must identify a tensor that describes the matter distribution at each event in spacetime.

We will use our discussion of the 4-current density in Chapter 6 as a guide. Thus, let us consider some general time-dependent distribution of (electrically neutral) *non-interacting* particles, each of rest mass m_0. This is commonly called *dust* in the literature. At each event P in spacetime we can characterise the distribution completely by giving the matter density ρ and 3-velocity $\vec{u}$ as measured in some inertial frame. For simplicity, let us consider the fluid in its *instantaneous rest frame* S at P, in which $\vec{u} = \vec{0}$. In this frame, the (proper) density is given by

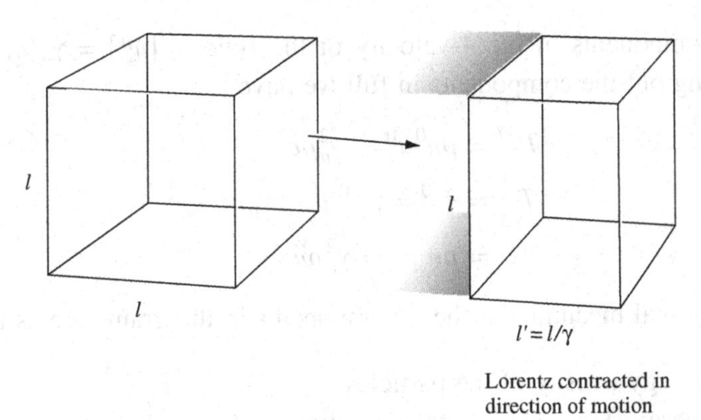

Figure 8.1 The Lorentz contraction of a fluid element in the direction of motion.

$\rho_0 = m_0 n_0$, where m_0 is the rest mass of each particle and n_0 is the number of particles in a unit volume. In some other frame S', moving with speed v relative to S, the volume containing a fixed number of particles is Lorentz contracted along the direction of motion (see Figure 8.1). Hence, in S' the number density of particles is $n' = \gamma_v n_0$. We now have an additional effect, however, since the mass of each particle in S' is $m' = \gamma_v m_0$. Thus, the matter density in S' is

$$\boxed{\rho' = \gamma_v^2 \rho_0.}$$

We may conclude that the matter density is *not* a scalar but *does* transform as the 00-component of a *rank-2 tensor*. This suggests that the source term in the gravitational field equations should be a rank-2 tensor. At each point in spacetime, the obvious choice is

$$\boxed{\boldsymbol{T}(x) = \rho_0(x)\boldsymbol{u}(x) \otimes \boldsymbol{u}(x),} \tag{8.1}$$

where $\rho_0(x)$ is the proper density of the fluid, i.e. that measured by an observer comoving with the local flow, and $\boldsymbol{u}(x)$ is its 4-velocity. The tensor $\boldsymbol{T}(x)$ is called the *energy–momentum* tensor (or the *stress–energy* tensor) of the matter distribution. We will see the reason for these names shortly. *Note that from now on we will denote the proper density simply by ρ, i.e. without the zero subscript.*

In some arbitrary coordinate system x^μ, in which the 4-velocity of the fluid is u^μ, the contravariant components of (8.1) are given simply by

$$\boxed{T^{\mu\nu} = \rho u^\mu u^\nu.} \tag{8.2}$$

To give a physical interpretation of the components of the energy–momentum tensor, it is convenient to consider a local Cartesian inertial frame at P in which the set of components of the 4-velocity of the fluid is $[u^\mu] = \gamma_u(c, \vec{u})$. In this frame, writing out the components in full we have

$$T^{00} = \rho u^0 u^0 = \gamma_u^2 \rho c^2,$$
$$T^{0i} = T^{i0} = \rho u^0 u^i = \gamma_u^2 \rho c \vec{u}^i,$$
$$T^{ij} = \rho u^i u^j = \gamma_u^2 \rho \vec{u}^i \vec{u}^j.$$

Thus the physical meanings of these components in this frame are as follows:

T^{00} is the energy density of the particles;
T^{0i} is the energy flux $\times c^{-1}$ in the i-direction;
T^{i0} is the momentum density $\times c$ in the i-direction;
T^{ij} is the rate of flow of the i-component of momentum per unit area in the j-direction.

It is because of these identifications that the tensor $\boldsymbol{T}$ is known as the energy–momentum or stress–energy tensor.

8.2 The energy–momentum tensor of a perfect fluid

To generalise our discussion to real fluids, we have to take account of the facts that (i) besides the bulk motion of the fluid, each particle has some random (thermal) velocity and (ii) there may be various forces between the particles that contribute potential energies to the total. The physical meanings of the components of the energy–momentum tensor $\boldsymbol{T}$ give us an insight into how to generalise its form to include these properties of real fluids.

Let us consider $\boldsymbol{T}$ at some event P and work in a local Cartesian inertial frame S that is the IRF of the fluid at P. For dust, the *only* non-zero component is T^{00}. However, let us consider the components of $\boldsymbol{T}$ in the IRF for a real fluid.

- T^{00} is the total energy density, including any potential energy contributions from forces between the particles and kinetic energy from their random thermal motions.
- T^{0i}: although there is no bulk motion, energy might be transmitted by heat conduction, so this is basically a heat conduction term in the IRF.
- T^{i0}: again, although the particles have no bulk motion, if heat is being conducted then the energy will carry momentum.
- T^{ij}: the random thermal motions of the particles will give rise to momentum flow, so that T^{ii} is the isotropic *pressure* in the i-direction and the T^{ij} (with $i \neq j$) are the *viscous stresses* in the fluid.

These identifications are valid for a general fluid. A *perfect fluid* is defined as one for which there is, and no heat conduction or viscosity in the IRF. Thus, in the IRF the components of $\boldsymbol{T}$ for a perfect fluid are given by

$$[T^{\mu\nu}] = \begin{pmatrix} \rho c^2 & 0 & 0 & 0 \\ 0 & p & 0 & 0 \\ 0 & 0 & p & 0 \\ 0 & 0 & 0 & p \end{pmatrix}. \tag{8.3}$$

It is not hard to show that

$$T^{\mu\nu} = (\rho + p/c^2)u^\mu u^\nu - p\eta^{\mu\nu}. \tag{8.4}$$

However, because of the way in which we have written this equation, it must be valid in *any* local Cartesian inertial frame at P. Moreover, we can obtain an expression that is valid in an *arbitrary* coordinate system simply by replacing $\eta^{\mu\nu}$ with the metric functions $g^{\mu\nu}$ in the arbitrary system. Thus, we arrive at a fully covariant expression for the components of the energy–momentum tensor of a perfect fluid:

$$\boxed{T^{\mu\nu} = (\rho + p/c^2)u^\mu u^\nu - pg^{\mu\nu}.} \tag{8.5}$$

We see that $T^{\mu\nu}$ is *symmetric* and is made up from the two scalar fields ρ and p and the vector field $\boldsymbol{u}$ that characterise the perfect fluid. We also see that in the limit $p \to 0$ a perfect fluid becomes dust.

Finally, we note that it is possible to give more complicated expressions representing the energy–momentum tensors for imperfect fluids, for charged fluids and even for the electromagnetic field. These tensors are all symmetric.

8.3 Conservation of energy and momentum for a perfect fluid

Let us investigate how to express energy and momentum conservation in a local Cartesian inertial frame S at some event P that is represented by the local inertial coordinates x^μ. In these coordinates, the energy–momentum tensor takes the form (8.4).

By analogy with the equation $\partial_\mu j^\mu = 0$ for the conservation of charge, which we derived in Chapter 6, the conservation of energy and momentum is represented by the equation

$$\boxed{\partial_\mu T^{\mu\nu} = 0.} \tag{8.6}$$

Rather than arriving at this result from first principles, which would take us into a lengthy discussion of relativistic fluid mechanics, let us instead reverse the process and justify our assertion by arguing that it produces the correct equations of motion and continuity for a fluid in the Newtonian limit.

Substituting the form (8.4) into (8.6) gives

$$\partial_\mu(\rho+p/c^2)u^\mu u^\nu + (\rho+p/c^2)[(\partial_\mu u^\mu)u^\nu + u^\mu(\partial_\mu u^\nu)] - (\partial_\mu p)\eta^{\mu\nu} = 0. \quad (8.7)$$

Now, the 4-velocity satisfies the normalisation condition $u^\nu u_\nu = c^2$ and differentation of this gives

$$(\partial_\mu u^\nu)u_\nu + u^\nu(\partial_\mu u_\nu) = 2(\partial_\mu u^\nu)u_\nu = 0.$$

Thus, contracting (8.7) with u_ν, dividing through by c^2 and collecting terms gives

$$\boxed{\partial_\mu(\rho u^\mu) + (p/c^2)\partial_\mu u^\mu = 0.} \quad (8.8)$$

Equation (8.7) therefore simplifies to

$$\boxed{(\rho+p/c^2)(\partial_\mu u^\nu)u^\mu = (\eta^{\mu\nu} - u^\mu u^\nu/c^2)\partial_\mu p.} \quad (8.9)$$

Equations (8.8) and (8.9) are, in fact, respectively the relativistic *equation of continuity* and the relativistic *equation of motion* for a perfect fluid in local inertial coordinates at some event P.[1] We will now show that for slowly moving fluids and small pressures they reduce to the classical equations of Newtonian theory.

By a slowly moving fluid, we mean one for which we may neglect u/c and so take $\gamma_u \approx 1$ and $[u^\mu] \approx (c, \vec{u})$; note that the difference between the proper density and the density disappears in this limit. By small pressures we mean that p/c^2 is negligible compared with ρ. In these limits, equation (8.8) then reduces to

$$\partial_\mu(\rho u^\mu) = 0,$$

or, in 3-vector notation,

$$\frac{\partial\rho}{\partial t} + \vec{\nabla}\cdot(\rho\vec{u}) = 0,$$

which is the classical equation of continuity for a fluid. In the limit of small pressures, equation (8.9) reduces to

$$\rho(\partial_\mu u^\nu)u^\mu = (\eta^{\mu\nu} - u^\mu u^\nu/c^2)\partial_\mu p.$$

Moreover, in our slowly-moving approximation, the zeroth components of the left- and right-hand sides are both zero. Thus the spatial components $i = 1, 2, 3$ satisfy

$$\rho(\partial_\mu u^i)u^\mu = -\delta^{ji}\partial_j p.$$

[1] As usual, these equations may be generalised to a form valid in arbitrary coordinates by replacing ∂_μ by ∇_μ and replacing $\eta^{\mu\nu}$ by $g^{\mu\nu}$.

In 3-vector notation this reads

$$\rho\left(\frac{\partial}{\partial t}+\vec{u}\cdot\vec{\nabla}\right)\vec{u}=-\vec{\nabla}p,$$

which is Euler's classical equation of motion for a perfect fluid. Hence we have shown that the relativistic continuity equation (8.8) and the equation of motion (8.9) for a perfect fluid reduce to the appropriate Newtonian equations. If we were to accept that a relativistic fluid were described by (8.8) and (8.9) then we could reverse our overall argument and derive the result $\partial_\mu T^{\mu\nu}=0$.

So far we have worked in local inertial coordinates in order to make contact with the Newtonian theory. Nevertheless, we can trivially obtain the condition for energy and momentum conservation in arbitrary coordinates by replacing ∂_μ by ∇_μ in (8.6), which then gives

$$\boxed{\nabla_\mu T^{\mu\nu}=0.} \tag{8.10}$$

This important equation is worthy of further comment. In our discussion so far, we have not been explicit about whether our spacetime is Minkowskian or curved. Although the form (8.10) is valid (in arbitrary coordinates) in both cases, its interpretation differs in the two cases. If we neglect gravity and assume a Minkowski spacetime, the relation (8.10) does indeed represent the conservation of energy and momentum. In the presence of a gravitational field (and hence a curved spacetime), however, the energy and momentum of the matter *alone* is not conserved. In this case, (8.10) represents the equation of motion of the matter under the influence of the gravitational field; this is discussed further in Section 8.8. As we will see below, the condition (8.10) places a tight restriction on the possible forms that the gravitational field equations may take.

8.4 The Einstein equations

We are now in a position to deduce the form of the gravitational field equations proposed by Einstein. Let us begin by recalling some of our previous results.

- The field equation of Newtonian gravity is

$$\vec{\nabla}^2\Phi=4\pi G\rho.$$

- If gravity is a manifestation of spacetime curvature, we showed in Chapter 7, equation (7.8), that for a weak gravitational field, in coordinates such that $g_{\mu\nu}=\eta_{\mu\nu}+h_{\mu\nu}$ (with $|h_{\mu\nu}|\ll 1$) and in which the metric is static, then

$$g_{00}=\left(1+\frac{2\Phi}{c^2}\right). \tag{8.11}$$

- The correct relativistic description of matter is provided by the energy–momentum tensor and, for a perfect fluid or dust, in the IRF we have

$$T_{00} = \rho c^2.$$

Combining these observations suggests that, for a weak static gravitational field in the low-velocity limit,

$$\vec{\nabla}^2 g_{00} = \frac{8\pi G}{c^4} T_{00}.$$

Einstein's fundamental intuition was that the curvature of spacetime at any event is related to the matter content at that event. The above considerations thus suggest that the gravitational field equations should be of the form

$$K_{\mu\nu} = \kappa T_{\mu\nu}, \tag{8.12}$$

where $K_{\mu\nu}$ is a rank-2 tensor related to the *curvature* of spacetime and we have set $\kappa = 8\pi G/c^4$. Since the curvature of spacetime is expressed by the curvature tensor $R_{\mu\nu\sigma\rho}$, the tensor $K_{\mu\nu}$ must be constructed from $R_{\mu\nu\sigma\rho}$ and the metric tensor $g_{\mu\nu}$. Moreover, $K_{\mu\nu}$ should have the following properties: (i) the Newtonian limit suggests that $K_{\mu\nu}$ should contain terms no higher than *linear* in the *second-order* derivatives of the metric tensor; and (ii) since $T_{\mu\nu}$ is symmetric then $K_{\mu\nu}$ should also be *symmetric*. The curvature tensor $R_{\mu\nu\sigma\rho}$ is already linear in the second derivatives of the metric, and so the most general form for $K_{\mu\nu}$ that satisfies (i) and (ii) is

$$K_{\mu\nu} = aR_{\mu\nu} + bRg_{\mu\nu} + \lambda g_{\mu\nu}, \tag{8.13}$$

where $R_{\mu\nu}$ is the Ricci tensor, R is the curvature scalar and a, b, λ are constants.

Let us now consider the constants a, b, λ. First, if we require that *every* term in $K_{\mu\nu}$ is linear in the second derivatives of $g_{\mu\nu}$ then we see immediately that $\lambda = 0$. We will relax this condition later, but for the moment we therefore have

$$K_{\mu\nu} = aR_{\mu\nu} + bRg_{\mu\nu}.$$

To find the constants a and b we recall that the energy–momentum tensor satisfies $\nabla_\mu T^{\mu\nu} = 0$; thus, from (8.10), we also require

$$\nabla_\mu K^{\mu\nu} = \nabla_\mu (aR^{\mu\nu} + bRg^{\mu\nu}) = 0.$$

However, in Section 7.11 we showed that

$$\nabla_\mu (R^{\mu\nu} - \tfrac{1}{2} g^{\mu\nu} R) = 0,$$

and so, remembering that $\nabla_\mu g^{\mu\nu} = 0$, we obtain

$$\nabla_\mu K^{\mu\nu} = (\tfrac{1}{2}a + b) g^{\mu\nu} \nabla_\mu R = 0.$$

The quantity $\nabla_\mu R$ will, in general, be non-zero throughout (a region of) spacetime unless the latter is flat and hence there is no gravitational field. Thus we find that $b = -a/2$, and so the gravitational field equations must take the form

$$a(R_{\mu\nu} - \tfrac{1}{2}g_{\mu\nu}R) = \kappa T_{\mu\nu}.$$

To fix the constant a, we must compare the weak-field limit of these equations with Poisson's equation in Newtonian gravity. The comparison is presented in the next section, where we show that, for consistency with the Newtonian theory, we require $a = -1$ and so

$$\boxed{R_{\mu\nu} - \tfrac{1}{2}g_{\mu\nu}R = -\kappa T_{\mu\nu},} \tag{8.14}$$

where $\kappa = 8\pi G/c^4$. Equation (8.14) constitutes *Einstein's gravitational field equations*, which form the mathematical basis of the theory of general relativity. We note that the left-hand side of (8.14) is simply the *Einstein tensor* $G_{\mu\nu}$, defined in Chapter 7.

We can obtain an alternative form of Einstein's equations by writing (8.14) in terms of mixed components,

$$R^\mu_{\ \nu} - \tfrac{1}{2}\delta^\mu_\nu R = -\kappa T^\mu_{\ \nu},$$

and contracting by setting $\mu = \nu$. We thus find that $R = \kappa T$, where $T \equiv T^\mu_{\ \mu}$. Hence we can write Einstein's equations (8.14) as

$$\boxed{R_{\mu\nu} = -\kappa(T_{\mu\nu} - \tfrac{1}{2}Tg_{\mu\nu}).} \tag{8.15}$$

In four-dimensional spacetime $g_{\mu\nu}$ has 10 independent components and so in general relativity we have 10 independent field equations. We may compare this with Newtonian gravity, in which there is only one gravitational field equation. Furthermore, the Einstein field equations are *non-linear* in the $g_{\mu\nu}$ whereas Newtonian gravity is linear in the field Φ. Einstein's theory thus involves numerous non-linear differential equations, and so it should come as no surprise that the theory is complicated.

8.5 The Einstein equations in empty space

In general, $T_{\mu\nu}$ contains *all* forms of energy and momentum. Of course, this includes any matter present but if there is also electromagnetic radiation, for example, then it too must be included in $T_{\mu\nu}$ (the resulting expression is somewhat complicated; see Exercise 8.3).

A region of spacetime in which $T_{\mu\nu} = 0$ is called *empty*, and such a region is therefore not only devoid of matter but also of radiative energy and momentum. It can be seen from (8.15) that the gravitational field equations for empty space are

$$\boxed{R_{\mu\nu} = 0.} \tag{8.16}$$

From this simple equation, we can immediately establish a profound result. Consider the number of field equations as a function of the number of spacetime dimensions; then, for two, three and four dimensions, the numbers of field equations and independent components of $R_{\mu\nu\sigma\rho}$ are as shown in the table.

No. of spacetime dimensions	2	3	4
No. of field equations	3	6	10
No. of independent components of $R_{\mu\nu\sigma\rho}$	1	6	20

Thus we see that in two or three dimensions the field equations in empty space *guarantee that the full curvature tensor must vanish*. In four dimensions, however, we have 10 field equations but 20 independent components of the curvature tensor. It is therefore possible to satisfy the field equations in empty space with a non-vanishing curvature tensor. Remembering that a non-vanishing curvature tensor represents a non-vanishing gravitational field, we conclude that *it is only in four dimensions or more that gravitational fields can exist in empty space*.

8.6 The weak-field limit of the Einstein equations

To determine the 'weak-field' limit of the Einstein equations our preliminary discussion in Section 8.4 suggests that we need only consider their 00-component. It is most convenient to use the form (8.15) of the equations, from which we have

$$R_{00} = -\kappa\left(T_{00} - \tfrac{1}{2}Tg_{00}\right). \tag{8.17}$$

In the weak-field approximation, spacetime is only 'slightly' curved and so there exist coordinates in which $g_{\mu\nu} = \eta_{\mu\nu} + h_{\mu\nu}$, with $|h_{\mu\nu}| \ll 1$, and the metric is stationary. Hence in this case $g_{00} \approx 1$. Moreover, from the definition of the curvature tensor we find that R_{00} is given by

$$R_{00} = \partial_0\Gamma^{\mu}{}_{0\mu} - \partial_\mu\Gamma^{\mu}{}_{00} + \Gamma^{\nu}{}_{0\mu}\Gamma^{\mu}{}_{\nu 0} - \Gamma^{\nu}{}_{00}\Gamma^{\mu}{}_{\nu\mu}.$$

In our coordinate system the $\Gamma^{\mu}{}_{\nu\sigma}$ are small, so we can neglect the last two terms to first order in $h_{\mu\nu}$. Also using the fact that the metric is stationary in our coordinate system, we then have

$$R_{00} \approx -\partial_i\Gamma^{i}{}_{00}.$$

In our discussion of the Newtonian limit in Chapter 7, however, we found that $\Gamma^{i}{}_{00} \approx \frac{1}{2}\delta^{ij}\partial_j h_{00}$ to first order in $h_{\mu\nu}$, and so

$$R_{00} \approx -\tfrac{1}{2}\delta^{ij}\partial_i\partial_j h_{00}.$$

Substituting our approximate expressions for g_{00} and R_{00} into (8.17), in the 'weak-field' limit we thus have

$$\tfrac{1}{2}\delta^{ij}\partial_i\partial_j h_{00} \approx \kappa(T_{00} - \tfrac{1}{2}T). \tag{8.18}$$

To proceed further we must assume a form for the matter producing the weak gravitational field and for simplicity we consider a perfect fluid. Most classical matter distributions have $p/c^2 \ll \rho$ and so we may in fact take the energy–momentum tensor to be that of dust, i.e.

$$T_{\mu\nu} = \rho u_\mu u_\nu,$$

which gives $T = \rho c^2$. In addition, let us assume that the particles making up the fluid have speeds u in our coordinate system that are small compared with c. We thus make the approximation $\gamma_u \approx 1$ and hence $u_0 \approx c$. Therefore equation (8.18) reduces to

$$\tfrac{1}{2}\delta^{ij}\partial_i\partial_j h_{00} \approx \tfrac{1}{2}\kappa\rho c^2.$$

We may, however, write $\delta^{ij}\partial_i\partial_j = \vec{\nabla}^2$; furthermore, from (8.11) we have $h_{00} = 2\Phi/c^2$, where Φ is the gravitational potential. Thus, remembering that $\kappa = 8\pi G/c^4$, we finally obtain

$$\boxed{\vec{\nabla}^2\Phi \approx 4\pi G\rho,}$$

which is Poisson's equation in Newtonian gravity. This identification verifies our earlier assertion that $a = -1$ in the derivation of Einstein's equations (8.14).

8.7 The cosmological-constant term

The standard Einstein gravitational field equations are

$$R_{\mu\nu} - \tfrac{1}{2}g_{\mu\nu}R = -\kappa T_{\mu\nu}. \tag{8.19}$$

However, these equations are *not* unique. In fact, shortly after Einstein derived them he proposed a modification known as the *cosmological term*.

In deriving the field equations (8.14), we assumed that the tensor $K_{\mu\nu}$ that makes up the left-hand side of the field equations,

$$K_{\mu\nu} = \kappa T_{\mu\nu},$$

should contain only terms that are linear in the second-order derivatives of $g_{\mu\nu}$. This led us to set $\lambda = 0$ in (8.13), i.e. to discard the term $\lambda g_{\mu\nu}$ in the tensor $K_{\mu\nu}$. Let us now relax this assumption.

Recalling that $\nabla_\mu T^{\mu\nu} = 0$ we still require $\nabla_\mu K^{\mu\nu} = 0$, but in Section 4.12 we showed that

$$\nabla_\mu g^{\mu\nu} = 0.$$

Thus, we can add *any constant multiple* of $g_{\mu\nu}$ to the left-hand side of (8.19) and still obtain a *consistent* set of field equations. It is usual to denote this multiple by Λ, so that the field equations become

$$\boxed{R_{\mu\nu} - \tfrac{1}{2}g_{\mu\nu}R + \Lambda g_{\mu\nu} = -\kappa T_{\mu\nu},} \tag{8.20}$$

where Λ is some new universal constant of nature known as the *cosmological constant*. By writing this equation in terms of the mixed components and contracting, as we did with the standard field equations, we find that $R = \kappa T + 4\Lambda$. Substituting this expression into (8.20), we obtain an alternative form of the field equations,

$$\boxed{R_{\mu\nu} = -\kappa\left(T_{\mu\nu} - \tfrac{1}{2}Tg_{\mu\nu}\right) + \Lambda g_{\mu\nu}.} \tag{8.21}$$

Following the procedure presented in Section 8.6, it is straightforward to show that, in the weak-field limit, the field equation of 'Newtonian' gravity becomes

$$\nabla^2\Phi = 4\pi G\rho - \Lambda c^2.$$

For a spherical mass M, the gravitational field strength is easily found to be

$$\vec{g} = -\vec{\nabla}\Phi = -\frac{GM}{r^2}\hat{\vec{r}} + \frac{c^2\Lambda r}{3}\hat{\vec{r}}.$$

Thus, in this case, we see that the cosmological constant term corresponds to a *gravitational repulsion* whose strength *increases linearly* with r.

The reason for calling Λ the cosmological constant is historical. Einstein first introduced this term because he was unable to construct static models of the universe from his standard field equations (8.19). What he found (and we will discuss this in detail in Chapter 15) was that the standard field equations predicted a universe that was either expanding or contracting. Einstein did this work in about 1916, when most people thought that our Milky Way Galaxy represented the whole universe, which Einstein represented as a uniform distribution of 'fixed stars'. By introducing Λ, Einstein constructed static models of the universe (which as we will see are actually unstable). It was later realised, however, that the Milky Way is just one of a great many galaxies. Moreover, in 1929 Edwin Hubble discovered the expansion of the universe by measuring distances and redshifts to nearby external galaxies. The universe was proved to be expanding and the need for a cosmological constant disappeared. Einstein is reputed to have said that the introduction of the cosmological constant was his 'biggest blunder'.

Nowadays we have a rather different view of the cosmological constant. Recall that the energy–momentum tensor of a perfect fluid is

$$T^{\mu\nu} = (\rho + p/c^2)u^\mu u^\nu - pg^{\mu\nu}.$$

Imagine some type of 'substance' with a strange equation of state $p = -\rho c^2$. This is unlike any kind of substance that you have ever encountered because it has a negative pressure! The energy–momentum tensor for this substance would be

$$\boxed{T_{\mu\nu} = -pg_{\mu\nu} = \rho c^2 g_{\mu\nu}.}$$

There are two points to note about this equation. First, the energy–momentum tensor of this strange substance depends only on the *metric* tensor – it is therefore a property of the vacuum itself and we can call ρ the energy density of the *vacuum*. Second, the form of $T_{\mu\nu}$ is the same as the cosmological-constant term in (8.20). We can therefore view the cosmological constant as a universal constant that fixes the energy density of the vacuum,

$$\boxed{\rho_{\text{vac}} c^2 = \frac{\Lambda c^4}{8\pi G}.} \tag{8.22}$$

Denoting the energy–momentum tensor of the vacuum by $T^{\text{vac}}_{\mu\nu} = \rho_{\text{vac}} c^2 g_{\mu\nu}$, we can thus write the modified gravitational field equations (8.20) as

$$R_{\mu\nu} - \tfrac{1}{2} g_{\mu\nu} R = -\kappa \left(T_{\mu\nu} + T^{\text{vac}}_{\mu\nu}\right),$$

where $T_{\mu\nu}$ is the energy–momentum tensor of any matter or radiation present.

How can we calculate the energy density of the vacuum? This is one of the major unsolved problems in physics. The simplest calculation involves summing the quantum mechanical zero-point energies of all the fields known in Nature. This gives an answer about 120 orders of magnitude higher than the upper limits on Λ set by cosmological observations. This is probably the worst theoretical prediction in the history of physics! Nobody knows how to make sense of this result. Some physical mechanism must exist that makes the cosmological constant very small.

Some physicists have thought that A mechanism must exist that makes Λ *exactly* equal to zero. But in the last few years there has been increasing evidence that the cosmological constant is small but *non-zero*. The strongest evidence comes from observations of distant Type Ia supernovae that indicate that the expansion of the universe is actually accelerating rather than decelerating. Normally, one would have thought that the gravity of matter in the universe would cause the expansion to slow down (perhaps even eventually halting the expansion and causing the universe to collapse). But if the cosmological constant is non-zero, the negative pressure of the vacuum can cause the universe to accelerate.

Whether these supernova observations are right or not is an area of active research, and the theoretical problem of explaining the value of the cosmological constant is one of the great challenges of theoretical physics. It is most likely that we require a fully developed theory of quantum gravity (perhaps superstring theory) before we can understand Λ.

8.8 Geodesic motion from the Einstein equations

The Einstein equations give a quantitative description of how the energy–momentum distribution of matter (or other fields) at any event determines the spacetime curvature at that event. We also know that, under the influence of gravity alone, matter moves along geodesics in the curved spacetime. We now show that it is, in fact, unnecessary to make the separate postulate of geodesic motion, since it follows directly from the Einstein equations themselves.

The field equations were derived partly from the requirement that the covariant divergence of the energy–momentum tensor vanishes,

$$\nabla_\mu T^{\mu\nu} = 0. \tag{8.23}$$

As noted in Section 8.3, this relation represents the equation of motion for matter in the curved spacetime, and in this section we explore this interpretation in more detail. For later convenience, we may also write (8.23) as

$$\begin{aligned}\nabla_\mu T^{\mu\nu} &= \partial_\mu T^{\mu\nu} + \Gamma^\mu{}_{\sigma\mu} T^{\sigma\nu} + \Gamma^\nu{}_{\sigma\mu} T^{\mu\sigma} \\ &= \frac{1}{\sqrt{-g}} \partial_\mu (\sqrt{-g}\, T^{\mu\nu}) + \Gamma^\nu{}_{\sigma\mu} T^{\mu\sigma},\end{aligned} \tag{8.24}$$

where in the last line we have used the expression (3.26) for the contracted connection coefficient $\Gamma^\mu{}_{\sigma\mu}$, and we note that $|g| = -g$ for a spacetime metric with signature -2.

Let us first consider directly the specific case of a single test particle of rest mass m. By analogy with (8.2), the energy–momentum tensor of the particle as a function of position x may be written as

$$T^{\mu\nu}(x) = \frac{m}{\sqrt{-g}} \int \frac{dz^\mu}{d\tau} \frac{dz^\nu}{d\tau} \delta^4(x - z(\tau))\, d\tau, \tag{8.25}$$

where $z^\mu(\tau)$ is the worldline of the particle and τ is its proper time.[2] Inserting (8.25) into (8.23) and using the result (8.24), we obtain

$$\int \dot{z}^\mu \dot{z}^\nu \frac{\partial}{\partial x^\mu}\delta^4(x-z(\tau))\,d\tau + \Gamma^\nu{}_{\sigma\mu}\int \dot{z}^\mu \dot{z}^\sigma \delta^4(x-z(\tau))\,d\tau = 0, \qquad (8.26)$$

where the dots denote differentiation with respect to τ. Since $\delta^4(x-z(\tau))$ depends only on the difference $x^\rho - z^\rho$, we can replace $\partial/\partial x^\mu$ by $-\partial/\partial z^\mu$ where it acts upon the delta function. Then, by noting that

$$\dot{z}^\mu \frac{\partial}{\partial z^\mu}\delta^4(x-z(\tau)) = \frac{d}{d\tau}\delta^4(x-z(\tau)),$$

we may write (8.26) as

$$-\int \dot{z}^\nu \frac{d}{d\tau}\delta^4(x-z(\tau))\,d\tau + \Gamma^\nu{}_{\sigma\mu}\int \dot{z}^\mu \dot{z}^\sigma \delta^4(x-z(\tau))\,d\tau = 0.$$

On performing the first integral on the left-hand side by parts and collecting together terms, this becomes

$$\int \left(\ddot{z}^\nu + \Gamma^\nu{}_{\sigma\mu}\dot{z}^\mu \dot{z}^\sigma\right)\delta^4(x-z(\tau))\,d\tau = 0.$$

For this integral to vanish, we clearly require the first factor in the integrand to equal zero, from which we recover directly the standard geodesic equation of motion.

The derivation above offers an entirely new insight into the equation of motion. The position of the particle is where the field equations become singular, but the solution of the field equations in the empty space surrounding the singularity determines how it should move, i.e. it obeys the same equation of motion as that of a 'test particle'. The fact that the Einstein equations predict the equation of motion is remarkable and should be contrasted with the situation in electrodynamics. In the latter case, the Maxwell equations for the electromagnetic field do *not* contain the corresponding equation of motion for a charged particle, which has to be postulated separately. The origin of this distinction between gravity and electromagnetism lies in the non-linear nature of the Einstein equations. The physical reason for this non-linearity is that the gravitational field itself carries energy–momentum and can therefore act as its own source, whereas electromagnetic field carries no charge and so cannot act as its own source.

[2] The four-dimensional delta function $\delta^4(x-y)$ is defined by the relation

$$\int \Phi(x)\delta^4(x-y)\,d^4x = \Phi(y),$$

where Φ is any scalar field. Since $\sqrt{-g}\,d^4x$ is the invariant volume element, it follows that $\delta^4(x-y)/\sqrt{-g}$ is the invariant scalar that must be used in (8.25).

It is worthwhile generalising the above discussion from a single point particle to a continuous matter distribution. As a simple example, we shall consider a distribution of dust (i.e. a pressureless perfect fluid), for which the energy–momentum tensor is given by

$$T^{\mu\nu} = \rho u^\mu u^\nu.$$

In this case, the equation of motion (8.23) thus reads

$$\nabla_\mu(\rho u^\mu u^\nu) = \nabla_\mu(\rho u^\mu)u^\nu + \rho u^\mu \nabla_\mu u^\nu = 0. \tag{8.27}$$

Contracting this expression with u_ν, we have

$$c^2\nabla_\mu(\rho u^\mu) + \rho u^\mu u_\nu \nabla_\mu u^\nu = 0, \tag{8.28}$$

where we have used the fact that $u_\nu u^\nu = c^2$. Using this result again, one finds that $u_\nu \nabla_\mu u^\nu = 0$, and so the second term in (8.28) vanishes. Thus, we obtain

$$\boxed{\nabla_\mu(\rho u^\mu) = 0,}$$

which is simply the general-relativistic conservation equation. Substituting this expression back into (8.27) gives

$$\boxed{u^\mu \nabla_\mu u^\nu = 0,} \tag{8.29}$$

which is the equation of motion for the dust distribution in a gravitational field. Moreover, let us consider the worldline $x^\mu(\tau)$ of a dust particle. From (3.38) the intrinsic derivative of the particle's 4-velocity u^μ along the worldline is given by

$$\frac{Du^\nu}{D\tau} = (\nabla_\mu u^\nu)u^\mu = 0,$$

where we have used (8.29) to obtain the last equality. Since the intrinsic derivative of the 4-velocity (i.e. the tangent vector to the worldline) is zero, the dust particle's worldline $x^\mu(\tau)$ is a *geodesic*. We can show this explicitly using the expression (3.37) for the intrinsic derivative, from which we immediately obtain the geodesic equation

$$\ddot{x}^\nu + \Gamma^\nu{}_{\sigma\mu}\dot{x}^\mu \dot{x}^\sigma = 0.$$

8.9 Concluding remarks

We have now completed the task commenced in Chapter 1 of formulating a consistent relativistic theory of gravity. This has led us to the interpretation of gravity as a manifestation of spacetime curvature induced by the presence of matter (and other fields). This principle is embodied mathematically in the Einstein field equations (8.20). In the remainder of this book, apart from the final chapter, we will explore the physical consequences of these equations in a wide variety of astrophysical and cosmological applications. In the final chapter we will return to the formulation of general relativity itself, rederiving the Einstein equations from a variational principle.

Appendix 8A: Alternative relativistic theories of gravity

In Section 8.7, we described a relatively simple (but theoretically profound) modification of the Einstein field equations. This shows that Einstein's field equations are not unique. It is also worth noting that it is possible to create more radically different theories of gravity, as follows.

Scalar theory of gravity

The simplest relativistic generalisation of Newtonian gravity is obtained by continuing to represent the gravitational field by the *scalar* Φ. Since matter is described relativistically by the energy–momentum tensor $T_{\mu\nu}$, the only scalar with the dimensions of a mass density is $T^{\mu}_{\ \mu}$. Thus a *consistent* scalar relativistic theory of gravity is given by the field equation

$$\boxed{\Box^2\Phi = -\frac{4\pi G}{c^2}T^{\mu}_{\ \mu}.} \tag{8.30}$$

However, this theory must be *rejected* since, when used with the appropriate equation of motion, it predicts a *retardation* of the perihelion of Mercury, in contradiction of observations. Moreover, it does not allow one to couple gravity to electromagnetism, since $(T^{\rm EM})^{\mu}_{\ \mu} = 0$; in such a theory we could have neither gravitational redshift nor the deflection of light by matter.

Brans–Dicke theory

A gravitational theory based on a *vector* field can be eliminated since such a theory predicts that two massive particles would repel one another, rather than attract. It is, of course, possible to construct relativistic theories of gravity in which combinations of the three kinds of field (scalar, vector and tensor) are used. The most important of these alternative theories is *Brans–Dicke theory*, which we now discuss briefly.

In deriving the Einstein field equations, we started with the principle of equivalence, which led us to consider gravitation as spacetime curvature, and we found a rank-2 tensor theory of gravity that agreed with Newton's theory in the limit of weak gravitational fields and small velocities. Brans and Dicke also took the principle of equivalence as a starting point, and thus again described gravity in terms of spacetime curvature. However, they set about finding a consistent *scalar–tensor* theory of gravity. Instead of treating the gravitational constant G as a constant of nature, Brans and Dicke introduced a scalar field ϕ that determines the strength of G, i.e. *the scalar field ϕ determines the coupling strength of matter to gravity.* The key ideas of the theory are thus:

- matter, represented by the energy–momentum tensor $(T^{\mathrm{M}})_{\mu\nu}$, and a coupling constant λ fix the scalar field ϕ;
- the scalar field ϕ fixes the value of G;
- the gravitational field equations relate the curvature to the energy–momentum tensors of the scalar field and matter.

The coupled equations for the scalar field and the gravitational field in this theory are therefore

$$\begin{aligned} \Box^2\phi &= -4\pi\lambda(T^{\mathrm{M}})^{\mu}_{\mu}, \\ R_{\mu\nu} - \frac{1}{2}g_{\mu\nu}R &= -\frac{8\pi}{c^4\phi}\left(T^{\mathrm{M}}_{\mu\nu} + T^{\phi}_{\mu\nu}\right), \end{aligned} \tag{8.31}$$

where $T^{\mathrm{M}}_{\mu\nu}$ is the energy–momentum tensor of the matter and $T^{\phi}_{\mu\nu}$ is the energy–momentum tensor of the scalar field ϕ (the form of $T^{\phi}_{\mu\nu}$ is rather complicated). It is usual (for historical reasons) to write the coupling constant as $\lambda = 2/(3+2\omega)$. In the limit $\omega \to \infty$ we have $\lambda \to 0$, so ϕ is not affected by the matter distribution and can be set equal to a constant $\phi = 1/G$. In this case, $T^{\phi}_{\mu\nu}$ vanishes, and hence Brans–Dicke theory reduces to Einstein's theory in the limit $\omega \to \infty$.

The Brans–Dicke theory is interesting because it shows that it is possible to construct alternative theories that are consistent with the principle of equivalence. Einstein's theory is beautiful and simple, but it is not unique. One must therefore look to experiment to find out which theory is correct. One of the features of the Brans–Dicke model is that the effective gravitational 'constant' G varies with time because it is determined by the scalar field ϕ. A variation in G would affect the orbits of the planets, altering, for example, the dates of solar eclipses (which can be checked against historical records). A reasonably conservative conclusion from experiments is that $\omega \geq 500$, so Einstein's theory does seem to be the correct theory of gravity, at least at low energies.

Torsion theories

Throughout our discussion of curved spacetimes we have assumed that the manifold is torsionless. This is not a requirement, and we can generalise our discussion to spacetimes with a *non-zero* torsion tensor,

$$T^{\mu}_{\nu\sigma} = \Gamma^{\mu}{}_{\nu\sigma} - \Gamma^{\mu}{}_{\sigma\nu}.$$

Typically, torsion is generated by the (quantum-mechanical) spin of particles. Such theories are rather complicated mathematically, since we must make the distinction between affine and metric connections and geodesics. Gravitational theories that include spacetime torsion are often described as *Einstein–Cartan* theories and have been extensively investigated. We will not discuss these theories any further, however.

Appendix 8B: Sign conventions

There is no accepted system of sign conventions in general relativity. Different books use different sign conventions for the metric tensor, for the curvature tensors and for the field equations. We can summarize these sign conventions in terms of three sign factors $S1$, $S2$ and $S3$. These are defined as follows:

$$\eta^{\mu\nu} = [S1]\,(-1,+1,+1,+1),$$
$$R^{\mu}{}_{\alpha\beta\gamma} = [S2]\left(\partial_\beta \Gamma^{\mu}{}_{\alpha\gamma} - \partial_\gamma \Gamma^{\mu}{}_{\alpha\beta} + \Gamma^{\mu}{}_{\sigma\beta}\Gamma^{\sigma}{}_{\gamma\alpha} - \Gamma^{\mu}{}_{\sigma\gamma}\Gamma^{\sigma}{}_{\beta\alpha}\right),$$
$$G_{\mu\nu} = [S3]\,\frac{8\pi G}{c^4} T_{\mu\nu},$$
$$R_{\mu\nu} = [S2]\,[S3]\,R^{\alpha}{}_{\mu\alpha\nu}.$$

In this text we have used a convention that matches that of both R. d'Inverno, *An Introduction to Einstein's Relativity*, Oxford University Press, 1992, and W. Rindler, *Relativity: Special, General and Cosmological*, Oxford University Press, 2001, but this differs from the convention used by, for example, Misner, Thorne and Wheeler, *Gravitation*, Freeman (1973) or Weinberg, *Gravitation and Cosmology*, Wiley, (1972). Here is a summary of the sign conventions used in the various books:

	Present text	d'Inverno, Rindler	MTW	Weinberg
[S1]	−	−	+	+
[S2]	+	+	+	−
[S3]	−	−	+	−

Exercises

8.1 Show that the components of the energy–momentum tensor of a perfect fluid in its instantaneous rest frame can be written as in (8.3):

$$T^{\mu\nu} = (\rho + p/c^2)u^{\mu}u^{\nu} - p\eta^{\mu\nu}.$$

Can the components be written in any other covariant form?

8.2 Show that, for any fluid,

$$u_{\nu}\nabla_{\mu}u^{\nu} = 0.$$

Hence show that a perfect fluid in a gravitational field must satisfy the equations

$$\nabla_\mu(\rho u^\mu) + \frac{p}{c^2}\nabla_\mu u^\mu = 0,$$

$$\left(\rho + \frac{p}{c^2}\right) u^\mu \nabla_\mu u^\nu = \left(g^{\mu\nu} - \frac{u^\mu u^\nu}{c^2}\right)\nabla_\mu p.$$

Obtain the equation of motion for the worldline $x^\mu(\tau)$ of a particle in a perfect fluid with pressure, and hence show that the particle is 'pushed off' geodesics by the pressure gradient.

8.3 The electromagnetic field *in vacuo* has an energy–momentum tensor $T^{\mu\nu}_{\text{em}}$. By analogy with the energy–momentum tensor for dust, we require that (i) $T^{\mu\nu}_{\text{em}}$ is symmetric; (ii) $\nabla_\mu T^{\mu\nu}_{\text{em}} = 0$; (iii) $T^{\mu\nu}_{\text{em}}$ must be quadratic in the dynamical variable $F^{\mu\nu}$. Hence show that

$$T^{\mu\nu}_{\text{em}} = \alpha(F^\mu{}_\sigma F^{\nu\sigma} - \tfrac{1}{4}g^{\mu\nu}F_{\sigma\rho}F^{\sigma\rho}),$$

where α is a constant. By examining the component T^{00}_{em} in local Cartesian inertial coordinates, show that the constant $\alpha = -1/\mu_0$.

8.4 Consider a cloud of *charged* dust particles. Show that the equation of motion of such a fluid is

$$\rho u^\nu \nabla_\nu u^\mu = \sigma F^\mu{}_\nu u^\nu,$$

where ρ is the proper matter density of the fluid, σ is its proper charge density and u^μ is the fluid 4-velocity. Define an energy–momentum tensor $T^{\mu\nu}_{\text{d}} = \rho u^\mu u^\nu$, where ρ is the proper density of the fluid. Hence show that

$$\nabla_\mu T^{\mu\nu}_{\text{d}} = F^\mu{}_\nu j^\nu,$$

$$\nabla_\mu T^{\mu\nu}_{\text{em}} = -F^\mu{}_\nu j^\nu,$$

where $j^\mu = \sigma u^\mu$ is the 4-current density. Thus write down the energy–momentum tensor for charged dust, $T^{\mu\nu} = T^{\mu\nu}_{\text{d}} + T^{\mu\nu}_{\text{em}}$.

8.5 The energy–momentum tensor of an electromagnetic field interacting with a source satisfies $\nabla_\mu T^{\mu\nu}_{\text{em}} = -F^{\nu\sigma}J_\sigma$, where J_σ is the 4-current density of the source. Hence show that the worldline of a particle of charge q in an electromagnetic field satisfies

$$\ddot{z}^\nu + \Gamma^\nu{}_{\sigma\mu}\dot{z}^\mu\dot{z}^\sigma = \frac{q}{m}F^\nu{}_\sigma\dot{z}^\sigma,$$

and interpret this result physically.

8.6 The *weak energy condition* (WEC) states that any energy–momentum tensor must satisfy

$$T_{\mu\nu}t^\mu t^\nu \geq 0$$

for all timelike vectors t^μ. Show that for a perfect fluid the WEC implies that

$$\rho \geq 0 \quad \text{and} \quad \rho c^2 + p \geq 0.$$

8.7 The *strong energy condition* (SEC) states that any energy–momentum tensor must satisfy

$$T_{\mu\nu}t^\mu t^\nu \geq \tfrac{1}{2}T^\rho_\rho t^\sigma t_\sigma$$

for all timelike vectors t^μ. Show that for a perfect fluid the SEC implies that

$$\rho c^2 + p \geq 0 \quad \text{and} \quad \rho c^2 + 3p \geq 0.$$

Does the SEC imply the WEC in Exercise 8.7? Show further that, from the Einstein equations, the SEC implies that $R_{\mu\nu}t^\mu t^\nu \geq 0$, where $R_{\mu\nu}$ is the Ricci tensor.

8.8 The *equation-of-state parameter* w is defined by $w = p/(\rho c^2)$. If one restricts oneself to sources for which $\rho \geq 0$, show that both the weak and strong energy conditions in Exercises 8.6 and 8.7 imply that $w \geq -1$.

8.9 Write down the form of the energy–momentum tensor for a perfect fluid with 4-velocity u^μ with respect to some Cartesian inertial frame S. Show that for the energy–momentum tensor to be invariant under a Lorentz transformation to any other inertial frame one requires $p = -\rho c^2$. Compare this result with that for the energy–momentum tensor of the vacuum.

8.10 Find the most general tensor which can be constructed from the curvature tensor and the metric tensor and which contains terms no higher than quadratic in the second-order derivatives of $g_{\mu\nu}$. Hence write down the most general form of the gravitational field equations in such a theory.

8.11 In the Newtonian limit of weak gravitational fields, for a slowly moving perfect fluid with pressure $p \ll \rho c^2$ show that the 00-component of the Einstein field equations with a non-zero cosmological constant Λ reduces to

$$\nabla^2\Phi = 4\pi G\rho - \Lambda c^2,$$

where $\nabla^2 = \delta^{ij}\partial_i\partial_j$ and ρ is the proper density of the fluid. Hence show that the corresponding Newtonian gravitational potential of a spherically symmetric mass M centred at the origin can be written as

$$\Phi = -\frac{GM}{r} - \frac{\Lambda c^2 r^2}{6}$$

where $r^2 = \delta^{ij}x_i x_j$. Give a physical interpretation of this result.

8.12 In the scalar theory of gravity (8.30) show that, in any inertial frame, the gravitational potential Φ produced by a perfect fluid at some event P satisfies

$$\frac{1}{c^2}\frac{\partial^2\Phi}{\partial t^2} - \vec{\nabla}^2\Phi = -4\pi G\left(\rho - \frac{3p}{c^2}\right),$$

where ρ and p are the density and isotropic pressure as measured in the instantaneous rest frame of the fluid at P. Hence show that the theory reduces to Newtonian gravity in the non-relativistic limit. How might a cosmological constant be included in the theory?

9

The Schwarzschild geometry

We now consider how to solve the Einstein field equations and so discover the metric functions $g_{\mu\nu}$ in any given physical situation. Clearly, the high degree of *non-linearity* in the field equations means that a general solution for an arbitrary matter distribution is analytically intractable. The problem becomes easier if we look for special solutions, for example those representing spacetimes possessing *symmetries*. The first exact solution to Einstein's equations was found by Karl Schwarzschild in 1916.[1] As we shall see, the Schwarzschild solution represents the spacetime geometry outside a spherically symmetric matter distribution.

9.1 The general static isotropic metric

Schwarzschild sought the metric $g_{\mu\nu}$ representing the static spherically symmetric gravitational field in the empty space surrounding some massive spherical object such as a star. Thus, a good starting point for us is to construct the most general form of the metric for a static spatially isotropic spacetime.

A *static* spacetime is one for which some timelike coordinate x^0 (say) with the following properties: (i) all the metric components $g_{\mu\nu}$ are independent of x^0; and (ii) the line element ds^2 is invariant under the transformation $x^0 \to -x^0$. Note that (i) does not necessarily imply (ii), as is made clear by the example of a rotating star: time reversal changes the sense of rotation, but the metric components are constant in time. A spacetime that satisfies (i) but not (ii) is called *stationary*.

Thus, starting from the general expression for the line element

$$ds^2 = g_{\mu\nu}\, dx^\mu\, dx^\nu,$$

we wish to find a set of coordinates x^μ in which the $g_{\mu\nu}$ do not depend on the timelike coordinate x^0 and the line element ds^2 is invariant under $x^0 \to -x^0$, i.e.

[1] Astonishingly, Schwarzschild derived the solution while in the trenches on the Eastern Front during the First World War but sadly he did not survive the conflict.

the metric is *static*, and in which ds^2 depends only on *rotational invariants* of the spacelike coordinates x^i and their differentials, i.e. the metric is *isotropic*.

In fact, it is only slightly more complicated to derive the general form of the spatially isotropic metric *without* insisting that it is static. We therefore begin by constructing this more general metric. Only after its derivation will we impose the additional constraint that the metric is static.

The only rotational invariants of the spacelike coordinates x^i and their differentials are

$$\vec{x}\cdot\vec{x}\equiv r^2, \qquad d\vec{x}\cdot d\vec{x}, \qquad \vec{x}\cdot d\vec{x},$$

where $\vec{x}\equiv(x^1, x^2, x^3)$ and we have defined the coordinate r. Denoting the timelike coordinate x^0 by t, we thus find that the most general form of a spatially isotropic metric must be

$$ds^2 = A(t, r)\,dt^2 - B(t, r)\,dt\,\vec{x}\cdot d\vec{x} - C(t, r)(\vec{x}\cdot d\vec{x})^2 - D(t, r)\,d\vec{x}^2, \qquad (9.1)$$

where A, B, C and D are *arbitrary* functions of the coordinates t and r.

Let us now transform to the (spherical polar) coordinates (t, r, θ, ϕ), defined by

$$x^1 = r\sin\theta\cos\phi, \qquad x^2 = r\sin\theta\sin\phi, \qquad x^3 = r\cos\theta.$$

In this case, we have

$$\vec{x}\cdot\vec{x} = r^2, \qquad \vec{x}\cdot d\vec{x} = r\,dr, \qquad d\vec{x}\cdot d\vec{x} = dr^2 + r^2 d\theta^2 + r^2\sin^2\theta\,d\phi^2,$$

and so the general metric (9.1) now takes the form

$$\begin{aligned} ds^2 = {} & A(t, r)\,dt^2 - B(t, r) r\,dt\,dr - C(t, r) r^2 dr^2 \\ & - D(t, r)\left(dr^2 + r^2 d\theta^2 + r^2\sin^2\theta\,d\phi^2\right). \end{aligned}$$

Collecting together terms and absorbing factors of r into our functions, thereby redefining A, B, C, D, the metric can be written

$$ds^2 = A(t, r)\,dt^2 - B(t, r)\,dt\,dr - C(t, r)\,dr^2 - D(t, r)(d\theta^2 + \sin^2\theta\,d\phi^2).$$

If we now define a new radial coordinate by $\bar{r}^2 = D(t, r)$ and collect together terms into new arbitrary functions of t and $\bar{r}$, thereby again redefining A, B, C, we can write the metric as

$$ds^2 = A(t, \bar{r})\,dt^2 - B(t, \bar{r})\,dt\,d\bar{r} - C(t, \bar{r})\,d\bar{r}^2 - \bar{r}^2(d\theta^2 + \sin^2\theta\,d\phi^2). \qquad (9.2)$$

Let us also introduce a new timelike coordinate $\bar{t}$ defined by the relation

$$d\bar{t} = \Phi(t, \bar{r}) \left[A(t, \bar{r})\, dt - \tfrac{1}{2}B(t, \bar{r})\, d\bar{r}\right],$$

where $\Phi(t, r)$ is an integrating factor that makes the right-hand side an exact differential. Squaring, we obtain

$$d\bar{t}^2 = \Phi^2 \left(A^2\, dt^2 - AB\, dt\, d\bar{r} + \tfrac{1}{4}B^2\, d\bar{r}^2\right),$$

from which we find

$$A\, dt^2 - B\, dt\, d\bar{r} = \frac{1}{A\Phi^2}\, d\bar{t}^2 - \frac{B^2}{4A}\, d\bar{r}^2.$$

Thus defining the new functions $\bar{A} = 1/(A\Phi^2)$ and $\bar{B} = C + B^2/(4A)$, our metric (9.2) becomes *diagonal* and takes the form

$$ds^2 = \bar{A}(\bar{t}, \bar{r})\, d\bar{t}^2 - \bar{B}(\bar{t}, \bar{r})\, d\bar{r}^2 - \bar{r}^2(d\theta^2 + \sin^2\theta\, d\phi^2).$$

There is no need to retain the bars on the variables, so we can write the metric as

$$\boxed{ds^2 = A(t, r)\, dt^2 - B(t, r)\, dr^2 - r^2(d\theta^2 + \sin^2\theta\, d\phi^2).} \tag{9.3}$$

Thus, the general isotropic metric is specified by *two* functions of t and r, namely $A(t, r)$ and $B(t, r)$. We will also see that, for surfaces given by t, r constant, the line element (9.3) describes the geometry of 2-spheres, which expresses the isotropy of the metric. In fact this line element shows that such a surface has a surface area $4\pi r^2$. However, because $B(t, r)$ is not necessarily equal to unity we cannot assume that r is the radial distance.

The final step in obtaining the most general *stationary* isotropic metric is now trivial. We require the metric functions $g_{\mu\nu}$ to be independent of the timelike coordinate, which means simply that A and B must be functions only of r. Thus, we have

$$\boxed{ds^2 = A(r)\, dt^2 - B(r)\, dr^2 - r^2(d\theta^2 + \sin^2\theta\, d\phi^2).} \tag{9.4}$$

Moreover, we see immediately that ds^2 is already invariant under $t \to -t$, and so this is the required form of the metric for a general static spatially isotropic spacetime.

9.2 Solution of the empty-space field equations

The functions $A(r)$ and $B(r)$ in the general static isotropic metric are determined by solving the Einstein field equations. We are interested in the spacetime geometry *outside* a spherical mass distribution, so we must solve the empty-space field equations, which simply require the Ricci tensor to vanish:

$$R_{\mu\nu} = 0. \tag{9.5}$$

From equation (7.13) we can write the Ricci tensor as

$$R_{\mu\nu} = \partial_\nu \Gamma^\sigma{}_{\mu\sigma} - \partial_\sigma \Gamma^\sigma{}_{\mu\nu} + \Gamma^\rho{}_{\mu\sigma}\Gamma^\sigma{}_{\rho\nu} - \Gamma^\rho{}_{\mu\nu}\Gamma^\sigma{}_{\rho\sigma}, \tag{9.6}$$

and, in turn, the connection is defined in terms of the metric $g_{\mu\nu}$ by

$$\Gamma^\sigma{}_{\mu\nu} = \tfrac{1}{2} g^{\sigma\rho}(\partial_\nu g_{\rho\mu} + \partial_\mu g_{\rho\nu} - \partial_\rho g_{\mu\nu}). \tag{9.7}$$

Thus, we see that the deceptively simple expression (9.5) in fact equates to a rather complicated set of differential equations for the components of the metric $g_{\mu\nu}$.

To proceed further, we must calculate the connection coefficients $\Gamma^\sigma{}_{\mu\nu}$ corresponding to our static isotropic metric. This can be done in two ways. The quicker route (with any metric) is to use the Lagrangian procedure for geodesics discussed in Section 3.19. This involves writing down the 'Lagrangian'

$$L = g_{\mu\nu}\dot{x}^\mu \dot{x}^\nu,$$

in which $\dot{x}^\mu$ denotes $dx^\mu/d\sigma$, where σ is some affine parameter along the geodesic. Substituting L into the Euler–Lagrange equations then yields the equations of an affinely parameterised geodesic, from which the connection coefficients can be identified. Since later we will discuss the motion of particles in the Schwarzschild geometry, this procedure would be doubly beneficial.

For illustration, however, we will adopt the more traditional (but slower) method, in which the $\Gamma^\sigma{}_{\mu\nu}$ are calculated directly from the metric $g_{\mu\nu}$ using (9.7). Thus we first need to identify the metric components from the line element (9.4). The non-zero elements of $g_{\mu\nu}$ and $g^{\mu\nu}$ are

$$\begin{aligned} g_{00} &= A(r), & g^{00} &= 1/A(r), \\ g_{11} &= -B(r), & g^{11} &= -1/B(r), \\ g_{22} &= -r^2, & g^{22} &= -1/r^2, \\ g_{33} &= -r^2\sin^2\theta, & g^{33} &= -1/(r^2\sin^2\theta), \end{aligned}$$

where we note that the contravariant components of the metric are simply the reciprocals of the covariant components, since the metric is diagonal.

Substituting the metric components into the expression (9.7) for the connection, we find the expressions given in Table 9.1 (with no sums on Latin indices) with all the other components equalling zero. Thus, summarising these results, we find

Table 9.1 *The connection coefficients of the general static isotropic metric*

$\Gamma^0{}_{00} = 0$		
$\Gamma^i{}_{00} = -\frac{1}{2}g^{i\rho}\partial_\rho g_{00}$	$\Rightarrow$	$\Gamma^1{}_{00} = \frac{1}{2B(r)}\frac{dA(r)}{dr}$
$\Gamma^0{}_{0i} = \frac{1}{2}g^{0\rho}(\partial_i g_{\rho 0} + \partial_0 g_{\rho i} - \partial_\rho g_{0i}) = \frac{1}{2}g^{00}\partial_i g_{00}$	$\Rightarrow$	$\Gamma^0{}_{01} = \frac{1}{2A(r)}\frac{dA(r)}{dr}$
$\Gamma^0{}_{ij} = 0$		
$\Gamma^i{}_{ii} = \frac{1}{2}g^{i\rho}(\partial_i g_{\rho i} + \partial_i g_{\rho i} - \partial_\rho g_{ii}) = \frac{1}{2}g^{ii}\partial_i g_{ii}$	$\Rightarrow$	$\Gamma^1{}_{11} = \frac{1}{2B(r)}\frac{dB(r)}{dr}$
$\Gamma^1{}_{22} = \frac{1}{2}g^{11}(\partial_2 g_{12} + \partial_2 g_{12} - \partial_1 g_{22})$	$\Rightarrow$	$\Gamma^1{}_{22} = -\frac{r}{B(r)}$
$\Gamma^1{}_{33} = -\frac{1}{2}g^{11}\partial_1 g_{33}$	$\Rightarrow$	$\Gamma^1{}_{33} = -\frac{r\sin^2\theta}{B(r)}$
$\Gamma^2{}_{21} = \frac{1}{2}g^{22}\partial_1 g_{22}$	$\Rightarrow$	$\Gamma^2{}_{21} = \frac{1}{r}$
$\Gamma^2{}_{33} = -\frac{1}{2}g^{22}\partial_2 g_{33}$	$\Rightarrow$	$\Gamma^2{}_{33} = -\sin\theta\cos\theta$
$\Gamma^3{}_{31} = \frac{1}{2}g^{33}\partial_1 g_{33}$	$\Rightarrow$	$\Gamma^3{}_{31} = \frac{1}{r}$
$\Gamma^3{}_{32} = \frac{1}{2}g^{33}\partial_2 g_{33}$	$\Rightarrow$	$\Gamma^3{}_{32} = \frac{\cos\theta}{\sin\theta}$

that only nine of the 40 independent connection coefficients are non-zero; they read as follows:

$$\begin{aligned}
&\Gamma^0{}_{01} = A'/(2A), && \Gamma^1{}_{00} = A'/(2B), && \Gamma^1{}_{11} = B'/(2B),\\
&\Gamma^1{}_{22} = -r/B, && \Gamma^1{}_{33} = -(r\sin^2\theta)/B, && \Gamma^2{}_{12} = 1/r,\\
&\Gamma^2{}_{33} = -\sin\theta\cos\theta, && \Gamma^3{}_{13} = 1/r, && \Gamma^3{}_{23} = \cot\theta.
\end{aligned}$$

We now substitute these connection coefficients into the expression (9.6) in order to obtain the components $R_{\mu\nu}$ of the Ricci tensor. This requires quite a lot of tedious (but simple) algebra. Fortunately the off-diagonal components $R_{\mu\nu}$ for $\mu \neq \nu$ are identically zero, and we find that the diagonal components are

$$R_{00} = -\frac{A''}{2B} + \frac{A'}{4B}\left(\frac{A'}{A} + \frac{B'}{B}\right) - \frac{A'}{rB}, \tag{9.8}$$

$$R_{11} = \frac{A''}{2A} - \frac{A'}{4A}\left(\frac{A'}{A} + \frac{B'}{B}\right) - \frac{B'}{rB}, \tag{9.9}$$

$$R_{22} = \frac{1}{B} - 1 + \frac{r}{2B}\left(\frac{A'}{A} - \frac{B'}{B}\right), \tag{9.10}$$

$$R_{33} = R_{22}\sin^2\theta. \tag{9.11}$$

The empty-space field equations (9.5) are thus obtained by setting each of the expressions (9.8–9.11) equal to zero. Of these four equations, only the first three are useful, since the fourth merely repeats the information contained in the third. Adding B/A times (9.8) to (9.9) and rearranging gives

$$A'B + AB' = 0,$$

which implies that $AB =$ constant. Let us denote this constant by α. Substituting $B = \alpha/A$ into (9.10) we obtain $A + rA' = \alpha$, which can be written as

$$\frac{d(rA)}{dr} = \alpha.$$

Integrating this equation gives $rA = \alpha(r + k)$, where k is another integration constant. Thus the functions $A(r)$ and $B(r)$ are given by

$$A(r) = \alpha\left(1 + \frac{k}{r}\right) \quad \text{and} \quad B(r) = \left(1 + \frac{k}{r}\right)^{-1}.$$

In solving for A and B we have used only the sum of equations (9.8) and (9.9), not the separate equations. It is, however, straightforward to check that, with these forms for A and B, the equations (9.8–9.11) are satisfied separately.

It can be seen that the integration constant k must in some way represent the mass of the object producing the gravitational field, as follows. We can identify k (and α) by considering the weak-field limit, in which we require that

$$\frac{A(r)}{c^2} \to 1 + \frac{2\Phi}{c^2},$$

where Φ is the Newtonian gravitational potential. Moreover, in the weak-field limit r can be identified as the radial distance, to a very good approximation. For a spherically symmetric mass M we thus have $\Phi = -GM/r$, and so we conclude that $k = -2GM/c^2$ and $\alpha = c^2$. Therefore, the *Schwarzschild metric* for the empty spacetime outside a spherical body of mass M is[2]

$$ds^2 = c^2\left(1 - \frac{2GM}{c^2 r}\right) dt^2 - \left(1 - \frac{2GM}{c^2 r}\right)^{-1} dr^2 - r^2\, d\theta^2 - r^2 \sin^2\theta\, d\phi^2. \tag{9.12}$$

We will use this metric to investigate the physics in the vicinity of a spherical object of mass M, in particular the trajectories of freely falling massive particles

[2] We note that the constant α could have been identified earlier by making the additional assumption that spacetime is *asymptotically flat*, i.e. that the line element (9.4) tends to the Minkowski line element in the limit $r \to \infty$. Thus we require that, in this limit, $A(r) \to c^2$ and $B(r) \to 1$ and so $AB = c^2$.

and photons. The Schwarzschild metric is valid down to the surface of the spherical object, at which point the empty-space field equations no longer hold. Clearly, the metric functions are infinite at $r = 2\mu$, which is known as the *Schwarzschild radius*. As we shall see, if the surface of the massive body contracts within this radius then the object becomes a *Schwarzschild black hole* (see Chapter 11). For the remainder of this chapter, however, we will restrict our attention to the region $r > 2GM/c^2$. We will often use the shorthand $\mu \equiv GM/c^2$ when writing down this metric.

9.3 Birkhoff's theorem

If we do not demand that our original metric is static (or stationary) but only that it is isotropic, then we would substitute the more general form (9.3),

$$ds^2 = A(t, r)\,dt^2 - B(t, r)\,dr^2 - r^2(d\theta^2 + \sin^2\theta\,d\phi^2),$$

into Einstein's empty-space field equations $R_{\mu\nu} = 0$ in order to determine the functions $A(t, r)$ and $B(t, r)$. On repeating our earlier analysis, we would find some additional non-zero connection coefficients and components of the Ricci tensor. However, on solving this new set of equations, one discovers that the resulting metric must still be the Schwarzschild metric (9.12). Thus, we obtain *Birkhoff's theorem*, which states that *the spacetime geometry outside a general spherically symmetric matter distribution is the Schwarzschild geometry.*

This is an unexpected result because in Newtonian theory spherical symmetry has nothing to do with time dependence. This highlights the special character of the empty-space Einstein equations and of the solutions they admit. In particular, Birkhoff's theorem implies that if a spherically symmetric star undergoes strictly radial pulsations then *it cannot propagate any disturbance into the surrounding space*. Looking ahead to Chapter 18, this means that a radially pulsating star cannot emit gravitational waves.

One can show that the converse of Birkhoff's theorem is *not* true, i.e. a matter distribution that gives rise to the Schwarzschild geometry outside it *need not be spherically symmetric*. Indeed, some specific counter-examples are known.

9.4 Gravitational redshift for a fixed emitter and receiver

We begin our discussion of the physics in the vicinity of a spherical mass M by considering the phenomenon of *gravitational redshift*. In particular, we consider the specific example of an emitter, at *fixed* spatial coordinates (r_E, θ_E, ϕ_E), which emits a photon that is received by an observer at *fixed* spatial coordinates (r_R, θ_R, ϕ_R). If t_E is the coordinate time of emission and t_R the coordinate time

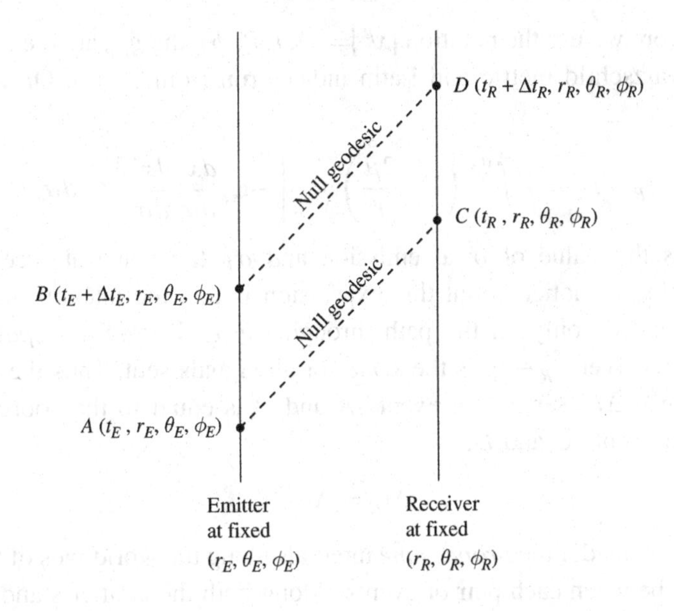

Figure 9.1 Schematic illustration of the emission and reception of two light signals.

of reception then the photon travels from the event $(t_E, r_E, \theta_E, \phi_E)$ to the event $(t_R, r_R, \theta_R, \phi_R)$ along a *null geodesic* in the Schwarzschild spacetime. This is illustrated schematically in Figure 9.1, which also shows a second photon, emitted at a later coordinate time $t_E + \Delta t_E$ and received at $t_R + \Delta t_R$.

In Appendix 9A we present an approach for calculating gravitational redshifts in more general situations. Nevertheless, in this simple case, it is instructive to derive the result in a more elementary manner: we need only use the fact that the photon geodesic is a null curve.[3] Thus $ds^2 = 0$ at all points along it, and from the Schwarzschild metric (9.12), we find that

$$c^2\left(1-\frac{2\mu}{r}\right) dt^2 = \left(1-\frac{2\mu}{r}\right)^{-1} dr^2 + r^2\,d\theta^2 + r^2\sin^2\theta\,d\phi^2,$$

where we have written $\mu \equiv GM/c^2$. Let us consider the first signal. Thus, if σ is some affine parameter along the null geodesic then we have

$$\frac{dt}{d\sigma} = \frac{1}{c}\left(1-\frac{2\mu}{r}\right)^{-1/2}\left[-g_{ij}\frac{dx^i}{d\sigma}\frac{dx^j}{d\sigma}\right]^{1/2},$$

[3] This approach is based on that presented in J. Foster & J. D. Nightingale, *A Short Course in General Relativity*, Springer-Verlag, 1995.

where as before we use the notation $[x^\mu] = (t, r, \theta, \phi)$, the $g_{\mu\nu}$ are the components of the Schwarzschild metric and Latin indices run from 1 to 3. On integrating, we obtain

$$t_R - t_E = \frac{1}{c}\int_{\sigma_E}^{\sigma_R}\left(1 - \frac{2\mu}{r}\right)^{-1/2}\left[-g_{ij}\frac{dx^i}{d\sigma}\frac{dx^j}{d\sigma}\right]^{1/2} d\sigma,$$

where σ_E is the value of σ at emission and σ_R the value at reception. The important thing to notice about this expression is that the integral on the right-hand side depends only on the path through *space*. Thus, for a *spatially fixed* emitter *and* receiver, $t_R - t_E$ is the *same* for all signals sent. Thus the coordinate time difference Δt_E separating events A and B is equal to the coordinate time Δt_R between events C and D,

$$\Delta t_R = \Delta t_E.$$

Now let us consider the *proper time* intervals along the worldlines of the emitter and receiver between each pair of events. Along both the emitter's and receiver's worldlines, $dr = d\theta = d\phi = 0$. Thus, from the Schwarzschild line element (9.12), in both cases

$$c^2\,d\tau^2 \equiv ds^2 = c^2\left(1 - \frac{2\mu}{r}\right) dt^2.$$

Moreover, in each case r is constant along the worldline, so we can immediately integrate this equation to obtain

$$\Delta\tau_E = \left(1 - \frac{2\mu}{r_E}\right)^{1/2}\Delta t_E \quad \text{and} \quad \Delta\tau_R = \left(1 - \frac{2\mu}{r_R}\right)^{1/2}\Delta t_R.$$

Thus, since $\Delta t_R = \Delta t_E$, we find that

$$\frac{\Delta\tau_R}{\Delta\tau_E} = \left(\frac{1 - 2\mu/r_R}{1 - 2\mu/r_E}\right)^{1/2},$$

which forms the basis of the formula for gravitational redshift. If we think of the two light signals as, for example, the two wavecrests of an electromagnetic wave, then it is clear that this ratio must also be the ratio of the *period* of the wave as observed by the receiver and emitter respectively. Thus the frequencies of the photon as measured by each observer are related by

$$\boxed{\frac{\nu_R}{\nu_E} = \left[\frac{1 - 2GM/(r_E c^2)}{1 - 2GM/(r_R c^2)}\right]^{1/2},} \tag{9.13}$$

which shows that $\nu_R < \nu_E$ if $r_R > r_E$. The photon *redshift* z is defined by

$$1 + z = (\nu_R/\nu_E)^{-1}.$$

There is an important point to notice about this derivation. It can be generalised very easily to any spacetime in which we can choose coordinates such that the spacetime is stationary ($\partial_0 g_{\mu\nu} = 0$) and $g_{0i}(\vec{x}) = 0$. In this case,

$$ds^2 = g_{00}(\vec{x})\, dt^2 + g_{ij}(\vec{x})\, dx^i\, dx^j,$$

where, as indicated, all the metric components are *independent* of t. By repeating the above derivation for an emitter and observer at fixed spatial coordinates in this more general spacetime, we easily find that

$$\boxed{\frac{\nu_R}{\nu_E} = \left[\frac{g_{00}(\vec{x}_E)}{g_{00}(\vec{x}_R)}\right]^{1/2}}. \tag{9.14}$$

The derivations presented here depend crucially upon the fact that the emitter and receiver are spatially fixed. However, this is not often physically realistic. For example, we might want to calculate the gravitational redshift of a photon if the emitter or receiver (or both) are in free fall or moving in some arbitrary manner. A method for calculating redshifts in such general situations is given in Appendix 9A. In order to use this formalism, however, we require knowledge of the paths followed by freely falling particles and photons. Therefore, we now consider geodesics in the Schwarzschild geometry.

9.5 Geodesics in the Schwarzschild geometry

In deriving the Schwarzschild line element,

$$ds^2 = c^2\left(1 - \frac{2\mu}{r}\right) dt^2 - \left(1 - \frac{2\mu}{r}\right)^{-1} dr^2 - r^2\, d\theta^2 - r^2 \sin^2\theta\, d\phi^2, \tag{9.15}$$

we also calculated the connection coefficients $\Gamma^\mu{}_{\nu\rho}$ for this metric. Thus we could now write down the geodesic equations for the Schwarzschild geometry in the form

$$\frac{d^2 x^\mu}{d\sigma^2} + \Gamma^\mu{}_{\nu\rho} \frac{dx^\nu}{d\sigma}\frac{dx^\rho}{d\sigma} = 0,$$

where σ is some affine parameter along the geodesic $x^\mu(\sigma)$. It is more instructive, however, to obtain the geodesic equations using the very neat 'Lagrangian' procedure discussed in Chapter 3.

Thus, let us consider the 'Lagrangian' $L = g_{\mu\nu}\dot{x}^\mu \dot{x}^\nu$, where $\dot{x}^\mu \equiv dx^\mu/d\sigma$. Using (9.15), L is given by

$$L = c^2\left(1 - \frac{2\mu}{r}\right)\dot{t}^2 - \left(1 - \frac{2\mu}{r}\right)^{-1}\dot{r}^2 - r^2\left(\dot{\theta}^2 + \sin^2\theta\, \dot{\phi}^2\right). \tag{9.16}$$

The geodesic equations are then obtained by substituting this form for L into the the Euler–Lagrange equations

$$\frac{d}{d\sigma}\left(\frac{\partial L}{\partial \dot{x}^{\mu}}\right) - \frac{\partial L}{\partial x^{\mu}} = 0.$$

Performing this calculation, we find that the four resulting geodesic equations (for $\mu = 0, 1, 2, 3$) are given by

$$\left(1 - \frac{2\mu}{r}\right)\dot{t} = k, \quad (9.17)$$

$$\left(1 - \frac{2\mu}{r}\right)^{-1}\ddot{r} + \frac{\mu c^2}{r^2}\dot{t}^2 - \left(1 - \frac{2\mu}{r}\right)^{-2}\frac{\mu}{r^2}\dot{r}^2 - r\left(\dot{\theta}^2 + \sin^2\theta\,\dot{\phi}^2\right) = 0, \quad (9.18)$$

$$\ddot{\theta} + \frac{2}{r}\dot{r}\dot{\theta} - \sin\theta\cos\theta\,\dot{\phi}^2 = 0, \quad (9.19)$$

$$r^2\sin^2\theta\,\dot{\phi} = h. \quad (9.20)$$

In (9.17) and (9.20) respectively, the quantities k and h are *constants*. These two equations are derived immediately since L is not an explicit function of t or ϕ.

We see immediately that $\theta = \pi/2$ satisfies the third geodesic equation (9.19). Because of the spherical symmetry of the Schwarzschild metric we can therefore, with no loss of generality, confine our attention to particles moving in the 'equatorial plane' given by $\theta = \pi/2$. In this case our set of geodesic equations reduces to

$$\left(1 - \frac{2\mu}{r}\right)\dot{t} = k, \quad (9.21)$$

$$\left(1 - \frac{2\mu}{r}\right)^{-1}\ddot{r} + \frac{\mu c^2}{r^2}\dot{t}^2 - \left(1 - \frac{2\mu}{r}\right)^{-2}\frac{\mu}{r^2}\dot{r}^2 - r\dot{\phi}^2 = 0, \quad (9.22)$$

$$r^2\dot{\phi} = h. \quad (9.23)$$

These equations are valid for both null and non-null affinely parameterised geodesics. In each of these cases, however, it is easier to replace the rather complicated r-equation (9.22) by a first integral of the geodesics equations. For a *non-null* geodesic this first integral is simply

$$g_{\mu\nu}\dot{x}^{\mu}\dot{x}^{\nu} = c^2, \quad (9.24)$$

whereas for a *null* geodesic it is

$$g_{\mu\nu}\dot{x}^{\mu}\dot{x}^{\nu} = 0. \quad (9.25)$$

Before going on to discuss separately non-null and null geodesics in the equatorial plane $\theta = \pi/2$, it is instructive to discuss the physical interpretation of the constants k and h. One can arrive at equations (9.21) and (9.23) simply by using the fact that the components p_0 and p_3 of a particle's 4-momentum are conserved along geodesics since L does not depend explicitly on t and ϕ (remember that $\boldsymbol{p}$ is proportional to the tangent vector to the geodesic at each point). For notational simplicity, for a massive particle, we shall take the particle to have unit rest mass and choose the affine parameter to be the particle's proper time τ, so that $p^\mu = \dot{x}^\mu$. Similarly, for a massless particle we are free to choose an appropriate affine parameter along the null geodesic, once again such that $p^\mu = \dot{x}^\mu$. Thus, for $\theta = \pi/2$ we may write

$$p_0 = g_{00}\dot{t} = c^2\left(1 - \frac{2\mu}{r}\right)\dot{t} = kc^2, \tag{9.26}$$

$$p_3 = g_{33}\dot{\phi} = -r^2\dot{\phi} = -h, \tag{9.27}$$

where in the last equality on each line we have defined the constants in a manner that coincides with (9.21) and (9.23). Let us first consider the constant k. If, at some event, an observer with 4-velocity $\boldsymbol{u}$ encountered a particle with 4-momentum $\boldsymbol{p}$ then he would measure the particle's energy to be

$$E = \boldsymbol{p}\cdot\boldsymbol{u} = p_\mu u^\mu.$$

For an observer at rest at infinity we have $[u^\mu] = (1, 0, 0, 0)$ and so $E = p_0 = kc^2$ (which is conserved along the particle geodesic). Thus we may take $k = E/c^2$, where E is the total energy of the particle in its orbit. Since for massive particles we have assumed unit mass, in the general case we have $k = E/(m_0c^2)$, where m_0 is the rest mass of the particle. For the constant h, we can see immediately from (9.27) that it equals the specific angular momentum of the particle and that (as result of the choice of signature for the metric) p_3 is equal to *minus* the specific angular momentum. Finally, we note that the results (9.17–9.20) can also be derived using the alternative form (3.56) of the geodesic equations, which may be written

$$\dot{p}_\mu = \tfrac{1}{2}(\partial_\mu g_{\nu\sigma})p^\nu p^\sigma. \tag{9.28}$$

9.6 Trajectories of massive particles

The trajectory of a massive particle is a timelike geodesic. Considering motion in the equatorial plane, we replace the geodesic equation (9.22) by (9.24), where $g_{\mu\nu}$ is taken from (9.15) with $\theta = \pi/2$. Moreover, since we are considering a timelike geodesic we can choose our affine parameter σ to be the *proper time* τ along the path. Thus we find that the worldline $x^\mu(\tau)$ of a massive particle moving in the equatorial plane of the Schwarzschild geometry must satisfy the equations

$$\left(1-\frac{2\mu}{r}\right)\dot{t} = k, \tag{9.29}$$

$$c^2\left(1-\frac{2\mu}{r}\right)\dot{t}^2 - \left(1-\frac{2\mu}{r}\right)^{-1}\dot{r}^2 - r^2\dot{\phi}^2 = c^2, \tag{9.30}$$

$$r^2\dot{\phi} = h. \tag{9.31}$$

By substituting (9.29) and (9.31) into (9.30), we obtain the combined 'energy' equation for the r-coordinate,

$$\boxed{\dot{r}^2 + \frac{h^2}{r^2}\left(1-\frac{2GM}{c^2 r}\right) - \frac{2GM}{r} = c^2(k^2-1),} \tag{9.32}$$

where we have written $\mu = GM/c^2$. We shall use this 'energy' equation to discuss radial free fall and the stability of orbits. Note that the right-hand side is a *constant* of the motion. We can verify the physical meaning of the constant k by noting that $E \propto k$. The constant of proportionality is fixed by requiring that, for a particle at rest at $r = \infty$, we have $E = m_0c^2$. Letting $r \to \infty$ and $\dot{r} = 0$ in (9.32), we thus require $k^2 = 1$. Hence, as previously, we must have $k = E/(m_0c^2)$, where E is the total energy of the particle in its orbit.

A second useful equation, which enables us to determine the *shape* of a particle orbit (i.e. r as a function of ϕ), may be found by using $h = r^2\dot{\phi}$ to express $\dot{r}$ in the energy equation (9.32) as

$$\frac{dr}{d\tau} = \frac{dr}{d\phi}\frac{d\phi}{d\tau} = \frac{h}{r^2}\frac{dr}{d\phi}.$$

We thus obtain

$$\left(\frac{h}{r^2}\frac{dr}{d\phi}\right)^2 + \frac{h^2}{r^2} = c^2(k^2-1) + \frac{2GM}{r} + \frac{2GMh^2}{c^2r^3}.$$

If we make the substitution $u \equiv 1/r$ that is usually employed in Newtonian orbit calculations, we find that

$$\left(\frac{du}{d\phi}\right)^2 + u^2 = \frac{c^2}{h^2}(k^2-1) + \frac{2GMu}{h^2} + \frac{2GMu^3}{c^2}.$$

We now differentiate this equation with respect to ϕ to obtain finally

$$\boxed{\frac{d^2u}{d\phi^2} + u = \frac{GM}{h^2} + \frac{3GM}{c^2}u^2.} \tag{9.33}$$

In Newtonian gravity, the equations of motion of a particle of mass m in the equatorial plane $\theta = \pi/2$ may be determined from the Lagrangian

$$L = \tfrac{1}{2}m(\dot{r}^2 + r^2\dot{\phi}^2) + \frac{GMm}{r}.$$

From the Euler–Lagrange equations we have

$$r^2\dot{\phi} = h,$$

$$\ddot{r} = \frac{h}{r^3} - \frac{GM}{r^2},$$

where the integration constant h is the specific angular momentum of the particle. If we now substitute $u = 1/r$ and eliminate the time variable, the Newtonian equation of motion for planetary orbits is obtained:

$$\frac{d^2u}{d\phi^2} + u = \frac{GM}{h^2}. \tag{9.34}$$

We must remember, however, that in this equation $u = 1/r$, where r is the radial *distance* from the mass, whereas in (9.33) r is a radial *coordinate* that is related to distance through the metric. Nevertheless, the forms of the two equations are very similar except for the extra term $3GMu^2/c^2$ in (9.33). We note that this term correctly tends to zero as $c \to \infty$.

Two interesting special cases of massive-particle orbits are worth investigating in detail, namely *radial motion* and *motion in a circle*.

9.7 Radial motion of massive particles

For radial motion ϕ is constant, which implies that $h = 0$. Thus, (9.32) reduces to

$$\dot{r}^2 = c^2(k^2 - 1) + \frac{2GM}{r}. \tag{9.35}$$

Differentiating this equation with respect to τ and dividing through by $\dot{r}$ gives

$$\ddot{r} = -\frac{GM}{r^2}, \tag{9.36}$$

which has precisely the same form as the corresponding equation of motion in Newtonian gravity. This does not imply, however, that general relativity and Newtonian gravity predict the same physical behaviour. It should be remembered that in (9.36) the coordinate r is not the radial distance, and dots indicate derivatives with respect to proper time rather than coordinate time.

As a specific example, consider a particle dropped from rest at $r = R$. From (9.35) we see immediately that $k^2 = 1 - 2GM/(c^2 R)$, so (9.35) can be written

$$\frac{\dot{r}^2}{2} = GM\left(\frac{1}{r} - \frac{1}{R}\right). \tag{9.37}$$

This has the same form as the Newtonian formula equating the gain in kinetic energy to the loss in gravitational potential energy for a particle (of unit mass) falling from rest at $r = R$. This provides a useful way to remember this equation, but the different meanings of r and of the dot should again be borne in mind.

We could continue our analysis of this quite general situation, but we can illustrate the main physical points by considering a particle dropped from rest at infinity. In this case $k = 1$ and the algebra is much less complicated. Thus, setting $k = 1$ in the geodesic equation (9.29) and in (9.35), we obtain

$$\frac{dt}{d\tau} = \left(1 - \frac{2\mu}{r}\right)^{-1}, \tag{9.38}$$

$$\frac{dr}{d\tau} = -\left(\frac{2\mu c^2}{r}\right)^{1/2}, \tag{9.39}$$

where in (9.39) we have taken the negative square root. These equations form the basis of our discussion of a radially *infalling* particle dropped from rest at infinity. From these equations we see immediately that the components of the 4-velocity of the particle in the (t, r, θ, ϕ) coordinate system are simply

$$[u^\mu] \equiv \left[\frac{dx^\mu}{d\tau}\right] = \left(\left(1 - \frac{2\mu}{r}\right)^{-1}, \quad -\left(\frac{2\mu c^2}{r}\right)^{1/2}, 0, 0\right).$$

Equation (9.39) determines the trajectory $r(\tau)$. On integrating (9.39) we immediately obtain

$$\tau = \frac{2}{3}\sqrt{\frac{r_0^3}{2\mu c^2}} - \frac{2}{3}\sqrt{\frac{r^3}{2\mu c^2}},$$

where we have written the integration constant in a form such that $\tau = 0$ at $r = r_0$. Thus τ is the proper time experienced by the particle in falling from $r = r_0$ to a coordinate radius r.

Instead of parameterising the worldline in terms of the proper time τ, we can alternatively describe the path as $r(t)$, thereby mapping out the trajectory of the particle in the (t, r) coordinate plane. This is easily achieved by writing

$$\frac{dr}{dt} = \frac{dr}{d\tau}\frac{d\tau}{dt} = -\left(\frac{2\mu c^2}{r}\right)^{1/2}\left(1 - \frac{2\mu}{r}\right). \tag{9.40}$$

On integrating, we find

$$t = \frac{2}{3}\left(\sqrt{\frac{r_0^3}{2\mu c^2}} - \sqrt{\frac{r^3}{2\mu c^2}}\right) + \frac{4\mu}{c}\left(\sqrt{\frac{r_0}{2\mu}} - \sqrt{\frac{r}{2\mu}}\right)$$
$$+ \frac{2\mu}{c}\ln\left|\left(\frac{\sqrt{r/(2\mu)}+1}{\sqrt{r/(2\mu)}-1}\right)\left(\frac{\sqrt{r_0/(2\mu)}-1}{\sqrt{r_0/(2\mu)}+1}\right)\right|,$$

where the choice of the integration constant gives $t = 0$ at $r = r_0$.

In particular we note that

$$\tau \to \frac{2}{3}\sqrt{\frac{r_0^3}{2\mu c^2}} \qquad \text{as } r \to 0,$$

$$t \to \infty \qquad \text{as } r \to 2\mu.$$

Evidently, the particle takes a *finite* proper time to reach $r = 0$. When the worldline is expressed in the form $r(t)$, however, we see that r asymptotically approaches 2μ as $t \to \infty$. Since the coordinate time t corresponds to the proper time experienced by a stationary observer at large radius, we must therefore conclude that, to such an observer, it takes an *infinite* time for the particle to reach $r = 2\mu$. We return to this point later.

It is interesting to ask what velocity a stationary observer at r measures for the infalling particle as it passes. From the Schwarzschild metric (9.15) we see that, for a stationary observer at coordinate radius r, a coordinate time interval dt corresponds to a proper time interval

$$dt' = \left(1 - \frac{2\mu}{r}\right)^{1/2} dt.$$

Similarly, a radial coordinate separation dr corresponds to a proper radial distance measured by the observer equal to

$$dr' = \left(1 - \frac{2\mu}{r}\right)^{-1/2} dr.$$

Thus the velocity of the radially infalling particle, as measured by a stationary observer at r, is given by

$$\frac{dr'}{dt'} = \left(1 - \frac{2\mu}{r}\right)^{-1} \frac{dr}{dt} = -\left(\frac{2\mu c^2}{r}\right)^{1/2}. \tag{9.41}$$

Thus we find the rather surprising result that, as the particle approaches $r = 2\mu$, a stationary observer at that radius observes that the particle's velocity tends to c. We note that the equation (9.41) is only physically valid for $r > 2\mu$ since, as we shall see, it is impossible to have a stationary observer at $r \leq 2\mu$.

9.8 Circular motion of massive particles

For circular motion in the equatorial plane we have $r = \text{constant}$, and so $\dot{r} = \ddot{r} = 0$. Setting $u = 1/r = \text{constant}$ in the 'shape' equation (9.33) we have

$$u = \frac{GM}{h^2} + \frac{3GM}{c^2}u^2,$$

from which we find that

$$h^2 = \frac{\mu c^2 r^2}{r - 3\mu}.$$

Putting $\dot{r} = 0$ in the energy equation (9.32) and substituting the above expression for h^2 allows us to identify the constant k:

$$k = \frac{1 - 2\mu/r}{(1 - 3\mu/r)^{1/2}}. \tag{9.42}$$

The energy of a particle of rest mass m_0 in a circular of radius r is then given by $E = km_0c^2$. We can use this result to determine which circular orbits are *bound*. For this we require $E < m_0c^2$, so the limits on r for the orbit to be bound are given by $k = 1$. This yields

$$(1 - 2\mu/r)^2 = 1 - 3\mu/r,$$

which is satisfied when $r = 4\mu$ or $r = \infty$. Thus, over the range $4\mu < r < \infty$, circular orbits are bound. A plot of $E/(m_0c^2)$ as a function of r/μ is shown in Figure 9.2.

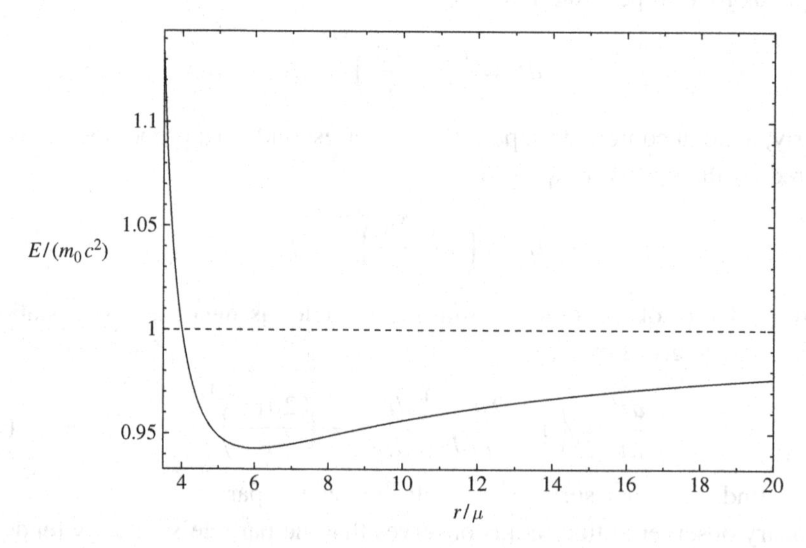

Figure 9.2 The variation of $k = E/(m_0c^2)$ as a function of r/μ for a circular orbit of a massive particle in the Schwarzschild geometry.

We can obtain another useful result by substituting our expression for h^2 into the geodesic equation $r^2\dot{\phi} = h$; then we can write

$$\left(\frac{d\phi}{d\tau}\right)^2 = \frac{\mu c^2}{r^2(r-3\mu)}.$$

The significance of this equation is that it *cannot* be satisfied for circular orbits with $r < 3\mu$. Such orbits cannot be geodesics (since they do not satisfy the geodesic equations) and so cannot be followed by freely falling particles. Thus, according to general relativity a free massive particle cannot maintain a circular orbit with $r < 3\mu$ around a spherical massive body, no matter how large the angular momentum of the particle. This is very different from Newtonian theory.

It is also useful to calculate the expression for $d\phi/dt$, which is given by

$$\left(\frac{d\phi}{dt}\right)^2 = \left(\frac{d\phi}{d\tau}\frac{d\tau}{dt}\right)^2 = \frac{(1-2\mu/r)^2}{k^2}\left(\frac{d\phi}{d\tau}\right)^2 = \frac{\mu c^2}{r^3} = \frac{GM}{r^3}.$$

This expression is exactly the same as the Newtonian expression for the period of a circular orbit of radius r. Although we cannot say that r is the radius of the orbit in the relativistic case, we see that the spatial distance travelled in one complete revolution is $2\pi r$, just as in the Newtonian case.

9.9 Stability of massive particle orbits

The above analysis appears to suggest that the closest bound circular orbit around a massive spherical body is at $r = 4\mu$. However, we have not yet determined whether this orbit is stable.

In Newtonian dynamics the equation of motion of a particle in a central potential can be written

$$\frac{1}{2}\left(\frac{dr}{dt}\right)^2 + V_{\text{eff}}(r) = E,$$

where $V_{\text{eff}}(r)$ is the effective potential and E is the total energy of the particle per unit mass. For an orbit around a spherical mass M, the effective potential is

$$V_{\text{eff}}(r) = -\frac{GM}{r} + \frac{h^2}{2r^2}, \tag{9.43}$$

where h is the specific angular momentum of the particle. This effective potential is shown in Figure 9.3. It can be seen that bound orbits have two turning points and that a circular orbit corresponds to the special case where the particle sits at the minimum of the effective potential. Furthermore one sees that, in Newtonian dynamics, a finite angular momentum provides an *angular momentum barrier* preventing a particle reaching $r = 0$. This is *not* true in general relativity.

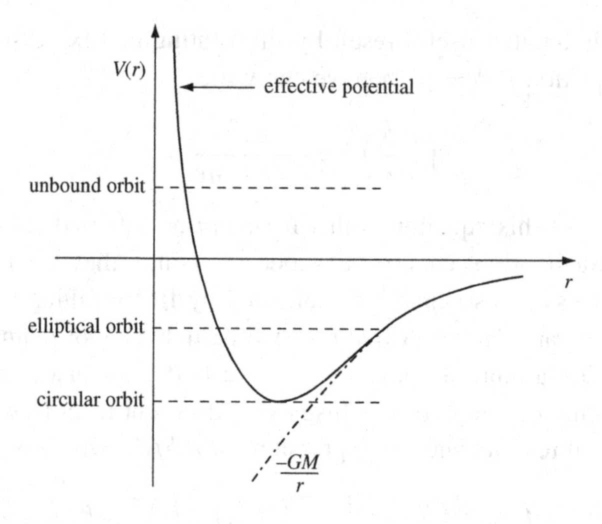

Figure 9.3 The Newtonian effective potential for $h \neq 0$, showing how an angular momentum barrier prevents particles reaching $r = 0$.

In general relativity, the 'energy' equation (9.32) for the motion of a particle around a central mass can be written

$$\frac{1}{2}\left(\frac{dr}{d\tau}\right)^2 + \frac{h^2}{2r^2}\left(1 - \frac{2\mu}{r}\right) - \frac{\mu c^2}{r} = \frac{c^2}{2}(k^2 - 1),$$

where we recall that the constant $k = E/(m_0 c^2)$. Thus in general relativity we identify the effective potential per unit mass as

$$V_{\text{eff}}(r) = -\frac{\mu c^2}{r} + \frac{h^2}{2r^2} - \frac{\mu h^2}{r^3}, \tag{9.44}$$

which has an additional term proportional to $1/r^3$ as compared with the Newtonian case (9.43). Remembering that $\mu = GM/c^2$, we see that (9.44) reduces to the form (9.43) in the non-relativistic limit $c \to \infty$.

Figure 9.4 shows the general relativistic effective potential for several values of $\bar{h} \equiv h/(c\mu)$. The dots indicate the locations of *stable* circular orbits, which occur at the local minimum of the potential. The local maxima in the potential curves are the locations of *unstable* circular orbits. For any given value of $\bar{h}$, circular orbits occur where $dV_{\text{eff}}/dr = 0$. Differentiating (9.44) gives

$$\frac{dV_{\text{eff}}}{dr} = \frac{\mu c^2}{r^2} - \frac{h^2}{r^3} + \frac{3\mu h^2}{r^4},$$

and so the extrema of the effective potential are located at the solutions of the quadratic equation

$$\mu c^2 r^2 - h^2 r + 3\mu h^2 = 0,$$

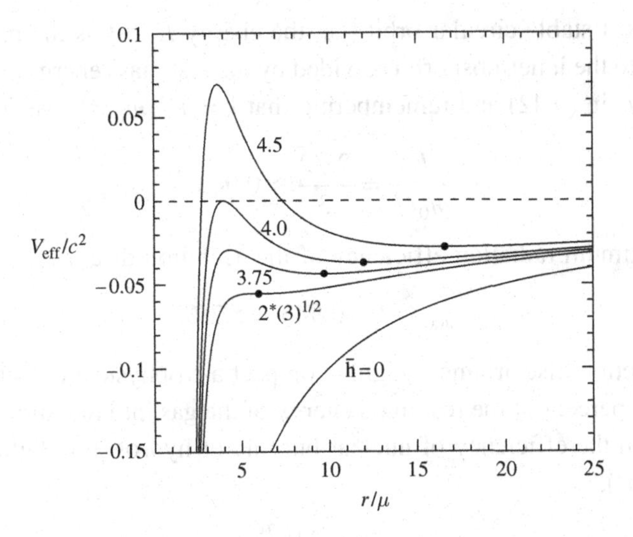

Figure 9.4 The general relativistic effective potential plotted for several values of the angular momentum parameter $\bar{h}$.

which occur at

$$r = \frac{h}{2\mu c^2}\left(h \pm \sqrt{h^2 - 12\mu^2 c^2}\right).$$

We note, in particular, that if $h = \sqrt{12}\mu c = 2\sqrt{3}\mu c$ then *there is only one extremum*, and there are no turning points in the orbit for lower values of h. The significance of this result is that the innermost stable circular orbit has

$$r_{\rm min} = 6\mu = \frac{6GM}{c^2}.$$

This orbit, with $r = 6\mu$ and $h/(\mu c) = 2\sqrt{3}$, is unique in satisfying both $dV_{\rm eff}/dr = 0$ and $d^2V_{\rm eff}/dr^2 = 0$, the latter being the condition for marginal stability of the orbit.

The existence of an innermost stable orbit has some interesting astrophysical consequences. Gas in an accretion disc around a massive compact central body settles into circular orbits around the compact object. However, the gas slowly loses angular momentum because of turbulent viscosity (the turbulence is thought to be generated by magnetohydrodynamic instabilities). As the gas loses angular momentum it moves slowly inwards, losing gravitational potential energy and heating up. Eventually it has lost enough angular momentum that it can no longer follow a stable circular orbit, and so it spirals rapidly inwards onto the central object.

We can make an estimate of the efficiency of energy radiation in an accretion disc. The maximum efficiency is of the order of the 'gravitational binding energy' at the innermost stable circular orbit (i.e. the energy E lost as the particle moves from infinity to the innermost orbit) divided by the rest mass energy of the particle. Setting $r = 6\mu$ in (9.42) and remembering that $k = E/(m_0c^2)$, we find that

$$\frac{E}{m_0c^2} = \frac{2\sqrt{2}}{3} \approx 0.943.$$

Thus the maximum radiation efficiency of the accretion disc is

$$\epsilon_{\text{acc}} \approx 1 - 0.943 = 5.7\%.$$

Thus, an accretion disc around a highly compact astrophysical object can convert perhaps a few percent of the rest mass energy of the gas into radiation; this may be compared with the efficiency of nuclear burning of hydrogen to helium (26 MeV per He nucleus),

$$\epsilon_{\text{nuclear}} \sim 0.7\%.$$

Accretion discs are therefore capable of converting rest mass energy into radiation with an efficiency that is about 10 times greater than the efficiency of the nuclear burning of hydrogen. The 'accretion power' of highly compact objects (such as black holes) cause some of the most energetic phenomena known in the universe.

A physically intuitive picture of a non-circular orbit and the capture of a particle with non-zero angular momentum h may be obtained by differentiating the energy

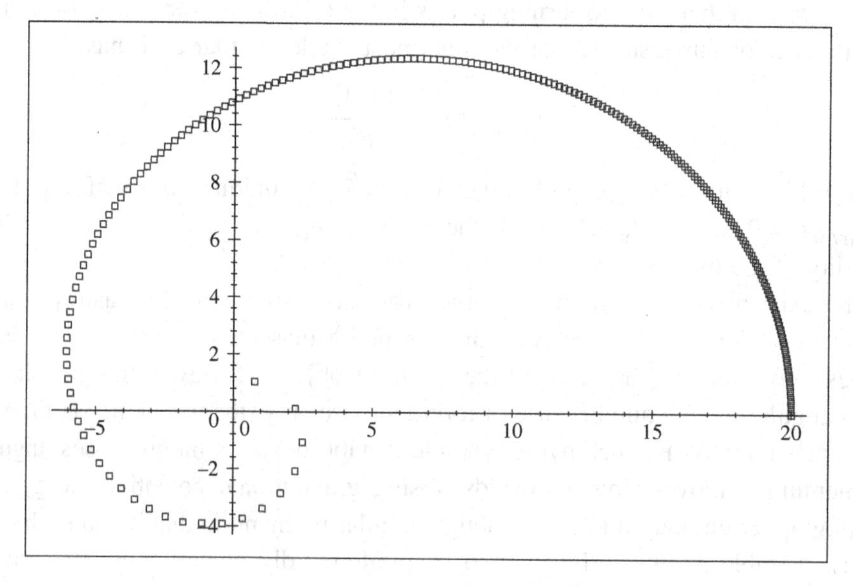

Figure 9.5 Orbit for a particle projected azimuthally from $r = 20GM/c^2$ with $h = 3.5GM/c$. A circular orbit would require $h = (20/\sqrt{17})GM/c$. The points are plotted at equal intervals of the particle's proper time.

equation (9.32) for massive particle orbits with respect to proper time τ. Using the original equation to remove the first derivative $dr/d\tau$, we find that

$$\frac{d^2r}{d\tau^2} = -\frac{GM}{r^2} + \frac{h^2}{r^3} - \frac{3h^2GM}{c^2r^4}.$$

As we might expect, the first two terms on the right-hand side are very like the Newtonian expressions corresponding to an inward gravitational force and a repulsive 'centrifugal force' proportional to h^2. The third term is new, however, and is also proportional to h^2 but this time acts *inwards*. This shows that close to a highly compact object, specifically within the radius $r = 3GM/c^2$, the centrifugal force 'changes sign' and is directed inwards, thus hastening the demise of any particle that strays too close to the object. This leads to spiral orbits of the type shown in Figure 9.5.

9.10 Trajectories of photons

The trajectory of a photon (and of any other particle having zero rest mass) is a null geodesic. We cannot use the proper time τ as a parameter, so instead we use some affine parameter σ along the geodesic. Considering motion in the equatorial plane, the equations of motion are given by the geodesic equations (9.21) and (9.23), and we replace the r-equation (9.22) by the condition $g_{\mu\nu}\dot{x}^\mu\dot{x}^\nu = 0$. Thus we have

$$\left(1 - \frac{2\mu}{r}\right)\dot{t} = k, \tag{9.45}$$

$$c^2\left(1 - \frac{2\mu}{r}\right)\dot{t}^2 - \left(1 - \frac{2\mu}{r}\right)^{-1}\dot{r}^2 - r^2\dot{\phi}^2 = 0, \tag{9.46}$$

$$r^2\dot{\phi} = h. \tag{9.47}$$

For photon trajectories, an analogue of the energy equation (9.32) can again be obtained by substituting (9.45) and (9.47) into (9.46), which gives

$$\boxed{\dot{r}^2 + \frac{h^2}{r^2}\left(1 - \frac{2\mu}{r}\right) = c^2k^2.} \tag{9.48}$$

Similarly, the analogue for photons of the shape equation (9.33) is obtained by substituting $h = r^2\dot{\phi}$ into (9.46) and using the fact that

$$\frac{dr}{d\sigma} = \frac{dr}{d\phi}\frac{d\phi}{d\sigma} = \frac{h}{r^2}\frac{dr}{d\phi}.$$

Making the usual substitution $u \equiv 1/r$ and differentiating with respect to ϕ we find

$$\boxed{\frac{d^2u}{d\phi^2} + u = \frac{3GM}{c^2}u^2.} \tag{9.49}$$

It is again worth mentioning the two special cases of radial motion and motion in a circle.

9.11 Radial motion of photons

For radial motion $\dot{\phi} = 0$ and (9.46) reduces to

$$c^2\left(1 - \frac{2\mu}{r}\right)\dot{t}^2 - \left(1 - \frac{2\mu}{r}\right)^{-1}\dot{r}^2 = 0,$$

from which we obtain

$$\boxed{\frac{dr}{dt} = \pm c\left(1 - \frac{2\mu}{r}\right).} \tag{9.50}$$

On integrating, we have

$$ct = r + 2\mu \ln\left|\frac{r}{2\mu} - 1\right| + \text{constant} \qquad \text{(outgoing photon)},$$

$$ct = -r - 2\mu \ln\left|\frac{r}{2\mu} - 1\right| + \text{constant} \qquad \text{(incoming photon)}.$$

Notice that under the transformation $t \to -t$, incoming and outgoing photon paths are interchanged, as we would expect. In fact for the moment the differential equation (9.50) is more useful. In a (ct, r)-diagram, we see that the photon worldlines will have slopes ± 1 as $r \to \infty$ (forming the standard special-relativistic lightcone), but their slopes approach $\pm\infty$ as $r \to 2\mu$. This means that they become more vertical; the cone 'closes up'.

Our knowledge of the lightcone structure allows us to construct the 'picture' behind our earlier algebraic result that a particle takes infinite coordinate time to reach the horizon; this is illustrated in Figure 9.6. The curved solid line is the worldline of a massive particle dropped from rest by an observer fixed at $r = R$. Since massive particle worldlines are confined within the forward lightcone in any event, the closing up of the lightcones forces the worldlines of massive particles to become more vertical as $r \to 2\mu$. Thus, the particle 'reaches' $r = 2\mu$ only at $t = \infty$. Further, suppose that at some point along its trajectory the particle emits a radially outgoing photon in the direction of the observer. The tangent to

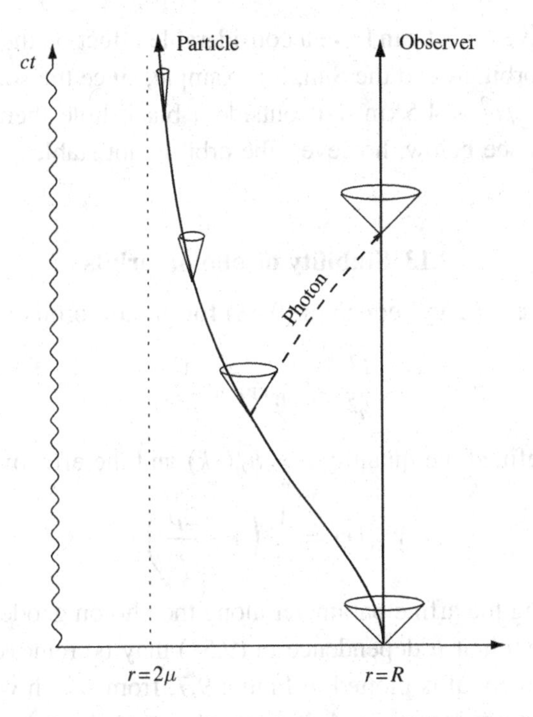

Figure 9.6 A radially infalling particle emitting a radially outgoing photon. The wavy line indicates the singularity at $r = 0$.

the resulting photon worldline must, at any event, lie along the outward-pointing forward lightcone at that point. This is illustrated by the broken line in Figure 9.6. Thus, in the limit where the particle approaches $r = 2\mu$, the initial direction of the photon worldline approaches the vertical and so the photon will be received by the observer only at $t = \infty$. Thus to an external observer the particle appears to take an infinite time to reach the horizon.

As discussed earlier, however, the *proper time* τ experienced by a massive particle in falling to $r = 2\mu$ is *finite*. Moreover, $dr/d\tau$ does not tend to zero at this point, so the particle has not 'run out of steam' and presumably passes beyond this threshold. Thus, our present coordinate system is inadequate for discussing what happens at and within $r = 2\mu$, and our (ct, r)-diagram is in some respects misleading in these regions. We discuss this further in Chapter 11.

9.12 Circular motion of photons

For motion in a circle we have $r =$ constant. Thus, from the shape equation (9.49), we see that the only possible radius for a circular photon orbit is

$$r = \frac{3GM}{c^2}.$$

Therefore a massive object can have a considerable effect on the path of a photon. There is no such orbit around the Sun, for example, since the solar radius is much larger than $3GM_\odot/c^2 \approx 4.5\,\text{km}$, but outside a black hole there can be such an orbit. As we shall see below, however, the orbit is not stable.

9.13 Stability of photon orbits

We can rewrite the 'energy' equation (9.48) for photon orbits as

$$\frac{\dot{r}^2}{h^2} + V_{\text{eff}}(r) = \frac{1}{b^2}, \tag{9.51}$$

where we have defined the quantity $b = h/(ck)$ and the effective potental

$$V_{\text{eff}}(r) = \frac{1}{r^2}\left(1 - \frac{2\mu}{r}\right).$$

In fact, by rescaling the affine parameter along the photon geodesic in such a way that $\lambda \to h\lambda$ the explicit h-dependence in (9.51) may be removed.

The effective potential is plotted in Figure 9.7, from which we see that $V_{\text{eff}}(r)$ has a single maximum at $r = 3\mu$, where the value of the potential is $1/(27\mu^2)$. Thus the circular orbit at $r = 3\mu$ is unstable. We conclude that there are no stable circular photon orbits in the Schwarzschild geometry.

The character of general photon orbits is determined by the value of the constant b. To find the physical meaning of b, we begin by using the geodesic equation (9.47) and the energy equation (9.51) to write

$$\frac{d\phi}{dr} = \frac{\dot{\phi}}{\dot{r}} = \frac{1}{r^2}\left[\frac{1}{b^2} - \frac{1}{r^2}\left(1 - \frac{2\mu}{r}\right)\right]^{-1/2}.$$

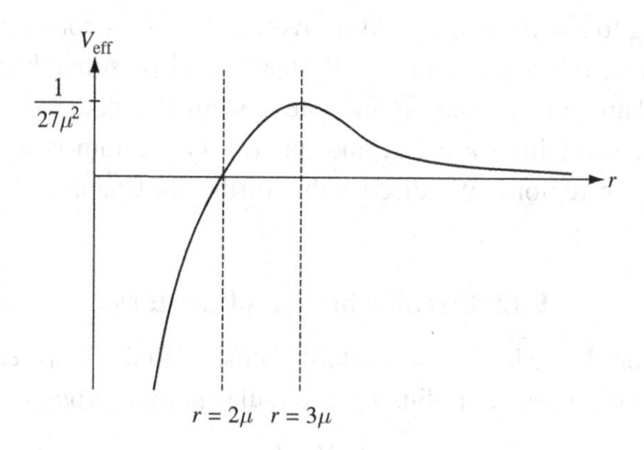

Figure 9.7 The effective potential for photon orbits.

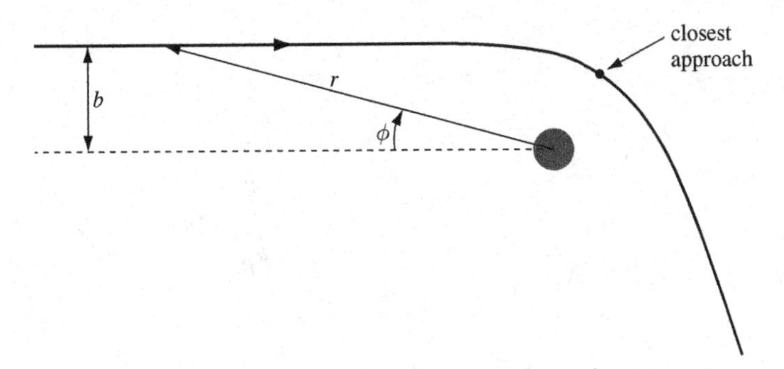

Figure 9.8 The shape of a photon orbit passing a spherical mass if $b > 3\sqrt{3}\mu$.

Thus, for a photon orbit, as $r \to \infty$ we have

$$r^2 \frac{d\phi}{dr} = \pm b.$$

Assuming that $\phi \to 0$ as $r \to \infty$, the solution to this equation is

$$r = \mp \frac{b}{\sin \phi},$$

which gives the equations of two straight lines with *impact parameter b* passing on either side of the origin.

The nature of the orbits depends very much on the value of the impact parameter b. Let us first consider inward-moving photons, i.e. photons for which r is initially decreasing. From (9.51) and Figure 9.7 we see that if $1/b^2 < 1/(27\mu^2)$, so that $b > 3\sqrt{3}\mu$, then the orbit will have a single turning point of closest approach and escape again to infinity. This situation is illustrated in Figure 9.8. If $b < 3\sqrt{3}\mu$, however, then the light ray will be captured by the massive body and spiral in towards the origin.

Similar considerations hold for trajectories that start at small radii. If $b > 3\sqrt{3}\mu$ then the photon will escape, and at infinity its straight-line path will have an impact parameter b. If $b < 3\sqrt{3}\mu$ then the photon path will have a turning point, and it will fall back towards the origin. In this case the particle does not reach infinity, so b cannot be interpreted simply as an impact parameter. It is straightfoward to show that if a photon is emitted from within the region $r = 2\mu$ to $r = 3\mu$ then the opening angle α from the radial direction for the photon to escape varies from $\alpha = 0$ at $r = 2\mu$ to $\alpha = \pi/2$ at $r = 3\mu$.

Appendix 9A: General approach to gravitational redshifts

Consider a *general* spacetime with metric $g_{\mu\nu}$ in some arbitrary coordinate system x^μ, where x^0 is a timelike coordinate and the x^i are spacelike. Suppose that an

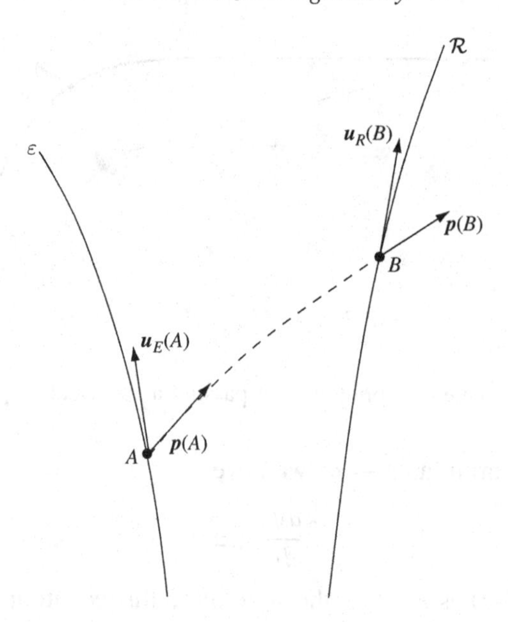

Figure 9.9 Schematic illustration of the emission and reception of a photon.

emitter $\mathcal{E}$ and a receiver $\mathcal{R}$ have worldlines $x_E^\mu(\tau_E)$ and $x_R^\mu(\tau_R)$ respectively, where τ_E and τ_R are the proper times of each observer. At some event A, $\mathcal{E}$ emits a photon with 4-momentum $\boldsymbol{p}(A)$ that is received by $\mathcal{R}$ at an event B. Furthermore, let us assume that at event A the emitter $\mathcal{E}$ has 4-velocity $\boldsymbol{u}_E(A)$ and that at event B the receiver has 4-velocity $\boldsymbol{u}_R(B)$. This is illustrated schematically in Figure 9.9.

The energies of the photon as observed by the emitter at A and by the receiver at B are respectively given by

$$E(A) = \boldsymbol{p}(A) \cdot \boldsymbol{u}_E(A) = p_\mu(A) u_E^\mu(A),$$
$$E(B) = \boldsymbol{p}(B) \cdot \boldsymbol{u}_R(B) = p_\mu(B) u_R^\mu(B).$$

Since in both cases $E = h\nu$, the ratio of the photon frequencies is given by the general result

$$\boxed{\frac{\nu_R}{\nu_E} = \frac{p_\mu(B) u_R^\mu(B)}{p_\mu(A) u_E^\mu(A)}.} \tag{9.52}$$

If we know the components of the 4-momentum $p_\mu(A)$ at emission then we can calculate the components $p_\mu(B)$ at reception, using the fact that the photon travels along a null geodesic. Since the photon 4-momentum $\boldsymbol{p}$ at any point is tangent to this geodesic, it is *parallel-transported* along the path. Thus, if the photon geodesic $x^\mu(\sigma)$ is described in terms of some affine parameter σ then

$$\frac{dp_\mu}{d\sigma} - \Gamma^\nu{}_{\mu\rho} p_\nu \frac{dx^\rho}{d\sigma} = 0.$$

Moreover, since $\boldsymbol{p}$ is tangent to the geodesics, we can choose the affine parameter σ so that $p^\mu = dx^\mu/d\sigma$, in which case

$$\frac{dp_\mu}{d\sigma} = \Gamma^\nu{}_{\mu\rho} p_\nu p^\rho.$$

It is also worth remembering that a first integral of the equation for a photon geodesic, which can prove very useful, is

$$p_\mu p^\mu = 0. \tag{9.53}$$

Let us now examine some special cases of the general formula (9.52). We begin by considering the case in which both the emitter $\mathcal{E}$ and the receiver $\mathcal{R}$ have *fixed spatial coordinates*. Thus, for $i = 1, 2, 3$ the spatial components of their 4-velocities are

$$u^i_E \equiv \frac{dx^i_E}{d\tau_E} = 0 \qquad \text{and} \qquad u^i_R \equiv \frac{dx^i_R}{d\tau_R} = 0.$$

Moreover, in each case, the squared length of the 4-velocity is $g_{\mu\nu}u^\mu u^\nu = c^2$. In our situation, this reduces to $g_{00}(u^0)^2 = c^2$, so we find that

$$u^0 = \frac{c}{(g_{00})^{1/2}}.$$

Hence the formula (9.52) reduces to

$$\boxed{\frac{\nu_R}{\nu_E} = \frac{p_0(B)}{p_0(A)} \left[\frac{g_{00}(A)}{g_{00}(B)}\right]^{1/2}}. \tag{9.54}$$

Let us now make the *additional* assumption that the metric is *stationary* in our chosen cooordinate system, i.e.

$$\partial_0 g_{\mu\nu} = 0.$$

Thus, the metric components $g_{\mu\nu}$ cannot depend explicitly on the coordinate x^0. As shown in Section 3.19, this means that the zeroth covariant component of the tangent vector is *constant* along an affinely parameterised geodesic. Since the photon 4-momentum is simply proportional to the tangent vector, this means that

p_0 *is constant along the photon's geodesic.* Thus, in this case, (9.54) reduces further to

$$\boxed{\frac{\nu_R}{\nu_E} = \left[\frac{g_{00}(A)}{g_{00}(B)}\right]^{1/2},} \tag{9.55}$$

and we have recovered the result (9.14) derived earlier.

Exercises

9.1 Show that surfaces of constant t and r in the general isotropic metric (9.3) have surface area $4\pi r^2$.

9.2 For the general static isotropic metric (9.4), show that the off-diagonal components of the Ricci tensor $R_{\mu\nu}$ are zero and that the diagonal components are given by (9.8–9.11).

9.3 The Schwarzschild line element is

$$ds^2 = c^2\left(1-\frac{2\mu}{r}\right)dt^2 - \left(1-\frac{2\mu}{r}\right)^{-1}dr^2 - r^2\,d\theta^2 - r^2\sin^2\theta\,d\phi^2.$$

By considering the 'Lagrangian' $L = g_{\mu\nu}\dot{x}^\mu\dot{x}^\nu$, where the dots denote differentiation with respect to an affine parameter λ, calculate the connection coefficients $\Gamma^\mu{}_{\nu\sigma}$. Hence verify that the geodesic equations are given by (9.17–9.20).

9.4 Derive the results (9.17–9.20) using the alternative form (9.28) of the geodesic equations.

9.5 Calculate the connection coefficients and the Ricci tensor for the general isotropic metric (9.3). Hence prove Birkoff's theorem.

9.6 Use Birkhoff's theorem to show that a particle inside a spherical shell of matter experiences no gravitational force.

9.7 Show that the 'Lemaitre' line element

$$ds^2 = c^2\,dw^2 - \frac{4}{9}\left[\frac{9\mu}{2(z-cw)}\right]^{2/3}dz^2 - \left[\frac{9\mu}{2}(z-cw)^2\right]^{2/3}d\Omega^2,$$

where $d\Omega^2 = d\theta^2 + \sin^2\theta\,d\phi^2$, describes the Schwarzschild geometry. Show that observers with fixed spatial coordinates (z, θ, ϕ) are in free fall and had zero velocity at infinity, and that the proper time of such observers is w.

9.8 For a general stationary spacetime with line element

$$ds^2 = g_{00}(\vec{x})\,dt^2 + g_{ij}(\vec{x})\,dx^i\,dx^j,$$

show that, for a fixed emitter and receiver, the ratio of the received photon frequency to the emitted frequency is

$$\frac{\nu_R}{\nu_E} = \left[\frac{g_{00}(\vec{x}_E)}{g_{00}(\vec{x}_R)}\right]^{1/2},$$

where $\vec{x}_E$ and $\vec{x}_R$ are the fixed spatial coordinates of the emitter and receiver respectively.

9.9 An isolated thin rigid spherical shell has mass M and radius R. Suppose that a small hole is drilled through the shell, so that an observer O at the shell's centre can observe the outside universe. Show that a photon emitted by a fixed observer E at $r = r_E$ (where $r_E > R$) and received by O is blueshifted by the amount

$$\frac{\nu_O}{\nu_E} = \left(\frac{1 - 2\mu/r_E}{1 - 2\mu/R}\right)^{1/2}.$$

9.10 Show that the quantity

$$L^2 = p_\theta^2 + \frac{p_\phi^2}{\sin^2\theta},$$

where $\boldsymbol{p}$ is the 4-momentum of a particle, is a constant of motion along any geodesic in the Schwarzschild geometry. Hence show that the particle orbits in a Schwarzschild geometry are stably planar.

9.11 For a particle dropped from rest at infinity in the Schwarzschild geometry, find expressions for $t(r)$ and $\tau(r)$, where t is the coordinate time and τ is the proper time of the particle.

9.12 A particle is dropped from rest at a coordinate radius $r = R$ in the Schwarzschild geometry. Obtain an expression for the 4-velocity of the particle in (t, r, θ, ϕ) coordinates when it passes coordinate radius r.

9.13 A particle at infinity in the Schwarzschild geometry is moving radially inwards with coordinate speed u_0. Show that at any coordinate radius r the coordinate velocity is given by

$$\left(\frac{dr}{dt}\right)^2 = \left(1 - \frac{2GM}{c^2 r}\right)^2 c^2\left[1 - \frac{1}{\gamma_0^2}\left(1 - \frac{2GM}{c^2 r}\right)\right],$$

where $\gamma_0 = \left(1 - u_0^2/c^2\right)^{-1/2}$. Determine the velocity relative to a stationary observer at r, and show that this velocity tends to c as r tends to $2GM/c^2$, irrespective of the value of u_0.

9.14 Suppose that the particle in Exercise 9.13 has rest mass m_0 and that it stopped at $r = r_1$. If its excess energy was converted to radiation that is observed at infinity, show that the energy released as seen by a stationary observer at r_1 is

$$E = m_0 c^2\left(\frac{\gamma_0}{\sqrt{1 - 2GM/(c^2 r_1)}} - 1\right).$$

What is the energy released as observed at infinity? Show that this tends to $\gamma_0 m_0 c^2$ as r_1 tends to $2GM/c^2$.

9.15 For a particle in a circular orbit of radius r in the Schwarzschild geometry, use the alternative form (9.28) of the geodesic equations to show that

$$\frac{d\phi}{dt} = \frac{GM}{r^3}.$$

9.16 In the Schwarzschild geometry, a photon is emitted from a coordinate radius $r = r_2$ and travels radially inwards until it is reflected by a fixed mirror at $r = r_1$, so that it travels radially outwards back to $r = r_2$. How long does the round trip take according to a stationary observer in infinity?

9.17 A photon moves in a circular orbit at $r = 3\mu$ in the Schwarzschild geometry. Show that the period of the orbit as measured by a stationary observer at this radius is $\tau = 6\pi\mu/c$. What is the period of the orbit as measured by a stationary observer infinity?

9.18 Show that a massive particle moving in the innermost stable circular orbit in the Schwarzschild geometry has speed $c/2$ as measured by a stationary observer at this radius. Hence calculate the period of the orbit as measured by the local observer. What is the period of the orbit as measured by a stationary observer infinity?

9.19 Alice is situated at a fixed position on the equator of the Earth (which is assumed to be spherical). In Schwarzschild coordinates (t, r, θ, ϕ), her worldline is described in terms of a parameter τ by

$$t = \gamma\tau, \qquad r = R, \qquad \theta = \pi/2, \qquad \phi = \omega\tau,$$

where γ and ω are constants and R is the coordinate radius of the Earth's surface. Bob is a distant stationary observer in space. Show that he will measure the orbital speed of Alice to be $v = R\omega/\gamma$. By considering the magnitude of Alice's 4-velocity, show that

$$\gamma = \left[1 - \left(\frac{v^2}{c^2} + \frac{GM}{c^2 r}\right)\right]^{-1/2},$$

where M is the mass of the Earth. Interpret this result physically.

9.20 All massive objects look larger than they really are. Show that a light ray grazing the surface of a massive sphere of coordinate radius $r > 3GM/c^2$ will arrive at infinity with impact parameter

$$b = r\left(\frac{r}{r - 2GM/c^2}\right)^{1/2}.$$

Hence show that the apparent diameter of the Sun ($M_\odot = 2 \times 10^{30}$ kg, $R_\odot = 7 \times 10^8$ m) exceeds the coordinate diameter by nearly 3 km.

9.21 The Hipparcos satellite can measure the positions of stars to an accuracy of 0.001 arcseconds. If it is measuring the position of a star in a direction perpendicular to the plane of the Earth's orbit, do Hipparcos observers need to account for the gravitational bending of light by the Sun?

9.22 A massive particle is moving in the equatorial plane of the Schwarzschild geometry. Show that at infinity the particle moves in a straight line with impact parameter $b = h/(c\sqrt{k^2 - 1})$.

9.23 An observer at rest at coordinate radius $r = R$ in the Schwarzschild geometry drops a massive particle which free-falls radially inwards. When the particle is at a coordinate radius $r = r_E$ it emits a photon radially outwards. Find an expression for the redshift z of the photon when it is received by the observer. Show that $z \to \infty$ as $r_E \to 2GM/c^2$.

9.24 Show that the geodesic equations for photon motion in the equatorial plane $\theta = \pi/2$ of the Schwarzschild geometry can be written in the form

$$\dot{t} = \frac{1}{bc}\left(1 - \frac{2\mu}{r}\right)^{-1}, \qquad \dot{\phi} = \frac{1}{r^2}, \qquad \dot{r}^2 = \frac{1}{b^2} - U(r),$$

where b is a constant, the dots correspond to differentiation with respect to some affine parameter and

$$U(r) = \frac{1}{r^2}\left(1 - \frac{2\mu}{r}\right).$$

Suppose that a photon moving in the equatorial plane passes an observer at rest at a coordinate radius in the range $2\mu < r < 3\mu$. Show that the observer measures the radial and azimuthal components of the photon's velocity to be

$$v_r = \pm c\left[1 - b^2 U(r)\right]^{1/2} \qquad \text{and} \qquad v_\phi = cb[U(r)]^{1/2}.$$

If the observer emits a photon that makes an angle α with the outward radial direction, show that the photon will escape to infinity provided that

$$\sin\alpha < 3\sqrt{3}\mu[U(r)]^{1/2}.$$

Find the values of α at $r = 2\mu$ and $r = 3\mu$.

9.25 Alice and Bob are astronauts in a space capsule, with no engine, in a circular orbit at $r = R$ (where $R > 3\mu$) in the equatorial plane of the Schwarzschild geometry. At some point in the orbit, Bob leaves the capsule, uses his rocket-pack to maintain a hovering position at that fixed point in space and then rejoins the capsule after it has completed one orbit. Show that the proper time interval measured by Alice while Bob is out of the capsule is

$$\Delta\tau_A = 2\pi\left[\frac{R^2}{\mu c^2}(R - 3\mu)\right]^{1/2}.$$

If $\Delta\tau_B$ is the corresponding proper time interval measured by Bob between these two events, show that

$$\frac{\Delta\tau_B}{\Delta\tau_A} = \left(\frac{R - 2\mu}{R - 3\mu}\right)^{1/2}.$$

Briefly compare this result with the 'twin paradox' result in special relativity. If Bob chooses not to rejoin the capsule but instead observes it fly past him, show that he will measure the capsule's speed as

$$v = \left(\frac{\mu c^2}{R - 2\mu}\right)^{1/2}.$$

9.26 A particle A and its antiparticle B are travelling in opposite senses in free circular orbits in the equatorial plane of the Schwarzschild geometry, one at coordinate radius r_A and the other at $r_B(> r_A)$. At some instant, A emits a photon of frequency ν_A that travels radially and is received by B with frequency ν_B. Show that

$$\frac{\nu_B}{\nu_A} = \left(\frac{1-3\mu/r_A}{1-3\mu/r_B}\right)^{1/2}.$$

Suppose now that $r_A = r_B = r$, so that the particles collide and annihilate each other. Show that the total radiated energy as measured by an observer at rest at the point of collision is given by

$$E = 2m_0c^2\left(\frac{1-2\mu/r}{1-3\mu/r}\right)^{1/2},$$

where m_0 is the rest mass of each particle.

9.27 If the cosmological constant Λ is non-zero, show that the line element outside a static spherically symmetric matter distribution is given by

$$ds^2 = c^2\left(1-\frac{2\mu}{r}-\frac{\Lambda r^2}{3}\right)dt^2 - \left(1-\frac{2\mu}{r}-\frac{\Lambda r^2}{3}\right)^{-1}dr^2 - r^2(d\theta^2+\sin^2\theta\,d\phi^2).$$

Hence show that, in the weak-field Newtonian limit, a spherically symmetric mass M produces a gravitational field strength $\vec{g}$ given by

$$\vec{g} = \left(-\frac{GM}{r^2}+\frac{c^2\Lambda r}{3}\right)\hat{r}.$$

Show that the shapes of massive particle orbits in the above geometry differ from those in the Schwarzschild geometry, but that the shapes of photon orbits do not.

9.28 Consider a static axisymmetric spacetime that is invariant under translations and reflections along the axis of symmetry. Show that, in general, the line element for such a spacetime can be written in the form

$$ds^2 = A(\rho)\,dt^2 - d\rho^2 - B(\rho)\,d\phi^2 - C(\rho)\,dz^2,$$

for arbitrary functions A, B and C. Show that the non-zero connection coefficients for this line element are given by

$$\Gamma^0{}_{01} = \frac{A'}{2A}, \qquad \Gamma^1{}_{22} = -\frac{B'}{2}, \qquad \Gamma^1{}_{33} = -\frac{C'}{2},$$

$$\Gamma^1{}_{00} = \frac{A'}{2}, \qquad \Gamma^2{}_{21} = \frac{B'}{2B}, \qquad \Gamma^3{}_{31} = \frac{C'}{2C},$$

where the primes denote $d/d\rho$. Hence show that the non-zero components of the Ricci tensor are given by

$$R_{00} = -\frac{A''}{2} + \frac{A'}{2}\left(\frac{A'}{2A} - \frac{B'}{2B} - \frac{C'}{2C}\right),$$

$$R_{11} = \frac{A''}{2A} - \frac{(A')^2}{4A^2} + \frac{B''}{2B} - \frac{(B')^2}{4B^2} + \frac{C''}{2C} - \frac{(C')^2}{4C^2},$$

$$R_{22} = \frac{B''}{2} - \frac{B'}{2}\left(\frac{B'}{2B} - \frac{A'}{2A} - \frac{C'}{2C}\right),$$

$$R_{33} = \frac{C''}{2} - \frac{C'}{2}\left(\frac{C'}{2C} - \frac{A'}{2A} - \frac{B'}{2B}\right).$$

9.29 Consider a static, infinitely long, cylindrically symmetric matter distribution of constant radius that is invariant to Lorentz boosts along the symmetry axis (a 'cosmic string'). Show that the line element outside the body can be written as

$$ds^2 = c^2\,dt^2 - d\rho^2 - (\alpha + \beta\rho)^2\,d\phi^2 - dz^2,$$

where α and β are constants. For the case $\alpha = 0$, consider the spacelike surfaces defined by $t = \text{constant}$ and $z = \text{constant}$ and calculate the circumference of a circle of constant coordinate radius ρ in such a surface. Hence show that, for $\beta < 1$, the geometry on the spacelike surface is that of a two-dimensional cone embedded in three-dimensional Euclidean space.

10

Experimental tests of general relativity

Most of the experimental tests of general relativity are based on the Schwarzschild geometry in the region $r > 2GM/c^2$. Some are based on the trajectories of massive particles and others on photon trajectories. Most of the 'classic' tests are in the weak-field limit, but more recent observations have begun to probe the more discriminative strong-field regime. We will now discuss both these 'classic' experimental tests and some of the more recent findings and proposals. Some later tests are in fact more closely linked to the Kerr geometry (see Chapter 13), which describes spacetime outside a rotating massive body, but the basic principles can still be understood in terms of the simpler Schwarzschild geometry.

10.1 Precession of planetary orbits

For a general non-circular orbit in Newtonian theory the equation of motion is

$$\frac{d^2u}{d\phi^2} + u = \frac{GM}{h^2},$$

where $u \equiv 1/r$ and h is the angular momentum per unit mass of the orbiting particle. For a bound orbit, the equation has the solution

$$u = \frac{GM}{h^2}(1 + e\cos\phi), \qquad (10.1)$$

which describes an ellipse; the parameter e measures the *ellipticity* of the orbit. Thus, for example, we can draw the orbit of a planet around the Sun as in Figure 10.1. We can write the distance of closest approach (*perihelion*) as $r_1 = a(1-e)$ and the distance of furthest approach (*aphelion*) as $r_2 = a(1+e)$. The equation of motion then requires that the semi-major axis is given by

$$a = \frac{h^2}{GM(1-e^2)}. \qquad (10.2)$$

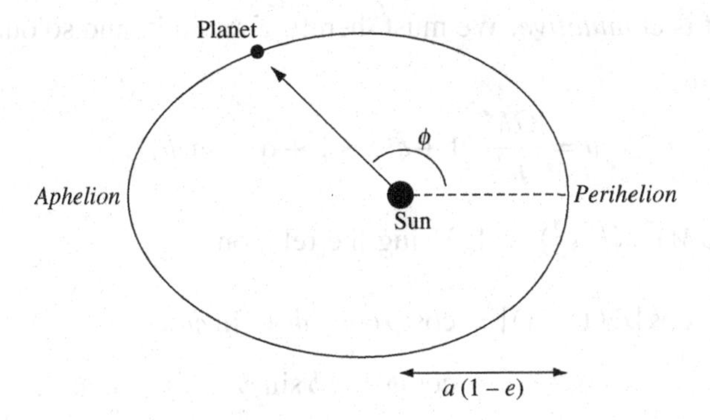

Figure 10.1 The elliptical orbit of a planet around the Sun; e is the ellipticity of the orbit.

The general-relativistic equation of motion is

$$\frac{d^2u}{d\phi^2}+u=\frac{GM}{h^2}+\frac{3GM}{c^2}u^2. \tag{10.3}$$

If the gravitational field is weak, as it is for planetary orbits around the Sun, then we expect Newtonian gravity to provide an excellent approximation to the motion in general relativity. We can therefore treat the Newtonian solution (10.1) as the zeroth-order solution to the general-relativistic equation of motion. Thus let us write the general-relativistic solution as

$$u=\frac{GM}{h^2}(1+e\cos\phi)+\Delta u,$$

where Δu is a perturbation. Substituting this expression into the general-relativistic equation (10.3) we find that, to first-order in Δu,

$$\frac{d^2\Delta u}{d\phi^2}+\Delta u=A\left(1+e^2\cos^2\phi+2e\cos\phi\right),$$

where the constant $A=3(GM)^3/(c^2h^4)$ is very small. A particular integral of this equation is easily found to be

$$\Delta u=A\left[1+e^2\left(\tfrac{1}{2}-\tfrac{1}{6}\cos 2\phi\right)+e\phi\sin\phi\right], \tag{10.4}$$

which can be checked by direct differentiation.

Since the constant A is very small, the first two terms on the right-hand side of (10.4) are tiny, and of no use in testing the theory. However, the last term, $Ae\phi\sin\phi$, might be tiny at first but will gradually grow with time, since the factor

ϕ means that it is *cumulative*. We must therefore retain it, and so our approximate solution reads

$$u = \frac{GM}{h^2}[1 + e(\cos\phi + \alpha\phi\sin\phi)], \quad (10.5)$$

where $\alpha = 3(GM)^2/(h^2c^2) \ll 1$. Using the relation

$$\cos[\phi(1-\alpha)] = \cos\phi\cos\alpha\phi + \sin\phi\sin\alpha\phi$$
$$\approx \cos\phi + \alpha\phi\sin\phi \quad \text{for} \quad \alpha \ll 1, \quad (10.6)$$

we can therefore write

$$u \approx \frac{GM}{h^2}\{1 + e\cos[\phi(1-\alpha)]\}. \quad (10.7)$$

From this expression, we see that the orbit is periodic, but with a period $2\pi/(1-\alpha)$, i.e. the r-values repeat on a cycle that is larger than 2π. The result is that the orbit cannot 'close', and so the ellipse *precesses* (see Figure 10.2). In one revolution, the ellipse will rotate about the focus by an amount

$$\Delta\phi = \frac{2\pi}{1-\alpha} - 2\pi \approx 2\pi\alpha = \frac{6\pi(GM)^2}{h^2c^2}.$$

Substituting for h from (10.2), we finally obtain

$$\boxed{\Delta\phi = \frac{6\pi GM}{a(1-e^2)c^2}.} \quad (10.8)$$

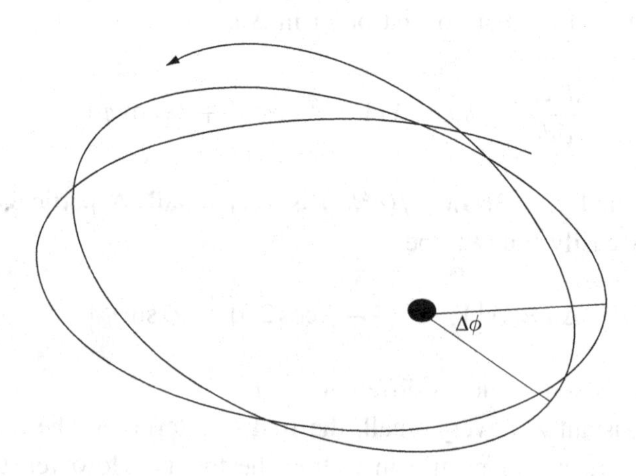

Figure 10.2 Precession of an elliptical orbit (greatly exaggerated).

Let us apply equation (10.8) to the orbit of Mercury, which has the following parameters: period = 88 days, $a = 5.8 \times 10^{10}$ m, $e = 0.2$. Using $M_\odot = 2 \times 10^{30}$ kg, we find

$$\Delta\phi = 43'' \text{ per century.}$$

In fact, the measured precession after correcting for the precession of equinoxes is

$$574''.1 \pm 0''.4 \text{ per century,}$$

but almost all of this is caused by perturbations from other planets. The residual, after taking perturbations into account, is in remarkable agreement with general relativity. The residuals for a number of planets (and Icarus, which is a large asteroid with a perihelion that lies within the orbit of Mercury) may also be calculated (in arcseconds per century):

	Observed residual	Predicted residual
Mercury	43.1 ± 0.5	43.03
Venus	8 ± 5	8.6
Earth	5 ± 1	3.8
Icarus	10 ± 1	10.3

In each case, the results are in excellent agreement with the predictions of general relativity. Einstein included this calculation regarding Mercury in his 1915 paper on general relativity. He had solved one of the major problems of celestial mechanics in the very first application of his complicated theory to an empirically testable problem. As you can imagine, this gave him tremendous confidence in his new theory.

10.2 The bending of light

We have already noted that a massive object can have a significant effect on the propagation of photons. For example, photons can travel in a circular orbit at $r = 3GM/c^2$. We do not, however, expect to observe this effect directly, but a more modest bending of light can be observed. For investigating the slight deflection of light by, for example, the Sun, it is easiest to follow an approximation technique analogous to that used in predicting the perihelion shift of Mercury.

As we showed in Chapter 9, the 'shape' equation for a photon trajectory in the equatorial plane of the Schwarzschild geometry is

$$\frac{d^2u}{d\phi^2} + u = \frac{3GM}{c^2}u^2, \tag{10.9}$$

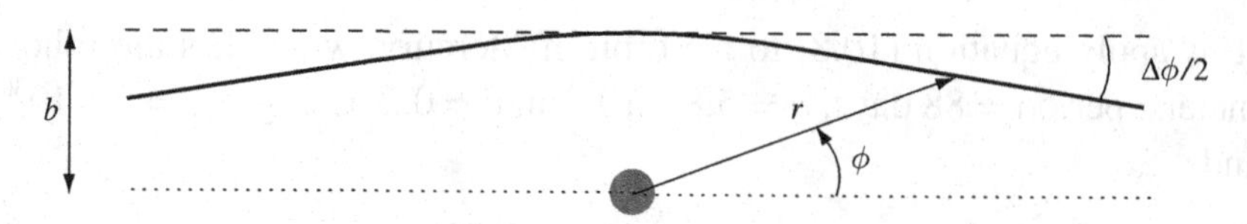

Figure 10.3 Angles and coordinates in the deflection of light by a spherical mass.

where $u \equiv 1/r$. In the absence of matter, the right-hand side vanishes and we may write the solution as

$$u = \frac{\sin\phi}{b}, \tag{10.10}$$

which represents a straight-line path with impact parameter b (see Figure 10.3). We again treat (10.10) as the zeroth-order solution to the equation of motion. Thus we write the general-relativistic solution as

$$u = \frac{\sin\phi}{b} + \Delta u,$$

where Δu is a perturbation. Substituting this expression into (10.9), we find that, to first order in Δu,

$$\frac{d^2\Delta u}{d\phi^2} + \Delta u = \frac{3GM}{c^2 b^2}\sin^2\phi.$$

This is satisfied by the particular integral

$$\Delta u = \frac{3GM}{2c^2b^2}\left(1 + \tfrac{1}{3}\cos 2\phi\right), \tag{10.11}$$

and adding (10.11) into the original solution yields

$$u = \frac{\sin\phi}{b} + \frac{3GM}{2c^2b^2}\left(1 + \tfrac{1}{3}\cos 2\phi\right). \tag{10.12}$$

Now consider the limit $r \to \infty$, i.e. $u \to 0$. Clearly, for a slight deflection we can take $\sin\phi \approx \phi$ and $\cos 2\phi \approx 1$ at infinity, to obtain $\phi = -2GM/(c^2b)$. Thus the total deflection (see Figure 10.3) is

$$\boxed{\Delta\phi = \frac{4GM}{c^2b}.} \tag{10.13}$$

This is the famous gravitational deflection formula (which incidentally is twice what had previously been worked out using a Newtonian approach). For light grazing the Sun it yields $\Delta\phi = 1''.75$. The 1919 eclipse expedition led by Eddington gave two sets of results:

$$\Delta\phi = 1''.98 \pm 0''.16,$$
$$\Delta\phi = 1''.61 \pm 0''.4,$$

both consistent with the theory. This provided the first experimental verification of a *prediction* of Einstein's theory (the 'anomalous' perihelion shift of Mercury had been known for many years) and turned Einstein into a scientific superstar.[1] Some historians have argued that Eddington 'fiddled' the results to agree with the theory. If Eddington did indeed massage the results, then he gambled correctly. Later high-precision tests using *radio sources*, which can be observed near the Sun even when there is no lunar eclipse, show there is now no doubt that the general-relativistic prediction is accurate to a fraction of a percent. Modern radio experiments using very long baseline interferometry (VLBI) have been performed to measure the gravitational deflection of the positions of radio quasars as they are eclipsed by the Sun. Such experiments can be performed to an accuracy of better than $\sim 10^{-4}$ arcseconds. Figure 10.4 summarizes the results of measurements of the deflection angle $\Delta\phi$ from experiments conducted over the period 1969–75. The results are in excellent agreement with the predictions of general relativity. Moreover, as one can see from the figure, the results constrain the parameter ω in the Brans–Dicke theory of gravity (see Appendix 8A): we have $\omega \geq 40$.

For more dramatic light deflection, our adopted approach of successive approximations is unsuitable. In this case, it is more appropriate to use the exact equation for $d\phi/dr$ derived in Chapter 9, which reads

$$\frac{d\phi}{dr} = \frac{1}{r^2}\left[\frac{1}{b^2} - \frac{1}{r^2}\left(1 - \frac{2\mu}{r}\right)\right]^{-1/2},$$

where b is the impact parameter at infinity. We also showed in Chapter 9 that if $b > 3\sqrt{3}\mu$ then the photon is not captured by the mass; the resulting general orbit shape is illustrated in Figure 10.5. From the figure, we see that the deflection angle is given by

$$\boxed{\Delta\phi = 2\int_{r_0}^{\infty} \frac{1}{r^2}\left[\frac{1}{b^2} - \frac{1}{r^2}\left(1 - \frac{2\mu}{r}\right)\right]^{-1/2} dr} \qquad (10.14)$$

where r_0 is the point of closest approach, at which the expression in the square brackets in (10.14) vanishes.

[1] The media had a great story. Remember that this was just after the end of the First World War, and so the headlines read something like 'Newton's theory of gravity overthrown by German physicist, verified by British scientists.'

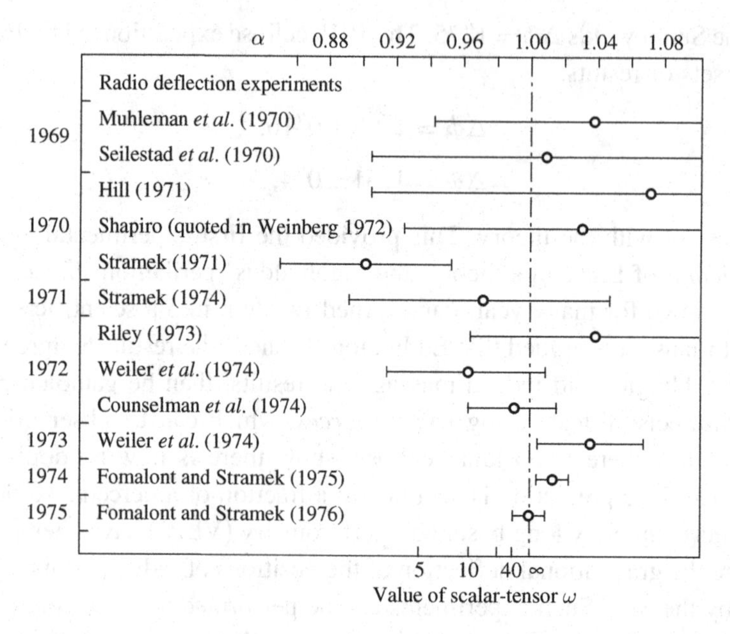

Figure 10.4 Results of radio-wave deflection measurements of the positions of quasars in the period 1969–75 (from C. Will, *Theory and Experiment in Gravitational Physics*, Cambridge University Press, 1981). The deflection angle is $\Delta\phi = \alpha 4GM/(R_{\odot}c^2)$ and the error bars are plotted on the parameter α. If general relativity is correct, we expect $\alpha = 1$. The abscissa scale gives the measured values of the parameter ω in the Brans–Dicke scalar-tensor theory of gravity, discussed in Appendix 8A.

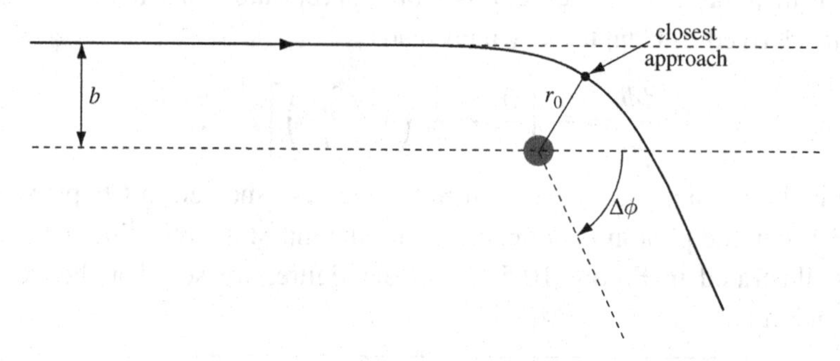

Figure 10.5 Angles and coordinates in the deflection of light by a spherical mass.

10.3 Radar echoes

In Chapter 9, we showed that the 'energy' equation for a photon orbit in the Schwarzschild geometry is

$$\dot{r}^2 + \frac{h^2}{r^2}\left(1 - \frac{2\mu}{r}\right) = c^2k^2.$$

Using the result

$$\left(\frac{dr}{d\sigma}\right)^2 = \left(\frac{dr}{dt}\frac{dt}{d\sigma}\right)^2 = \frac{k^2}{(1-2\mu/r)^2}\left(\frac{dr}{dt}\right)^2,$$

we can rewrite the energy equation as

$$\frac{1}{(1-2\mu/r)^3}\left(\frac{dr}{dt}\right)^2 + \frac{h^2}{k^2r^2} - \frac{c^2}{1-2\mu/r} = 0. \tag{10.15}$$

Now consider a photon path from Earth to another planet (say Venus), as shown in Figure 10.6. Evidently the photon path will be deflected by the gravitational field of the Sun (assuming that the planets are in a configuration like that shown in the figure, where the photon has to pass close to the Sun in order to reach Venus). Let r_0 be the coordinate distance of closest approach of the photon to the Sun; then

$$\left(\frac{dr}{dt}\right)_{r_0} = 0,$$

and so from (10.15) we have

$$\frac{h^2}{k^2r_0^2} = \frac{c^2}{1-2\mu/r_0}.$$

Thus, after rearrangement, we can write (10.15) as

$$\frac{dr}{dt} = c(1-2\mu/r)\left[1 - \frac{r_0^2(1-2\mu/r)}{r^2(1-2\mu/r_0)}\right]^{1/2},$$

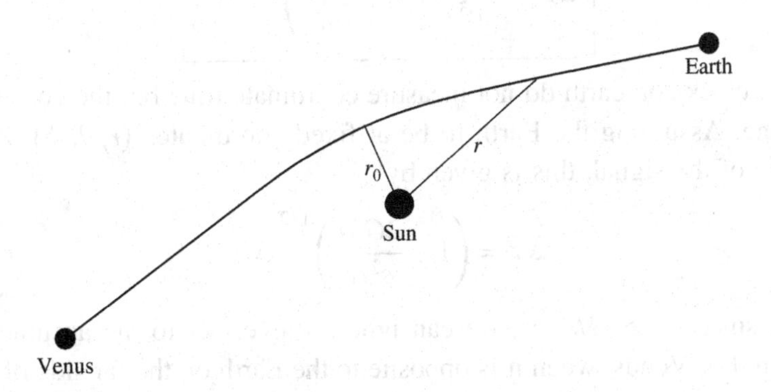

Figure 10.6 Photon path from Earth to Venus deflected by the Sun.

which can be integrated to give for the time taken to travel between points r_0 and r

$$t(r, r_0) = \int_{r_0}^{r} \frac{1}{c(1-2\mu/r)}\left[1 - \frac{r_0^2(1-2\mu/r)}{r^2(1-2\mu/r_0)}\right]^{-1/2} dr.$$

The integrand can be expanded to first order in μ/r to obtain

$$t(r, r_0) = \int_{r_0}^{r} \frac{r}{c(r^2-r_0^2)^{1/2}}\left[1+\frac{2\mu}{r}+\frac{\mu r_0}{r(r+r_0)}\right] dr,$$

which can be evaluated to give

$$t(r, r_0) = \frac{(r^2-r_0^2)^{1/2}}{c} + \frac{2\mu}{c}\ln\left[\frac{r+(r^2-r_0^2)^{1/2}}{r_0}\right] + \frac{\mu}{c}\left(\frac{r-r_0}{r+r_0}\right)^{1/2}. \quad (10.16)$$

The first term on the right-hand side is just what we would have expected if the light had been travelling in a straight line. The second and third terms give us the extra coordinate time taken for the photon to travel along the *curved* path to the point r. So, you can see from Figure 10.6 that if we bounce a radar beam to Venus and back then the excess coordinate-time delay over a straight-line path is

$$\Delta t = 2\left[t(r_E, r_0) + t(r_V, r_0) - \frac{\left(r_E^2 - r_0^2\right)^{1/2}}{c} - \frac{\left(r_V^2 - r_0^2\right)^{1/2}}{c}\right],$$

where the factor 2 is included because the photon has to go to Venus and back. Since $r_E \gg r_0$ and $r_V \gg r_0$ we have

$$t(r_E, r_0) - \frac{\left(r_E^2 - r_0^2\right)^{1/2}}{c} \approx \frac{2\mu}{c}\ln\left(\frac{r_E}{r_0}\right) + \frac{\mu}{c},$$

and likewise for t_V and r_V. Thus, the excess coordinate-time delay is

$$\boxed{\Delta t \approx \frac{4GM}{c^3}\left[\ln\left(\frac{r_E r_V}{r_0^2}\right) + 1\right].}$$

Of course, clocks on earth do not measure coordinate time but the corresponding proper time. Assuming the Earth to be at fixed coordinates (r, θ, ϕ) during the travel time of the signal, this is given by

$$\Delta\tau = \left(1 - \frac{2GM}{c^2 r_E}\right)^{1/2} \Delta t.$$

However, since $r_E \gg GM/c^2$, we can ignore this effect to the accuracy of our calculation. For Venus, when it is opposite to the Earth on the far side of the Sun,

$$\Delta\tau \approx 220\,\mu s.$$

The idea of the experiment is as follows. Fire an intense radar beam towards Venus when it is almost opposite to the Earth on the far side of the Sun and measure the time delay of the radar echo with a sensitive radio telescope. The excess time delay gives us a test of the principle of equivalence. This sounds straightforward, but the time delay is very small and depends on the values of r_E, r_V, r_0. How can one determine these parameters to the required precision? The answer is to fit the measured delays over a long period of time to a curve chosen by varying r_E, r_V, μ etc. as free parameters (see Figure 10.7). There are a number of technical problems that limit the accuracy of this method. Firstly, we must correct for the motion of Venus and the Earth in their orbits and for their individual gravitational fields. Also, in practice, the radar beam is reflected from different points on the surface of Venus (mountain peaks, valleys, etc.) and this introduces a dispersion in the time delay of several hundred μs. This problem can be solved by bouncing the radar beams from a mirror – as has since been done using the Viking landers on Mars. Another, more complicated, problem is correcting for refraction by the Solar corona – this can be important for photon paths that graze the surface of the Sun. Nevertheless, Figure 10.7 confirms that the corrected measurements are in excellent agreement with the general-relativistic prediction.

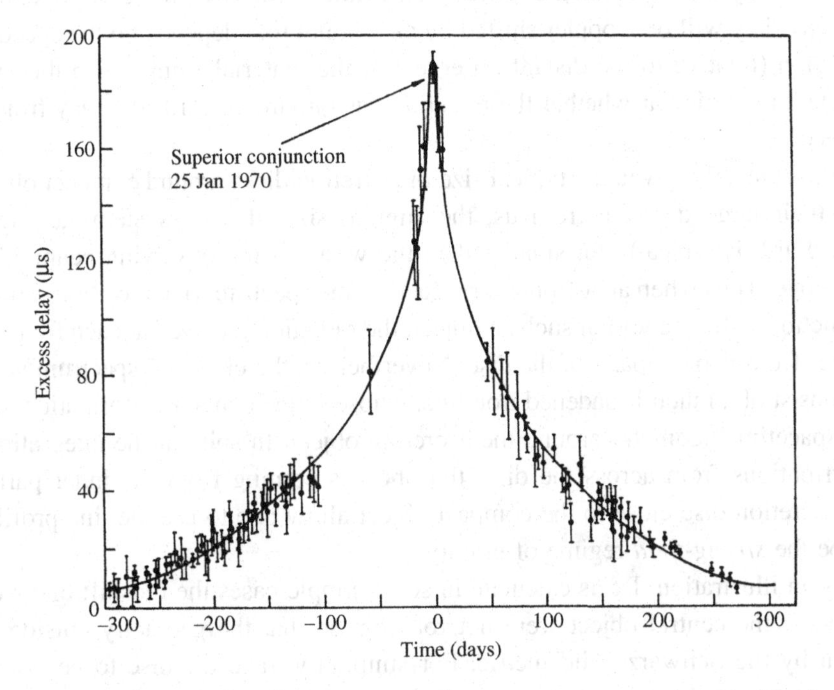

Figure 10.7 The Earth–Venus time-delay measurement compared with the general-relativistic prediction.

10.4 Accretion discs around compact objects

As we have seen, the orbits of particles and photons are probes of the geometry of spacetime. Information about the geometry produced by compact massive objects or black holes can be obtained from observations of the orbits of particles in the accretion disc that often surrounds them. As we saw in Chapter 9, the radiation efficiency of the accretion disc around a Schwarzschild black hole is 10 times greater than the efficiency of the nuclear burning of hydrogen, and such disks are very strong emitters in X-rays.

Even at the temperatures of $\sim 10^7$ K that characterise an accretion disc, some heavy nuclei retain bound electrons. The small trace of iron found in the accreting matter is such a nucleus. Incident radiation from X-ray flares above and below the disc can lead to fluorescence from the highly ionised atoms in the disc; in this process an electron in the atom is de-excited from a higher energy level to a lower one and emits a photon. For iron atoms, this results in photons of energy 6.4 keV, giving a spectral line roughly in the middle of the X-ray band. As one might expect, however, the frequency of the emitted photons as measured by some observer at infinity (i.e. an astronomer on Earth) will differ from the frequency with which the photons were emitted. Qualitatively, there are two effects that cause this frequency shift. First, the photons will be gravitationally redshifted by an amount that depends on the radius from which they were emitted. Second, they will be Doppler shifted by an amount that depends on the speed and direction (relative to the distant observer) of the material from which they were emitted, in particular whether the material was moving towards or away from the observer.

Unfortunately, given the typical size of accretion disks around compact objects, and their large distance from us, the angular size of such systems as viewed from Earth is typically far smaller than the width of the observing beam of any telescope. Thus when an astronomer measures the spectrum (i.e. the photon flux as a function of frequency) of such an object, the radiation received at each frequency comes from various parts of the disc. Nevertheless, the observed spectrum is seen to consist of a much-broadened iron line, whose shape contains information about the spacetime geometry around the accreting object. In spite of the integration of contributions from across the disc, the photons coming from the inner parts of the accretion disc close to the compact object allow one to use the line profile to probe the *strong-field* regime of gravity.

As an illustration, let us calculate in some simple cases the redshift one would expect if the central object were not rotating, so that the geometry outside it is given by the Schwarzschild metric. For simplicity, take the disc to be oriented edge-on to the observer, as shown in Figure 10.8. All orbits are then in the plane of the observer and the disc, which we take to be the equatorial plane $\theta = \pi/2$.

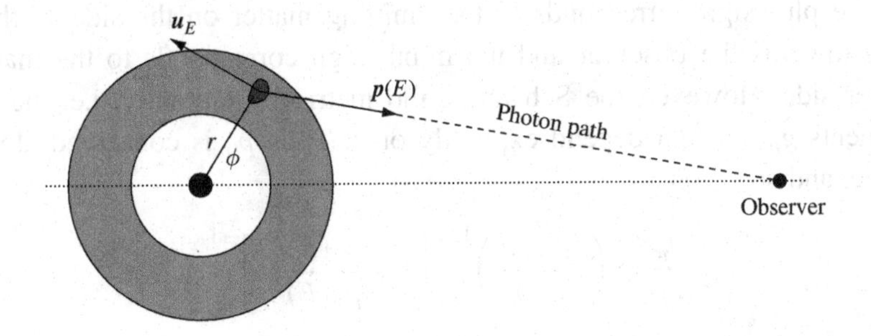

Figure 10.8 The emission of a photon by matter in an accretion disc around a compact object. The observer is viewing the disc edge-on.

The ratio of the photon's frequency at reception to that at emission is given by

$$\frac{\nu_R}{\nu_E} = \frac{\boldsymbol{p}(R)\cdot\boldsymbol{u}_R}{\boldsymbol{p}(E)\cdot\boldsymbol{u}_E} = \frac{p_\mu(R)u_R^\mu}{p_\mu(E)u_E^\mu}, \tag{10.17}$$

where $\boldsymbol{p}(E)$ and $\boldsymbol{p}(R)$ are the photon 4-momenta at emission and reception respectively, $\boldsymbol{u}_E$ is the 4-velocity of the material at emission and $\boldsymbol{u}_R$ is the 4-velocity of the observer at reception. Assuming the observer to be fixed at infinity, the components of his 4-velocity in the (t, r, θ, ϕ) coordinate system are

$$[u_R^\mu] = (1, 0, 0, 0).$$

Now consider the 4-velocity of the emitting material. Since we are assuming that this material is moving in a circular orbit it must have a 4-velocity of the form

$$[u_E^\mu] = \left(u_E^0, 0, 0, u_E^3\right).$$

Using the fact that

$$u_E^3 = \frac{d\phi}{d\tau} = \frac{d\phi}{dt}\frac{dt}{d\tau} = \frac{d\phi}{dt}u_E^0,$$

we can write the emitter's 4-velocity as

$$[u_E^\mu] = u_E^0(1, 0, 0, \Omega),$$

where, for circular motion, $\Omega \equiv d\phi/dt = (GM/r^3)^{1/2}$, which we derived in Chapter 9. We can now fix u_E^0 by using the fact that $g_{\mu\nu}u^\mu u^\nu = c^2$. If the emitting material is at a coordinate radius r, we have

$$u_E^0 = c\left[c^2\left(1 - \frac{2\mu}{r}\right) - r^2\Omega^2\right]^{-1/2} = \left(1 - \frac{3\mu}{r}\right)^{-1/2}.$$

Our general expression (10.17) therefore yields

$$\frac{\nu_R}{\nu_E} = \frac{p_0(R)}{p_0(E)u_E^0 + p_3(E)u_E^3} = \left(1 - \frac{3\mu}{r}\right)^{1/2}\frac{p_0(R)}{p_0(E)}\left[1 + \frac{p_3(E)}{p_0(E)}\Omega\right]^{-1},$$

where the plus sign corresponds to the emitting matter on the side of the disc moving towards the observer and the minus sign corresponds to the matter on the other side. However, the Schwarzschild metric is stationary, i.e. the metric components $g_{\mu\nu}$ do not depend explicitly on t. Thus p_0 is conserved along the geodesic, and so

$$\frac{\nu_R}{\nu_E} = \left(1 - \frac{3\mu}{r}\right)^{1/2} \left[1 + \frac{p_3(E)}{p_0(E)}\Omega\right]^{-1},$$

It therefore remains only to fix the ratio $p_3(E)/p_0(E)$ in order to determine the observed redshift. In general, we must use the fact that the photon worldline is null, and so $g^{\mu\nu}p_\mu p_\nu = 0$. As we are working in the equatorial plane $\theta = \pi/2$, this yields

$$\frac{1}{c^2}\left(1 - \frac{2\mu}{r}\right)^{-1}(p_0)^2 - \left(1 - \frac{2\mu}{r}\right)(p_1)^2 - \frac{1}{r^2}(p_3)^2 = 0. \tag{10.18}$$

For photons emitted from material at a general position angle ϕ, one would now need to use the geodesic equations for the photon worldline in order to eliminate p_1. There are, however, two special cases for which this is not necessary.

The simplest case occurs when the photon is emitted from matter moving transverse to the observer, i.e. when $\phi = 0$ or $\phi = \pi$. We then have $p_3(E) = 0$, and so the observed frequency ratio is

$$\boxed{\frac{\nu_R}{\nu_E} = \left(1 - \frac{3\mu}{r}\right)^{1/2}.} \tag{10.19}$$

The other simple cases occur when the matter is moving either directly towards or away from the observer, i.e. when $\phi = -\pi/2$ or $\phi = \pi/2$. Then the radial components of the photon 4-momentum, $p_1(E)$, will be zero. From (10.18) we obtain

$$\frac{p_3(E)}{p_0(E)} = \frac{r}{c(1 - 2\mu/r)^{1/2}},$$

so that the photon frequency shift for $\phi = \mp\pi/2$ is given by

$$\boxed{\frac{\nu_R}{\nu_E} = \frac{(1 - 3\mu/r)^{1/2}}{1 \pm (r/\mu - 2)^{-1/2}}.} \tag{10.20}$$

The above discussion has been for a disc viewed edge-on. The other limiting case, when the disc is viewed face-on, is easier to analyse. Since the motion of the emitting matter is always transverse to the observer, the frequency shift is given by (10.19).

Although the observed iron line consists of photons coming from different radii in the disc, we may still calculate the smallest possible frequency (or largest redshift) present in the observed spectrum. It is clear that such photons must be emitted from the smallest possible value of r. As discussed in the previous chapter, the innermost stable circular orbit for the Schwarzschild metric is at $r = 6\mu$. Thus the smallest frequency represented is therefore given by

$$\nu_R/\nu_E = \begin{cases} \sqrt{2}/3 = 0.47 & \text{for a disc viewed edge-on,} \\ 1/\sqrt{2} = 0.71 & \text{for a disc viewed face-on.} \end{cases}$$

If the central object were rotating (so that the exterior geometry is given by the Kerr metric – see Chapter 13), then the smallest frequencies could be even lower. Figure 10.9 shows the iron spectral line measured in the galaxy MCG-6-30-15.

In general the detailed shape of the line profile depends on the mass and rotation of the central object, the inclination of the disc to the line of sight and relativistic beaming effects. It is hoped that, in the future, line profiles can be measured to

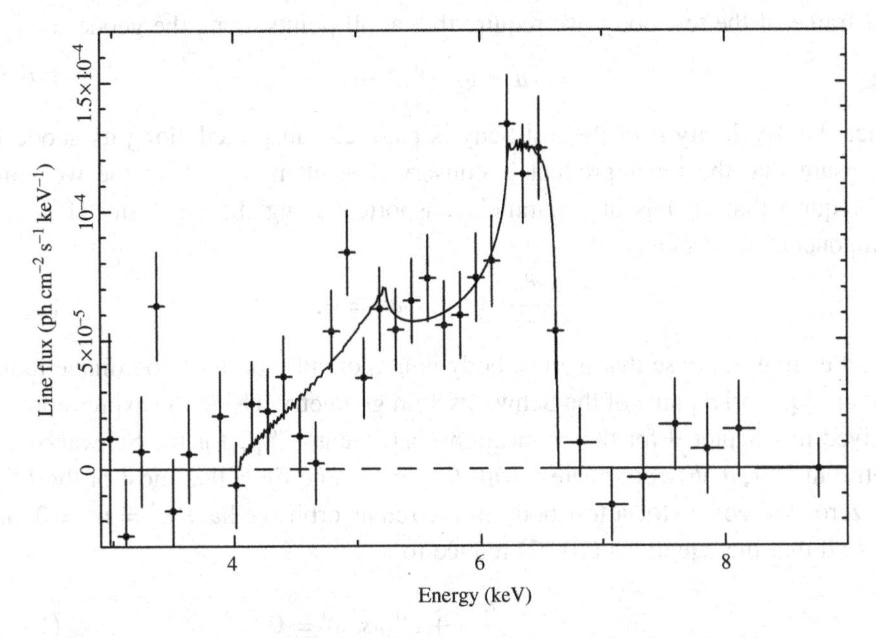

Figure 10.9 The line profile of the iron 6.4 keV spectral line from MCG-6-30-15 observed by the ASCA satellite (Y. Tanaka *et al.*, *Nature* **375**, 659, 1995). The emission line is extremely broad, the width indicating velocities of order one-third the speed of light. There is a marked asymmetry towards energies lower than the rest energy of the emission line, with a smallest energy of about 4 keV. The solid line shows a fit to the data assuming a disc around a non-rotating Schwarzschild black hole, extending between 3 and 10 Schwarzschild radii and inclined at an angle of 30° to the line of sight. Certain features suggest that the central object may in fact be rapidly rotating.

sufficient accuracy to determine the mass and angular momentum of the central compact object.

10.5 The geodesic precession of gyroscopes

We have seen how the motion of test bodies can be used to explore the geometry of a curved spacetime. If the test body has spin then the motion of its spin vector can also be used to probe the spacetime geometry. Here we discuss the idealised case of an infinitesimally small test body with spin, such as a small gyroscope.

The test body moves along a timelike geodesic curve, so its 4-velocity $\boldsymbol{u}(\tau)$ is parallel-transported along its worldline. Thus, in some coordinate system, its components satisfy

$$\frac{du^\mu}{d\tau} + \Gamma^\mu{}_{\nu\sigma} u^\nu u^\sigma = 0.$$

Suppose that the spin of the test body is described by the 4-vector $\boldsymbol{s}(\tau)$ along the geodesic. Since this vector can have no timelike component in the instantaneous rest frame of the test body, we require that at all points along the geodesic

$$\boldsymbol{s} \cdot \boldsymbol{u} = g_{\mu\nu} s^\mu u^\nu = 0. \qquad (10.21)$$

Since the 4-velocity $\boldsymbol{u}$ of the test body is parallel-transported along its geodesic, to ensure that the inner product is conserved at all points along the worldline we require that $\boldsymbol{s}(\tau)$ is also parallel-transported along the geodesic. Hence its components must satisfy

$$\frac{ds^\mu}{d\tau} + \Gamma^\mu{}_{\nu\sigma} s^\nu u^\sigma = 0. \qquad (10.22)$$

Let us now suppose that the test body is in a circular orbit of coordinate radius r in the equatorial plane of the Schwarzschild geometry. Using the expressions we derived in Chapter 9 for the connection coefficients $\Gamma^\mu{}_{\nu\sigma}$ for the Schwarzschild metric in (t, r, θ, ϕ) coordinates (with $\theta = \pi/2$), one finds that most of the $\Gamma^\mu{}_{\nu\sigma}$ are zero. Moreover, for a test body in a circular orbit we have $u^1 = u^2 = 0$, and we find that the equations (10.22) reduce to

$$\frac{ds^0}{d\tau} + \Gamma^0{}_{10} s^1 u^0 = 0, \qquad (10.23)$$

$$\frac{ds^1}{d\tau} + \Gamma^1{}_{00} s^0 u^0 + \Gamma^1{}_{33} s^3 u^3 = 0, \qquad (10.24)$$

$$\frac{ds^2}{d\tau} = 0, \qquad (10.25)$$

$$\frac{ds^3}{d\tau} + \Gamma^3{}_{13} s^1 u^3 = 0, \qquad (10.26)$$

where the connection coefficients are given by

$$\Gamma^0{}_{10} = \frac{\mu}{r^2}\left(1-\frac{2\mu}{r}\right)^{-1}, \quad \Gamma^1{}_{00} = \frac{c^2\mu}{r^2}\left(1-\frac{2\mu}{r}\right),$$
$$\Gamma^1{}_{33} = -r\left(1-\frac{2\mu}{r}\right), \quad \Gamma^3{}_{13} = \frac{1}{r}.$$

Moreover, from our discussion in the previous section, we can write the test body's 4-velocity as

$$[u^\mu] = u^0(1,0,0,\Omega),$$

where $u^0 = dt/d\tau = (1-3\mu/r)^{-1/2}$ and $\Omega = d\phi/dt = (\mu c^2/r^3)^{1/2}$ are both constants.

Since $u^1 = u^2 = 0$ the orthogonality condition (10.21) reduces to

$$c^2(1-2\mu/r)s^0u^0 - r^2s^3u^3 = 0$$

and noting that $u^3/u^0 = \Omega$ we may express s^0 in terms of s^3:

$$s^0 = \frac{\Omega r^2}{c^2(1-2\mu/r)}s^3.$$

Using this result it is straightforward to show that equation (10.23) is equivalent to equation (10.26). Thus the system of equations reduces to

$$\frac{ds^1}{d\tau} - \frac{r\Omega}{u^0}s^3 = 0, \qquad \frac{ds^2}{d\tau} = 0, \qquad \frac{ds^3}{d\tau} + \frac{u^0\Omega}{r}s^1 = 0. \tag{10.27}$$

It is more convenient to convert the τ-derivatives to t-derivatives using $u^0 = dt/d\tau$. Then, on using the third equation to eliminate s^3 from the first, the system of equations becomes

$$\frac{d^2s^1}{dt^2} + \left(\frac{\Omega}{u^0}\right)^2 s^1 = 0, \qquad \frac{ds^2}{dt} = 0, \qquad \frac{ds^3}{dt} + \frac{\Omega}{r}s^1 = 0.$$

Let us take the initial spatial direction $\vec{s}$ of the spin vector to be radial, so that $s^2(0) = s^3(0) = 0$. The corresponding solution to our system of equations is easily shown to be

$$s^1(t) = s^1(0)\cos\Omega' t, \qquad s^2(t) = 0, \qquad s^3(t) = -\frac{\Omega}{r\Omega'}s^1(0)\sin\Omega' t, \tag{10.28}$$

where $\Omega' = \Omega/u^0 = \Omega(1-3\mu/r)^{1/2}$. This solution shows that the spatial part $\vec{s}$ of the spin vector *rotates* relative to the radial direction with a coordinate angular speed Ω' in the negative ϕ-direction. However, the radial direction itself rotates with coordinate angular speed Ω in the positive ϕ-direction, and it is the difference between these two speeds that gives rise to the *geodesic precession*

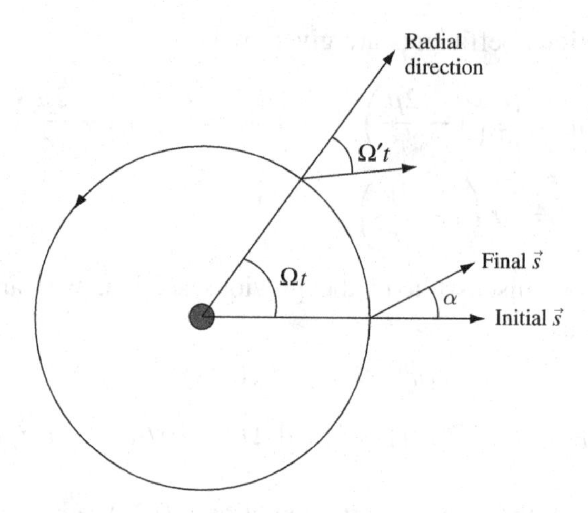

Figure 10.10 The geodesic precession effect for a spinning object in a circular orbit in the equatorial plane of the Schwarzschild geometry. Here the initial direction ($t = 0$) is radial.

effect. This is illustrated in Figure 10.10. Since one revolution is completed in a coordinate time $t = 2\pi/\Omega$, the final direction of $\vec{s}$ is therefore $2\pi + \alpha$, where $\alpha = (2\pi/\Omega)(\Omega - \Omega')$. Thus, after one revolution the spatial spin vector is rotated in the direction of the orbital motion by an angle

$$\boxed{\alpha = 2\pi[1 - (1 - 3\mu/r)^{1/2}].}$$

The geodesic precession effect may be observable experimentally by measuring the spacelike spin vector of a gyroscope in an orbiting spacecraft. Although the effect is small, it is cumulative. Thus, for a gyroscope in a near-Earth orbit, the precession rate is about $8''$ per year, which should be measurable. (In fact there is an additional very small effect, which may also be measurable, due to the fact that the Earth is slowly rotating and so the geometry outside it is correctly described by the Kerr metric). In April 2004, NASA launched the Gravity Probe B (GP-B) satellite to carry out this experiment and it is currently recording measurements; the results are eagerly awaited.

Exercises

10.1 Show that the equation of motion for planetary orbits in Newtonian gravity is

$$\frac{d^2u}{d\phi^2} + u = \frac{GM}{h^2},$$

where $u = 1/r$ and r is the radial distance from the centre of mass of the central object.

10.2 Show that the equation of motion in Exercise 10.1 has the solution

$$u = \frac{GM}{h^2}(1 + e\cos\phi),$$

and that this describes an ellipse. Show further that

$$a = \frac{h^2}{GM(1-e^2)},$$

where $r_1 = a(1-e)$ and $r_2 = a(1+e)$ are the distances of closest and furthest approach respectively.

10.3 Verify that the general-relativistic equation of motion for planetary orbits (10.3) has the solution

$$u = \frac{GM}{h^2}(1 + e\cos\phi) + \frac{3(GM)^2}{c^2h^4}\left[1 + e^2\left(\frac{1}{2} - \frac{1}{6}\cos 2\phi\right) + e\phi\sin\phi\right],$$

to first order in the relativistic perturbation to the Newtonian solution.

10.4 Verify that the general-relativistic equation of motion for a photon trajectory (10.9) has the solution

$$u = \frac{\sin\phi}{b} + \frac{3GM}{2c^2b^2}\left(1 + \frac{1}{3}\cos 2\phi\right),$$

to first order in the relativistic perturbation to a straight-line path.

10.5 Show that the gravitational deflection of light by the Sun predicted in the Newtonian theory of gravity is exactly half the value predicted in general relativity.

10.6 In the radar-echoes test, a photon travels from radial coordinate r to r_0, which is the radial coordinate of the closest approach of the photon to the Sun. Verify that, to first order in μ/r, the elapsed coordinate time is given by

$$t(r, r_0) = \frac{(r^2 - r_0^2)^{1/2}}{c} + \frac{2\mu}{c}\ln\left[\frac{r + (r^2 - r_0^2)^{1/2}}{r_0}\right] + \frac{\mu}{c}\left(\frac{r - r_0}{r + r_0}\right)^{1/2}.$$

10.7 An accretion disc extends from $r = 6\mu$ to $r = 20\mu$ in the equatorial plane of the Schwarzschild geometry. A photon is emitted radially outwards by a particle on the inner edge of the disc and is absorbed by a particle on the outer edge of the disc. Find the ratio of the energy absorbed to that emitted.

10.8 For a gyroscope in a circular orbit in the equatorial plane of the Schwarzschild geometry, show that the components s^μ of its spin 4-vector satisfy equations (10.23–10.26).

10.9 Show that the system of equations (10.27) has the solution (10.28).

10.10 A gyroscope in a circular orbit of radius r in the equatorial plane of the Schwarzschild geometry has its spatial spin vector $\vec{s}$ also lying in the equatorial plane. Show that, after one complete orbit, the angle between the initial and final directions of the spatial spin vector is given by

$$\alpha = 2\pi\left[1 - (1 - 3\mu/r)^{1/2}\right],$$

irrespective of the initial direction of the spin vector. Does this still hold if the original spatial spin vector does not lie in the plane of the orbit?

11

Schwarzschild black holes

In our discussion of the Schwarzschild *geometry*, we have thus far used the coordinates (t, r, θ, ϕ) to label events in the spacetime. In this context, (t, r, θ, ϕ) are called the *Schwarzschild coordinates*. Moreover, until now we have been concerned only with the *exterior region* $r > 2\mu$. We now turn to the discussion of the Schwarzschild geometry in the *interior region* $r < 2\mu$, and the significance of the hypersurface $r = 2\mu$. We shall see that, in order to understand the entire Schwarzschild geometry, we must relabel the events in spacetime using different sets of coordinates.

11.1 The characterisation of coordinates

Before discussing the Schwarzschild geometry in detail, let us briefly consider the characterisation of coordinates. In general, if we wish to write down a solution of Einstein's field equations then we need to do so in some particular coordinate system. But what, if any, is the significance of any such system? For example, suppose we take the Schwarzschild solution and apply some complicated coordinate transformation $x^\mu \to x'^\mu$. The resulting metric will still be a solution of the empty-space field equations, of course, but there is likely to be little or no physical or geometrical significance attached to the new coordinates x'^μ.

One thing we can do, however, is to establish whether at some event P a coordinate x^μ is *timelike*, *null* or *spacelike*. This corresponds directly to the nature of the tangent vector $\boldsymbol{e}_\mu$ to the coordinate curve at P. The easiest way to determine this property of the coordinate is to fix the other coordinates at their values at P and consider an infinitesimal variation dx^μ in the coordinate of interest. If the corresponding change in the interval ds^2 is positive, zero or negative then x^μ is timelike, null or spacelike respectively. This, in turn, corresponds simply to the sign of the relevant diagonal element $g_{\mu\mu}$ (no sum) of the metric.

11.2 Singularities in the Schwarzschild metric

With these ideas in mind, let us look at the Schwarzschild metric in the traditional (t, r, θ, ϕ) coordinate system. We have

$$ds^2 = c^2\left(1-\frac{2GM}{c^2 r}\right)dt^2 - \left(1-\frac{2GM}{c^2 r}\right)^{-1} dr^2 - r^2\,d\theta^2 - r^2\sin^2\theta\,d\phi^2. \quad (11.1)$$

Inspection of this line element shows immediately that the metric is *singular* at $r = 0$ and $r = 2GM/c^2$. The latter value is known as the *Schwarzschild radius* and is often denoted r_S, so that

$$\boxed{r_S = \frac{2GM}{c^2}.}$$

We must remember, however, that we derived the Schwarzschild solution by solving the *vacuum* field equations $R_{\mu\nu} = 0$, and so the metric given by (11.1) is only valid down to the surface of the spherical matter distribution. For example, the Schwarzschild radius for the Sun is

$$r_S = \frac{2GM_\odot}{c^2} = 2.95\,\text{km},$$

which is *much* smaller than the radius of the Sun ($R_\odot = 7 \times 10^5$ km). Similarly, the Schwarzschild radius for a proton is

$$r_S = \frac{2GM_p}{c^2} = 2.5 \times 10^{-54}\,\text{m},$$

again much smaller than the characteristic radius of a proton ($R_p = 10^{-15}$ m). In fact, for *most* real objects the Schwarzschild radius lies deep within the object, where the vacuum field equations do not apply. But *what if* there exist objects so compact that they lie well within the Schwarzschild radius? For such an object, the Schwarzschild solution looks very odd. Ignoring for the moment the singularity in the metric at $r = r_S$, let us denote the region $r > r_S$ as region I, and $r < r_S$ as region II.

From the Schwarzschild metric (11.1) we see that, in region I, the metric coefficient g_{00} is positive and the g_{ii} (for $i = 1, 2, 3$) are negative. It therefore follows that for $r > r_S$ the coordinate t is timelike and the coordinates r, θ, ϕ are spacelike. Indeed, in region I we may attach simple physical meanings to the coordinates. For example, t is the proper time measured by an observer at rest at infinity. Similarly, r is a radial coordinate with the property that the surface area of a 2-sphere $t =$ constant, $r =$ constant is $4\pi r^2$. In region II, however, the metric coefficients g_{00} and g_{11} change sign. Hence, for $r < r_S$, t is a *spacelike* coordinate and r is *timelike*. Thus 'time' and 'radial' coordinates swap character on either side of $r = r_S$. It is natural to ask what this means, and, indeed, whether it is physically meaningful.

Let us therefore consider in more detail the singularities in the metric at $r = 0$ and $r = r_S$. We must remember that coordinates are simply a way of labelling events in spacetime. The physically meaningful geometric quantites are the 4-tensors defined at any point on the spacetime manifold. Spacetime curvature is described covariantly by the components of the curvature tensor $R_{\mu\nu\rho\sigma}$ (and its contractions), which we may easily calculate for the Schwarzschild metric (11.1). For example, the curvature scalar at any point is given by

$$R_{\mu\nu\rho\sigma}R^{\mu\nu\rho\sigma} = \frac{48\mu^2}{r^6}, \tag{11.2}$$

which we see is *finite* at $r = r_S$. Moreover, since it is a scalar, its value remains the same in all coordinate systems. Thus the spacetime curvature at $r = r_S$ is perfectly well behaved, and so we see that $r = r_S$ is a *coordinate singularity*. By the same token, (11.2) is *singular* at $r = 0$ and so this point is a true *intrinsic singularity* of the Schwarzschild geometry.

We may illustrate the idea of coordinate singularities with a simple example. As discussed in Chapter 2, one may write the line element for the surface of a 2-sphere as

$$ds^2 = \frac{a^2 d\rho^2}{a^2 - \rho^2} + \rho^2 d\phi^2.$$

This line element has a singularity at $\rho = a$. Embedding this manifold, for the moment, in three-dimensional Euclidean space, we know that $\rho = a$ corresponds simply to the equator of the sphere (relative to the origin of the coordinate system) and it is clear why the (ρ, ϕ) coordinates cover the surface of the sphere uniquely only up to this point. There is nothing pathological occurring in the intrinsic geometry of the 2-sphere at the equator, i.e. there is no 'real' (or intrinsic) singularity in the metric. As shown in Appendix 7A, the Gaussian curvature of a 2-sphere is simply $K = 1/a^2$, which does not 'blow up' anywhere. Thus, $\rho = a$ is only a *coordinate singularity*, which has resulted simply from choosing coordinates with a restricted domain of validity. In an analogous way, the coordinate singularity of the Schwarzschild metric is simply a result of the coordinate system that we have chosen to use. We can *remove* it by making appropriate transformations of coordinates, which we will discuss later. For the time being, however, let us continue our investigation of the Schwarzschild geometry using the Schwarzschild coordinates (t, r, θ, ϕ).

11.3 Radial photon worldlines in Schwarzschild coordinates

Let us investigate the spacetime diagram of the Schwarzschild solution in (t, r, θ, ϕ) coordinates. The metric reads

$$ds^2 = c^2\left(1 - \frac{2\mu}{r}\right)dt^2 - \left(1 - \frac{2\mu}{r}\right)^{-1} dr^2 - r^2\, d\Omega^2,$$

where $d\Omega$ is an element of solid angle. We have written it in this form because we shall usually ignore the angular coordinates in drawing spacetime diagrams, i.e. these diagrams will show the (r, ct)-plane for *fixed* values of θ and ϕ.

We begin by determining the *lightcone structure* in the diagram, by considering the paths of radially incoming and outgoing photons; these were discussed briefly in Section 9.11. From the metric, for a radially moving photon we have

$$\frac{dt}{dr} = \pm\frac{1}{c}\left(1 - \frac{2\mu}{r}\right)^{-1},$$

where the plus sign corresponds to a photon that is *outgoing* (in that dr/dt is positive in the region $r > 2\mu$) and the minus sign corresponds to a photon that is *incoming* (in that dr/dt is negative in the region $r > 2\mu$). On integrating, we obtain

$$ct = r + 2\mu \ln\left|\frac{r}{2\mu} - 1\right| + \text{constant} \qquad \text{(outgoing photon)},$$

$$ct = -r - 2\mu \ln\left|\frac{r}{2\mu} - 1\right| + \text{constant} \qquad \text{(incoming photon)}.$$

Notice that under the transformation $t \to -t$ the incoming and outgoing photon paths are interchanged, as we would expect. We can now plot these curves in the (r, ct)-plane, as shown in Figure 11.1. The diagram is drawn for fixed θ and ϕ. Since the diagram will be the same for all other θ and ϕ, we should think of each point (r, ct) in the diagram as representing a 2-sphere of area $4\pi r^2$.

Figure 11.1 requires some words of explanation. At large radii in region I the gravitational field becomes weak and the metric tends to the Minkowski metric of special relativity. Thus, as expected, the lightcone structure becomes that of Minkowski spacetime, where incoming and outgoing light rays define straight lines of slope ± 1 in the diagram. As we approach the Schwarzschild radius, the ingoing light rays tend to the ordinate $t \to +\infty$ and outgoing light rays tend to $t \to -\infty$. This seems to suggest that it takes an infinite time for an incoming signal to cross the Schwarzschild radius, but in this respect the diagram is misleading, as we shall see shortly (we discussed this point briefly in Section 9.11).

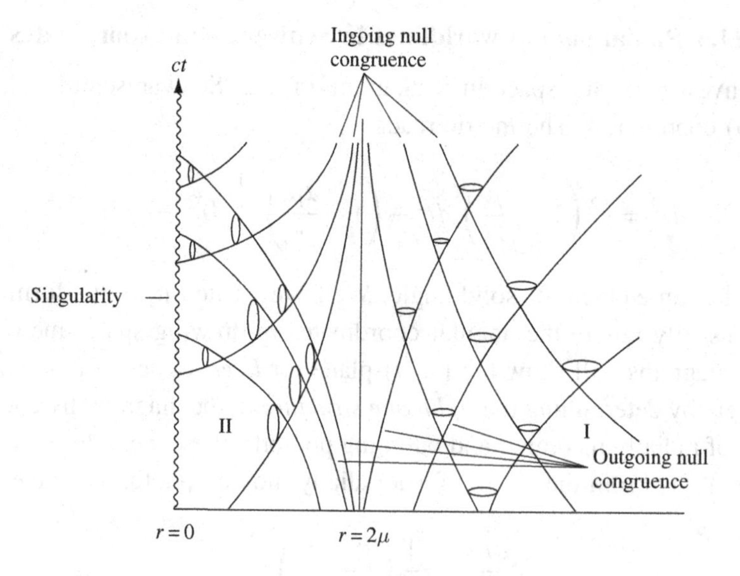

Figure 11.1 Lightcone structure of the Schwarzschild solution.

In region II the lightcones flip their orientation by 90°, since the coordinates t and r reverse their character. We see that all photons in this region *must* end up at $r = 0$. At this point there is *real* singularity, where the curvature of the Schwarzschild solution diverges. Moreover, any massive particle in region II must also end up at the singularity, since a timelike worldline must lie within the forward light-cone at each point. Thus we conclude that *once within the Schwarzschild radius you necessarily end up at a spacelike singularity at* $r = 0$. To escape would require a violation of causality.

11.4 Radial particle worldlines in Schwarzschild coordinates

The causal structure in Figure 11.1 is determined by radially moving photons. It is also of interest to determine the worldlines of radially moving massive particles in Schwarzschild coordinates. For simplicity let us consider an infalling particle released from rest at infinity, which we investigated in detail in Chapter 9. Parameterising the particle worldline in terms of the proper time τ, we found that the trajectory $r(\tau)$ could be written implicitly as

$$\tau = \frac{2}{3}\sqrt{\frac{r_0^3}{2\mu c^2}} - \frac{2}{3}\sqrt{\frac{r^3}{2\mu c^2}}, \tag{11.3}$$

taking $\tau = 0$ at $r = r_0$. Alternatively, if the trajectory is described as $r(t)$, where t is the coordinate time, we found that

$$t = \frac{2}{3}\left(\sqrt{\frac{r_0^3}{2\mu c^2}} - \sqrt{\frac{r^3}{2\mu c^2}}\right) + \frac{4\mu}{c}\left(\sqrt{\frac{r_0}{2\mu}} - \sqrt{\frac{r}{2\mu}}\right)$$
$$+ \frac{2\mu}{c}\ln\left|\left(\frac{\sqrt{r/(2\mu)}+1}{\sqrt{r/(2\mu)}-1}\right)\left(\frac{\sqrt{r_0/(2\mu)}-1}{\sqrt{r_0/(2\mu)}+1}\right)\right|, \tag{11.4}$$

where $t = 0$ at $r = r_0$. Using equations (11.3) and (11.4), we can associate a given value of the particle's proper time τ with a point in a (r, ct)-diagram. Thus, as τ increases, we can plot out the particle trajectory in the (r, ct)-plane.

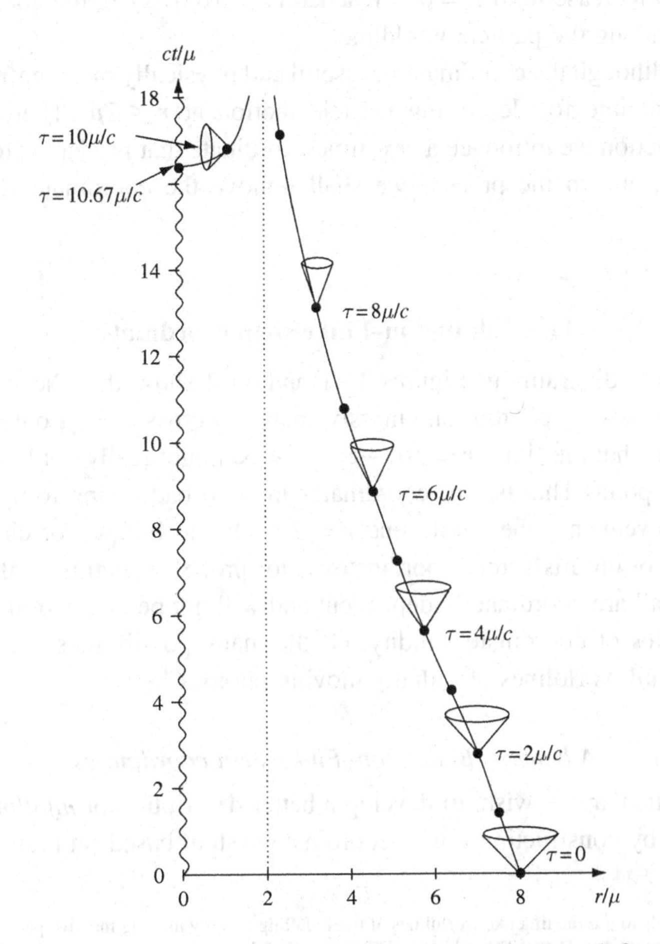

Figure 11.2 Trajectory of a radially infalling particle released from rest at infinity. The dots correspond to unit intervals of $c\tau/\mu$, where τ is the particle's proper time and we have taken $\tau = t = 0$ at $r_0 = 8\mu$.

The corresponding curve is shown in Figure 11.2, which is a more quantitative version of Figure 9.6; we have taken $\tau = t = 0$ at $r_0 = 8\mu$. Also plotted are dots showing unit intervals of $c\tau/\mu$, together with the light-cone structure at particular points on the trajectory. We see from the plot that the particle worldline has a singularity at $r = 2\mu$ and that it takes an *infinite* coordinate time t for the particle to travel from $r = 8\mu$ to $r = 2\mu$. Since t is the time experienced by a stationary observer at large radius, to such an observer it thus takes an *infinite* time for the particle to reach $r = 2\mu$. However, the proper time taken by the particle to reach $r = 2\mu$ is *finite* ($\tau = 9.33\mu/c$). Moreover, we see that for later values of τ the particle worldline lies in the region $r < 2\mu$, which was not plotted in Figure 9.6. In this region the coordinates t and r swap character, as indicated by the fact that the light-cone is flipped by 90°. For $r < 2\mu$, we also note that, although τ continues to increase until $r = 0$ is reached ($\tau = 10.67\mu/c$), the coordinate time t *decreases* along the particle worldline.

Clearly, although the coordinate t is useful and physically meaningful as $r \to \infty$, it is inappropriate for describing particle motion at $r \leq 2\mu$. Therefore, in the following section we introduce a new time coordinate that is adapted to describing radial infall, and in the process we shall remove the coordinate singularity at $r = 2\mu$.

11.5 Eddington–Finkelstein coordinates

The spacetime diagrams in Figures 11.1 and 11.2 show that the worldlines of both radially moving photons and massive particles cross $r = 2\mu$ only at $t = \pm\infty$. This suggests that the 'line' $r = 2\mu$, $-\infty < t < \infty$ might really not be a line at all, but a single point. That is, our coordinates may go bad owing to the expansion of a single event into the whole line $r = 2\mu$. One technique for circumventing the problem of unsatisfactory coordinates is to 'probe' spacetime with geodesics, which after all are coordinate independent and will not be affected in any way by the boundaries of coordinate validity. Of the many possibilities, we will use as probes the null worldlines of radially moving photons.[1]

Advanced Eddington–Finkelstein coordinates

Since in particular, we wish, to develop a better description of *infalling* particles, let us begin by constructing a new coordinate system based on radially infalling

[1] It is also possible to use the timelike worldlines of freely falling radially moving massive particles as probes of the spacetime geometry. The traditional approach leads to useful new coordinates, called *Novikov coordinates*, but they are related to Schwarzschild coordinates by transformations that are algebraically very complicated. A more physically meaningful set of new coordinates that are also based on radially moving massive particle geodesics is discussed in Exercise 11.9.

photons. Recall that the worldline of a radially ingoing photon is given by

$$ct = -r - 2\mu \ln \left| \frac{r}{2\mu} - 1 \right| + \text{constant}.$$

The trick is to use the *integration constant* as the new coordinate, which we denote by p. Thus, we make the coordinate transformation

$$p = ct + r + 2\mu \ln \left| \frac{r}{2\mu} - 1 \right|, \tag{11.5}$$

where p, for historical reasons, is known as the *advanced time parameter* and is clearly a null coordinate (see Section 11.1). Since p is constant along the entire worldline of the radially ingoing photon, it will be a 'good' coordinate wherever that worldline penetrates.

Differentiating (11.5), we obtain

$$dp = c\,dt + \frac{r}{r - 2\mu}\,dr,$$

and, on substituting for dt in the Schwarzschild line element, we find that in terms of the parameter p the line element takes the simple form

$$ds^2 = \left(1 - \frac{2\mu}{r}\right) dp^2 - 2\,dp\,dr - r^2\left(d\theta^2 + \sin^2\theta\, d\phi^2\right). \tag{11.6}$$

We see immediately from (11.6) that ds^2 is now *regular* at $r = 2\mu$; indeed it is regular for the whole range $0 < r < \infty$, which is the range of r-values probed by an infalling photon geodesic. Thus, in some sense, the transformation (11.5) has extended the coordinate range of the solution in a way reminiscent of the *analytic continuation* of a complex function.

One might object that the coordinate transformation (11.5) cannot be used at $r = 2\mu$ because it becomes singular. This must happen, however, if one is to remove the coordinate singularity there. In any case, this transformation takes the standard form (11.1) for the Schwarzschild line element to the form (11.6). Given these two solutions, we can simply ask, what is the largest range of coordinates for which each solution is regular? For the standard form this is $2\mu < r < \infty$, whereas for the new form (11.6) it is $0 < r < \infty$. In the overlap region $2\mu < r < \infty$ the two solutions are related by (11.5), and hence they must represent the *same* solution in this region.

As one might expect, the metric (11.6) is especially convenient for calculating the paths of null geodesics. In particular, we see that radial null geodesics (for which $ds = d\theta = d\phi = 0$) are given by

$$\left(1-\frac{2\mu}{r}\right)\left(\frac{dp}{dr}\right)^2 - 2\frac{dp}{dr} = 0,$$

which has the two solutions

$$\frac{dp}{dr} = 0 \qquad \Rightarrow \qquad p = \text{constant},$$

$$\frac{dp}{dr} = 2\left(1-\frac{2\mu}{r}\right)^{-1} \qquad \Rightarrow \qquad p = 2r + 4\mu \ln\left|\frac{r}{2\mu} - 1\right| + \text{constant}, \tag{11.7}$$

which correspond to incoming and outgoing radial null geodesics respectively (the former being valid by construction).

Since p is a null coordinate, which might be intuitively unfamiliar, it is common practice to work instead with the related *timelike* coordinate t', defined by

$$ct' \equiv p - r = ct + 2\mu \ln\left|\frac{r}{2\mu} - 1\right|. \tag{11.8}$$

The line element then takes the form

$$\boxed{ds^2 = c^2\left(1-\frac{2\mu}{r}\right)dt'^2 - \frac{4\mu c}{r}\,dt'dr - \left(1+\frac{2\mu}{r}\right)dr^2 - r^2\left(d\theta^2 + \sin^2\theta\, d\phi^2\right),} \tag{11.9}$$

which is again regular for the whole range $0 < r < \infty$. The coordinates (t', r, θ, ϕ) are called *advanced Eddington–Finkelstein* coordinates. We note that the line element (11.9) is not invariant with respect to the transformation $t' \to -t'$, under which the second term on the right-hand side changes sign. From (11.7), we see that incoming and outgoing photon worldlines are given by

$$ct' = -r + \text{constant}, \tag{11.10}$$

$$ct' = r + 4\mu \ln\left|\frac{r}{2\mu} - 1\right| + \text{constant}. \tag{11.11}$$

The first equation, for ingoing photons, corresponds to a straight line making an angle of 45° with the r-axis and is valid for $0 < r < \infty$. Thus the photon geodesics are continuous straight lines across $r = 2\mu$. The spacetime diagram

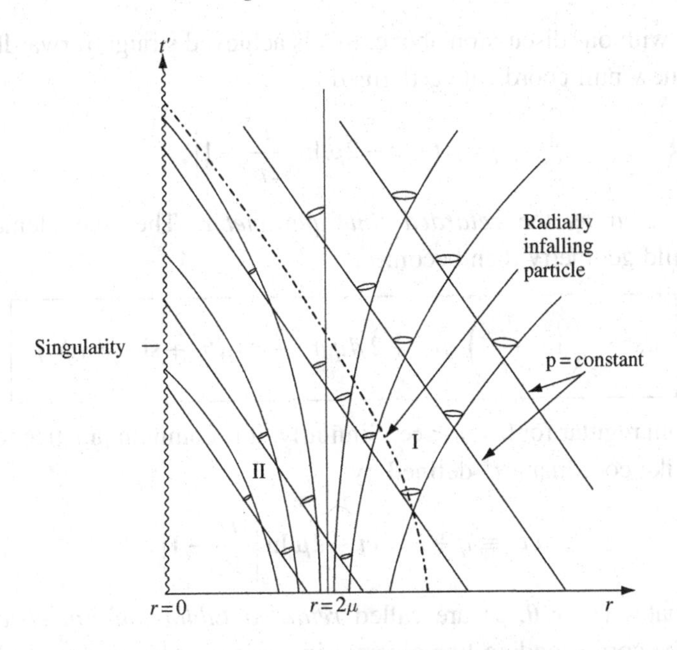

Figure 11.3 Lightcone structure in advanced Eddington–Finkelstein coordinates.

of the Schwarzschild geometry in advanced Eddington–Finkelstein coordinates is shown in Figure 11.3.

The spacetime diagram now appears more sensible. It is straightforward to see that the radial trajectory of an infalling particle or photon is continuous at the Schwarzschild radius $r = 2\mu$. The lightcone structure changes at the Schwarzschild radius and, as you can see from the diagram, once you have crossed the boundary $r = 2\mu$ your future is directed towards the singularity. Similarly, it can be seen that a photon (or particle) starting at $r < 2\mu$ cannot escape to the region $r > 2\mu$. The Schwarzschild radius $r = 2\mu$ defines an *event horizon*, a boundary of no return. Once a particle crosses the event horizon it must fall to the singularity at $r = 0$. Moreover, from the paths of the 'outgoing' null geodesics, we see that any photons emitted by the infalling particle at $r < 2\mu$ will not reach an observer in region I. Thus to such an observer the particle appears never to cross the event horizon. A compact object that has an event horizon is called a *black hole*.

Retarded Eddington–Finkelstein coordinates

One may reasonably ask what occurs if one instead chooses to construct a new coordinate system based on the worldlines of *radially outgoing* photons.

By analogy with our discussion above, this is achieved straightforwardly by introducing the new null coordinate q defined by

$$q = ct - r - 2\mu \ln \left| \frac{r}{2\mu} - 1 \right|,$$

which is known as the *retarded time parameter*. The line element of the Schwarzschild geometry then becomes

$$\boxed{ds^2 = \left(1 - \frac{2\mu}{r}\right) dq^2 + 2\, dq\, dr - r^2(d\theta^2 + \sin^2\theta\, d\phi^2).}$$

which is again regular for $0 < r < \infty$. Similarly, it is common practice to introduce a new timelike coordinate t^* defined by

$$ct^* \equiv q + r = ct - 2\mu \ln \left| \frac{r}{2\mu} - 1 \right|.$$

The coordinates (t^*, r, θ, ϕ) are called *retarded Eddington–Finkelstein* coordinates, and the corresponding line element in these coordinates is simply the time reversal of the advanced Eddington–Finkelstein line element (11.9).

It is straightforward to draw an spacetime diagram analogous to Figure 11.3 in retarded Eddington–Finkelstein coordinates, and one finds that (by construction) the *outgoing* radial null geodesics are continuous straight lines at 45° but the *ingoing* null rays are discontinuous, tending to $t^* = +\infty$ at $r = 2\mu$. In this case, the surface $r = 2\mu$ again acts as a one-way membrane, but this time letting only outgoing timelike or null geodesics cross from inside to outside. Indeed, particles must move away from the singularity at $r = 0$ and are forcibly expelled from the region $r < 2\mu$. Such an object is called a *white hole*.

This behaviour appears completely at odds with our intuition regarding the gravitational attraction of a massive body. Moreover, how can the physical processes that occur be so radically different depending on one's choice of coordinates, since we have maintained throughout that coordinates are merely arbitrary labels of spacetime events? The key to resolving this apparent paradox is to realise that our original coordinates (t, r, θ, ϕ) covered only a part of the 'full' Schwarzschild geometry. This topic is discussed fully in Section 11.9, in which we introduce Kruskal coordinates, which cover the entire geometry and which show that it possesses both a black-hole and a white-hole singularity. The advanced Eddington–Finkelstein coordinates 'extend' the solution into the (more familiar) part of the manifold that constitutes a black hole, whereas the retarded Eddington–Finkelstein coordinates extend the solution into a different part of the manifold, corresponding to a white hole. As we will discuss in Section 11.9, the existence of white holes as a physical reality (as opposed to a mathematical curiosity) is rather doubtful. Black holes, however, *are* likely to occur physically, as we now go on to discuss.

11.6 Gravitational collapse and black-hole formation

Our investigation of the properties of a black hole would be largely academic unless there were reasons for believing that they might exist in Nature. The possibility of their existence arises from the idea of *gravitational collapse*.

A star is held up by a mixture of gas and radiation pressure, the relative contributions depending on its mass. The energy to provide this pressure support is derived from the fusion of light nuclei into heavier ones, predominantly hydrogen into helium, which releases about 26 MeV for each atom of He that is formed. When all the nuclear fuel is used up, however, the star begins to cool and collapse under its own gravity. For most stars, the collapse ends in a high-density stellar remnant known as a *white dwarf*. In fact, we expect that in around 5 billion years the Sun will collapse to a form a white dwarf with a radius of about 5000 km and a spectacularly high mean density of about $10^9\,\mathrm{kg\,m^{-3}}$.

Astronomers have known about white dwarfs since as long ago as 1915 (the earliest example being the companion to the bright star Sirius, known as Sirius B), but nobody at the time knew how to explain them. The physical mechanism providing the internal pressure to support such a dense object was a mystery. The answer had to await the development of quantum mechanics and the formulation of Fermi–Dirac statistics. Fowler realised in 1926 that white dwarfs were held up by *electron degeneracy pressure*. The electrons in a white dwarf behave like the free electrons in a metal, but the electron states are widely spaced in energy because of the small size of the star in its white-dwarf form. Because of the Pauli exclusion principle, the electrons completely fill these states up to a high characteristic Fermi energy. It is these high electron energies that save the star from collapse.

In 1930, Chandrasekhar realized that the more massive a white dwarf, the denser it must be and so the stronger the gravitational field. For white dwarfs over a critical mass of about $1.4\,M_\odot$ (now called the Chandrasekhar limit), gravity would overwhelm the degeneracy pressure and no stable solution would be possible. Thus, the gravitational collapse of the object must continue. At first it was thought that the white dwarf must collapse to a point. After the discovery of the neutron, however, it was realized that at some stage in the collapse the extremely high densities occurring would cause the electrons to interact with the protons via inverse β-decay to form neutrons (and neutrinos, which simply escape). A new stable configuration – a *neutron star* – was therefore possible in which the pressure support is provided by degenerate neutrons. A neutron star of one solar mass would have a radius of only 30 km, with a density of around $10^{16}\,\mathrm{kg\,m^{-3}}$. Since the matter in a neutron star is at nuclear density, the gravitational forces inside the star are extremely strong. In fact, it is the first point in the evolution of a stellar object at which general relativistic effects are expected to be important (we will discuss relativistic stars in Chapter 12).

Given the extreme densities inside a neutron star, there remain uncertainties in the equation of state of matter. Nevertheless, it is believed that (as for white dwarfs), there exists a maximum mass above which no stable neutron-star configuration is possible. This maximum mass is believed to be about $3\,M_{\odot}$ (which is known as the Oppenheimer–Volkoff limit). Thus, we believe that stars more massive than this limit should collapse to form *black holes*. Moreover if the collapse is spherically symmetric then it must produce a Schwarzschild black hole.

Some theorists were very sceptical about the formation of black holes. The Schwarzschild solution in particular is very special – it is exactly spherically symmetric by construction. In reality, a star will not be perfectly symmetric and so perhaps, as it collapses, the asymmetries will amplify and avoid the formation of an event horizon. In the early 1960s, however, Penrose applied global geometrical techniques to prove a famous series of 'singularity theorems'. These showed that in realistic situations an event horizon (a closed trapped surface) will be formed and that there must exist a singularity within this surface, i.e. a point at which the curvature diverges and general relativity ceases to be valid. The singularity theorems were important in convincing people that black holes must form in Nature. In Appendices 11A and 11B, we discuss some of the observational evidence for the existence of black holes. As we will see, there is compelling evidence that black holes do indeed exist. Furthermore, as mentioned in Section 10.4, it should become possible within the next few years not only to measure the masses of black holes but also to measure their angular momenta, using powerful X-ray telescopes! Direct experimental probes of the strong-gravity regime are now possible.

11.7 Spherically symmetric collapse of dust

Let us consider the spherically symmetric collapse of a massive star to form a Schwarzschild black hole and also the view this process seen by a stationary observer at large radius. For simplicity, we consider the case in which the star has a uniform density and the internal pressure is assumed to be zero. In the absence of pressure gradients to deflect their motion, the particles on the outer surface of this 'ball of dust' will simply follow radial geodesics. In order to simplify our analysis still further, we will assume that initially the surface of the 'star' is at rest at infinity.[2] In this case, the particles on the surface will follow the radial geodesics we discussed earlier.

Consider two observers participating in the gravitational collapse of the spherical star. One observer rides the surface of the star down to $r = 0$, and the other observer remains fixed at a large radius. Moreover, suppose that the infalling observer carries a clock and communicates with the distant one by sending out radial light signals at equal intervals according to this clock. Figure 11.4 shows the relevant spacetime diagram in advanced Eddington–Finkelstein coordinates (ct', r), with θ and ϕ suppressed. The dots denote unit intervals of ct/μ and we have chosen $\tau = t' = 0$ at $r = 8\mu$. This diagram is easily constructed from the results that were used to obtain Figure 11.2.

For a distant observer at fixed r, we know that the standard Schwarzschild coordinate time t measures proper time. From (11.8), however, we see that if r is fixed then $dt' = dt$. Thus, a unit interval of t' corresponds to a unit interval of

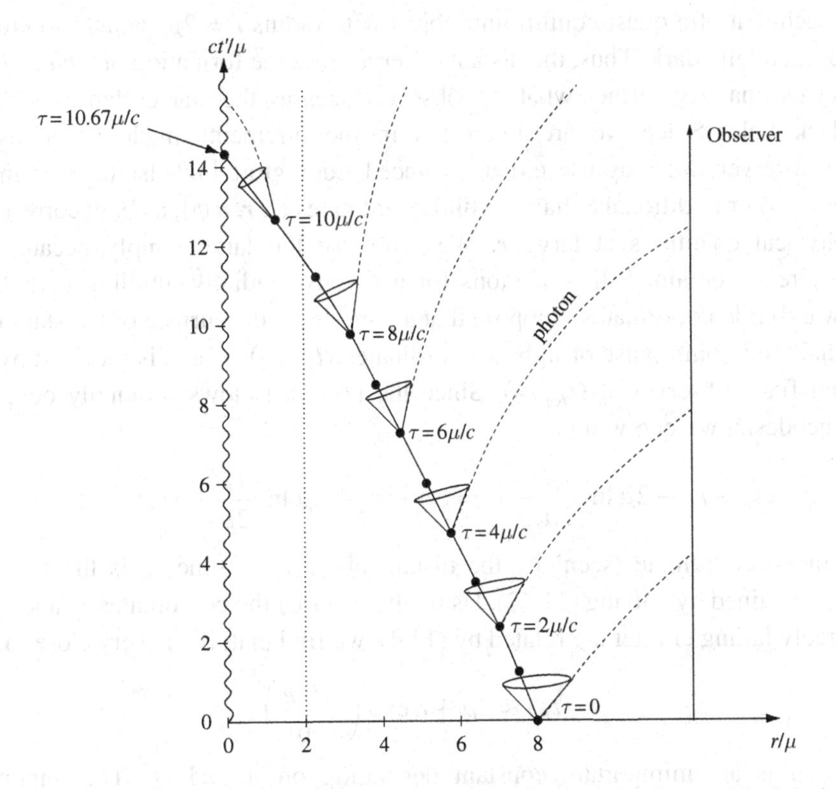

Figure 11.4 Collapse of the surface of a pressureless star to form a black hole in advanced Eddington–Finkelstein coordinates. The star's surface started at rest at infinity, and we have chosen $\tau = t' = 0$ at $r = 8\mu$.

[2] This is equivalent to the collapse commencing with the star's surface at some finite radius $r = r_0$ with some finite inwards velocity.

proper time for a distant fixed observer. From the diagram, we see that the light pulses are not received at equal intervals of t'. Rather, the proper time interval measured by the distant observer between each pulse steadily increases. Indeed, the last light pulse to reach this observer is the one emitted just before the surface of the star crosses $r = 2\mu$. The worldine of this photon is simply the vertical line $r = 2\mu$, and so this pulse would only 'reach' the distant observer at $t' = \infty$. Pulses emitted after the surface of the star has crossed the event horizon do not progress to larger r but instead progress to smaller r and end up at the singularity at $r = 0$.

Thus, the distant observer never sees the star's surface cross the radius $r = 2\mu$. Furthermore, the pulses emitted at equal intervals by the falling observer's clock arrive at the distant observer at increasingly longer intervals. Correspondingly, the photons received by the distant observer are increasingly redshifted, the redshift tending to infinity as the star's surface approaches $r = 2\mu$. Both these effects mean that the distant observer sees the luminosity of the star fall to zero. To summarise, the distant observer sees the collapse slow down and the star's state approach that of a quasi-equilibrium object with radius $r = 2\mu$, which eventually becomes totally dark. Thus, the distant observer sees the formation of a *black hole*.

Let us quantify further what the observer sees as the star collapses to form a black hole. Since we are interested in measurements made by a distant fixed observer, we may use either advanced Eddington–Finkelstein coordinates (t', r, θ, ϕ) or traditional Scharwzschild coordinates $(t, r, \theta\, \phi)$, as both correspond to physical quantities at large r. We shall use the latter simply because we have already obtained the equations for a massive radially infalling particle in Schwarzschild coordinates. Suppose that a particle on the surface of the star emits a radially outgoing pulse of light at coordinates (t_E, r_E), which is received by the distant fixed observer at (t_R, r_R). Since the photon follows a radially outgoing null geodesic, we can write

$$ct_E - r_E - 2\mu \ln\left|\frac{r_E}{2\mu} - 1\right| = ct_R - r_R - 2\mu \ln\left|\frac{r_R}{2\mu} - 1\right|. \tag{11.12}$$

The radial coordinate 'seen' by the distant observer at time t_R is the function $r_E(t_R)$ obtained by solving (11.12). Using the fact that the coordinates t_E and r_E of the freely falling emitter are related by (11.4), we find that, if r is very close to 2μ,

$$r_E(t_R) \approx 2\mu + a \exp\left(-\frac{ct_R}{4\mu}\right), \tag{11.13}$$

where a is an unimportant constant depending on μ and r_R. The important consequence of this result is that the radius $r = 2\mu$ is approached *exponentially*, as seen by the distant observer, with a characteristic time $4\mu/c$. Since

$$\frac{\mu}{c} = \frac{GM}{c^3} = 5 \times 10^{-6} \left(\frac{M}{M_\odot}\right) \text{seconds},$$

the time scale for stellar-size objects is very small by the usual astrophysical standards. Thus for any collapse even approximately like the free-fall collapse described here, the approach to a black hole is extremely rapid.

Let us work out the redshift seen by the distant observer as a function of time t. The ratio of the frequencies of a photon at emission and reception is

$$\frac{\nu_R}{\nu_E} = \frac{u_R^\mu p_\mu(R)}{u_E^\mu p_\mu(E)}, \tag{11.14}$$

where $\boldsymbol{u}_E$ and $\boldsymbol{u}_R$ are the 4-velocities of the emitter and receiver respectively and $\boldsymbol{p}$ is the photon 4-momentum. The 4-velocity of our emitter riding on the star's surface is

$$[u_E^\mu] = [(1-2\mu/r)^{-1}, -(2\mu c^2/r)^{1/2}, 0, 0],$$

whereas the 4-velocity of the stationary observer at infinity is

$$[u_R^\mu] = [1, 0, 0, 0].$$

Hence (11.14) reduces to

$$\frac{\nu_R}{\nu_E} = \frac{p_0(R)}{u_E^0 p_0(E) + u_E^1 p_1(E)} = \left[u_E^0 + \frac{p_1(E)}{p_0(E)} u_E^1\right]^{-1},$$

where we have used that fact that the Schwarzschild metric is stationary and so p_0 is conserved along the photon geodesic. Moreover, since $\boldsymbol{p}$ is null we require $g^{\mu\nu} p_\mu p_\nu = 0$, which in our case reduces to

$$\frac{1}{c^2}\left(1 - \frac{2\mu}{r}\right)^{-1} (p_0)^2 - \left(1 - \frac{2\mu}{r}\right)(p_1)^2 = 0.$$

So, for a radially outgoing photon, $p_1 = -(1-2\mu/r)^{-1} p_0/c$ and we find that

$$\frac{\nu_R}{\nu_E} = \left(1 - \frac{2\mu}{r}\right)\left[1 + \left(\frac{2\mu}{r}\right)^{1/2}\right]^{-1} = 1 - \left(\frac{2\mu}{r}\right)^{1/2}. \tag{11.15}$$

As $r \to 2\mu$ we see that $\nu_R \to 0$, so the redshift is infinite. By Taylor-expanding (11.15) about $r = 2\mu$, we find that for r_E close to 2μ we can write

$$\frac{\nu_R}{\nu_E} \approx \frac{r - 2\mu}{4\mu};$$

however, near the event horizon the time of reception is given by (11.13). Hence

$$\frac{\nu_R}{\nu_E} \sim \exp\left(-\frac{ct}{4\mu}\right),$$

so that the redshift goes exponentially to infinity with a characteristic time $4\mu/c$. The computation of the luminosity is more complicated since it involves non-radial photon geodesics also. Nevertheless, using the above analysis we see that the time intervals between successive photons will also decrease as $\sim \exp[-ct/(4\mu)]$ and so we expect the luminosity to decay exponentially as $\sim \exp[-ct/(2\mu)]$.

11.8 Tidal forces near a black hole

As discussed in Section 7.14, in Newtonian gravity a distribution of non-interacting particles freely falling towards the Earth will be elongated in the direction of motion and compressed in the transverse directions, as a result of gravitational tidal forces. The same effect occurs in a body falling towards a spherical object in general relativity, but if the object is a black hole then the effect becomes infinite at $r = 0$.

We may calculate the tidal forces in the Schwarzschild geometry, working in traditional Schwarzschild coordinates (t, r, θ, ϕ). At any particular point in space, the tidal forces have the same form for *any* (close) pair of particles that are in free fall. Thus, it is easiest to calculate the tidal forces at some coordinate radius r for the case in which the two particles are released from rest at r. In this case, a frame of orthonormal basis vectors defining the inertial instantaneous rest frame of one of the particles may be taken as

$$(\hat{e}_0)^\mu = \frac{1}{c}u^\mu = \frac{1}{c}\left(1 - \frac{2\mu}{r}\right)^{-1/2}\delta^\mu_0, \qquad (\hat{e}_1)^\mu = \left(1 - \frac{2\mu}{r}\right)^{1/2}\delta^\mu_1,$$

$$(\hat{e}_2)^\mu = \frac{1}{r}\delta^\mu_2, \qquad (\hat{e}_3)^\mu = \frac{1}{r\sin\theta}\delta^\mu_3.$$

Substituting these expressions into (7.28), together with the appropriate expressions for the components of the Riemann tensor in Schwarzschild coordinates, from (7.27) we obtain (after some algebra) that the spatial components of the orthogonal connecting vector between the two particles satisfy

$$\frac{d^2\zeta^{\hat{r}}}{d\tau^2} = +\frac{2\mu c^2}{r^3}\zeta^{\hat{r}}, \qquad \frac{d^2\zeta^{\hat{\theta}}}{d\tau^2} = -\frac{\mu c^2}{r^3}\zeta^{\hat{\theta}}, \qquad \frac{d^2\zeta^{\hat{\phi}}}{d\tau^2} = -\frac{\mu c^2}{r^3}\zeta^{\hat{\phi}}.$$

The positive sign in the $\zeta^{\hat{r}}$-equation indicates a tension or stretching in the radial direction and the negative signs in the $\zeta^{\hat{\theta}}$- and $\zeta^{\hat{\phi}}$- equations indicate a pressure or compression in the transverse directions. Note the $1/r^3$ radial dependence in each case, which is characteristic of tidal gravitational forces. Moreover, the equations reveal that the tidal forces do not undergo any 'transition' at $r = 2\mu$ but become infinite at $r = 0$.

Let us consider an intrepid astronaut falling feet first into a black hole. The equations derived above will not hold exactly, since there will exist forces between the particles (atoms) that comprise the astronaut. Nevertheless, when the tidal gravitational forces become strong we can neglect the interatomic forces, and the equations derived above will be valid to an excellent approximation. Thus the unfortunate astronaut would be stretched out like a piece of spaghetti (!), as illustrated in Figure 11.5. In fact, not only do the tidal forces tear the astronaut to pieces, but the very atoms of which the astronaut is composed must ultimately suffer the same fate! Assuming that the limit of tolerance to stretching or compression of a human body is an acceleration gradient of $\sim 400\,\mathrm{m\,s^{-2}}$ per metre, for a human to survive the tidal forces at the Schwarzschild radius requires a very massive black hole with

$$M \gtrsim 10^4 M_\odot.$$

If you fell towards a supermassive black hole, with say $M \sim 10^9 M_\odot$ (such black holes are believed to lie at the centres of some galaxies; see Appendix 11B) you would cross the event horizon without feeling a thing. However, your fate will have been sealed – you will end up shredded by the tidal forces of the black hole as you approach the singularity, from which there is no escape. If you fell towards a 'small' black hole, of mass say $10\,M_\odot$, you would be shredded by the tidal forces of the hole well before you reached the event horizon.

Figure 11.5 An astronaut stretched by the tidal forces of a black hole. For a human to survive this stretching at the Schwarzschild radius requires a very massive black hole, with $M \gtrsim 10^4 M_\odot$

11.9 Kruskal coordinates

In our discussion of advanced and retarded Eddington–Finkelstein coordinates, we found that neither coordinate system was completely satisfactory. In the advanced coordinates the outgoing null rays are discontinuous, and in the retarded coordinates the ingoing null rays are discontinuous. It is natural to ask whether it is possible to find a system of coordinates in which both the incoming and outgoing radial photon geodesics are continuous straight lines. Such a coordinate system was indeed discovered in 1961 by Martin Kruskal, and it serves also to clarify the structure of the complete Schwarzschild geometry.

An obvious way to begin is to introduce both the advanced null coordinate p and the retarded null coordinate q that we met during our discussion of Eddington–Finkelstein coordinates. In the coordinates (p, q, θ, ϕ) the Schwarzschild metric becomes

$$ds^2 = \left(1 - \frac{2\mu}{r}\right) dp\,dq - r^2(d\theta^2 + \sin^2\theta\, d\phi^2), \qquad (11.16)$$

where r is considered as a function of p and q, defined implicitly by

$$\tfrac{1}{2}(p-q) = r + 2\mu \ln\left|\frac{r}{2\mu} - 1\right|.$$

Our new system of coordinates has some appealing properties. Most importantly, the 2-space defined by $\theta = \text{constant}$, $\phi = \text{constant}$ has the simple metric

$$ds^2 = \left(1 - \frac{2\mu}{r}\right) dp\,dq. \qquad (11.17)$$

Transforming from the null coordinates p and q to the new coordinates

$$ct = \tfrac{1}{2}(p+q), \qquad (11.18)$$

$$\tilde{r} = \tfrac{1}{2}(p-q) = r + 2\mu \ln\left|\frac{r}{2\mu} - 1\right|, \qquad (11.19)$$

where t is the standard Schwarzschild timelike coordinate and $\tilde{r}$ is a radial spacelike coordinate (sometimes called the *tortoise coordinate*!), the 2-space metric then becomes

$$ds^2 = \left(1 - \frac{2\mu}{r}\right)(c^2\,dt^2 - d\tilde{r}^2) \qquad (11.20)$$

$$= \Omega^2(x)\eta_{\mu\nu}\,dx^\mu\,dx^\nu, \qquad (11.21)$$

where $x^0 = ct$ and $x^1 = \tilde{r}$. This line element has the same form as that of a Minkowski 2-space (which is spatially flat) but it is multiplied by what mathematicians call a *conformal* scaling factor, $\Omega^2(x)$, which is a function of position. The

2-space itself is curved, because the derivatives of the function $\Omega(x)$ enter into the components of the curvature tensor, but the line element (11.21) of the 2-space is manifestly *conformally flat*. In fact, *any* two-dimensional (pseudo-)Riemannian manifold is conformally flat (see Appendix 11C), in that a coordinate system always exists in which the line element takes the form (11.21). We have thus succeeded in finding such a coordinate system for the 2-space (11.17).

The form of the line element (11.21) has an important consequence for studying the paths of radially moving photons (for which $d\theta = d\phi = 0$). Since the conformal factor $\Omega^2(x)$ is just a scaling, it does not change the lightcone structure and so the latter should just look like that in Minkowski space. Thus, in a spacetime diagram in $(ct, \tilde{r})$ coordinates, both ingoing and outgoing radial null geodesics are straight lines with slope ± 1, as is easily seen by setting $ds^2 = 0$ in (11.20).

Unfortunately, however, the coordinates $(ct, \tilde{r})$ are pathological when $r = 2\mu$, as is easily seen from (11.19). This suggests that, instead of using the parameters p and q directly, we should look for a coordinate transformation that preserves the manifest conformal nature of the 2-space defined by (11.17) but removes the offending factor $1 - 2\mu/r$, which is the cause of the pathological behaviour. It is straightforward to see that a transformation of the form $\tilde{p}(p)$ and $\tilde{q}(q)$ will achieve this goal, since, in this case, the metric becomes

$$ds^2 = \left(1 - \frac{2\mu}{r}\right) \frac{dp}{d\tilde{p}} \frac{dq}{d\tilde{q}} \, d\tilde{p} \, d\tilde{q},$$

which has the same general form as (11.17). An appropriate choice of the functions $\tilde{p}(p)$ and $\tilde{q}(q)$ that removes the factor $(1 - 2\mu/r)$ in the line element is (as suggested by Kruskal)

$$\tilde{p} = \exp\left(\frac{p}{4\mu}\right), \qquad \tilde{q} = -\exp\left(-\frac{q}{4\mu}\right),$$

for which we find that

$$ds^2 = \frac{32\mu^3}{r} \exp\left(-\frac{r}{2\mu}\right) d\tilde{p} \, d\tilde{q}.$$

The usual form of the metric is then obtained by defining a timelike variable v and a spacelike variable u by

$$v = \tfrac{1}{2}(\tilde{p} + \tilde{q}), \qquad u = \tfrac{1}{2}(\tilde{p} - \tilde{q}).$$

Thus, the full line element for the Schwarzschild geometry in *Kruskal coordinates* (v, u, θ, ϕ) is given by

$$\boxed{ds^2 = \frac{32\mu^3}{r} \exp\left(-\frac{r}{2\mu}\right) (dv^2 - du^2) - r^2 (d\theta^2 + \sin^2\theta \, d\phi^2),} \tag{11.22}$$

where r is considered as a function of v and u that is defined implicitly by

$$u^2 - v^2 = \left(\frac{r}{2\mu} - 1\right) \exp\left(\frac{r}{2\mu}\right). \tag{11.23}$$

It is straightforward to show that the coordinates v and u are related to the original Schwarzschild coordinates t and r by the following transformations. For $r > 2\mu$ we have

$$v = \left(\frac{r}{2\mu} - 1\right)^{1/2} \exp\left(\frac{r}{4\mu}\right) \sinh\left(\frac{ct}{4\mu}\right),$$

$$u = \left(\frac{r}{2\mu} - 1\right)^{1/2} \exp\left(\frac{r}{4\mu}\right) \cosh\left(\frac{ct}{4\mu}\right),$$

whereas, for $r < 2\mu$,

$$v = \left(1 - \frac{r}{2\mu}\right)^{1/2} \exp\left(\frac{r}{4\mu}\right) \cosh\left(\frac{ct}{4\mu}\right),$$

$$u = \left(1 - \frac{r}{2\mu}\right)^{1/2} \exp\left(\frac{r}{4\mu}\right) \sinh\left(\frac{ct}{4\mu}\right).$$

Considerable insight into the nature of the Schwarzschild geometry can be obtained by plotting its spacetime diagram in Kruskal coordinates. The causal structure defined by radial light rays is (by construction) particularly easy to analyse in Kruskal coordinates. From the metric (11.22), we see that for $ds = d\theta = d\phi = 0$ we have

$$v = \pm u + \text{constant},$$

which represents straight lines at $\pm 45°$ to the axes. This is a direct consequence of the fact that the 2-space with $d\theta = d\phi = 0$ is manifestly conformally flat in (v, u) coordinates. Thus, the lightcone structure should look like that in Minkowski space. Also, we note that a massive particle worldline must always lie within the future light-cone at each point.

It is also instructive to plot lines of constant t and r. From (11.23) we see that lines of constant r are curves of constant $u^2 - v^2$ and are hence hyperbolae. In particular, the value $r = 2\mu$ correpsonds to either of the straight lines $u = \pm v$, which are the asymptotes to the set of constant-r hyperbolae, and the value $r = 0$ corresponds to the hyperbolae $v = \pm\sqrt{u^2 + 1}$. Thus the 'point' in space $r = 0$ is mapped into *two lines*. However, not too much can be made of this since it is a singularity of the geometry. We should not glibly speak of it as a part of spacetime with a well-defined dimensionality. Similarly, lines of constant t may be mapped out. It is straightforward to show that

$$\tanh[ct/(4\mu)] = \begin{cases} v/u & \text{for } r > 2\mu, \\ u/v & \text{for } r < 2\mu, \end{cases}$$

so fixed values of t correspond to lines of constant u/v, i.e. straight lines through the origin. The value $t = -\infty$ corresponds to $u = -v$, while $t = \infty$ corresponds to $u = v$. The value $t = 0$ for $r > 2\mu$ corresponds to the line $v = 0$, whereas for $r < 2\mu$ it is the line $u = 0$.

We note that the entire region covered by the Schwarzschild coordinates $-\infty < t < \infty$, $0 < r < \infty$ is mapped onto the regions I and II in Figure 11.6. Thus, we would require *two* Schwarzschild coordinate patches (I, II) and (I′, II′) to cover the entire Schwarzschild geometry, but a single Kruskal coordinate system suffices. The diagonal lines $r = 2\mu$, $t = \infty$ and $r = 2\mu$, $t = -\infty$ define event horizons separating the regions of spacetime II and II′ from the other regions, I and I′.

The Kruskal diagram has some curious features. There are two 'Minkowski' regions, I and I′, so apparently there are two universes. We can identify region I as the spacetime region outside a Schwarzschild black hole and region II as the interior of the black-hole event horizon. Any particle that travels from region I to

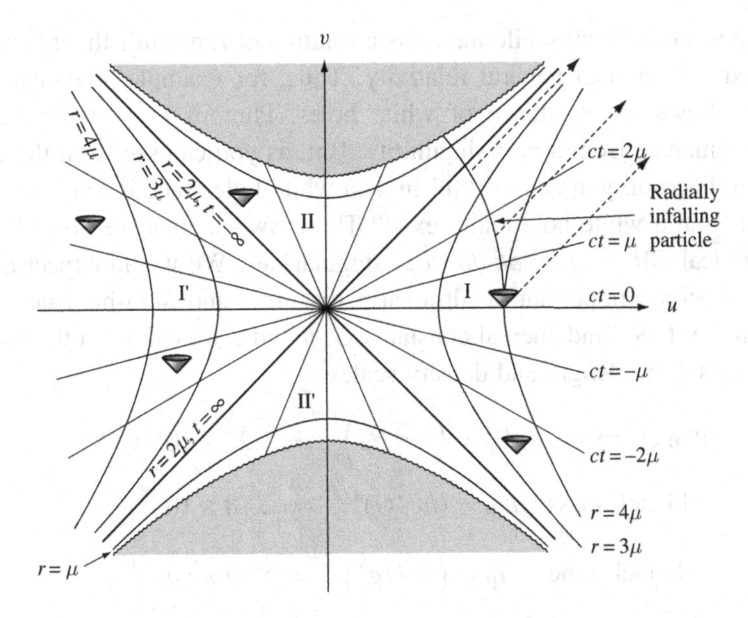

Figure 11.6 Spacetime diagram of the Schwarzschild geometry in Kruskal coordinates. The lower and upper wavy lines at the boundaries of the shaded regions are respectively the past singularity and the future singularity at $r = 0$. The broken-line arrows show escaping signals.

region II can never return and, moreover, must eventually reach the singularity $r = 0$. Regions I′ and II′ are completely inaccessible from regions I or II. Region II′ is similar to region II but in reverse: it is a part of spacetime from which a particle can escape (into regions I and I′) but not enter. Moreover, there is a singularity in the past – a *white hole* – from which particles can emanate. Indeed, we may now understand more clearly our discussion of the advanced and retarded Eddington–Finkelstein coordinates in Section 11.5: the advanced coordinates describe the Schwarzschild geometry in regions I and II, whereas the retarded coordinates cover the regions I′ and II′. The two universes I and I′ are actually connected by a *wormhole* at the origin, which we discuss in more detail in the next section, but, as we will show, no particle can travel between regions I and I′.

It is worth asking what has happened here. How can a few simple coordinate transformations lead to what is apparently new physics? What we have done amounts to mathematically extending the Schwarzschild solution. Mathematicians would call this a *maximal extension* of the Schwarzschild solution because *all* geodesics either extend to infinite values of their affine parameter or end at a past or future singularity. Thus Kruskal coordinates probe all the Schwarzschild geometry. Hence, we find that the complete Schwarzschild geometry consists of a black hole and white hole and two universes connected at their horizons by a wormhole.

The extended Schwarzschild metric is a solution of Einstein's theory and hence is allowed by classical general relativity. Thus, for example, classical general relativity allows the existence of white holes. Photons or particles could, in principle, emanate from a past singularity. But, as you can see from the Kruskal spacetime diagram, you cannot 'fall into' a white hole since it can only exist in your past. Can a white hole really exist? The answer is that we don't know for sure. Classical GR *must break down* at singularities. We would expect quantum effects to become important at ultra-short distances and ultra-high energies. In fact, from the three fundamental constants G, $\hbar$ and c we can form the following energy, mass, time, length and density scales:

$$\text{Planck energy} \quad E_P = \left(\hbar c^5/G\right)^{1/2} = 1.22 \times 10^{19}\,\text{GeV},$$

$$\text{Planck mass} \quad m_P = (\hbar c/G)^{1/2} = 2.18 \times 10^{-5}\,\text{g},$$

$$\text{Planck time} \quad t_P = \left(\hbar G/c^5\right)^{1/2} = 5.39 \times 10^{-44}\,\text{s},$$

$$\text{Planck length} \quad l_P = \left(\hbar G/c^3\right)^{1/2} = 1.62 \times 10^{-33}\,\text{cm},$$

$$\text{Planck density} \quad \rho_P = \left(c^5/\hbar G^2\right) = 5.16 \times 10^{93}\,\text{g}\,\text{cm}^{-3}.$$

These *Planck scales* define the characteristic energies, lengths, times, etc. *at which we expect quantum gravitational effects to become important.* To put it into some kind of perspective, an elementary particle with the Planck mass would weigh about the same as a small bacterium.

Nobody really expects the centres of black holes to harbour true singularities. Instead, it is expected that, close to the classical singularity, quantum gravitational effects will occur that will prevent the divergences of classical general relativity. We do not yet have a complete theory of quantum gravity, though many people hope that M-theory (formerly known as superstring theory) might one day provide such a theory. Theorists have developed semi-classical theories, however, which might (or might not) contain some of the features of a complete theory of quantum gravity. Such calculations suggest that white holes would be unstable and could not exist for more than about a Planck time. It is interesting that within a few pages we have pushed Einstein's theory of gravity to the edge of known physics.

11.10 Wormholes and the Einstein–Rosen bridge

Although it is not obvious from Figure 11.6, the two universes I and I′ are actually connected by a *wormhole* at the origin. To understand the structure at the origin, you must realize that the coordinates θ and ϕ have been suppressed in this figure; each point in Figure 11.6 actually represents a 2-sphere.

We can gain some intuitive insight into wormholes by considering the geometry of the spacelike hypersurface $v = 0$, which extends from $u = +\infty$ to $u = -\infty$. The line element for this hypersurface is

$$ds^2 = -\frac{32\mu^3}{r}\exp\left(-\frac{r}{2\mu}\right)du^2 - r^2(d\theta^2 + \sin^2\theta\, d\phi^2).$$

We can draw a cross-section of this hypersurface corresponding to the equatorial plane $\theta = \pi/2$, in which the line element reduces further to

$$ds^2 = -\frac{32\mu^3}{r}\exp\left(-\frac{r}{2\mu}\right)du^2 - r^2\, d\phi^2. \tag{11.24}$$

To interpret this, we may consider a two-dimensional surface possessing a line element $d\sigma^2$ given by minus (11.24) and embed it in a three-dimensional Euclidean space.

This embedding is most easily performed by re-expressing $d\sigma^2$ in terms of the coordinates r and ϕ, which is easily shown to yield the familiar form

$$d\sigma^2 = \left(1 - \frac{2\mu}{r}\right)^{-1} dr^2 + r^2\, d\phi^2. \tag{11.25}$$

However, we must remember that, in the spacelike hypersurface $v=0$, as we move along the u-axis from $+\infty$ to $-\infty$ the value of r decreases to a minimum value $r=2\mu$ (at $u=0$) and then increases again. In general, in Euclidean space, a 2-surface parameterised by arbitrary coordinates (ξ, η) can be specified by giving three functions $x^a(\xi, \eta)(a=1,2,3)$, where the x^a define some coordinate system in the three-dimensional Euclidean space. In our particular case, it will be useful to use cylindrical polar coordinates (ρ, ψ, z), in which case the line element of the three-dimensional space is

$$d\sigma^2 = d\rho^2 + \rho^2\, d\psi^2 + dz^2. \tag{11.26}$$

Moreover, since the 2-surface we wish to embed (which is parameterised by the coordinates r and ϕ) is clearly axisymmetric, we may take the three functions specifying this surface to have the form

$$\rho = \rho(r), \qquad \psi = \phi, \qquad z = z(r).$$

Substituting these forms into (11.26), we may thus write the line element on the embedded 2-surface as

$$d\sigma^2 = \left[\left(\frac{d\rho}{dr}\right)^2 + \left(\frac{dz}{dr}\right)^2\right] dr^2 + \rho^2\, d\phi^2. \tag{11.27}$$

For the geometry of the embedded 2-surface to be identical to the geometry of the 2-space of interest, we require the line elements (11.25) and (11.27) to be identical, and so we require $\rho(r) = r$ and thus

$$1 + \left(\frac{dz}{dr}\right)^2 = \left(1 - \frac{2\mu}{r}\right)^{-1}.$$

The solution to this differential equation is easily found to be

$$z(r) = \sqrt{8\mu(r-2\mu)} + \text{constant},$$

and substituting $r=\rho$ gives us the equation of the cross-section of the embedded 2-surface in the (ρ, z)-plane of the Euclidean 3-space. Taking the constant of integration to be zero, and remembering that r (and hence ρ) is never less than 2μ, we find that the surface has the form shown in Figure 11.7. Thus, the geometry of the spacelike hypersurface at $v=0$ can be thought of as two distinct, but identical, asymptotically flat Schwarzschild manifolds joined at the 'throat' $r=2\mu$ by an *Einstein–Rosen bridge*. If one so wishes, one can also connect the two asymptotically flat regions together in a region distant from the throat. In this case, the wormhole connects two distant regions of a single universe.

In either case, the structure of the wormhole is *dynamic*. One is used to thinking of the Schwarzschild geometry as 'static'. However, working for the moment in

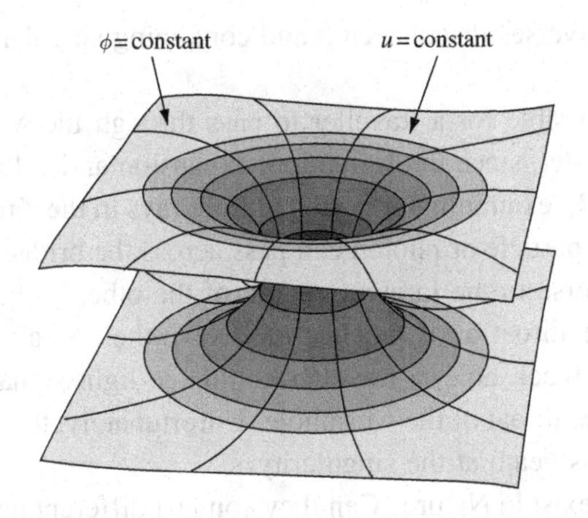

Figure 11.7 The structure of the Einstein–Rosen bridge.

terms of the traditional Schwarzschild coordinate, it is only in regions I and I′ that t is timelike and the fact that the metric coefficients are independent of t means that spacetime is static. In regions II and II′, the t-coordinate is spacelike and the r-coordinate is timelike. Since the metric coefficients do depend explicitly on r, the spacetime in these regions is no longer static but evolves with respect to this timelike coordinate. Returning to Kruskal coordinates, consider the spacelike hypersurface $v=0$. As this surface is pushed forwards in time (in the $+v$ direction in the Kruskal diagram), part of it enters region II and begins to evolve.

As v increases, the picture of the geometry of the hypersurface is qualitatively the same as that illustrated in Figure 11.7, but the bridge narrows, the universes now joining at $r<2\mu$. At $v=1$, the bridge pinches off completely and the two universes simply touch at the singularity $r=0$. For larger values of v the two universes, each containing a singularity at $r=0$, are completely separate. Since the Kruskal solution is symmetric in v, the same things happen for negative values of v. The full time evolution is shown schematically in Figure 11.8. Thus, the two universes are disconnected at the beginning, each containing a singularity of infinite curvature ($r=0$). As they evolve in time, their singularities join each other and form a non-singular bridge. The bridge enlarges until at $v=0$ it reaches a maximum radius at the throat equal to $r=2\mu$. It then contracts and pinches off,

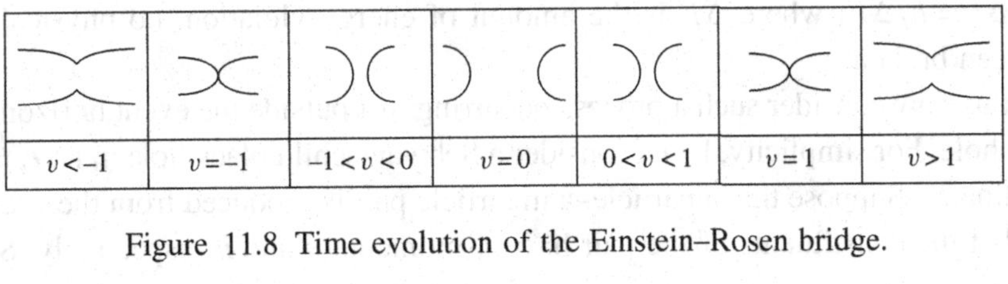

$v<-1$	$v=-1$	$-1<v<0$	$v=0$	$0<v<1$	$v=1$	$v>1$

Figure 11.8 Time evolution of the Einstein–Rosen bridge.

leaving the two universes disconnected and containing singularities ($r = 0$) once again.

Sadly, it is impossible for a traveller to pass through the wormhole from one universe into the other, since the formation, expansion and collapse of the bridge occur too rapidly. By examining the paths of light rays in the Kruskal diagram, we can deduce that no particle or photon can pass across the bridge from the faraway region of one universe to the faraway region of the other without getting caught and crushed in the throat as it pinches off. Nevertheless, after falling through the horizon of the black hole, a traveller could see light signals from the other universe through the throat of the wormhole. Unfortunately, the penalty for seeing the other universe is death at the singularity.

Can wormholes exist in Nature? Can they connect different universes, or different parts of the same universe? Again, nobody knows for sure. Many theorists would argue that we need to understand quantum gravity to understand wormholes. Wormholes are probably unstable, but 'virtual' wormholes are a feature of some formulations of quantum gravity.

11.11 The Hawking effect

So far our discussion of black holes has been purely classical. Indeed, we have found that classically nothing can escape from the within the event horizon of a black hole; that is why they are called black holes! However, in 1974, Stephen Hawking applied the principles of quantum mechanics to electromagnetic fields near a black hole and found the amazing result that black holes radiate continuously as a blackbody with a temperature inversely proportional to their mass! Hawking's original calculation uses the techniques of quantum field theory, but we can derive the main results very simply from elementary arguments.

According to quantum theory, even the vacuum of empty space exhibits quantum fluctuations, in which particle–antiparticle pairs are created at one event only to annihilate one another at some other event. Pair creation violates the conservation of energy and so is classically forbidden. In quantum mechanics, however, one form of Heisenberg's uncertainty principle is $\Delta t \Delta E = \hbar$, where ΔE is the minimum uncertainty in the energy of a particle that resides in a quantum mechanical state for a time Δt. Thus, provided the pair annihilates in a time less than $\Delta t = \hbar/\Delta E$, where ΔE is the amount of energy violation, no physical law has been broken.

Let us now consider such a process occurring just outside the event horizon of a black hole. For simplicity, let us consider a Schwarzschild black hole in (t, r, θ, ϕ) coordinates. Suppose that a particle–antiparticle pair is produced from the vacuum and that the constituents of the pair have 4-momenta $\boldsymbol{p}$ and $\bar{\boldsymbol{p}}$ respectively. Since the spacetime is stationary ($\partial_0 g_{\mu\nu} = 0$), the quantities $p_0 = \boldsymbol{e}_0 \cdot \boldsymbol{p}$ and $\bar{p}_0 = \boldsymbol{e}_0 \cdot \bar{\boldsymbol{p}}$ are conserved along the particle worldlines; here $\boldsymbol{e}_0$ is the t-coordinate basis vector. Thus, for a fluctuation from the vacuum, classical conservation requires

$$\boldsymbol{e}_0 \cdot \boldsymbol{p} + \boldsymbol{e}_0 \cdot \bar{\boldsymbol{p}} = 0. \tag{11.28}$$

The squared 'length' of the coordinate basis vector $\boldsymbol{e}_0$ is given by

$$\boldsymbol{e}_0 \cdot \boldsymbol{e}_0 = g_{00} = c^2(1 - 2\mu/r). \tag{11.29}$$

Thus, outside the horizon ($r > 2\mu$), $\boldsymbol{e}_0$ is *timelike*. The components $\boldsymbol{e}_0 \cdot \boldsymbol{p}$ and $\boldsymbol{e}_0 \cdot \bar{\boldsymbol{p}}$ are therefore proportional to the particle energies as measured by an observer whose 4-velocity is along the $\boldsymbol{e}_0$-direction. Hence both must be positive, so the conservation condition (11.28) cannot be satisfied.

However, if the fluctuation occurs near the event horizon then the inward-moving particle may travel to the region $r < 2\mu$. Inside the event horizon $\boldsymbol{e}_0$ is *spacelike*, as shown by (11.29). Thus $\boldsymbol{e}_0 \cdot \boldsymbol{p}$ is a component of the *spatial momentum* for some observer and so may be negative. Hence the conservation condition (11.28) *can* be satisfied if the antiparticle (say) crosses the horizon with negative $\boldsymbol{e}_0 \cdot \bar{\boldsymbol{p}}$ and the particle escapes to infinity with positive $\boldsymbol{e}_0 \cdot \boldsymbol{p}$. As seen by an observer at infinity, the black hole has emitted a particle of energy $\boldsymbol{e}_0 \cdot \boldsymbol{p}$ and the black hole's mass has decreased by $(\boldsymbol{e}_0 \cdot \bar{\boldsymbol{p}})c^2$ as a consequence of the particle falling into it. This is the *Hawking effect*. Of course, the argument is equally valid if it is the particle that falls into the black hole and the antiparticle that escapes to infinity. The black hole emits particles and antiparticles in equal numbers.

For a fluctuation near the horizon, the inward-travelling particle needs to endure in a prohibited negative $\boldsymbol{e}_0 \cdot \boldsymbol{p}$ condition only for a short proper time, as measured by some locally free-falling observer, before reaching the inside of the horizon where negative $\boldsymbol{e}_0 \cdot \boldsymbol{p}$ is allowed. The particle has, in fact, *tunnelled* quantum mechanically through a region outside the horizon, where negative $\boldsymbol{e}_0 \cdot \boldsymbol{p}$ is classically forbidden, to a region inside the event horizon where it is classically allowed. The process works best where the proper time in the forbidden region is smallest, i.e. close to the horizon.

The distant observer sees a steady flux of particles and antiparticles. The flux must be steady, since the geometry is independent of t and so the rate of particle emission must also be independent of t. Let us calculate the typical energy of such a particle as measured by the distant observer. Suppose that the particle–antiparticle pair is created at some event P with coordinate radius $R = 2\mu + \epsilon$. Let us consider this event as viewed by a freely falling observer, starting from rest at this point. Since the observer is in free fall, the rules of special relativity apply in his frame. A typical measure of the proper time $\Delta\tau$ elapsed before the observer reaches the horizon may be obtained by considering a radially free-falling particle that starts from rest at $r = R$. In this case,

$$\dot{t} = \frac{(1-2\mu/R)^{1/2}}{1-2\mu/r},$$

$$\dot{r} = -\left[2\mu c^2\left(\frac{1}{r}-\frac{1}{R}\right)\right]^{1/2}.$$

Thus the required proper time interval is

$$\Delta\tau = -\int_{2\mu+\epsilon}^{2\mu}\left(\frac{2\mu c^2}{r} - \frac{2\mu c^2}{2\mu+\epsilon}\right)^{-1/2} dr \approx \frac{2(2\mu\epsilon)^{1/2}}{c},$$

where the final result is quoted to first order in ϵ. From the uncertainty principle, the typical energy $\mathcal{E}$ of the particle, as measured by a freely falling observer, is given by

$$\mathcal{E} = \frac{\hbar}{\Delta\tau} = \frac{\hbar c}{2(2\mu\epsilon)^{1/2}}.$$

However, this can also be written as

$$\mathcal{E} = \boldsymbol{p}\cdot\boldsymbol{u} \approx p_0 u^0,$$

where $\boldsymbol{u}$ is the observer's 4-velocity and the approximation holds since $u^1 \ll u^0$. Now, $u^0 = \dot{t} \approx (2\mu/\epsilon)^{1/2}$ to first order in ϵ. Moreover, p_0 is conserved along the particle's worldline and is equal to the energy E of the particle as measured by the distant observer, whose 4-velocity is simply $[u^\mu] = (1, 0, 0, 0)$. Thus, we finally obtain

$$E = \mathcal{E}\left(\frac{\epsilon}{2\mu}\right)^{1/2} = \frac{\hbar c^3}{4GM}. \tag{11.30}$$

Remarkably, this result does not depend on ϵ; the particle always emerges with this characteristic energy.

The full quantum field theory calculation shows that the particles are in fact received with a *blackbody* energy spectrum characterised by the *Hawking temperature*

$$\boxed{T = \frac{\hbar c^3}{8\pi k_B GM}.}$$

The typical particle energy is thus $E = k_B T = \hbar c^3/(8\pi GM)$, which is only a factor 2π smaller than our crude estimate (11.30). Putting in numbers, we find that

$$T = 6\times 10^{-8}\left(\frac{M}{M_\odot}\right)^{-1}\ \text{K}.$$

Thus the radiation from a solar-mass black hole, such as might be formed by the gravitational collapse of a massive star, is negligibly small.

It is straightforward to calculate the rate dM/dt at which the black hole loses mass, as determined by a stationary distant observer whose proper time is t. Since the black-hole event horizon emits radiation as a blackbody of temperature T, the black-hole mass must decrease at a rate

$$\frac{dM}{dt} = -\frac{\sigma T^4 A}{c^2},$$

where $\sigma = \pi^2 k_B^4/(60\hbar^3 c^2)$ is the Stefan–Boltzmann constant and A is the proper area of the event horizon. From the Schwarzschild metric we find that $A = 16\pi\mu^2$, and so we obtain

$$\frac{dM}{dt} = -\frac{\alpha\hbar}{M^2}, \tag{11.31}$$

where the constant $\alpha = c^4/(15\,360\pi G^2) = 3.76 \times 10^{49}\text{kg}^2\text{m}^{-2}$. The solution $M(t)$ to (11.31) is easily calculated. For a black hole whose evaporation is complete at time t_0, we find that

$$\boxed{M(t) = [3\alpha\hbar(t_0 - t)]^{1/3}.} \tag{11.32}$$

This result shows that a burst of energy is emitted right at the end of a black hole's life. For example, in the final second it should emit $\sim 10^{22}$ J of energy, primarily as γ-rays. No such events have yet been identified.

Appendix 11A: Compact binary systems

One of the best ways of finding candidate black holes is to search for luminous compact X-ray sources. The reason is that if a black hole has a stellar companion then the intense tidal field can pull gas from the companion, producing an accretion disc around the black hole. A schematic picture is shown in Figure 11.9. As we showed in Chapter 10, accretion discs can radiate very efficiently and we would expect to observe high-energy (X-ray) photons emitted from a small region of space.

Table 11.1 summarizes the two common classes of compact binaries. The compact object can be a white dwarf, neutron star or black hole. If you find a compact binary system then you can set limits on the mass of the compact object from the dynamics of the binary orbit. If you find evidence for a compact object that is more massive than the Oppenheimer–Volkoff limit then you have good evidence that the object might be a black hole.

Table 11.1 *Compact accreting binary systems*

Companion star	Compact object		
	White dwarf	Neutron star	Black hole
Early type, massive	None known	Massive X-ray binaries	Cyg X-I
Late type, low mass	Cataclysmic variables (e.g., dwarf novae)	Low mass X-ray binaries	A0620 – 00

Figure 11.9 Schematic picture of a compact binary system.

In fact it is not so straightforward. What observers actually measure is the *mass function*

$$f(M) = \frac{PK^3}{2\pi G},$$

where P is the orbital period, and K is the radial velocity amplitude. For example, for the low-mass X-ray binary A0620-00 the period is $P = 7.7$ hours and $K = 457\,\mathrm{km\,s^{-1}}$. From Kepler's laws we can show that the mass function is related to the masses M_1 and M_2 of the compact object and the companion star and the

Table 11.2 *Derived parameters and dynamical mass measurements of SXTs*

Source	$f(M)(M_\odot)$	$\rho(\mathrm{g\,cm^{-3}})$	$q(=M_1/M_2)$	i (deg)	$M_1(M_\odot)$	$M_2(M_\odot)$
V404 Cyg	6.08 ± 0.06	0.005	17 ± 1	55 ± 4	12 ± 2	0.6
G2000+25	5.01 ± 0.12	1.6	24 ± 10	56 ± 15	10 ± 4	0.5
N Oph 77	4.86 ± 0.13	0.7	>19	60 ± 10	6 ± 2	0.3
N Mus 91	3.01 ± 0.15	1.0	8 ± 2	54^{+20}_{-15}	6^{+5}_{-2}	0.8
A0620−00	2.91 ± 0.08	1.8	15 ± 1	37 ± 5	10 ± 5	0.6
J0422+32	1.21 ± 0.06	4.2	>12	20−40	10 ± 5	0.3
J1655−40	3.24 ± 0.14	0.03	3.6 ± 0.9	67 ± 3	6.9 ± 1	2.1
4U1543−47	0.22 ± 0.02	0.2	—	20−40	5.0 ± 2.5	2.5
Cen X-4	0.21 ± 0.08	0.5	5 ± 1	43 ± 11	1.3 ± 0.6	0.4

inclination angle i of the orbit to the plane of the sky by

$$f(M) = \frac{M_1^3 \sin^3 i}{(M_1 + M_2)^2}.$$

You can see from this equation that the mass function is a strict lower limit on the mass M_1 of the compact object. It is equal to the latter, $f = M_1$, *only* if $M_2 = 0$ and the orbit is viewed edge on (so that $\sin i = 1$). For example, for A0620 − 00 the lower limit on the mass of the compact object is $2.9M_\odot$, and this makes it a very good black hole candidate because this mass limit is very close to the theoretical *upper* limit for the mass of a neutron star. In fact, it is possible to make reasonable estimates[3] for M_2 and $\sin i$ in this system, leading to a probable mass of $\approx 10M_\odot$ for the compact object – well into the black-hole regime.

Table 11.2 summarises the dynamical mass limits on some good black-hole candidates (so-called short X-ray transients). As you can see, in several systems, such as V404 Cyg, G2000 + 25 and N Oph 77, the minimum mass inferred from the mass function is *well above* the theoretical maximum mass limit for a neutron star. As we understand things at present there can be no other explanation than that the compact objects are black holes.

Appendix 11B: Supermassive black holes

The first quasar[4] (3C273) was discovered in 1963 by Maarten Schmidt. He measured a cosmological redshift of $z = 0.15$ for this object, which was

[3] An estimate of the mass M_2 can be made by measuring the spectral type and luminosity of the companion star. The inclination angle can be estimated from the shape of the star's light curve by searching for evidence of eclipsing by the compact object.

[4] *Quasi-stellar radio source.* We now know that the majority of quasars are radio quiet, and so they are often called QSOs for *quasi-stellar object*.

unprecedently high at the time (quasars have since been discovered with redshifts as high as $z = 5.8$). Quasars are very luminous, typically 100–1000 times brighter than a large galaxy. However, they are *compact*, so compact, in fact, that quasars look like stars in photographs. In fact, from variability and other studies one can infer that the size of the continuum-emitting region of a quasar is of order a few parsecs or less. How can we explain such a phenomenon? Imagine an object radiating many times the luminosity of an entire galaxy from a region smaller than the Solar System. Donald Lynden-Bell was one of the first to suggest that the quasar phenomenon is caused by accretion of gas onto a *supermassive* black hole residing at the centre of a galaxy. The black-hole masses required to explain the high luminosities of quasars are truly spectacular – we require black holes with masses a few million to a few billion times the mass of the Sun.

Do such supermassive black holes exist? The evidence in recent years has become extremely strong. Using the Hubble Space Telescope it is possible to probe the velocity dispersions of stars in the central regions of galaxies. According to Newtonian dynamics, we would expect the characteristic velocities to vary as

$$v^2 \sim \frac{GM}{r}.$$

If the central mass is dominated by a supermassive black hole then we expect the typical velocities of stars to *increase* as we go closer to the centre. This is indeed what is found in a number of galaxies. From the rate of increase of the velocities with radius, we can estimate the mass of the central object, which seems to be correlated with the mass of the bulge component of the galaxy:

$$M_{\text{bh}} \approx 0.006 M_{\text{bulge}}.$$

It seems as though, at the time of galaxy formation, about half a percent of the mass of the bulge material collapses right to the very centre of a galaxy to form a supermassive black hole. During this phase the infalling gas radiates efficiently, producing a quasar. When the gas supply is used up, the quasar quickly fades away leaving a dormant massive black hole that is starved of fuel. Nobody has yet developed a convincing theory of how this happens, or of what determines the mass of the central black hole.

A sceptic might argue that these observations merely prove that a dense compact object exists at the centre of a galaxy that is not necessarily a black hole. But there are two beautiful observational results that probe compact objects on parsecond scales – making it almost certain that the central objects are black holes. In our own Milky Way Galaxy it is possible to measure the *proper motions* of stars in the Galactic centre (using infrared wavelengths to penetrate through the dense dust that obscures optical light). This has allowed astronomers to see the stars actually moving and so infer their three-dimensional motions. These observations imply that there exists a black hole of mass $2.5 \times 10^6 M_\odot$ at the centre of our Galaxy.

In a remarkable set of observations, a disc of H_2O masers has been detected in the galaxy NGC 4258 using very long baseline interferometry (VLBI). The VLBI observations measure the velocities of the masing clouds on scales of ~ 0.03–0.2 parsecs and are well fitted by a thin (actually slightly warped) disc in circular motion (see Figure 11.10). The mass of the central black hole is estimated to be $4 \times 10^7 M_\odot$.

Table 11.3 lists the masses of some potential supermassive black holes, with a five-star rating. The masing disc of NGC 4258 gets a full five stars – this is the strongest observational evidence for a supermassive black hole. The stellar

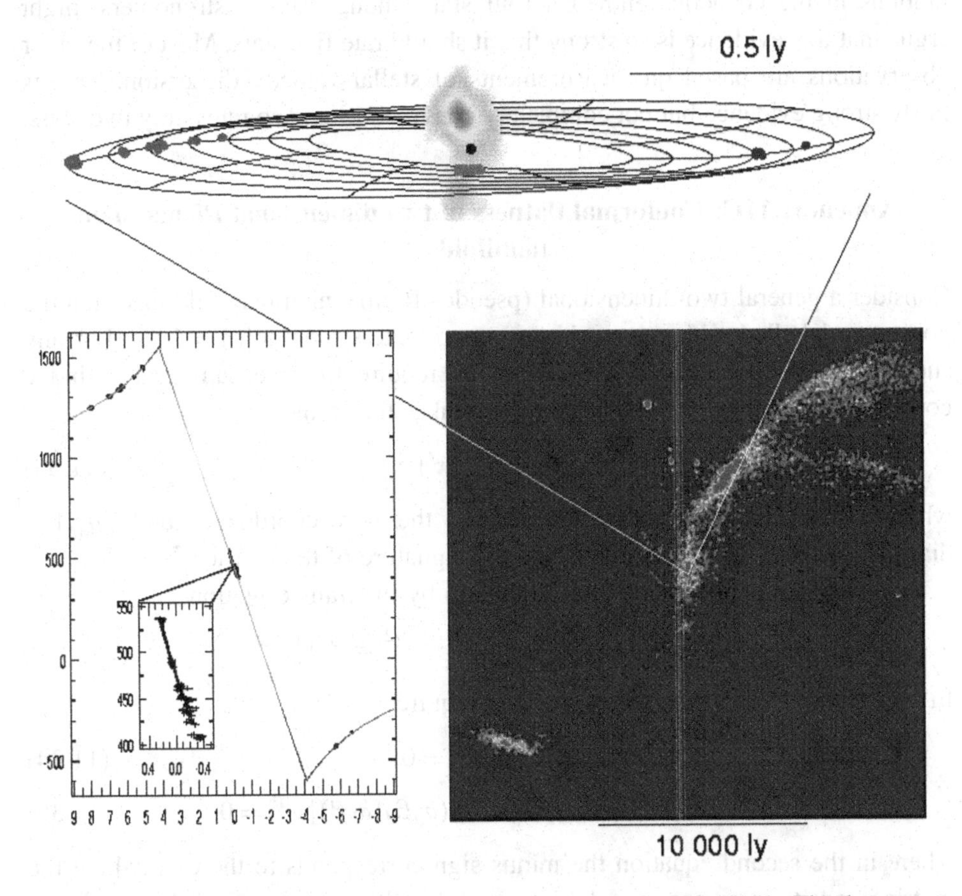

Figure 11.10 The masing H_2O disc in the centre of NGC 4258. The lower left-hand panel shows the variation in the line-of-sight velocity in km s^{-1} of the material in the disc as a function of the distance along its major axis in milliarcseconds. In the upper panel and lower right-hand panel the distance scales are given in light years.

Table 11.3 *Potential supermassive black holes*

Rating	Source	$M_{bh}/M_\odot$	Evidence
***	M87	2×10^9	stars and optical disc
**	NGC 3115	1×10^9	stars
**	NGC 4594 (Sombrero)	5×10^8	stars
**	NGC 3377	1×10^9	stars
*****	NGC 4258	4×10^7	masing H_2O disc
**	M31 (Andromeda)	3×10^7	stars
**	M32	3×10^6	stars
****	Galactic centre	2.5×10^6	stars and 3D motions

motions in the Galactic centre get four stars, though some astronomers might argue that this evidence is so strong that it should rate five stars. Most of the other observations are based on measurements of stellar-velocity dispersion. This is fairly strong evidence but not completely convincing[5] and so rates only two stars.

Appendix 11C: Conformal flatness of two-dimensional Riemannian manifolds

Consider a general two-dimensional (pseudo-)Riemannian manifold in which the points are labelled with some arbitrary coordinate system $x^a (a = 1, 2)$. For any such manifold to be conformally flat, we require that we can always find a coordinate system x'^a in which the metric takes the form

$$g'_{ab}(x') = \Omega^2(x')\eta_{ab}, \tag{11.33}$$

where $\Omega^2(x')$ is an arbitrary function of the new coordinates and $[\eta_{ab}] = \text{diag}(\pm 1, \pm 1)$, the signs depending on the signature of the metric.

Suppose the primed coordinates are given by the transformation

$$x'^1 = \alpha(x^1, x^2) \qquad \text{and} \qquad x'^2 = \beta(x^1, x^2).$$

In order that (11.33) is satisfied, we thus require

$$g'^{12} = (\partial_a\alpha)(\partial_b\beta)g^{ab} = 0, \tag{11.34}$$

$$g'^{11} \mp g'^{22} = [(\partial_a\alpha)(\partial_b\alpha) \mp (\partial_a\beta)(\partial_b\beta)]\, g^{ab} = 0, \tag{11.35}$$

where in the second equation the minus sign corresponds to the case where the metric is positive- or negative-definite, and the plus sign corresponds to the case where the metric is indeterminate.

[5] The interpretation of velocity dispersion measurements requires some assumptions about the degree of velocity anisotropy.

It is straightforward to verify that (11.34) is satisfied identically if

$$\partial_a \alpha = \kappa \epsilon_{ab} g^{bc} \partial_c \beta,$$

where κ is an arbitrary function of the coordinates and ϵ_{ab} is the alternating symbol, for which $\epsilon_{11} = \epsilon_{22} = 0$ and $\epsilon_{12} = -\epsilon_{21} = 1$. Moreover, substituting this expression into (11.35), we find that

$$\left(\frac{\kappa^2}{g} \mp 1\right) [(\partial_a \beta)(\partial_b \beta) g^{ab}] = 0,$$

where $g = \det[g_{ab}]$. For a positive- or negative-definite metric the factor in square brackets cannot be zero and, moreover, we can guarantee that $g \neq 0$. Thus, in this case, we can satisfy our requirements by choosing $\kappa^2 = g$. For an indeterminate metric, however, we must require that the above factor is zero, i.e. β must not be a null coordinate. In this case, we can again guarantee that $g \neq 0$, and so we choose $\kappa^2 = -g$. Thus we have shown explicitly that any two-dimensional (pseudo-)Riemannian manifold is conformally flat.

Exercises

11.1 In the Schwarzschild geometry, we introduce the new coordinates

$$x = r \sin\theta \cos\phi, \qquad y = r \sin\theta \sin\phi, \qquad z = r\cos\theta.$$

Find the form of the line element in these coordinates.

11.2 By introducing the new coordinate ρ defined by

$$r = \rho\left(1 + \frac{\mu}{2\rho}\right)^2,$$

show that the line element for the Schwarzschild geometry can be written in the *isotropic form*

$$ds^2 = c^2\left(1 - \frac{\mu}{2\rho}\right)^2 \left(1 + \frac{\mu}{2\rho}\right)^{-2} dt^2 - \left(1 + \frac{\mu}{2\rho}\right)^4 (d\rho^2 + \rho^2\, d\theta^2 + \rho^2 \sin^2\theta\, d\phi^2).$$

Show that $g_{00} \approx 1 - 2\mu/\rho$ in the weak-field limit $\mu \ll \rho$.

11.3 Show that the worldlines of radially moving photons in the Schwarzschild geometry are given by

$$ct = r + 2\mu \ln\left|\frac{r}{2\mu} - 1\right| + \text{constant} \qquad \text{(outgoing photon)},$$

$$ct = -r - 2\mu \ln\left|\frac{r}{2\mu} - 1\right| + \text{constant} \qquad \text{(incoming photon)}.$$

11.4 Show that, on introduction of the advanced Eddington–Finkelstein timelike coordinate $t' = ct + 2\mu \ln|r/(2\mu) - 1|$, the Schwarzschild line element takes the form

$$ds^2 = c^2\left(1 - \frac{2\mu}{r}\right)dt'^2 - \frac{4\mu c}{r}\,dt'dr - \left(1 + \frac{2\mu}{r}\right)dr^2 - r^2\left(d\theta^2 + \sin^2\theta\, d\phi^2\right).$$

Hence show that the worldlines of radially moving photons in advanced Eddington–Finkelstein coordinates are given by

$$ct' = r + 4\mu \ln\left|\frac{r}{2\mu} - 1\right| + \text{constant} \quad \text{(outgoing photon)},$$

$$ct' = -r + \text{constant} \quad \text{(incoming photon)}.$$

11.5 Show that, on introduction of the retarded Eddington–Finkelstein timelike coordinate $t^* = ct + 2\mu \ln|r/(2\mu) - 1|$, the Schwarzschild line element takes the form

$$ds^2 = c^2\left(1 - \frac{2\mu}{r}\right)dt^{*2} + \frac{4\mu c}{r}\,dt^*\,dr - \left(1 + \frac{2\mu}{r}\right)dr^2 - r^2\left(d\theta^2 + \sin^2\theta\, d\phi^2\right).$$

Hence find the equations for the worldlines of radially moving photons in retarded Eddington–Finkelstein coordinates. Use this result to sketch the spacetime diagram showing the light-cone structure in this coordinate system.

11.6 A particle in the Schwarzschild geometry emits a radially outgoing photon at coordinates (t_E, r_E), which is received by the distant fixed observer at (t_R, r_R). Show that, if r_E lies just outside the horizon $r = 2\mu$, the radial coordinate 'seen' by the distant observer at the time t_R is given by

$$r_E(t_R) \approx 2\mu + 2\mu e^{-c(t_R - r_R)/4\mu}.$$

11.7 An observer sits on the surface of a star as it collapses to form a black hole. Once an event horizon forms would the observer see any light from the star?

11.8 A spherical distribution of dust of coordinate radius R and total mass M collapses from rest under its own gravity. Show that, as the collapse progresses, the coordinate radius r of the star's surface and the elapsed proper time τ of an observer sitting on the surface are related by

$$\tau(r) = -\frac{1}{(2GM)^{1/2}} \int_R^r \left(\frac{r}{1 - r/R}\right)^{1/2} dr.$$

By making the substitution $r = R\cos^2(\psi/2)$, or otherwise, show that the solution can be expressed parametrically as

$$r = \frac{R}{2}(1 + \cos\phi), \qquad \tau = \frac{R}{2}\left(\frac{R}{2GM}\right)^{1/2}(\psi + \sin\psi).$$

Calculate the proper time experienced by the observer before the star collapses to a point.

11.9 A massive particle is released from rest at infinity in the Schwarzschild geometry. Show that the covariant components of its subsequent 4-velocity at coordinate radius r can be written as $u_\mu = c^2 \partial_\mu T$, where

$$T = t + \tfrac{1}{c}\int_\infty^r \left(\frac{2\mu}{r'}\right)^{1/2}\left(1 - \frac{2\mu}{r'}\right)^{-1} dr'.$$

Hence show that the line element of the Schwarzschild geometry in (T, r, θ, ϕ) coordinates is given by

$$ds^2 = c^2 dT^2 - \left(dr + \sqrt{2\mu c^2/r}\, dT\right)^2 - r^2\left(d\theta^2 + sin^2\theta\, d\theta^2\right).$$

Is this new form singular at $r = 2\mu$? What can you say about the hypersurface $T =$ constant? Show that observes infalling radially from rest at infinity have $\dot{T} = 1$ and hence give a physical interpretation of the T coordinate.

11.10 A massive particle is released from rest at coordinate radius r in the Schwarzschild geometry. Show that a frame of orthonormal basis vectors defining the inertial instantaneous rest frame of the particle may be taken as

$$(\hat{e}_0)^\mu = \frac{1}{c}u^\mu = \frac{1}{c}\left(1 - \frac{2\mu}{r}\right)^{-1/2}\delta_0^\mu, \qquad (\hat{e}_1)^\mu = \left(1 - \frac{2\mu}{r}\right)^{1/2}\delta_1^\mu,$$

$$(\hat{e}_2)^\mu = \frac{1}{r}\delta_2^\mu, \qquad (\hat{e}_3)^\mu = \frac{1}{r\sin\theta}\delta_3^\mu.$$

Hence show that the spatial components of the orthogonal connecting vector between two such nearby particles satisfy

$$\frac{d^2\zeta^{\hat{r}}}{d\tau^2} = +\frac{2\mu c^2}{r^3}\zeta^{\hat{r}}, \qquad \frac{d^2\zeta^{\hat{\theta}}}{d\tau^2} = -\frac{\mu c^2}{r^3}\zeta^{\hat{\theta}}, \qquad \frac{d^2\zeta^{\hat{\phi}}}{d\tau^2} = -\frac{\mu c^2}{r^3}\zeta^{\hat{\phi}}.$$

11.11 Two compact masses, each of mass m, are connected by a light strong wire of length l. The system is aligned in such a way that the two masses lie along a radial line from a Schwarzschild black hole, and it is released from rest at coordinate radius r. Obtain an expression for the tension in the wire immediately after the system is released.

11.12 An astronaut, starting from rest at infinity, falls radially inwards towards a Schwarzschild black hole with $M = 10^5 M_\odot$. Calculate the radial coordinate from the centre of the black hole at which the astronaut first experiences a lateral tidal force of $400\,\mathrm{m\,s^{-2}\,m^{-1}}$ and is therefore crushed. How does this radial coordinate compare with the position of the event horizon?

11.13 An unpowered satellite is in radial free fall towards a Schwarzschild black hole. Show that the principal stresses in the satellite are given by

$$\frac{2\mu c^2}{r^3}, \qquad -\frac{\mu c^2}{r^3}, \qquad -\frac{\mu c^2}{r^3}.$$

In what directions do these principal stresses act? Compare your answer with that obtained in Exercise 11.10.

11.14 An unpowered satellite follows a circular orbit of radius r around a Schwarzschild black hole. Show that the principal stresses in the satellite are given by

$$\frac{\mu c^2}{r^3}\frac{2r-3\mu}{r-3\mu}, \qquad -\frac{\mu c^2}{r^3}\frac{r}{r-3\mu}, \qquad -\frac{\mu c^2}{r^3}.$$

In what directions do these principal stresses act?

11.15 Suppose that p and q are respectively the advanced and retarded Eddington–Finkelstein time parameters, defined in terms of Schwarzschild coordinates by

$$p = ct + r + 2\mu \ln\left|\frac{r}{2\mu} - 1\right|,$$

$$q = ct - r - 2\mu \ln\left|\frac{r}{2\mu} - 1\right|.$$

A new set of (Kruskal) coordinates is defined by

$$v = \tfrac{1}{2}(e^{p/4\mu} - e^{-q/4\mu}) \qquad \text{and} \qquad u = \tfrac{1}{2}(e^{p/4\mu} + e^{-q/4\mu}).$$

Show that these new coordinates are related to Schwarzschild coordinates for $r > 2\mu$ by

$$v = \left(\frac{r}{2\mu} - 1\right)^{1/2} \exp\left(\frac{r}{4\mu}\right) \sinh\left(\frac{ct}{4\mu}\right),$$

$$u = \left(\frac{r}{2\mu} - 1\right)^{1/2} \exp\left(\frac{r}{4\mu}\right) \cosh\left(\frac{ct}{4\mu}\right),$$

and for $r < 2\mu$ by

$$v = \left(1 - \frac{r}{2\mu}\right)^{1/2} \exp\left(\frac{r}{4\mu}\right) \cosh\left(\frac{ct}{4\mu}\right),$$

$$u = \left(1 - \frac{r}{2\mu}\right)^{1/2} \exp\left(\frac{r}{4\mu}\right) \sinh\left(\frac{ct}{4\mu}\right).$$

11.16 Show that, in terms of the Kruskal coordinates u and v defined in Exercise 11.15, the Schwarzschild line element takes the form

$$ds^2 = \frac{32\mu^3}{r} \exp\left(-\frac{r}{2\mu}\right)(dv^2 - du^2) - r^2(d\theta^2 + \sin^2\theta\, d\phi^2),$$

where r is considered as a function of v and u and is defined implicitly by

$$u^2 - v^2 = \left(\frac{r}{2\mu} - 1\right) \exp\left(\frac{r}{2\mu}\right).$$

Show further that v is timelike and u is spacelike throughout the Schwarzschild geometry.

11.17 Perform an embedding into three-dimensional Euclidean space of the 2-space with line element

$$d\sigma^2 = dr^2 + (r^2 + a^2)\, d\phi^2,$$

and hence show that the resulting 2-surface has a geometry reminiscent of a wormhole.

11.18 By examining the paths of light rays in the Kruskal diagram, deduce that no particle can pass through the Einstein–Rosen wormhole from region I to region I′ or vice versa, before the throat of the wormhole pinches off.

11.19 A Schwarzschild black hole of mass M radiates as a blackbody of temperature $T = \hbar c^3/(8\pi k_B GM)$. Show from first principles that the black hole has a lifetime $\tau = M^3/(3\alpha\hbar)$, where $\alpha = c^4/(15360\pi G^2)$. In its last second, calculate the total energy radiated and estimate the typical energy of each radiated particle.

11.20 From observations of a compact binary system, one may calculate the mass function

$$f(M) = \frac{PK^3}{2\pi G},$$

where P is the orbital period and K is the radial velocity amplitude. From Kepler's laws in Newtonian gravity, show that $f(M)$ is related to the masses M_1 and M_2 of the compact object and the companion star and the inclination angle i of the orbit to the plane of the sky by

$$f(M) = \frac{M_1^3 \sin^3 i}{(M_1 + M_2)^2}.$$

12

Further spherically symmetric geometries

In the preceding three chapters, we have considered in some detail the Schwarzschild geometry, which represents the gravitational field outside a static spherically symmetric object. We also considered the structure of the Schwarzschild black hole, in which the empty-space field equations are satisfied everywhere except at the central intrinsic singularity. In this chapter, we consider solving the Einstein equations for a static spherically symmetric spacetime in regions where the presence of other fields means that the energy–momentum tensor is non-zero. In particular, we will concentrate on two physically interesting situations. First, we discuss the relativistic gravitational equations for the *interior* of a spherically symmetric matter distribution (or *star*); in this case the energy–momentum tensor of the matter making up the star must be included in the Einstein field equations. Second, we consider the spacetime geometry *outside* a static spherically symmetric *charged* object; once again this is not a vacuum, since it is filled with a static electric field whose energy–momentum must be included in the field equations.

12.1 The form of the metric for a stellar interior

Most stars in the sky are nowhere near dense enough for general-relativistic effects to be important in determining their structure. This is true for main sequence stars (of which our Sun is an example), red giants and even such high-density objects as white dwarfs. Thus, most stars will never even evolve into an object that is not adequately described by the Newtonian theory of stellar structure.[1] For neutron stars, however, the extremely high densities involved (see Section 11.6) mean that the internal gravitational forces will be very strong, and so we expect general-relativistic effects to play a significant role in determining their structure and their stability to collapse. As a result, it is of practical (as well as theoretical) interest

[1] See, for example, S. Chandrasekhar, *An Introduction to the Study of Stellar Structure*, Dover, 1958.

to consider the relativistic equations governing the equilibrium of a centrally symmetric self-gravitating distribution of matter.

Since we are assuming spherical symmetry and a static matter distribution, the appropriate general form of the metric is that derived in Section 9.1, namely

$$\boxed{ds^2 = A(r)\,dt^2 - B(r)\,dr^2 - r^2(d\theta^2 + \sin^2\theta\,d\phi^2).} \tag{12.1}$$

As in our derivation of the Schwarzschild metric, the two functions $A(r)$ and $B(r)$ are determined by solving the Einstein equations. For our present discussion, however, we shall not solve the empty-space field equations $R_{\mu\nu} = 0$, which are valid outside the spherical object, but instead solve the full field equations that hold in the interior of the object. These are most conveniently written in the form (8.15), namely

$$R_{\mu\nu} = -\kappa\left(T_{\mu\nu} - \tfrac{1}{2}Tg_{\mu\nu}\right), \tag{12.2}$$

where $T_{\mu\nu}$ is the energy–momentum tensor of the matter of which the object is composed, $T \equiv T^{\mu}_{\mu}$ and $\kappa = 8\pi G/c^4$. For the discussion in this chapter we will assume the matter to be described by a perfect fluid, so that

$$T_{\mu\nu} = \left(\rho + \frac{p}{c^2}\right)u_\mu u_\nu - pg_{\mu\nu}, \tag{12.3}$$

where $\rho(r)$ is the proper mass density and $p(r)$ is the isotropic pressure in the instantaneous rest frame of the fluid, both of which may be taken as functions only of the radial coordinate r for a static matter distribution. Using the fact that $u_\mu u^\mu = c^2$, we find that

$$T = \left(\rho + \frac{p}{c^2}\right)c^2 - p\delta^{\mu}_{\mu} = \rho c^2 - 3p,$$

and so the field equations (12.2) read

$$R_{\mu\nu} = -\kappa\left[\left(\rho + \frac{p}{c^2}\right)u_\mu u_\nu - \tfrac{1}{2}(\rho c^2 - p)g_{\mu\nu}\right]. \tag{12.4}$$

As shown in Section 9.2, the off-diagonal components of the Ricci tensor $R_{\mu\nu}$ for the metric (12.1) are all zero and the diagonal components are given by

$$R_{00} = -\frac{A''}{2B} + \frac{A'}{4B}\left(\frac{A'}{A} + \frac{B'}{B}\right) - \frac{A'}{rB}, \tag{12.5}$$

$$R_{11} = \frac{A''}{2A} - \frac{A'}{4A}\left(\frac{A'}{A} + \frac{B'}{B}\right) - \frac{B'}{rB}, \tag{12.6}$$

$$R_{22} = \frac{1}{B} - 1 + \frac{r}{2B}\left(\frac{A'}{A} - \frac{B'}{B}\right), \tag{12.7}$$

$$R_{33} = R_{22}\sin^2\theta. \tag{12.8}$$

It is of interest first to determine the consequences of the vanishing off-diagonal components of the Ricci tensor, $R_{0i} = 0$ for $i = 1, 2, 3$. From the field equations (12.4), and using the fact that $g_{0i} = 0$, we see immediately that we require $u_i u_0 = 0$. Combining this with $u_\mu u^\mu = c^2$, we find that the covariant components of the fluid 4-velocity are given by

$$[u_\mu] = c\sqrt{A}(1, 0, 0, 0), \tag{12.9}$$

and thus the spatial 3-velocity of the fluid must *vanish* everywhere. In particular, we note that this conclusion holds *without* our assuming in advance that the proper density ρ and the pressure p are independent of t. Thus, the fact the metric (12.1) is independent of t automatically implies that the matter distribution itself is static and so the object is in a state of hydrostatic equilibrium. This is another illustration how the equations of motion for matter follow directly from the field equations (see Section 8.8).

Let us now use the diagonal ($\mu = \nu$) components of the field equations (12.2) to obtain the differential equations that the functions $A(r)$ and $B(r)$ must satisfy. Inserting the expression (12.3) into the right-hand side of the field equations and using the metric (12.1), we find that

$$R_{00} = -\tfrac{1}{2}\kappa(\rho c^2 + 3p)A, \tag{12.10}$$

$$R_{11} = -\tfrac{1}{2}\kappa(\rho c^2 - p)B, \tag{12.11}$$

$$R_{22} = -\tfrac{1}{2}\kappa(\rho c^2 - p)r^2, \tag{12.12}$$

$$R_{33} = R_{22}\sin^2\theta. \tag{12.13}$$

From these equations, one quickly obtains

$$\frac{R_{00}}{A} + \frac{R_{11}}{B} + \frac{2R_{22}}{r^2} = -2\kappa\rho c^2.$$

On substituting the expressions (12.5–12.8) for the Ricci tensor components and simplifying, one finds that

$$\left(1 - \frac{1}{B}\right) + \frac{rB'}{B^2} = \kappa r^2 \rho c^2, \tag{12.14}$$

which can be rewritten in the form

$$\frac{d}{dr}\left[r\left(1 - \frac{1}{B}\right)\right] = \kappa r^2 \rho c^2.$$

Integrating this expression with respect to r, and noting that the associated constant of integration must be zero in order for $B(r)$ to be non-zero at the origin (as demanded by (12.14)), we find that the solution for $B(r)$ is given by

$$\boxed{B(r) = \left[1 - \frac{2Gm(r)}{c^2 r}\right]^{-1},} \tag{12.15}$$

where we have defined the function

$$m(r) = 4\pi \int_0^r \rho(\bar{r})\bar{r}^2\, d\bar{r}. \tag{12.16}$$

This function is worthy of further comment, since it has the appearance of being the mass contained within a coordinate radius r. This interpretation is not quite correct, however, since the proper spatial volume element for the metric (12.1) is given by

$$d^3V = \sqrt{B(r)}\, r^2 \sin\theta\, dr\, d\theta\, d\phi.$$

Thus the proper integrated 'mass' (i.e. energy/c^2) within a coordinate radius r is

$$\widetilde{m}(r) = 4\pi \int_0^r \rho(\bar{r})\sqrt{B(\bar{r})}\,\bar{r}^2\, d\bar{r} = 4\pi \int_0^r \rho(\bar{r})\left[1 - \frac{2Gm(\bar{r})}{c^2\bar{r}}\right]^{-1/2} \bar{r}^2\, d\bar{r}.$$

Nevertheless, we note that it is $m(r)$, not $\widetilde{m}(r)$, that appears in the radial metric coefficient $B(r)$ in (12.15). In particular, if the object extends to a coordinate radius $r = R$, beyond which there is empty space, then the spacetime geometry outside this radius is described by the Schwarzschild metric with mass parameter $M = m(R)$, rather than $\widetilde{M} = \widetilde{m}(R)$. The difference $E = \widetilde{M} - M$ corresponds to the gravitational binding energy of the object, which is the amount of energy required to disperse the material of which the object consists to infinite spatial separation.

We now turn to determining the differential equation that must be satisfied by the function $A(r)$ in (12.1). In principle, this could be obtained by substituting for $B(r)$ using (12.15) in any of the equations (12.10–12.13). It is more convenient and instructive, however, to use the conservation equation $\nabla_\mu T^{\mu\nu} = 0$ directly, from which the fluid equations of motion may be derived (as discussed in Section 8.3). Using (12.3), we may write

$$\begin{aligned}\nabla_\mu T^{\mu\nu} &= \nabla_\mu\left[\left(\rho + \frac{p}{c^2}\right)u^\mu u^\nu\right] - \nabla_\mu(pg^{\mu\nu})\\ &= \frac{1}{\sqrt{-g}}\partial_\mu\left[\sqrt{-g}\left(\rho + \frac{p}{c^2}\right)u^\mu u^\nu\right] + \left(\rho + \frac{p}{c^2}\right)\Gamma^\nu{}_{\sigma\mu}u^\mu u^\sigma - g^{\mu\nu}\partial_\mu p,\end{aligned} \tag{12.17}$$

where, in going from the first to the second line, the first term has been rewritten using the expression (8.24) for the covariant divergence and the second term has been manipulated by noting that $\nabla_\mu g^{\mu\nu} = 0$ and that p is a scalar function.

From (12.9), however, we have $u^0 = c/\sqrt{A}$ and $u^i = 0$, and since ρ and p do not depend on t the first term on the right-hand side of (12.17) must vanish. For the same reason, the second term becomes equal to $(\rho c^2 + p)\Gamma^{\nu}{}_{00}/A$. From (3.21) and (12.1), we have

$$\Gamma^{\nu}{}_{00} = -\tfrac{1}{2} g^{\mu\nu} \partial_{\mu} g_{00} = -\tfrac{1}{2} g^{\mu\nu} \partial_{\mu} A,$$

and so the conservation equation $\nabla_{\mu} T^{\mu\nu} = 0$ can be written as

$$\frac{\rho c^2 + p}{2A} g^{\mu\nu} \partial_{\mu} A + g^{\mu\nu} \partial_{\mu} p = 0.$$

Multiplying through by $g_{\nu\sigma}$ and simplifying, one obtains

$$\frac{\rho c^2 + p}{2A} \partial_{\sigma} A + \partial_{\sigma} p = 0. \tag{12.18}$$

Since A is a function only of r, the above equation is trivial for $\sigma = 0$, in which case we recover the fact that p is independent of t. Similarly, for $\sigma = 2$ and $\sigma = 3$ we find that the corresponding (tangential) derivatives of p also vanish, as dictated by spherical symmetry. For $\sigma = 1$, however, the relation (12.18) is non-trivial and reads (where primes denote d/dr)

$$\boxed{\frac{A'}{A} = -\frac{2p'}{\rho c^2 + p},} \tag{12.19}$$

which gives a differential equation, in terms of $\rho(r)$ and $p(r)$, that $A(r)$ must satisfy in hydrostatic equilibrium.

12.2 The relativistic equations of stellar structure

The equations (12.15) and (12.19) show how to calculate the functions $A(r)$ and $B(r)$ in the metric (12.1), given particular functions of $\rho(r)$ and $p(r)$. Specifying these two functions does, however, imply an *equation of state* $p = p(\rho)$ by elimination of r, and this is likely to be physically unrealistic for arbitrary choices of $\rho(r)$ and $p(r)$. For astrophysical investigations, one is more interested in building models of the density and pressure distribution inside a star under the assumption of some (quasi-)realistic equation of state. Thus, it is usual to recast the results obtained in the previous section into an alternative form.

In this approach, the first equation of stellar structure is taken simply from (12.16) and written as

$$\boxed{\frac{dm(r)}{dr} = 4\pi r^2 \rho(r),} \tag{12.20}$$

which clearly relates the functions $m(r)$ and $\rho(r)$. The next step is to obtain an equation linking $m(r)$ and $p(r)$. This is most conveniently achieved by using (12.7) and (12.12):

$$\frac{1}{B} - 1 + \frac{r}{2B}\left(\frac{A'}{A} - \frac{B'}{B}\right) = -\frac{1}{2}\kappa(\rho c^2 - p)r^2.$$

Eliminating the functions A and B using (12.19) and (12.15) and simplifying, one obtains the second equation of stellar structure,

$$\boxed{\frac{dp}{dr} = -\frac{1}{r^2}(\rho c^2 + p)\left[\frac{4\pi G}{c^4}pr^3 + \frac{Gm(r)}{c^2}\right]\left[1 - \frac{2Gm(r)}{c^2 r}\right]^{-1},} \tag{12.21}$$

which is also known as the *Oppenheimer–Volkoff equation.* As mentioned above, to obtain a closed system of equations we need to define the equation of state for the matter, which gives the pressure in terms of the density, namely

$$\boxed{p = p(\rho).} \tag{12.22}$$

This provides the third (and final) equation of stellar structure. We note that, for many astrophysical systems, the matter obeys a *polytropic* equation of state of the form $p = K\rho^\gamma$, where K and γ are both constants. In the usual notation used in this field, $\gamma = 1 + 1/n$, where n is known as the *polytropic index.*

The closed system of three equations (12.20–12.22) contains two coupled first-order differential equations, and so to obtain a unique solution one must specify two boundary conditions. The first is straightforward, since we must have $m(0) = 0$, leaving just one further adjustable boundary condition to be specified. It is most common to choose this adjustable parameter to be the central pressure $p(0)$, or equivalently the central density $\rho(0)$, which can be obtained immediately from the equation of state (12.22). Very few exact solutions are known for realistic equations of state, and so in practice the system of equations is integrated numerically on a computer. The procedure is to 'integrate outwards' from $r = 0$ (in practice in small radial steps of size Δr) until the pressure drops to zero. This condition defines the surface ($r = R$) of the star, since otherwise there would be an infinite pressure gradient, and hence an infinite force, on the material elements constituting the outer layer of the star. For $r > R$, $\rho(r)$ and $p(r)$ are both zero and $m(r) = m(R) = M$, and the spacetime geometry is described by the Schwarzschild metric with mass parameter M.

Before looking for particular solutions to the set of equations (12.20–12.22), it is worthwhile considering briefly their Newtonian limit. In fact, the forms of (12.20) and (12.22) remain unchanged in this limit, and it is only the equation (12.21) for the pressure gradient that is simplified. In the Newtonian limit we have $p \ll \rho$ and therefore $4\pi r^3 p \ll mc^2$. Moreover, we require the metric (12.1) to be close to Minkowski, and so we require $2Gm/(c^2 r) \ll 1$. Thus, the Oppenheimer–Volkoff equation reduces to

$$\frac{dp}{dr} = -\frac{Gm(r)\rho(r)}{r^2}, \tag{12.23}$$

which is simply the Newtonian equation for hydrostatic equilibrium. Comparing (12.23) with (12.21), we see that all the relativistic effects serve to steepen the pressure gradient relative to the Newtonian case. Thus, for an object to remain in hydrostatic equilibrium, the fluid of which it consists must experience stronger internal forces when general-relativistic effects are taken into account.

12.3 The Schwarzschild constant-density interior solution

The simplest analytic interior solution for a relativistic star is obtained by making the assumption that, throughout the star,

$$\boxed{\rho = \text{constant},}$$

which constitutes an equation of state. There is no physical justification for this assumption, but it is on the borderline of being realistic. It corresponds to an *ultra-stiff* equation of state that represents an incompressible fluid. Consequently, the speed of sound in the fluid, which is proportional to $(dp/d\rho)^{1/2}$, is infinite (which is clearly not allowed relativistically). Nevertheless, it is believed that the interiors of dense neutron stars are of nearly uniform density, and so this simple case is of some practical interest.

Equation (12.20) immediately integrates to give

$$\boxed{m(r) = \begin{cases} \frac{4}{3}\pi\rho r^3 & \text{for } r \leq R \\ \frac{4}{3}\pi\rho R^3 \equiv M & \text{for } r > R, \end{cases}} \tag{12.24}$$

where R is the radius of the star, as yet undetermined, and M is the mass parameter for the Schwarzschild metric describing the spacetime geometry outside the star. Moreover, the Oppenheimer–Volkoff equation (12.21) becomes

$$\frac{dp}{dr} = -\frac{4\pi G}{3c^4} r(\rho c^2 + p)(\rho c^2 + 3p)\left(1 - \frac{8\pi G}{3c^2}\rho r^2\right)^{-1}.$$

This equation is separable and we may write

$$\int_{p_0}^{p(r)} \frac{d\bar{p}}{(\rho c^2 + \bar{p})(\rho c^2 + 3\bar{p})} = -\frac{4\pi G}{3c^4}\int_0^r \frac{\bar{r}\, d\bar{r}}{1 - 8\pi G\rho\bar{r}^2/(3c^2)},$$

where $p_0 = p(0)$ is the central pressure of the star. Performing these standard integrals, one finds that

$$\frac{\rho c^2 + 3p}{\rho c^2 + p} = \frac{\rho c^2 + 3p_0}{\rho c^2 + p_0}\left(1 - \frac{8\pi G}{3c^2}\rho r^2\right)^{1/2}. \tag{12.25}$$

At the surface $r = R$ of the star, the pressure p is zero and so the left-hand side of the above equation equals unity. Thus, we obtain

$$R^2 = \frac{3c^2}{8\pi G\rho}\left[1 - \left(\frac{\rho c^2 + p_0}{\rho c^2 + 3p_0}\right)^2\right],$$

which gives the radius of a star of uniform density ρ with a central pressure p_0. Alternatively, we may rearrange this result and use (12.24) to obtain a useful expression for the central pressure,

$$p_0 = \rho c^2 \frac{1 - (1 - 2\mu/R)^{1/2}}{3(1 - 2\mu/R)^{1/2} - 1}, \tag{12.26}$$

where $\mu = GM/c^2$. Using this expression to replace p_0 in (12.25) gives

$$\boxed{p(r) = \rho c^2 \frac{(1 - 2\mu r^2/R^3)^{1/2} - (1 - 2\mu/R)^{1/2}}{3(1 - 2\mu/R)^{1/2} - (1 - 2\mu r^2/R^3)^{1/2}} \quad \text{for } r \leq R.} \tag{12.27}$$

To obtain the complete solution to the problem, it remains to determine the functions $A(r)$ and $B(r)$ in the metric (12.1). From (12.15) and (12.24), we immediately find that

$$B(r) = \left(1 - \frac{2\mu r^2}{R^3}\right)^{-1}. \tag{12.28}$$

In particular, we note that at the star's surface, where $r = R$, the above solution matches with the corresponding expression from the Schwarzschild metric for the exterior solution. The function $A(r)$ is obtained from (12.19), (12.24) and (12.27). One may fix the integration constant arising from (12.19) by imposing the boundary condition that $A(r)$ matches the corresponding expression in the Schwarzschild metric at $r = R$. One then finds

$$A(r) = \frac{c^2}{4}\left[3\left(1 - \frac{2\mu}{R}\right)^{1/2} - \left(1 - \frac{2\mu r^2}{R^3}\right)^{1/2}\right]^2. \tag{12.29}$$

The expressions (12.28) and (12.29) constitute Schwarzschild's interior solution for a constant-density object.

12.4 Buchdahl's theorem

The most important feature of the Schwarzschild constant-density solution discussed above is that it imposes a constraint connecting the star's 'mass' M and its (coordinate) radius R. To derive this constraint, one notices that (12.26) implies that $p_0 \to \infty$ as $\mu/R \to 4/9$. Since pressure is a general scalar, this infinity will persist in any coordinate system, and so one can only avoid this behaviour by demanding that

$$\boxed{\frac{GM}{c^2 R} < \frac{4}{9}.} \tag{12.30}$$

Although we have only shown that this constraint holds for an object of constant density, *Buchdahl's theorem* states that (12.30) is in fact valid for *any* equation of state. This theorem can be proved directly from the Einstein equations but requires considerable care and lies outside the scope of our discussion.

Equation (12.30) can be regarded as providing an upper limit on the mass of a star for a fixed radius. If one attempts to pack more mass inside R than is allowed by (12.30), general relativity admits no static solution: the hydrostatic equilibrium is destroyed by the increased gravitational attraction. Such a star must therefore collapse inwards without stopping. Throughout the collapse, the exterior geometry is described by the Schwarzschild metric, and so eventually one obtains a Schwarzschild black hole. The limit (12.30) is, in fact, quite easily reached. For example, the density of a neutron star is around 10^{16} kg m^{-3} and, assuming it to be of uniform density, we find from (12.30) and (12.24) that $M < 7 \times 10^{31}$ kg. This is approximately 35 solar masses, which is of same order as the most massive stars in our Galaxy.

12.5 The metric outside a spherically symmetric charged mass

We now turn to our second physical application, namely the form of the metric *outside* a static spherically symmetric *charged* body. The exterior of such an object is not a vacuum, since it is filled with a static electric field. We must therefore once again solve the Einstein field equations for a static spherically symmetric spacetime in the presence of a non-zero energy–momentum tensor, this time representing the electromagnetic field of the object.

Since we are assuming spherical symmetry and a static object, the general form of the metric is once more given by (12.1). The two functions $A(r)$ and $B(r)$ are determined by solving the full Einstein equations outside the spherical object; these equations are again most conveniently written in the form (12.2). In this case, however, $T_{\mu\nu}$ is the energy–momentum tensor of the electromagnetic field of the charged object, which from Exercise 8.3 has the general form

$$T_{\mu\nu} = -\mu_0^{-1}(F_{\mu\rho}F_\nu{}^\rho - \tfrac{1}{4}g_{\mu\nu}F_{\rho\sigma}F^{\rho\sigma}), \tag{12.31}$$

where $F_{\mu\nu} = \partial_\mu A_\nu - \partial_\mu A_\nu$ is the electromagnetic field strength tensor and A_μ is the electromagnetic 4-potential. The first point to note about this energy–momentum tensor is that it has zero trace,

$$T \equiv T^\mu_\mu = -\mu_0^{-1}(F_{\mu\rho}F^{\mu\rho} - \tfrac{1}{4}\delta^\mu_\mu F_{\rho\sigma}F^{\rho\sigma}) = 0.$$

Thus, in this case, the Einstein field equations (12.2) take the simplified form

$$R_{\mu\nu} = -\kappa T_{\mu\nu}. \tag{12.32}$$

In addition to the Einstein field equations, our solution must also satisfy the Maxwell equations. In the region outside the charged object, the 4-current density j^μ is zero and so the Maxwell equations read

$$\nabla_\mu F^{\mu\nu} = 0, \tag{12.33}$$

$$\nabla_\sigma F_{\mu\nu} + \nabla_\nu F_{\sigma\mu} + \nabla_\mu F_{\nu\sigma} = 0. \tag{12.34}$$

The Einstein and Maxwell equations are coupled together, since $F_{\mu\nu}$ enters the gravitational field equations through the energy–momentum tensor (12.31) and the metric $g_{\mu\nu}$ enters the electromagnetic field equations through the covariant derivative.

The constraint imposed on the metric coefficients $g_{\mu\nu}$ (or gravitational fields) by requiring the solution to be spherically symmetric and static is embodied in the choice of line element (12.1). We thus begin by considering the corresponding consequences of these symmetry constraints for the form of the electromagnetic field. In this case, the electromagnetic 4-potential in (t, r, θ, ϕ) coordinates takes the form

$$[A^\mu] = \left(\frac{\phi(r)}{c^2}, a(r), 0, 0\right), \tag{12.35}$$

where $\phi(r)$ and $a(r)$ depend only on r and may be interpreted respectively as the electrostatic potential and the radial component of the 3-vector potential as $r \to \infty$ (the extra factor of $1/c$ multiplying $\phi(r)$ in (12.35), as compared with the usual form in Minkowski coordinates, is a result of taking $x^0 = t$ rather than $x^0 = ct$; also, note that the 3-vector potential $a(r)$ should not be confused with the function $A(r)$ in the metric (12.1)). From (12.35), the field-strength tensor has the form

$$[F_{\mu\nu}] = E(r)\begin{pmatrix} 0 & -1 & 0 & 0 \\ 1 & 0 & 0 & 0 \\ 0 & 0 & 0 & 0 \\ 0 & 0 & 0 & 0 \end{pmatrix}, \tag{12.36}$$

where $E(r)$ is an arbitrary function of r only and may be interpreted as the radial component of the static electric field as $r \to \infty$. Thus our task is to use the Einstein equations (12.32) and the Maxwell equations (12.33–12.34) to determine the three unknown functions $A(r)$, $B(r)$ and $E(r)$.

Let us begin by using the Maxwell equations. As discussed in Section 6.6, the equations (12.34) are automatically satisfied by the definition of $F_{\mu\nu}$. Moreover, from Exercise 4.10, since $F_{\mu\nu}$ is antisymmetric we may rewrite the covariant divergence in the first Maxwell equation (12.33) to obtain

$$\nabla_\mu F^{\mu\nu} = \frac{1}{\sqrt{-g}}\partial_\mu\left(\sqrt{-g}F^{\mu\nu}\right) = 0, \tag{12.37}$$

where g is the determinant of the metric. For a diagonal line element such as (12.1), the determinant is simply the product of the diagonal elements, so that $g = -A(r)B(r)r^4\sin^2\theta$. Given the form of $F_{\mu\nu}$ in (12.36), the expression (12.37) yields the single equation

$$\partial_1\left(\sqrt{AB}r^2F^{10}\right) = 0.$$

Writing the required contravariant component as $F^{10} = g^{1\mu}g^{0\nu}F_{\mu\nu} = g^{11}g^{00}F_{10} = -E/(AB)$, we thus obtain the equation

$$\frac{d}{dr}\left(\frac{r^2E}{\sqrt{AB}}\right) = 0.$$

This integrates to give

$$E(r) = \frac{k\sqrt{A(r)B(r)}}{r^2}, \tag{12.38}$$

where k is a constant of integration. If we make the assumption that the metric is asymptotically flat then $A(r) \to c^2$ and $B(r) \to 1$ as $r \to \infty$. Identifying $E(r)$ with the radial electric field component at infinity, we thus require $k = Q/(4\pi\epsilon_0 c)$, where Q is the total charge of the object.

We now turn to the Einstein equations (12.32). The Ricci tensor components for the metric (12.1) are given in (12.5–12.8), and the form of the electromagnetic field energy–momentum tensor $T_{\mu\nu}$ may be found by substituting the form (12.36) for $F_{\mu\nu}$ into the expression (12.31). On performing this substitution, one quickly finds that the off-diagonal components of $T_{\mu\nu}$ are zero, and so the Einstein equations for $\mu \neq \nu$ are satisfied identically. For the diagonal components of the Einstein equations, one finds

$$R_{00} = -\tfrac{1}{2}\kappa c^2 \epsilon_0 E^2/B, \tag{12.39}$$

$$R_{11} = \tfrac{1}{2}\kappa c^2 \epsilon_0 E^2/A, \tag{12.40}$$

$$R_{22} = -\tfrac{1}{2}\kappa c^2 \epsilon_0 r^2 E^2/(AB), \tag{12.41}$$

$$R_{33} = R_{22}\sin^2\theta, \tag{12.42}$$

where we have used the facts that $F_0{}^1 = g^{11}F_{01} = E/B$ and $F_1{}^0 = g^{00}F_{10} = E/A$; we have also made use of the relation $\mu_0\epsilon_0 = 1/c^2$. From (12.39) and (12.40), we immediately obtain

$$BR_{00} + AR_{11} = 0.$$

On substituting the expressions (12.5, 12.6) for the Ricci tensor coefficients and rearranging, this yields

$$A'B + B'A = 0,$$

which implies that $AB =$ constant. We may fix this constant from the requirement that the metric is asymptotically flat as $r \to \infty$, and so we have

$$A(r)B(r) = c^2. \tag{12.43}$$

A further independent equation may be obtained from the 22-component (12.41) of the Einstein equations. Inserting the expression (12.7) for the Ricci tensor component and using (12.38), one finds that

$$A + rA' = c^2\left(1 - \frac{GQ^2}{4\pi\epsilon_0 c^4 r^2}\right).$$

Noting that $A + rA' = (rA)'$ and integrating, one thus obtains

$$A(r) = c^2\left(1 - \frac{2GM}{c^2 r} + \frac{GQ^2}{4\pi\epsilon_0 c^4 r^2}\right),$$

where we have identified the integration constant as $-2GM/c^2$, M being the mass of the object, since the line element must reduce to the Schwarzschild case when $Q = 0$. The solutions for $B(r)$ and $E(r)$ are then found immediately from (12.43) and (12.38) respectively.

Thus, collecting our results together and defining the constants $\mu = GM/c^2$ and $q^2 = GQ^2/(4\pi\epsilon_0 c^4)$, the line element for the spacetime outside a static spherically symmetric body of mass M and charge Q has the form

$$\boxed{ds^2 = c^2\left(1 - \frac{2\mu}{r} + \frac{q^2}{r^2}\right) dt^2 - \left(1 - \frac{2\mu}{r} + \frac{q^2}{r^2}\right)^{-1} dr^2 - r^2(d\theta^2 + \sin^2\theta d\phi^2),} \tag{12.44}$$

from which one may read off the metric coefficients $g_{\mu\nu}$ that determine the gravitational field of the object. The resulting solution is known as the *Reissner–Nordström* geometry. The electromagnetic $F_{\mu\nu}$ of the field of the object is given by (12.36) with

$$\boxed{E(r) = \frac{Q}{4\pi\epsilon_0 r^2}.}$$

12.6 The Reissner–Nordström geometry: charged black holes

The Reissner–Nordström (RN) metric (12.44) is only valid down to the surface of the charged object. As in our discussion of the Schwarzschild solution, however, it is of interest to consider the structure of the full RN geometry, namely the solution to the coupled Einstein–Maxwell field equations for a charged *point* mass located at the origin $r = 0$, in which case the RN metric is valid for all positive r.

Calculation of the invariant curvature scalar $R_{\mu\nu\sigma\rho}R^{\mu\nu\sigma\rho}$ shows that the only intrinsic singularity in the RN metric occurs at $r = 0$. In the 'Schwarzschild-like' coordinates (t, r, θ, ϕ), however, the RN metric also possesses a coordinate singularity wherever r satisfies

$$\Delta(r) \equiv 1 - \frac{2\mu}{r} + \frac{q^2}{r^2} = 0, \tag{12.45}$$

with $\Delta(r) = -1/g_{11}(r) = g_{00}(r)/c^2$. Multiplying (12.45) through by r^2 and solving the resulting quadratic equation, we find that the coordinate singularities occur on the surfaces $r = r_\pm$, where

$$\boxed{r_\pm = \mu \pm (\mu^2 - q^2)^{1/2}.} \tag{12.46}$$

It is clear that there exist three distinct cases, depending on the relative values of μ^2 and q^2; we now discuss these in turn.

- *Case 1*: $\mu^2 < q^2$ In this case $r_\pm$ are both imaginary, and so no coordinate singularities exist. The metric is therefore regular for all positive values of r. Since the function $\Delta(r)$ always remains positive, the coordinate t is always timelike and r is always spacelike. Thus, the intrinsic singularity at $r = 0$ is a timelike line, as opposed to a spacelike line in the Schwarzschild case. This means that the singularity does not necessarily lie in the future of timelike trajectories and so, in principle, can be avoided. In the absence of any event horizons, however, $r = 0$ is a *naked singularity*, which is visible to the outside world. The physical consequences of a naked singularity, such as the existence of closed timelike curves, appear so extreme that Penrose has suggested the existence of a *cosmic censorship hypothesis*, which would only allow singularities that are hidden behind an event horizon. As a result, the case $\mu^2 < q^2$ is not considered physically realistic.

- *Case 2*: $\mu^2 > q^2$ In this case, $r_\pm$ are both real and so there exist *two* coordinate singularities, occurring on the surfaces $r = r_\pm$. The situation at $r = r_+$ is very similar to the Schwarzschild case at $r = 2\mu$. For $r > r_+$, the function $\Delta(r)$ is positive and so the coordinates t and r are timelike and spacelike respectively. In the region $r_- < r < r_+$, however, $\Delta(r)$ becomes negative and so the physical natures of the coordinates t and r are interchanged. Thus, a massive particle or photon that enters the surface $r = r_+$ from outside must necessarily move in the direction of decreasing r, and thus $r = r_+$ is an event horizon. The major difference from the Schwarzschild geometry is that the irreversible infall of the particle need only continue to the surface $r = r_-$, since for $r < r_-$ the function $\Delta(r)$ is again positive and so t and r recover their timelike and spacelike properties. Within $r = r_-$, one may (with a rocket engine) move in the direction of either positive or negative r, or stand still. Thus, one may avoid the intrinsic singularity at $r = 0$, which is consistent with the fact that $r = 0$ is a timelike line. Perhaps even more astonishing is what happens if one then chooses to travel back in the direction of positive r in the region $r < r_-$. On performing a *maximal analytic extension* of the RN geometry, in analogy with the Kruskal extension for the Schwarzschild geometry discussed in Section 11.9, one finds that one may *re-cross* the surface $r = r_-$, but this time from the inside. Once again one is moving from a region in which r is spacelike to a region in which it is timelike, but this time the sense is reversed and one is forced to move in the direction of *increasing* r. Thus $r = r_-$ acts as an 'inside-out' event horizon. Moreover, one is eventually forceably ejected from the surface $r = r_+$ but, according to the maximum analytic extension, the particle emerges into a asymptotically flat spacetime *different* from that from which it first entered the black hole. As discussed in Section 11.9, however, such matters are at best highly speculative, and we shall not pursue them further here.

- *Case 3*: $\mu^2 = q^2$ In this case, called the *extreme* Reissner–Nordström black hole, the function $\Delta(r)$ is positive everywhere *except* at $r = \mu$, where it equals zero. Thus, the coordinate r is everywhere spacelike except at $r = \mu$, where it becomes null, and hence $r = \mu$ is an event horizon. The extreme case is basically the same as that considered in case 2, but with the region $r_- < r < r_+$ removed.

We may illustrate the properties of the RN spacetime in more detail by considering the paths of photons and massive particles in the geometry, which we now go on to discuss. Since the case $\mu^2 > q^2$ is the most physically reasonable RN spacetime, we shall restrict our discussion to this situation.

12.7 Radial photon trajectories in the RN geometry

Let us begin by investigating the paths of radially incoming and outgoing photons in the RN metric for the case $\mu^2 > q^2$. Since $ds = d\theta = d\phi = 0$ for a radially moving photon, we have immediately from (12.44) that

$$\frac{dt}{dr} = \pm\frac{1}{c}\left(1 - \frac{2\mu}{r} + \frac{q^2}{r^2}\right)^{-1} = \pm\frac{1}{c}\frac{r^2}{(r-r_-)(r-r_+)}, \tag{12.47}$$

where, in the second equality, we have used the result (12.46); the plus sign corresponds to an *outgoing* photon and the minus sign to an *incoming* photon. On integrating, we obtain

$$ct = r - \frac{r_-^2}{r_+ - r_-}\ln\left|\frac{r}{r_-} - 1\right| + \frac{r_+^2}{r_+ - r_-}\ln\left|\frac{r}{r_+} - 1\right| + \text{constant} \qquad \text{(outgoing)},$$

$$ct = -r + \frac{r_-^2}{r_+ - r_-}\ln\left|\frac{r}{r_-} - 1\right| - \frac{r_+^2}{r_+ - r_-}\ln\left|\frac{r}{r_+} - 1\right| + \text{constant} \qquad \text{(ingoing)}.$$

We will concentrate in particular on the ingoing radial photons. To develop a better description of infalling particles in general, we may construct the equivalent of the advanced Eddington–Finkelstein coordinates derived for the Schwarzschild metric in Section 11.5. Once again this coordinate system is based on radially infalling photons, and the trick is to use the integration constant as the new coordinate, which we denote by p. As before, p is a null coordinate and it is more convenient to work instead with the timelike coordinate t' defined by $ct' = p - r$. Thus, we have

$$ct' = ct - \frac{r_-^2}{r_+ - r_-}\ln\left|\frac{r}{r_-} - 1\right| + \frac{r_+^2}{r_+ - r_-}\ln\left|\frac{r}{r_+} - 1\right|. \tag{12.48}$$

On differentiating, or from (12.47) directly, one obtains

$$c\,dt' = dp - dr = c\,dt + \left[\frac{1}{\Delta(r)} - 1\right] dr, \tag{12.49}$$

where $\Delta(r)$ is defined in (12.45). Using the above expression to substitute for t in (12.44), one quickly finds that

$$\boxed{ds^2 = c^2\Delta\, dt'^2 - 2c(1-\Delta)\, dt'\, dr - (2-\Delta)\, dr^2 - r^2(d\theta^2 + \sin^2\theta\, d\phi^2),}$$

which is the RN metric in advanced Eddington–Finkelstein coordinates. In particular, we note that this form is regular for all positive values of r and has an instrinsic singularity at $r = 0$.

From (12.47) and (12.49), one finds that, in advanced Eddington–Finkelstein coordinates, the equation for ingoing radial photon trajectories is

$$ct' + r = \text{constant}, \tag{12.50}$$

whereas the trajectories for outgoing radial photons satisfy the differential equation

$$c\frac{dt'}{dr} = \frac{2-\Delta}{\Delta}. \tag{12.51}$$

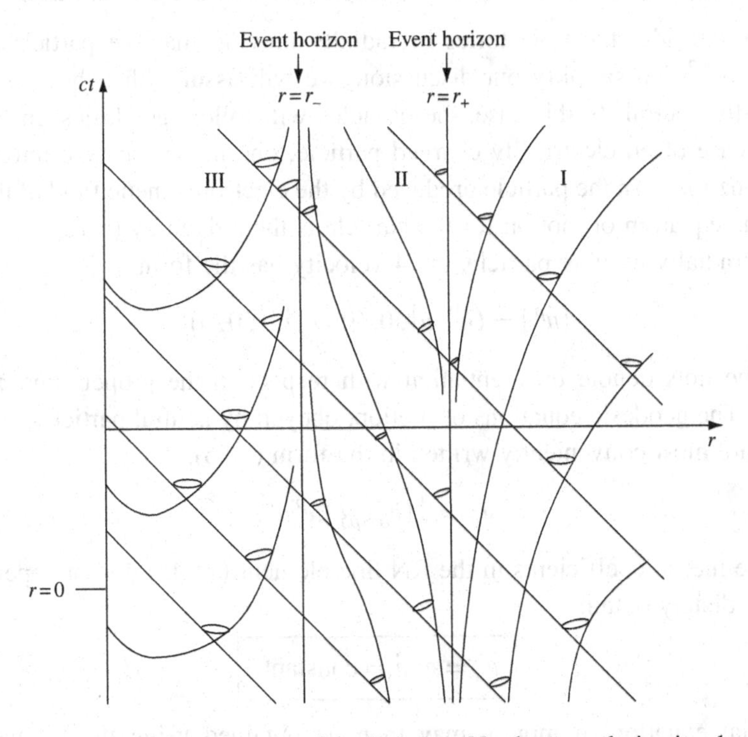

Figure 12.1 Spacetime diagram of the Reissner–Nordström solution in advanced Eddington–Finkelstein coordinates. The straight diagonal lines are ingoing photon worldlines whereas the curved lines correspond to outgoing photon worldlines.

We may use these equations to determine the light-cone structure of the RN metric in these coordinates. For ingoing radial photons, the trajectories (12.50) are simply straight lines at 45° in a spacetime diagram. For outgoing radial photons, (12.51) gives the gradient of the trajectory at any point in the spacetime diagram, and so one may sketch these without solving (12.51) explicitly. This resulting spacetime diagram is shown in Figure 12.1. It is worth noting that the light-cone structure depicted confirms the nature of the event horizon at $r = r_+$. Moreover, the light-cones remain tilted over in the region $r_- < r < r_+$, indicating that any particle falling into this region must move inwards until it reaches $r = r_-$. Once in the region $r < r_-$, the lightcones are no longer tilted and so particles need not fall into the singularity $r = 0$. As was the case in Section 11.5 for the Schwarzschild metric, however, this spacetime diagram may be somewhat misleading. For an outward-moving particle in the region $r < r_-$, Figure 12.1 suggests that it can only reach $r = r_-$ asymptotically, but by peforming an analytic extension of the RN solution one can show that the particle can cross the surface $r = r_-$ in finite proper time.

12.8 Radial massive particle trajectories in the RN geometry

We now consider the trajectories of radially moving massive particles for the case $\mu^2 > q^2$. To simplify our discussion, we will assume that the particles are electrically neutral. In this case, the particles will follow geodesics. In the more general case of an electrically charged particle, one must also take into account the Lorenz force on the particle produced by the electromagnetic field of the black hole. The equation of motion for the particle is then given by (6.13).

For a radially moving particle, the 4-velocity has the form

$$[u^\mu] = (u^0, u^1, 0, 0) = (\dot{t}, \dot{r}, 0, 0),$$

where the dots denote differentiation with respect to the proper time τ of the particle. The geodesic equations of motion, obeyed by neutral particles in the RN metric, are most conveniently written in the form (3.56):

$$\dot{u}_\sigma = \tfrac{1}{2}(\partial_\sigma g_{\mu\nu})u^\mu u^\nu.$$

Since the metric coefficients in the RN line element (12.44) do not depend on t, we immediately obtain

$$\boxed{u_0 = g_{00}\dot{t} = \text{constant}.}$$

The radial equation of motion may then be obtained using the normalisation condition $g_{\mu\nu}u^\mu u^\nu = c^2$, which gives

$$g_{00}(u^0)^2 + g_{11}(u^1)^2 = c^2.$$

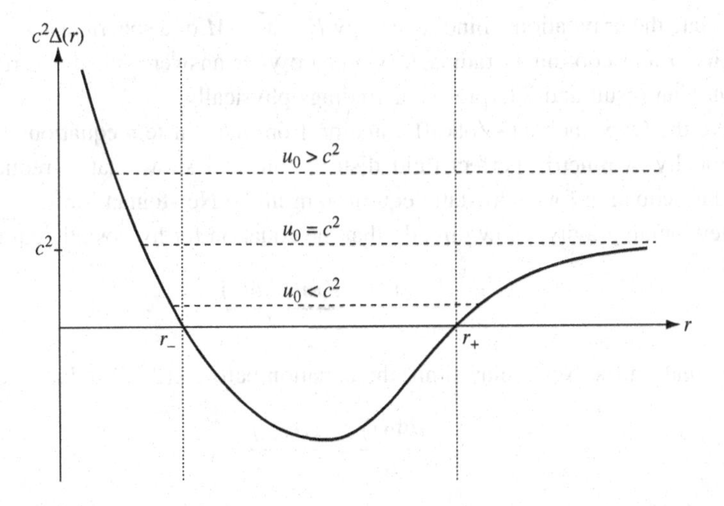

Figure 12.2 The limits of radial motion for a neutral massive particle in the Reissner–Nordström geometry.

Using the fact, from (12.45), that $\Delta(r) = g_{00}/c^2 = -1/g_{11}$, one finds that

$$\boxed{\dot{r}^2 + c^2\Delta(r) = \frac{u_0^2}{c^2}.} \tag{12.52}$$

This clearly has the form of an 'energy' equation, in which $c^2\Delta(r)$ plays the role of a potential. Qualitative information on the properties of the radial trajectories can be obtained directly from (12.52) by simply plotting the function $c^2\Delta(r)$; this plot is shown in Figure 12.2. The radial limits of the motion depend on the choice of the constant u_0, as indicated. The case $u_0 = c^2$ corresponds to the particle being released from rest at infinity. In all cases, there exists an inner radial limit that is greater than zero. This indicates that a neutral particle moving freely under gravity *cannot reach* the central intrinisic singularity at $r = 0$ but is instead repelled once it has approached to within some finite distance. As mentioned in Section 12.6. performing a maximum analytic extension the RN metric suggests that the particle passes back through $r = r_-$ and $r = r_+$ and ultimately emerges in a different asymptotically flat spacetime.

Exercises

12.1 For a general static diagonal metric, show that the 4-velocity of a perfect fluid in the spacetime must have the form

$$[u^\mu] = \frac{c}{\sqrt{g_{00}}}(1, 0, 0, 0).$$

12.2 Calculate the gravitational binding energy $E = \widetilde{M} - M$ of a spherical star of constant density ρ and coordinate radius R. Compare your answer with the corresponding Newtonian result and interpret your findings physically.

12.3 Derive the Oppenheimer–Volkoff equation from the Einstein equations for a static spherically symmetric perfect-fluid distribution, and show that it reduces to the standard equation for hydrostatic equilibrium in the Newtonian limit.

12.4 In Newtonian gravity, show directly that the equation for hydrostatic equilibrium is

$$\frac{dp(r)}{dr} = -\frac{Gm(r)\rho(r)}{r^2}.$$

12.5 Show that, in the Newtonian limit, the equation before (12.15) reduces to

$$\frac{d\Phi(r)}{dr} = \frac{Gm(r)}{r},$$

where $\Phi(r)$ is the Newtonian gravitational potential.

12.6 For a spherical star of uniform density ρ and central pressure p_0, verify that the Oppenheimer–Volkoff equation requires $p(r)$ to satisfy

$$\frac{\rho c^2 + 3p(r)}{\rho c^2 + p(r)} = \frac{\rho c^2 + 3p_0}{\rho c^2 + p_0}\left(1 - \frac{8\pi G}{3c^2}\rho r^2\right)^{1/2},$$

and hence show that

$$p(r) = \rho c^2 \frac{(1 - 2\mu r^2/R^3)^{1/2} - (1 - 2\mu/R)^{1/2}}{3(1 - 2\mu/R)^{1/2} - (1 - 2\mu r^2/R^3)^{1/2}},$$

where R is the coordinate radius of the star.

12.7 In Newtonian gravity, obtain the expression for $p(r)$ for a spherical star of uniform density ρ, central pressure p_0 and radius R. Compare your result with that obtained in Exercise 12.6.

12.8 Show that, for a spherical star of uniform density ρ,

$$R^2 < \frac{c^2}{3\pi G\rho} \quad \text{and} \quad M^2 < \frac{16c^6}{243\pi\rho G^3}.$$

If a photon is emitted from the star's surface and received by a stationary observer at infinity, show that the observed redshift must obey the constraint $z < 2$. Show also, however, that the observed redshift for a photon emitted from the star's centre can be arbitrarily large.

12.9 For a spherical star of uniform density ρ, show that in order for the star not to lie within its own Schwarzschild radius, one requires

$$M^2 < \frac{3c^6}{32\pi\rho G^3}.$$

Compare this limit with that derived in Exercise 12.8.

12.10 For a spherical uniform-density star of mass M and coordinate radius R, show that the line element of spatial sections with $t = $ constant can be written in the form

$$d\sigma^2 = \frac{Rc^2}{2GM}\left[d\chi^2 + \sin^2\chi\,(d\theta^2 + \sin^2\theta\, d\phi^2)\right].$$

12.11 Consider a static infinitely long cylindrical configuration of matter that is invariant to translations and Lorentz boosts along the axis of symmetry (a cosmic string). Adopting 'cylindrical polar' coordinates (ct, r, ϕ, z), show that a self-consistent solution to the Einstein field equations may be obtained if the stress-energy tensor for the matter is of the form

$$[T^{\mu\nu}] = \mathrm{diag}(\rho c^2, 0, 0, -\rho c^2),$$

such that there is a negative pressure (or tension) along the string, and the line element is of the form

$$ds^2 = c^2\,dt^2 - dr^2 - B(r)\,d\phi^2 - dz^2,$$

where $B(r)$ satisfies

$$\frac{B''}{2B} - \frac{(B')^2}{4B^2} = -\kappa\rho c^2.$$

Show further that $b(r) = \sqrt{B(r)}$ satisfies $b'' = -\kappa c^2 \rho b$.
Hint: You may find your answers to Exercises 8.9, 9.28 and 9.29 useful.

12.12 Suppose that the matter distribution in a cosmic string has a uniform density across the string, such that

$$\rho(r) = \begin{cases} \rho_0 & \text{for } r \le r_0, \\ 0 & \text{for } r > r_0. \end{cases}$$

By demanding that $g_{\phi\phi} \to -r^2$ as $r \to 0$, so that the spacetime geometry is regular on the axis of the string, show that the line element for $r \le r_0$ is

$$ds^2 = c^2\,dt^2 - dr^2 - \left(\frac{\sin\lambda r}{\lambda r}\right)^2 d\phi^2 - dz^2,$$

where $\lambda = \sqrt{\kappa\rho_0 c^2}$. By demanding that $g_{\phi\phi}$ and its derivative with respect to r are both continuous at $r = r_0$, show that the line element for $r > r_0$ is

$$ds^2 = c^2\,dt^2 - dr^2 - \left[\frac{\sin\lambda r_0}{\lambda r} + (r - r_0)\cos\lambda r_0\right]^2 d\phi^2 - dz^2.$$

For the interesting case in which $\lambda r_0 \ll 1$, show that for $r \gg r_0$ the line element takes the form

$$ds^2 = c^2\,dt^2 - dr^2 - \left(1 - \frac{8G\mu}{c^2}\right) r^2\,d\phi^2 - dz^2,$$

where $\mu = \pi r_0^2 \rho_0$ is the 'mass per unit length' of the string. Interpret this line element physically.

12.13 Show that the electromagnetic field tensor outside a static spherically symmetric charged matter distribution has the form

$$[F_{\mu\nu}] = E(r) \begin{pmatrix} 0 & -1 & 0 & 0 \\ 1 & 0 & 0 & 0 \\ 0 & 0 & 0 & 0 \\ 0 & 0 & 0 & 0 \end{pmatrix},$$

where $E(r)$ is some arbitrary function. Hence show that, if the line element outside the matter distribution has the form

$$ds^2 = A(r)\,dt^2 - B(r)\,dr^2 - r^2(d\theta^2 + \sin^2\theta\,d\phi^2),$$

the energy–momentum tensor of the electromagnetic field in this region is given by

$$[T_{\mu\nu}] = \frac{1}{2}c^2\epsilon_0 E^2 \,\mathrm{diag}\left(\frac{1}{B}, -\frac{1}{A}, \frac{r^2E^2}{AB}, \frac{r^2E^2\sin^2\theta}{AB}\right).$$

12.14 Calculate the invariant curvature scalar $R_{\mu\nu\rho\sigma}R^{\mu\nu\rho\sigma}$ for the Reissner–Nordström geometry and hence show that the only intrinsic singularity occurs at $r = 0$.

12.15 Show that the worldlines of radially moving photons in the Reissner–Nordström geometry are given by

$$ct = r - \frac{r_-^2}{r_+ - r_-}\ln\left|\frac{r}{r_-} - 1\right| + \frac{r_+^2}{r_+ - r_-}\ln\left|\frac{r}{r_+} - 1\right| + \text{constant} \qquad \text{(outgoing)},$$

$$ct = -r + \frac{r_-^2}{r_+ - r_-}\ln\left|\frac{r}{r_-} - 1\right| - \frac{r_+^2}{r_+ - r_-}\ln\left|\frac{r}{r_+} - 1\right| + \text{constant} \qquad \text{(ingoing)}.$$

12.16 Show that, by introducing the advanced Eddington–Finkelstein timelike coordinate

$$ct' = ct - \frac{r_-^2}{r_+ - r_-}\ln\left|\frac{r}{r_-} - 1\right| + \frac{r_+^2}{r_+ - r_-}\ln\left|\frac{r}{r_+} - 1\right|,$$

the Reissner–Nordström line element takes the form

$$ds^2 = c^2\Delta\,dt'^2 - 2(1-\Delta)\,dt'\,dr - (2-\Delta)\,dr^2 - r^2(d\theta^2 + \sin^2\theta\,d\phi^2),$$

where $\Delta \equiv \Delta(r) = 1 - 2\mu/r + q^2/r^2$. Hence show that the worldlines of radially moving photons in advanced Eddington–Finkelstein coordinates are given by

$$ct' + r = \text{constant} \quad \text{(incoming)}, \qquad c\frac{dt'}{dr} = \frac{2-\Delta}{\Delta} \quad \text{(outgoing)}.$$

What is the significance, if any, of the fact that $c\,dt'/dr = 0$ at $\Delta(r) = 2$ for outgoing radially moving photons?

12.17 For a particle of mass m and charge e in geodesic motion in the Reissner–Nordström geometry, show that the quantity

$$k = m\left(1 - \frac{2\mu}{r} + \frac{q^2}{r^2}\right)\frac{dt}{d\tau} + \frac{eq}{r}$$

is conserved, and interpret this result physically.

12.18 An observer is in a circular orbit of coordinate radius $r = R$ in the Reissner–Nordström geometry. Find the components of the magnetic field measured by the observer.

13

The Kerr geometry

The Schwarzschild solution describes the spacetime geometry outside a spherically symmetric massive object, characterised only by its mass M. In the previous chapter we derived further spherically symmetric solutions. Most real astrophysical objects, however, are *rotating*. In this case, a spherically symmetric solution cannot apply because the rotation axis of the object defines a special direction, so destroying the isotropy of the solution. For this reason, in general relativity it is *not* possible to find a coordinate system that reduces the spacetime geometry outside a rotating (uncharged) body to the Schwarzschild geometry. The non-linear field equations couple the source to the exterior geometry. Moreover, a rotating body is characterised not only by its mass M but also by its angular momentum J, and so we would expect the corresponding spacetime metric to depend upon these two parameters.

We now consider how to derive the metric describing the spacetime geometry outside a rotating body. Since the mathematical complexity in this case is far greater than that encountered in deriving the Schwarzschild metric (or the other spherically symmetric geometries discussed in the previous chapter), we shall content ourselves with just an outline of how the solution may be obtained.

13.1 The general stationary axisymmetric metric

In our derivation of the Schwarzschild solution, we began by constructing the general form of the static isotropic metric. We are now interested in deriving the spacetime geometry outside a steadily rotating massive body. Thus we begin by constructing the general form of the stationary axisymmetric metric.

For the description of such a spacetime, it is convenient to introduce the timelike coordinate $t(= x^0)$ and the azimuthal angle $\phi(= x^3)$ about the axis of symmetry. The stationary and axisymmetric character of the spacetime requires that the metric coefficients $g_{\mu\nu}$ be independent of t and ϕ, so that

$$g_{\mu\nu} = g_{\mu\nu}(x^1, x^2),$$

where x^1 and x^2 are the two remaining spacelike coordinates.

Besides stationarity and axisymmetry, we shall also require that the line element is *invariant* to simultaneous inversion of the coordinates t and ϕ, i.e. the transformations

$$t \to -t \qquad \text{and} \qquad \phi \to -\phi.$$

The physical meaning of this additional requirement is that the source of the gravitational field, whatever it may be, has motions that are purely rotational about the axis of symmetry, i.e. we are considering the spacetime associated with a *rotating body*. This assumed invariance requires that

$$g_{01} = g_{02} = g_{13} = g_{23} = 0,$$

since the corresponding terms in the line element would change sign under the simultaneous inversion of t and ϕ. Therefore, under the assumptions made thus far, the line element must have the form

$$ds^2 = g_{00}\, dt^2 + 2g_{03}\, dt\, d\phi + g_{33}\, d\phi^2 + \left[g_{11}(dx^1)^2 + 2g_{12}\, dx^1\, dx^2 + g_{22}(dx^2)^2\right]. \tag{13.1}$$

We note that, since the metric coefficients $g_{\mu\nu}$ are functions only of x^1 and x^2, the expression in square brackets in (13.1) can be considered as a separate two-dimensional submanifold. A further reduction in the form of the metric can thus be achieved by using the fact that *any* two-dimensional (pseudo-)Riemannian manifold is conformally flat, i.e. it is always possible to find a coordinate system in which the metric takes the form

$$g_{ab} = \Omega^2(x)\eta_{ab}, \tag{13.2}$$

where $\Omega^2(x)$ is an arbitrary function of the coordinates and $[\eta_{ab}] = \text{diag}(\pm 1, \pm 1)$; the signs depend on the signature of the manifold. We proved this result in Appendix 11C. Thus, taking advantage of this fact, and writing the result in way suggestive of a rotating body, we can express the line element (13.1) in the form

$$ds^2 = A\, dt^2 - B(d\phi - \omega\, dt)^2 - C\left[(dx^1)^2 + (dx^2)^2\right], \tag{13.3}$$

where A, B, C and ω are arbitrary functions of the spacelike coordinates x^1 and x^2.

For definiteness, let us denote the coordinates x^1 and x^2 by r and θ respectively. For our axisymmetric metric, these coordinates are not so readily associated with any geometrical meaning. Nevertheless, in order that they can be chosen later to be as similar as possible to the spherically symmetric r and θ, it is useful to allow some extra freedom in the metric by not demanding that the metric coefficients g_{22} and g_{33} be identical. Thus, from now on we will work with the metric

$$\boxed{ds^2 = A\,dt^2 - B(d\phi - \omega\,dt)^2 - C\,dr^2 - D\,d\theta^2,} \tag{13.4}$$

where A, B, C, D and ω are arbitrary functions of the spacelike coordinates r and θ but we have the freedom to relate C and D in such a way so that the physical meanings of r and θ are as close as possible to the spherically symmetric case. The functions in (13.4) are related to the metric coefficients $g_{\mu\nu}$ by

$$g_{tt} = A - B\omega^2, \qquad g_{t\phi} = B\omega, \qquad g_{\phi\phi} = -B, \qquad g_{rr} = -C, \qquad g_{\theta\theta} = -D,$$

where, from now on, we use coordinate names rather than numbers to denote the components. Note that $\omega = -g_{t\phi}/g_{\phi\phi}$ and, if the body is *not* rotating, we can set $\omega = 0$ since in this case we would require that the metric is invariant under the single transformation $t \to -t$ and consequently $g_{t\phi} = 0$.

For later convenience, let us also calculate the contravariant components $g^{\mu\nu}$ of the metric corresponding to the line element (13.4). The only off-diagonal terms involve t and ϕ, and so immediately we have

$$g^{rr} = -1/C, \qquad g^{\theta\theta} = -1/D.$$

To find the remaining contravariant components, we must invert the matrix

$$\mathsf{G} = \begin{pmatrix} g_{tt} & g_{t\phi} \\ g_{t\phi} & g_{\phi\phi} \end{pmatrix} \qquad \Rightarrow \qquad \mathsf{G}^{-1} = \frac{1}{|\mathsf{G}|}\begin{pmatrix} g_{\phi\phi} & -g_{t\phi} \\ -g_{t\phi} & g_{tt} \end{pmatrix},$$

where the determinant $|\mathsf{G}| = g_{tt}g_{\phi\phi} - (g_{t\phi})^2 = -AB$. Thus

$$g^{tt} = \frac{g_{\phi\phi}}{|\mathsf{G}|} = \frac{1}{A}, \qquad g^{t\phi} = -\frac{g_{t\phi}}{|\mathsf{G}|} = \frac{\omega}{A}, \qquad g^{\phi\phi} = \frac{g_{tt}}{|\mathsf{G}|} = \frac{B\omega^2 - A}{AB}. \tag{13.5}$$

Shortly we will show that a metric of the form (13.4) can indeed be made to satisfy the empty-space field equations $R_{\mu\nu} = 0$ by suitable choice of the functions A, B, C, D and ω. Before specialising to any particular solution, however, we investigate three particularly interesting generic properties of such spacetimes: the *dragging of inertial frames* and the existence of *stationary limit surfaces* and *event horizons*.

13.2 The dragging of inertial frames

The presence of $g_{t\phi} \neq 0$ in the metric (13.4) introduces qualitatively new effects into particle trajectories. Since $g_{\mu\nu}$ is independent of ϕ, the covariant component p_ϕ of a particle's 4-momentum is still conserved along its geodesic. Indeed

$p_\phi = -L$, where L is the component of angular momentum of the particle along the rotation axis, which is conserved (note the minus sign, which also occurred in the Schwarzschild case discussed in Chapter 9). This conservation law is a direct consequence of the axisymmetry of the spacetime. Note, however, that the total angular momentum of a particle is not a conserved quantity, since the spacetime is not spherically symmetric about any point.

The corresponding contravariant component p^ϕ of the particle's 4-momentum is given by

$$p^\phi = g^{\phi\mu} p_\mu = g^{\phi t} p_t + g^{\phi\phi} p_\phi,$$

and similarly the contravariant time component of the 4-momentum is

$$p^t = g^{t\mu} p_\mu = g^{tt} p_t + g^{t\phi} p_\phi.$$

Let is now consider a particle (or photon) with zero angular momentum, so that $p_\phi = 0$ along its geodesic. Using the definition of the 4-momentum, for either a massive particle or a photon we have

$$p^t \propto \frac{dt}{d\sigma} \quad \text{and} \quad p^\phi \propto \frac{d\phi}{d\sigma},$$

where σ is an affine parameter along the geodesic and the constants of proportionality in each case are equal. Thus the particle's trajectory is such that

$$\frac{d\phi}{dt} = \frac{p^\phi}{p^t} = \frac{g^{t\phi}}{g^{tt}} = \omega(r, \theta).$$

This equation defines what we mean by ω: it is the coordinate angular velocity of a zero-angular-momentum particle.

We shall find the explicit form for ω for the Kerr geometry later, but it is clear that this effect is present in any metric for which $g_{t\phi} \neq 0$, which in turn happens whenever the source of the gravitational field is rotating. So we have the remarkable result that a particle dropped 'straight in' from infinity ($p_\phi = 0$) is 'dragged' just by the influence of gravity so that it acquires an angular velocity in the same sense as that of the source of the metric. This effect weakens with distance (roughly as $\sim 1/r^3$ for the Kerr metric) and makes the angular momentum of the source measurable in practice.

The effect is called the *dragging of inertial frames*. Remember that inertial frames are defined as those in which free-falling test bodies are stationary or move along straight lines at constant speed. Consider the freely falling particle discussed above. At any spatial point (r, θ, ϕ), in order for the particle to be at rest in some (inertial) frame the frame must be moving with an angular speed $\omega(r, \theta)$. Any other inertial frame is then related to this instantaneous rest frame by a Lorentz transformation. Thus the inertial frames are 'dragged' by the rotating source. A schematic

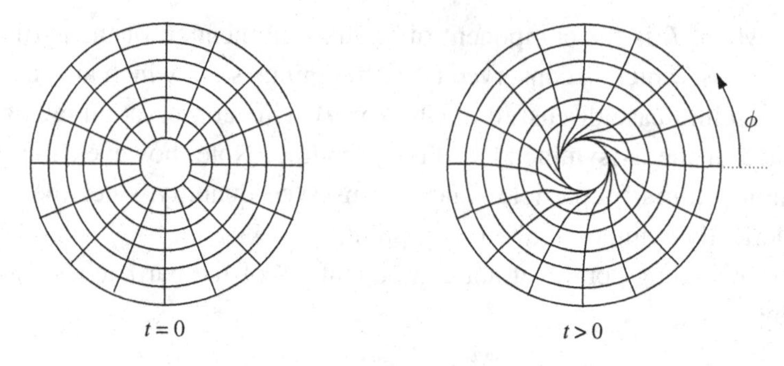

Figure 13.1 A schematic illustration of the dragging of inertial frames around a rotating source.

illustration of this effect in a plane $\theta =$ constant is shown in Figure 13.1, where the spacetime around the source is viewed along the rotation axis.

13.3 Stationary limit surfaces

A second generic property of spacetimes outside a rotating source is the existence of *stationary limit surfaces*; this is related to the dragging of inertial frames. This effect may be illustrated by considering, for example, photons emitted from a position with fixed spatial coordinates (r, θ, ϕ) in the spacetime. In particular, consider those photons emitted in the $\pm\phi$ directions so that, at first, only dt and $d\phi$ are non-zero along the path. Since $ds^2 = 0$ for a photon trajectory, we have

$$g_{tt}\, dt^2 + 2g_{t\phi}\, dt\, d\phi + g_{\phi\phi}\, d\phi^2 = 0,$$

from which we obtain

$$\frac{d\phi}{dt} = -\frac{g_{t\phi}}{g_{\phi\phi}} \pm \left[\left(\frac{g_{t\phi}}{g_{\phi\phi}}\right)^2 - \frac{g_{tt}}{g_{\phi\phi}}\right]^{1/2}.$$

Now, provided that $g_{tt}(r, \theta) > 0$ at the point of emission, we see that $d\phi/dt$ is positive (negative) for a photon emitted in the positive (negative) ϕ-direction, as we would expect, although the *value* of $d\phi/dt$ is different for the two directions. On any surface defined by $g_{tt}(r, \theta) = 0$, however, a remarkable thing happens. The two solutions of the above equation in this case are

$$\frac{d\phi}{dt} = -\frac{2g_{t\phi}}{g_{\phi\phi}} = 2\omega \qquad \text{and} \qquad \frac{d\phi}{dt} = 0.$$

The first solution represents the photon sent off in the same direction as the source rotation, and the second solution corresponds to the photon sent in the opposite direction. For this second case, we see that when $g_{tt} = 0$ the dragging of orbits is so severe that the photon initially does not move at all! Clearly, any massive particle, which must move more slowly than a photon, will therefore have to rotate with the source, even if it has an angular momentum arbitrarily large in the opposite sense. Any surface defined by $g_{tt}(r, \theta) = 0$ is called a stationary limit surface. Inside the surface, where $g_{tt} < 0$, no particle can remain at fixed (r, θ, ϕ) but must instead rotate around the source in the same sense as the source's rotation. This is consistent with our discussion of the Schwarzschild metric, for which $g_{tt} = 0$ occurs at $r = 2\mu$, within which no particle can remain at fixed spatial coordinates.

The fact that a particle (or observer) cannot remain at a fixed (r, θ, ϕ) inside a stationary limit surface, where $g_{tt} < 0$, may also be shown directly by considering the 4-velocity of an observer at fixed (r, θ, ϕ), which is given by

$$[u^\mu] = (u^t, 0, 0, 0). \tag{13.6}$$

We require, however, that $\boldsymbol{u} \cdot \boldsymbol{u} = g_{tt}(u^t)^2 = c^2$, but this cannot be satisfied if $g_{tt} < 0$, hence showing that a 4-velocity of the form (13.6) is not possible in such a region.

Any surface defined by $g_{tt} = 0$ is also physically interesting in another way. In Appendix 9A, we presented a general approach to the calculation of gravitational redshifts. In particular, we showed that, for an emitter E and receiver R with *fixed spatial coordinates* in a *stationary spacetime* (i.e. one for which $\partial_t g_{\mu\nu} = 0$), the gravitational frequency shift of a photon is, quite generally,

$$\frac{\nu_R}{\nu_E} = \left[\frac{g_{tt}(A)}{g_{tt}(B)}\right]^{1/2},$$

where A is the event at which the photon is emitted and B the event at which it is received. Thus, we see that if the photon is emitted from a point with fixed spatial coordinates, then $\nu_R \to 0$ in the limit $g_{tt} \to 0$, so that the photon suffers an *infinite redshift*. Thus a surface defined by $g_{tt}(r, \theta) = 0$ is also often called an *infinite redshift surface*. This is again consistent with our discussion of the Schwarzschild metric, for which the surface $r = 2\mu$ (where $g_{tt} = 0$) is indeed an infinite redshift surface.

13.4 Event horizons

In the Schwarzschild metric, the surface $r = 2\mu$ is both a surface of infinite redshift and an event horizon, but in our more general axisymmetric spacetime these surfaces need *not* coincide. In general, as we shall see below, the defining property of an event horizon is that it is a *null* 3-surface, i.e. a surface whose normal at every point is a null vector.

Before discussing the particular case of a stationary axisymmetric spacetime, let us briefly consider null 3-surfaces in general. Suppose that such a surface is defined by the equation

$$f(x^\mu) = 0.$$

The normal to the surface is directed along the 4-gradient $n_\mu = \nabla_\mu f = \partial_\mu f$ (remembering that f is a scalar quantity), and for a null surface we have

$$g^{\mu\nu} n_\mu n_\nu = 0. \tag{13.7}$$

This last property means that the direction of the normal lies in the surface itself; along the surface $df = n_\mu \, dx^\mu = 0$, and this equation is satisfied when the directions of the 4-vectors dx^μ and n^μ coincide. In this same direction, from the property (13.7) we see that the element of length in the 3-surface is $ds = 0$. In other words, along this direction the 3-surface is tangent, at any given point, to the lightcone at that point. Thus, the lightcone at each point of a null 3-surface (say, in the future direction) lies entirely on one side of the surface and is tangent to the 3-surface at that point. This means that the (future-directed) worldline of a particle or photon can cross a null 3-surface in only one direction, and hence the latter forms an event horizon.

In a stationary axisymmetric spacetime the equation of the surface must take the form

$$f(r, \theta) = 0.$$

Moreover, the condition that the surface is null means that

$$g^{\mu\nu}(\partial_\mu f)(\partial_\nu f) = 0,$$

which, for a metric of the form (13.4), reduces to

$$g^{rr}(\partial_r f)^2 + g^{\theta\theta}(\partial_\theta f)^2 = 0. \tag{13.8}$$

This is therefore the general condition for a surface $f(r, \theta)$ to be an event horizon.

We may, however, choose our coordinates r and θ in such a way that we can write the equation of the surface as $f(r) = 0$, i.e. as a function of r alone. In this case, the condition (13.8) reduces to

$$g^{rr}(\partial_r f)^2 = 0,$$

from which we see that an event horizon occurs when $g^{rr} = 0$, or equivalently $g_{rr} = \infty$. This is consistent with our analysis of the Schwarzschild metric, for which $g_{rr} = \infty$ at $r = 2\mu$.

13.5 The Kerr metric

So far our discussion has been limited to using symmetry arguments to restrict the possible form of the stationary axisymmetric line element, which we assumed to be

$$ds^2 = g_{tt}\,dt^2 + 2g_{t\phi}\,dt\,d\phi + g_{\phi\phi}\,d\phi^2 + g_{rr}\,dr^2 + g_{\theta\theta}\,d\theta^2 \tag{13.9}$$

or, equivalently,

$$ds^2 = A\,dt^2 - B(d\phi - \omega\,dt)^2 - C\,dr^2 - D\,d\theta^2, \tag{13.10}$$

where the arbitrary functions in either form depend only on r and θ. As we have seen, the general form of this line element leads to some interesting new physical phenomena in such spacetimes. Nevertheless, we must now verify that such a line element does indeed satisfy Einstein's gravitational field equations and thus obtain explicit forms for the metric functions appearing in ds^2.

The general approach to performing this calculation is the same as that used in deriving the Schwarzschild metric. We first calculate the connection coefficients $\Gamma^{\mu}{}_{\nu\sigma}$ for the metric (13.9) or (13.10) and then use these coefficients to obtain expressions for the components $R_{\mu\nu}$ of the Ricci tensor in terms of the unknown functions in the line element. Since we are again interested in the spacetime geometry *outside* the rotating matter distribution, we must then solve the empty-space field equations

$$R_{\mu\nu} = 0.$$

Although this process is conceptually straightforward, it is algebraically *very* complicated, and the full calculation is extremely lengthy.[1]

In fact, one finds that the Einstein equations alone are *insufficient* to determine all the unknown functions uniquely. This should not come as a surprise since the requirement of axisymmetry is far less restrictive than that of spherical symmetry, used in the derivation of the Schwarzschild geometry. Although we are envisaging a 'compact' rotating body, such as a star or planet, the general form of the metric (13.10) would also be valid outside a rotating 'extended' axisymmetric body, such as a rotating cosmic string. To obtain the Kerr metric, we must therefore impose some additional conditions on the solution. It transpires that if we demand that the spacetime geometry tends to the Minkowski form as $r \to \infty$ and that somewhere there exists a *smooth closed convex event horizon* outside which the geometry is non-singular, then the solution is unique.

[1] For a full derivation, see (for example) S. Chandrasekhar, *The Mathematical Theory of Black Holes*, Oxford University Press, 1983.

In this case, in terms of our 'Schwarzschild-like' coordinates (t, r, θ, ϕ), the line element for the *Kerr geometry* takes the form

$$\begin{aligned} ds^2 = {} & c^2\left(1 - \frac{2\mu r}{\rho^2}\right) dt^2 + \frac{4\mu acr\sin^2\theta}{\rho^2}\,dt\,d\phi - \frac{\rho^2}{\Delta}\,dr^2 - \rho^2\,d\theta^2 \\ & - \left(r^2 + a^2 + \frac{2\mu ra^2\sin^2\theta}{\rho^2}\right)\sin^2\theta\,d\phi^2, \end{aligned} \tag{13.11}$$

where μ and a are *constants* and we have introduced the functions ρ^2 and Δ, defined by

$$\rho^2 = r^2 + a^2\cos^2\theta, \qquad \Delta = r^2 - 2\mu r + a^2.$$

This standard expression for ds^2 is known as the *Boyer–Lindquist form* and (t, r, θ, ϕ) as *Boyer–Lindquist coordinates*. The dedicated student may wish to verify that this metric does indeed satisfy the empty-space field equations.

We can write the metric (13.11) in several other useful forms. In particular, it is common also to define the function

$$\Sigma^2 = (r^2 + a^2)^2 - a^2\Delta\sin^2\theta$$

and write the metric as

$$\begin{aligned} ds^2 = {} & \frac{\Delta - a^2\sin^2\theta}{\rho^2}\,c^2\,dt^2 + \frac{4\mu ar\sin^2\theta}{\rho^2}\,c\,dt\,d\phi \\ & - \frac{\rho^2}{\Delta}\,dr^2 - \rho^2\,d\theta^2 - \frac{\Sigma^2\sin^2\theta}{\rho^2}\,d\phi^2. \end{aligned} \tag{13.12}$$

This form can be rearranged in a manner that is more suggestive of a rotating object, to give

$$ds^2 = \frac{\rho^2\Delta}{\Sigma^2}\,c^2\,dt^2 - \frac{\Sigma^2\sin^2\theta}{\rho^2}(d\phi - w\,dt)^2 - \frac{\rho^2}{\Delta}\,dr^2 - \rho^2\,d\theta^2, \tag{13.13}$$

where the physically meaningful function ω is given by $\omega = 2\mu cra/\Sigma^2$.

For later convenience, it is useful to calculate the contravariant components $g^{\mu\nu}$ of the Kerr metric in Boyer–Lindquist coordinates. Using our earlier calculations for the general stationary axisymmetric metric, we find that g^{rr} and $g^{\theta\theta}$ are simply the reciprocals of g_{rr} and $g_{\theta\theta}$ respectively,

$$g^{rr} = -\frac{\Delta}{\rho^2}, \qquad g^{\theta\theta} = -\frac{1}{\rho^2},$$

whereas the remaining contravariant components are given by

$$g^{tt} = \frac{\Sigma^2}{c^2\rho^2\Delta}, \qquad g^{t\phi} = \frac{2\mu a r}{c\rho^2\Delta}, \qquad g^{\phi\phi} = \frac{a^2\sin^2\theta - \Delta}{\rho^2\Delta\sin^2\theta}.$$

13.6 Limits of the Kerr metric

We see that the Kerr metric depends on *two* parameters μ and a, as we might expect for a rotating body. Moreover, in the limit $a \to 0$,

$$\Delta \to r^2\left(1 - \frac{2\mu}{r}\right),$$
$$\rho^2 \to r^2,$$
$$\Sigma^2 \to r^4,$$

and so any of the forms for the Kerr metric above tends to the Schwarzschild form,

$$ds^2 \to c^2\left(1 - \frac{2\mu}{r}\right)dt^2 - \left(1 - \frac{2\mu}{r}\right)^{-1} dr^2 - r^2\,d\theta^2 - r^2\sin^2\theta\,d\phi^2.$$

Thus suggests that we should make the identification $\mu = GM/c^2$, where M is the mass of the body, and also that a corresponds in some way to the angular velocity of the body. In fact, by investigating the slow-rotation weak-field limit (see Section 13.20), one can show that the angular momentum J of the body about its rotation axis is given by $J = Mac$.

The fact that the Kerr metric tends to the Schwarzschild metric as $a \to 0$ allows us to give some geometrical meaning to the coordinates r and θ in the limit of a slowly rotating body. In the general case, however, r and θ are not the standard Schwarzschild polar coordinates. In particular, from (13.11) we see that surfaces $t =$ constant, $r =$ constant do not have the metric of 2-spheres.

The geometrical nature of Boyer–Lindquist coordinates is elucidated further by considering the Kerr metric in the limit $\mu \to 0$, i.e. in the absence of a gravitating mass, in which case the spacetime should be Minkowski. One quickly finds that, in this limit, the line element becomes

$$ds^2 = c^2\,dt^2 - \frac{\rho^2}{r^2 + a^2}\,dr^2 - \rho^2\,d\theta^2 - (r^2 + a^2)\sin^2\theta\,d\phi^2.$$

This is indeed the Minkowski metric $ds^2 = c^2\,dt^2 - dx^2 - dy^2 - dz^2$, but written in terms of spatial coordinates (r, θ, ϕ) that are related to Cartesian coordinates by[2]

$$x = \sqrt{r^2 + a^2}\sin\theta\cos\phi,$$
$$y = \sqrt{r^2 + a^2}\sin\theta\sin\phi,$$
$$z = r\cos\theta,$$

where $r \geq 0$, $0 \leq \theta \leq \pi$ and $0 \leq \phi < 2\pi$ (see Figure 13.2).

In this case (with $\mu = 0$), the surfaces $r =$ constant are oblate ellipsoids of rotation about the z-axis, given by

$$\frac{x^2 + y^2}{r^2 + a^2} + \frac{z^2}{r^2} = 1.$$

The special case $r = 0$ corresponds to the disc of radius a in the equatorial plane, centred on the origin of the Cartesian coordinates. The surfaces $\theta =$ constant correspond to hyperbolae of revolution about the z-axis given by

$$\frac{x^2 + y^2}{a^2\sin^2\theta} - \frac{z^2}{a^2\cos^2\theta} = 1.$$

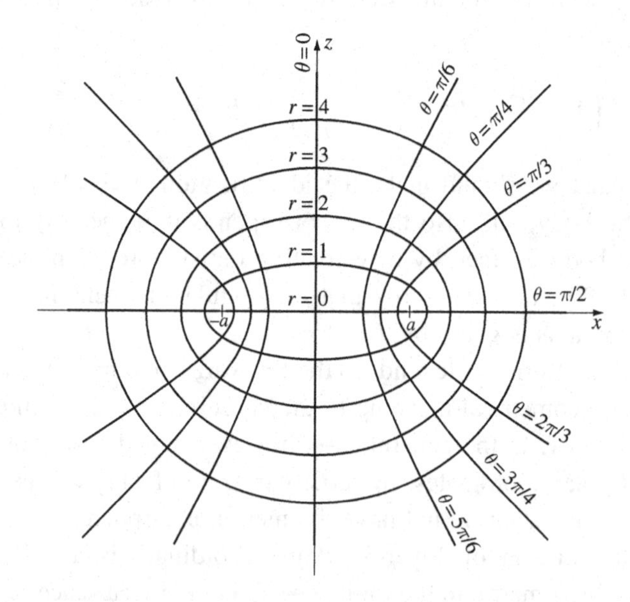

Figure 13.2 Boyer–Lindquist coordinates in the $\phi = 0$ plane in Euclidean space.

[2] The coordinates (r, θ, ϕ) are related to the standard oblate spheroidal coordinates (ξ, η, ϕ) by $r = a\sinh\xi$ and $\theta = \eta - \pi/2$; see, for example, M. Abramowitz & I. Stegun, *Handbook of Mathematical Functions*, Dover (1972).

The asymptote for large values of r is a cone, with its vertex at the origin, that subtends a half-angle θ. The angle ϕ is the standard azimuthal angle. Clearly, in the limit $a \to 0$ the coordinates (r, θ, ϕ) correspond to standard spherical polar coordinates. It should be remembered, however, that the simple interpretation of the coordinates given above no longer holds in the general case of the Kerr metric, when $\mu \neq 0$.

13.7 The Kerr–Schild form of the metric

The form (13.11) for the line element is *not* in fact the form originally discovered by Roy Kerr in 1963. Indeed, Kerr himself followed an approach to the derivation very different from that presented here. His original interest was in line elements of the general form

$$ds^2 = \eta_{\mu\nu}\, dx^\mu\, dx^\nu - \lambda l_\mu l_\nu\, dx^\mu\, dx^\nu,$$

where the vector l^μ is null with respect to the Minkowski metric $\eta_{\mu\nu}$, i.e.

$$\eta_{\mu\nu} l^\mu l^\nu = 0.$$

This form for a line element is now known as the *Kerr–Schild form.* Kerr showed that a line element of this form satisfied the empty-space field equations (together with our additional conditions on the solution mentioned above), provided that

$$\lambda = \frac{2\mu r^3}{r^4 + a^2 z^2},$$

$$[l_\mu] = \left(c, \frac{rx + ay}{a^2 + z^2}, \frac{ry - ax}{a^2 + z^2}, \frac{z}{r}\right),$$

where $[x^\mu] = (\bar{t}, x, y, z)$ and r is defined implicitly in terms of x, y and z by

$$r^4 - r^2(x^2 + y^2 + z^2 - a^2) - a^2 z^2 = 0. \tag{13.14}$$

The corresponding form for the line element is given by

$$\boxed{\begin{aligned} ds^2 = c^2\, d\bar{t}^2 - dx^2 - dy^2 - dz^2 - \frac{2\mu r^3}{r^4 + a^2 z^2} \\ \times \left[c\, d\bar{t} - \frac{r}{r^2 + a^2}(x\, dx + y\, dy) - \frac{a}{r^2 + a^2}(x\, dy - y\, dx) - \frac{z}{r}\, dz\right]^2. \end{aligned}} \tag{13.15}$$

It is straightforward, but lengthy, to show that the two forms (13.15) and (13.11) for the line element are identical if the two sets of coordinates are related by

$$c\,d\bar{t} = c\,dt - \frac{2\mu r}{\Delta}\,dr, \tag{13.16}$$

$$x = (r\cos\phi' + a\sin\phi')\sin\theta, \tag{13.17}$$

$$y = (r\sin\phi' - a\cos\phi')\sin\theta, \tag{13.18}$$

$$z = r\cos\theta, \tag{13.19}$$

where $d\phi' = d\phi - (a/\Delta)\,dr$.

13.8 The structure of a Kerr black hole

The Kerr metric is the solution to the empty-space field equations outside a rotating massive object and so is only valid down to the surface of the object. As in our discussion of the Schwarzschild solution, however, it is of interest to consider the structure of the full Kerr geometry as a vacuum solution to the field equations.

Singularities and horizons

The Kerr metric in Boyer–Lindquist coordinates is singular when $\rho = 0$ and when $\Delta = 0$. Calculation of the invariant curvature scalar $R_{\mu\nu\sigma\rho}R^{\mu\nu\sigma\rho}$ reveals that only $\rho = 0$ is an intrinsic singularity. Since

$$\rho^2 = r^2 + a^2\cos^2\theta = 0,$$

it follows that this occurs when

$$\boxed{r = 0, \qquad \theta = \pi/2.}$$

From our earlier discussion of Boyer–Lindquist coordinates, we recall that $r = 0$ represents a disc of coordinate radius a in the equatorial plane. Moreover, the collection of points with $r = 0$ and $\theta = \pi/2$ constitutes the outer edge of this disc. Thus, rather surprisingly, the singularity has the form of a *ring*, of coordinate radius a, lying in the equatorial plane. Similarly, using (13.14) and (13.19), we see that, in terms of the 'Cartesian' coordinates $(\bar{t}, x, y, z)$, the singularity occurs when $x^2 + y^2 = a^2$ and $z = 0$.

The points where $\Delta = 0$ are coordinate singularities, which occur on the surfaces

$$\boxed{r_{\pm} = \mu \pm \left(\mu^2 - a^2\right)^{1/2}.} \tag{13.20}$$

As discussed above, event horizons in the Kerr metric will occur where $r =$ constant is a null 3-surface, and this is given by the condition $g^{rr} = 0$ or, equivalently, $g_{rr} = \infty$. From (13.11), we have

$$g_{rr} = -\frac{\rho^2}{\Delta},$$

from which we see that the surfaces $r = r_+$ and $r = r_-$, for which $\Delta = 0$, are in fact event horizons. Thus, the Kerr metric has *two* event horizons. In the Schwarzschild limit $a \to 0$, these reduce to $r = 2\mu$ and $r = 0$. The surfaces $r = r_\pm$ are axially symmetric, but their intrinsic geometries are not spherically symmetric. Setting $r = r_\pm$ and $t =$ constant in the Kerr metric and noting from (13.20) that $r_\pm^2 + a^2 = 2\mu r_\pm$, we obtain two-dimensional surfaces with the line elements

$$d\sigma^2 = \rho_\pm^2\, d\theta^2 + \left(\frac{2\mu r_\pm}{\rho_\pm}\right)^2 \sin^2\theta\, d\phi^2, \tag{13.21}$$

which do not describe the geometry of a sphere. If one embeds a 2-surface with geometry given by (13.21) in three-dimensional Euclidean space, one obtains a surface resembling an axisymmetric ellipsoid, flattened along the rotation axis.

The existence of the outer horizon $r = r_+$, in particular, shows that the Kerr geometry represents a (rotating) black hole. It is a one-way surface, like $r = 2\mu$ in the Schwarzschild geometry. Particles and photons can cross it once, from the outside, but not in the opposite direction. It is common practice to define three distinct regions of a Kerr black hole, bounded by the event horizons, in which the solution is regular: region I, $r_+ < r < \infty$; region II, $r_- < r < r_+$; and region III, $0 < r < r_-$.

Not all values of μ and a correspond to a black hole, however. From (13.20), we see that horizons (at real values of r) exist only for

$$a^2 < \mu^2. \tag{13.22}$$

Thus the magnitude of the angular momentum $J = Mac$ of a rotating black hole is limited by its squared mass. Moreover, if the condition (13.22) is satisfied then the intrinsic singularity at $\rho = 0$ is contained safely within the outer horizon $r = r_+$. An *extreme* Kerr black hole is one that has the limiting value $a^2 = \mu^2$. In this case, the event horizons r_+ and r_- coincide at $r = \mu$. It may be that near-extreme Kerr black holes develop naturally in many astrophysical situations. Matter falling towards a rotating black hole forms an accretion disc that rotates in the same sense as the hole. As matter from the disc spirals inwards and falls into the black hole, it carries angular momentum with it and hence increases the angular momentum of the hole. The process is limited by the fact that radiation from the infalling matter carries away angular momentum. Detailed calculations suggest that the limiting value is $a \approx 0.998\mu$, which is very close to the extreme value.

For $a^2 > \mu^2$ we find that $\Delta > 0$ throughout, and so the Kerr metric is regular everywhere except $\rho = 0$, where there is a ring singularity. Since the horizons have disappeared, this means that the ring singularity is visible to the outside world. In fact, one can show explicitly that timelike and null geodesics in the equatorial plane can start at the singularity and reach infinity, thereby making the singularity visible to the outside world. Such a singularity is called a *naked singularity* (as mentioned in Section 12.6) and opens up an enormous realm for some truly wild speculation. However, Penrose's *cosmic censorship hypothesis* only allows singularities that are hidden behind an event horizon.

Stationary limit surfaces

As we showed earlier, in a general stationary axisymmetric spacetime the condition $g_{tt} = 0$ defines a surface that is both a stationary limit surface and a surface of infinite redshift. For the Kerr metric, we have

$$g_{tt} = c^2\left(1 - \frac{2\mu r}{\rho^2}\right) = c^2\frac{r^2 - 2\mu r + a^2\cos^2\theta}{\rho^2},$$

so that (for $a^2 \leq \mu^2$) these surfaces, S^+ and S^-, occur at

$$\boxed{r_{S\pm} = \mu \pm \left(\mu^2 - a^2\cos^2\theta\right)^{1/2}.}$$

The two surfaces are axisymmetric, but setting $r = r_{S\pm}$ and $t =$ constant in the Kerr metric, and noting that $r_{S\pm}^2 + a^2 = 2\mu r_{S\pm} + a^2\sin^2\theta$, we obtain two-dimensional surfaces with line elements

$$d\sigma^2 = \rho_{S\pm}^2\, d\theta^2 + \left[\frac{2\mu r_{S\pm}(2\mu r_{S\pm} + 2a^2\sin^2\theta)}{\rho_{S\pm}^2}\right]\sin^2\theta\, d\phi^2, \qquad (13.23)$$

which again do not describe the geometry of a sphere. If one embeds a 2-surface with geometry given by (13.23) in three-dimensional Euclidean space then a surface resembling an axisymmetric ellipsoid, flattened along the rotation axis, is once more obtained. In the Schwarzschild limit $a \to 0$, the surface S^+ reduces to $r = 2\mu$ and S^- to $r = 0$. As anticipated we see that, in the Schwarzschild solution, the surfaces of infinite redshift and the event horizons coincide.

The surface S^- coincides with the ring singularity in the equatorial plane. Moreover, S^- lies completely within the inner horizon $r = r_-$ (except at the poles, where they touch). The surface S^+ has coordinate radius 2μ at the equator and, for all θ, it completely encloses the outer horizon $r = r_+$ (except at the poles, where they touch), giving rise to a region between the two called the *ergoregion*. The structure of a Kerr black hole is illustrated in Figure 13.3.

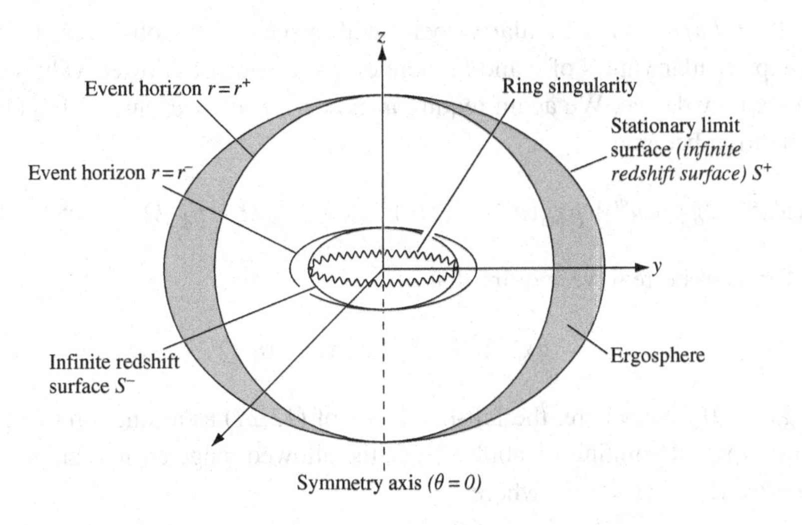

Figure 13.3 The structure of a Kerr black hole.

The ergoregion

The ergoregion gets its name from the Greek word *ergo* meaning work. The key property of an ergoregion (which can occur in other spacetime geometries) is that it is a region for which $g_{tt} < 0$ *and* from which particles can escape. Clearly, the Schwarzschild geometry does not possess an ergoregion, since $g_{tt} < 0$ is only satisfied within its event horizon. As we will discuss in Section 13.9, Roger Penrose has shown that it is possible to extract the rotational energy of a Kerr black hole from within the ergoregion. To assist in that discussion, it is useful here to consider the constraints induced by the spacetime geometry on the motion of observers within the ergoregion.

Since $g_{tt} < 0$ at all points within the ergoregion, an immediate consequence (as already discussed in Section 13.3) is that an observer (even in a spaceship with an arbitrarily powerful rocket) cannot remain at a fixed (r, θ, ϕ) position. The 4-velocity of such an observer would be given by

$$[u^\mu] = (u^t, 0, 0, 0), \tag{13.24}$$

but the requirement that $\boldsymbol{u} \cdot \boldsymbol{u} = g_{tt}(u^t)^2 = c^2$ cannot be satisfied if $g_{tt} < 0$, showing that a 4-velocity of the form (13.24) is not possible.

It is possible, however, for a rocket-powered observer to remain at fixed r and θ coordinates by rotating around the black hole (with respect to an observer at infinity) in the same sense as the hole's rotation; this is an illustration of the frame-dragging phenomenon discussed in Section 13.3. The 4-velocity of such an observer is

$$[u^\mu] = u^t(1, 0, 0, \Omega), \tag{13.25}$$

where $\Omega = d\phi/dt$ is his angular velocity with respect to the observer at infinity. For any particular values of r and θ, there exists a range of allowed values for Ω, which we now derive. We again require $\boldsymbol{u} \cdot \boldsymbol{u} = g_{\mu\nu}u^\mu u^\nu = c^2$ and, using (13.25), this condition becomes

$$g_{tt}(u^t)^2 + 2g_{t\phi}u^t u^\phi + g_{\phi\phi}(u^\phi)^2 = (u^t)^2(g_{tt} + 2g_{t\phi}\Omega + g_{\phi\phi}\Omega^2) = c^2. \quad (13.26)$$

Thus, for u^t to be real we require that

$$g_{\phi\phi}\Omega^2 + 2g_{t\phi}\Omega + g_{tt} > 0. \quad (13.27)$$

Since $g_{\phi\phi} < 0$ everywhere, the left-hand side of (13.27) as a function of Ω gives rise to an upward pointing parabola. Thus, the allowed range of angular velocities is given by $\Omega_- < \Omega < \Omega_+$, where

$$\boxed{\Omega_\pm = -\frac{g_{t\phi}}{g_{\phi\phi}} \pm \left[\left(\frac{g_{t\phi}}{g_{\phi\phi}}\right)^2 - \frac{g_{tt}}{g_{\phi\phi}}\right]^{1/2} = \omega \pm \left(\omega^2 - \frac{g_{tt}}{g_{\phi\phi}}\right)^{1/2}.} \quad (13.28)$$

There are clearly two special cases to be considered. First, when $g_{tt} = 0$ we have $\Omega_- = 0$ and $\Omega_+ = 2\omega$. This occurs on the stationary limit surface $r = r_{S+}$, which is the outer defining surface of the ergoregion. The lower limit $\Omega_- = 0$ is precisely the physical meaning of a stationary limit surface: within it an observer must rotate in the same direction as the black hole and so Ω must be positive. For larger values of r, however, Ω can be negative. The second special case to consider is when $\omega^2 = g_{tt}/g_{\phi\phi}$, in which case $\Omega_\pm = \omega$. Thus, at points where this condition holds, every observer on a circular orbit is forced to rotate with angular velocity $\Omega = \omega$. Where (if anywhere) does this condition hold? Upon inserting the appropriate expressions for ω, g_{tt} and $g_{\phi\phi}$ from the Kerr metric (13.13) into (13.28), one finds, after some careful algebra, that our condition holds where $\Delta = 0$, i.e. at the outer event horizon $r = r_+$, which is the inner defining surface of the ergoregion.

Putting our results together we find that, for an observer at fixed r and θ coordinates within the ergoregion, the allowed range of angular velocities $\Omega_- < \Omega < \Omega_+$ becomes progressively narrower as the observer is located closer and closer to the horizon $r = r_+$, and at the horizon itself the angular velocity is limited to the single value

$$\boxed{\Omega_\text{H} \equiv \omega(r_+, \theta) = \frac{ac}{2\mu r_+},} \quad (13.29)$$

which is, in fact, independent of θ. We also note that Ω_H is the maximum allowed value of the angular velocity for any observer at fixed r and θ within the ergoregion.

Extension of the Kerr metric

So far we have not discussed the disc region *interior* to the ring singularity. Although beyond the scope of our discussion, it may be shown that if a particle passes through the interior of the ring singularity then it emerges into another asymptotically flat spacetime, but not a copy of the original one. The new spacetime is described by the Kerr metric with $r < 0$ and hence Δ never vanishes, so there are no event horizons.[3]

In the new spacetime, the region in the vicinity of the ring singularity has the very strange property that it allows the existence of *closed timelike curves*. For example, consider a trajectory in the equatorial plane that winds around in ϕ while keeping t and r constant. The line element along such a path is

$$ds^2 = -\left(r^2 + a^2 + \frac{2\mu a^2}{r}\right) d\phi^2,$$

which is *positive* if r is negative and small. These are then closed timelike curves, which violate causality and would seem highly unphysical. If they represent worldlines of observers, then these observers would travel back and meet themselves in the past! It must be remembered, however, that the analytic extension of the Kerr metric to negative values of r is subject a number of caveats and may not be physically meaningful. It seems highly improbable that in practice the gravitational collapse of a real rotating object would lead to such a strange spacetime.

13.9 The Penrose process

We now discuss the Penrose process, by which energy may be extracted from the rotation of a Kerr black hole (or, indeed, from any spacetime possessing an ergoregion). Suppose that an observer, with a fixed position at infinity, for simplicity, fires a particle A into the ergoregion of a Kerr black hole. The energy of particle A, as measured by the observer at the emission event $\mathcal{E}$, is given by

$$E^{(A)} = \boldsymbol{p}^{(A)}(\mathcal{E}) \cdot \boldsymbol{u}_{\text{obs}} = p_t^{(A)}(\mathcal{E}), \qquad (13.30)$$

where $\boldsymbol{p}^{(A)}(\mathcal{E})$ is the 4-momentum of the particle at this event and $\boldsymbol{u}_{\text{obs}}$ is the 4-velocity of the observer, which has components $[u^\mu_{\text{obs}}] = (1, 0, 0, 0)$.

[3] In the extended Kerr solution it is common to define region III to cover the coordinate range $-\infty < r < r_-$.

Suppose now that, at some point in the ergoregion, particle A decays into two particles B and C. By the conservation of momentum, at the decay event $\mathcal{D}$ one has

$$\boldsymbol{p}^{(A)}(\mathcal{D}) = \boldsymbol{p}^{(B)}(\mathcal{D}) + \boldsymbol{p}^{(C)}(\mathcal{D}). \tag{13.31}$$

If the decay occurs in such a way that particle C (say) eventually reaches infinity, a stationary observer there would measure the particle's energy at the reception event $\mathcal{R}$ to be

$$E^{(C)} = p_t^{(C)}(\mathcal{R}) = p_t^{(C)}(\mathcal{D}),$$

where, in the second equality, we have made use of the fact that the covariant time component of a particle's 4-momentum is conserved along geodesics in the Kerr geometry, since the metric is stationary ($\partial_t g_{\mu\nu} = 0$). Similarly, for the original particle we have $p_t^{(A)}(\mathcal{D}) = p_t^{(A)}(\mathcal{E})$. Thus, the time component of the momentum conservation condition (13.31) may be written in the form

$$E^{(C)} = E^{(A)} - p_t^{(B)}(\mathcal{D}), \tag{13.32}$$

where $p_t^{(B)}$ is also conserved along the geodesic followed by particle B.

The key step is now to note that $p_t^{(B)} = \boldsymbol{e}_t \cdot \boldsymbol{p}^{(B)}$, where $\boldsymbol{e}_t$ is the t-coordinate basis vector, whose squared 'length' is given by

$$\boldsymbol{e}_t \cdot \boldsymbol{e}_t = g_{tt}.$$

If particle B were ever to escape beyond the outer surface of the ergoregion, i.e. to a region where $g_{tt} > 0$, then $\boldsymbol{e}_t$ would be timelike. Thus, $p_t^{(B)}$ would be proportional to the particle energy as measured by an observer with 4-velocity along the $\boldsymbol{e}_t$-direction. In this case $p_t^{(B)}$ must therefore be positive, and so (13.32) shows that $E^{(C)} < E^{(A)}$, i.e. we get less energy out than we put in. However, if the particle B were never to escape the ergoregion but instead fall into the black hole, then it would remain in a region where $g_{tt} < 0$ and so $\boldsymbol{e}_t$ is spacelike. In this case $p_t^{(B)}$ would be a component of spatial momentum, which might be positive or negative. For decays where it is negative, from (13.32) we see that $E^{(C)} > E^{(A)}$ and so we have extracted energy from the black hole. This is the Penrose process.

What are the consequences of the Penrose process for the black hole? Once the particle has fallen inside the event horizon, the mass M and angular momentum $J = Mac$ of the black hole are changed:

$$M \to M + p_t^{(B)}/c^2, \tag{13.33}$$

$$J \to J - p_\phi^{(B)}, \tag{13.34}$$

where in the last equation we must remember that, for particle orbits in general, p_ϕ is *minus* the component of angular momentum of the particle along the rotation axis of the black hole. From (13.33), we see that the negative value of $p_t^{(B)}$ for the infalling particle in the Penrose process reduces the total mass of the black hole. As we now show, however, the Penrose process also reduces the angular momentum of the black hole. This is what is meant by saying that the Penrose process extracts rotational energy from the black hole.

To show that the angular momentum of the black hole is reduced by the infalling particle, it is useful to consider an observer in the ergoregion at fixed r and θ coordinates, who observes the particle B as it passes him. As shown in Section 13.8, the 4-velocity of such an observer is

$$[u^\mu] = u^t(1, 0, 0, \Omega), \tag{13.35}$$

where $\Omega = d\phi/dt$ is the observer's angular velocity with respect to infinity. This observer would measure the energy of particle B to be

$$E^{(B)} = p_\mu^{(B)} u^\mu = u^t \left(p_t^{(B)} + p_\phi^{(B)} \Omega \right).$$

Since this energy must be positive, we require

$$L < \frac{p_t^{(B)}}{\Omega} \tag{13.36}$$

where $L = -p_\phi^{(B)}$ is the component of the angular momentum of the particle along the rotation axis of the hole. Since $p_t^{(B)}$ is negative in the Penrose process and Ω must be positive for an observer in the ergoregion, we see that $L < 0$. Thus the infalling particle must have *negative* angular momentum, which therefore *reduces* the net angular momentum of the black hole. Rotational energy can continue to be extracted until the angular momentum of the black hole is reduced to zero and it becomes a Schwarzschild black hole.

We can, in fact, go slightly further and set a strict upper limit on L (which, since L is negative, is equivalent to a lower limit on its magnitude). We actually require (13.36) to hold for *any* observer at fixed r and θ in the ergoregion. From our earlier discussion of the ergoregion, the maximum value of the angular velocity occurs for an observer at the horizon $r = r_+$, in which case $\Omega = \Omega_{\rm H}$, (13.29). Thus, denoting the changes in the mass and angular momentum of the black hole by δM and δJ respectively, the condition (13.36) becomes

$$\boxed{\delta J < \frac{c^2 \delta M}{\Omega_{\rm H}},}$$

where it should be remembered that both δM and δJ are negative.

13.10 Geodesics in the equatorial plane

As one might expect, the general equations for non-null and null geodesics in the Kerr geometry are much less tractable than in the Schwarzschild case, and particle trajectories exhibit complicated behaviour. For example, in general the trajectory of a massive particle or photon is not constrained to lie in a plane. This is a direct consequence of the fact that the spacetime is not spherically symmetric and so, in general, the angular momentum of a test particle is not conserved. Since the Kerr geometry is stationary and axisymmetric, the conserved quantities along particle trajectories are p_t and p_ϕ. The latter corresponds to the conservation of only the component of angular momentum along the rotation axis. Nevertheless, since the metric is reflection-symmetric through the equatorial plane, particles for which $p^\theta = 0$, will always have $p^\theta = 0$ and so the trajectory remains in this plane. We shall therefore confine our attention to this simpler special case.

Setting $\theta = \pi/2$ in the Kerr metric (13.11), we obtain

$$ds^2 = c^2\left(1 - \frac{2\mu}{r}\right) dt^2 + \frac{4\mu ac}{r}\, dt\, d\phi - \frac{r^2}{\Delta}\, dr^2 - \left(r^2 + a^2 + \frac{2\mu a^2}{r}\right) d\phi^2, \tag{13.37}$$

from which the covariant metric components $g_{\mu\nu}$ in the equatorial plane can be read off. Following the method described in Section 13.1, the corresponding contravariant metric components are found to be

$$g^{tt} = \frac{1}{c^2\Delta}\left(r^2 + a^2 + \frac{2\mu a^2}{r}\right), \qquad g^{t\phi} = \frac{2\mu a}{cr\Delta},$$

$$g^{rr} = -\frac{\Delta}{r^2}, \qquad g^{\phi\phi} = -\frac{1}{\Delta}\left(1 - \frac{2\mu}{r}\right).$$

From (13.37) one can immediately write down the corresponding 'Lagrangian' $\mathcal{L} = g_{\mu\nu}\dot{x}^\mu \dot{x}^\nu$. In the interests of notational simplicity, for a massive particle we shall take the particle to have unit rest mass and for a photon we shall choose an appropriate affine parameter along the null geodesic such that, in both cases, $p^\mu = \dot{x}^\mu$. One may obtain the geodesic equations by writing down the appropriate Euler–Lagrange equations. It is quicker, however, simply to use the fact that p_t and p_ϕ are conserved along geodesics (since the metric does not depend explicitly on t and ϕ), which leads immediately to the first integrals of the t- and ϕ- equations. These are given by

$$p_t = g_{tt}\dot{t} + g_{t\phi}\dot{\phi} = c^2\left(1 - \frac{2\mu}{r}\right)\dot{t} + \frac{2\mu ac}{r}\dot{\phi} = kc^2, \tag{13.38}$$

$$p_\phi = g_{\phi t}\dot{t} + g_{\phi\phi}\dot{\phi} = \frac{2\mu ac}{r}\dot{t} - \left(r^2 + a^2 + \frac{2\mu a^2}{r}\right)\dot{\phi} = -h, \tag{13.39}$$

where we have defined the *constants* k and h so that in the Schwarzschild limit $a \to 0$ they coincide with the constants introduced in Chapter 9. This pair of simultaneous equations for $\dot{t}$ and $\dot{\phi}$ is straightforwardly solved to give

$$\boxed{\begin{aligned} \dot{t} &= \frac{1}{\Delta}\left[\left(r^2+a^2+\frac{2\mu a^2}{r}\right)k-\frac{2\mu a}{cr}h\right], \\ \dot{\phi} &= \frac{1}{\Delta}\left[\frac{2\mu ac}{r}k+\left(1-\frac{2\mu}{r}\right)h\right]. \end{aligned}} \tag{13.40}$$

Instead of using the complicated Euler–Lagrange equation for r, we may use the first integral provided by the invariant length of the 4-momentum $\boldsymbol{p}$. Since the covariant components of $\boldsymbol{p}$ are particularly simple, the most convenient form to use is $g^{\mu\nu}p_\mu p_\nu = \epsilon^2$, where $\epsilon^2 = c^2$ for a massive particle and $\epsilon^2 = 0$ for a photon.[4] Since $p_\theta = 0$ this gives

$$g^{tt}(p_t)^2 + 2g^{t\phi}p_t p_\phi + g^{\phi\phi}(p_\phi)^2 + g^{rr}(p_r)^2 = \epsilon^2, \tag{13.41}$$

where, for the moment, it is simpler not to write out the contravariant metric components in full. By substituting $p_t = kc^2$ and $p_\phi = -h$ into (13.41) and remembering that $p_r = g_{rr}\dot{r}$ and $g^{rr} = 1/g_{rr}$, we may then obtain the 'energy' equation for equatorial trajectories, which gives $\dot{r}$ in terms of only r and a set of constants. This yields

$$\dot{r}^2 = g^{rr}(\epsilon^2 - g^{tt}c^4k^2 + 2g^{t\phi}c^2kh - g^{\phi\phi}h^2). \tag{13.42}$$

At this stage, we may (if we wish) substitute the explicit forms for the contravariant metric coefficients to obtain

$$\boxed{\dot{r}^2 = c^2k^2 - \epsilon^2 + \frac{2\epsilon^2\mu}{r} + \frac{a^2(c^2k^2-\epsilon^2)-h^2}{r^2} + \frac{2\mu(h-ack)^2}{r^3}.} \tag{13.43}$$

In the limit $a \to 0$, the energy equation reduces to those derived in Chapter 9 for massive-particle ($\epsilon^2 = c^2$) and photon ($\epsilon^2 = 0$) orbits in the Schwarzschild geometry.

Since we are restricting our attention to the equatorial plane, we need not consider the Euler–Lagrange equation for θ, since it will not yield an independent equation of motion. Thus, equations (13.40) and (13.43) completely determine the null and non-null geodesics in the equatorial plane for given values of the constants k and h.

[4] The device of working in terms of ϵ^2 allows one to calculate the null and non-null geodesic equations simultaneously; one simply sets ϵ^2 to the appropriate value at the end of the calculation.

The null and non-null geodesics in the equatorial plane can be delineated in much the same way as for the Schwarzschild case in Chapter 9, albeit requiring significantly more complicated algebra. Before moving on to discuss particular examples, however, it is worth noting two essential differences from the Schwarzschild case. First, in the Kerr equatorial geometry trajectories will depend upon whether the particle or photon is in a co-rotating (prograde) or counter-rotating (retrograde) orbit, i.e. rotating about the symmetry axis in the same sense or the opposite sense to that of the rotating gravitational source. Second, *both t and ϕ* are 'bad' coordinates near the horizons. Expressed in terms of these coordinates, a trajectory approaching an horizon (at r_+ or r_-) will spiral around the black hole an infinite number of times, just as it takes an infinite coordinate time t to cross the horizon; neither behaviour is experienced by an observer comoving with the particle.

13.11 Equatorial trajectories of massive particles

For a massive particle, the timelike geodesics in the equatorial plane are governed by (13.40), and the 'energy' equation (13.43) with $\epsilon^2 = c^2$, which reads

$$\boxed{\dot{r}^2 = c^2(k^2-1) + \frac{2\mu c^2}{r} + \frac{a^2c^2(k^2-1) - h^2}{r^2} + \frac{2\mu(h-ack)^2}{r^3}.} \tag{13.44}$$

The interpretation of the constants k and h may be obtained by considering the limit $r \to \infty$, in the same way as for the Schwarzschild geometry. One thus finds that kc^2 and h are, respectively, the energy and angular momentum per unit rest mass of the particle describing the trajectory.

One may rewrite the energy equation (13.44) in the form

$$\tfrac{1}{2}\dot{r}^2 + V_{\text{eff}}(r; h, k) = \tfrac{1}{2}c^2(k^2-1), \tag{13.45}$$

where we have identified the effective potential per unit mass as

$$V_{\text{eff}}(r; h, k) = -\frac{\mu c^2}{r} + \frac{h^2 - a^2c^2(k^2-1)}{2r^2} - \frac{\mu(h-ack)^2}{r^3}. \tag{13.46}$$

There are several points to note here. First, V_{eff} has the same r-dependence as the corresponding expression for the Schwarzschild case, derived in Chapter 9; it is only that the coefficients of the last two terms are more complicated in the Kerr case. The graph of V_{eff} therefore has the same general shape as those shown in Figure 9.4. Indeed, as one would expect, in the limit $a \to 0$ equation (13.46) reduces to the Schwarzschild result. When $a \neq 0$, however, one must be careful in interpreting (13.46) as an effective potential, since it depends on the energy k

of the particle (as well as the usual dependence on the angular momentum h). Nevertheless, by differentiating (13.45) with respect to τ, one finds that the radial acceleration of a particle is still given by $\ddot{r} = -dV_{\text{eff}}/dr$. Similarly, the stability of circular orbits, for example, may be deduced by considering the sign of d^2V_{eff}/dr^2 in the usual manner. Also, an incoming particle will fall into the black hole only if the parameters h and k defining its trajectory are such that the maximum value of $V_{\text{eff}}(r; h, k)$ exceeds $\frac{1}{2}c^2(k^2 - 1)$.

In our discussion of the Schwarzschild geometry in Chapter 9, in addition to the energy equation it was reasonably straightforward also to derive the 'shape' equation for a general massive particle orbit and, equivalently, a simple expression for $d\phi/dr$. Unfortunately, it is algebraically very complicated (and unilluminating) to obtain the equivalent expressions for the Kerr geometry, even in the case of equatorial orbits. It is therefore natural to confine our attention to special cases in which the symmetry of the orbit makes the algebra more manageable. We are once again unfortunate, however, since the Kerr solution does not admit radial geodesics (either null or non-null). In a loose sense, the reason is that the rotating object 'drags' the surrounding space and the geodesics with it. Nevertheless, it is still reasonably straightforward to consider *motion with zero angular momentum* and *motion in a circle*.

13.12 Equatorial motion of massive particles with zero angular momentum

For a particle falling into a Kerr black hole whose angular momentum about the black hole is zero, we have $h = 0$. Setting $h = 0$ reduces the complexity of the geodesic equations (13.40–13.44) somewhat. To simplify the equations still further, however, we will also consider the limit in which the particle starts at rest from infinity, in which case $k = 1$. In this case the particle will initially be moving radially.

Using these values of h and k, the geodesic equations become

$$\dot{t} = \frac{1}{\Delta}\left(r^2 + a^2 + \frac{2\mu a^2}{r}\right),$$

$$\dot{\phi} = \frac{2\mu ac}{r\Delta},$$

$$\dot{r}^2 = \frac{2\mu c^2}{r}\left(1 + \frac{a^2}{r^2}\right).$$

From these expression, we see that both $\dot{t}$ and $\dot{\phi}$ are infinite at the horizons (when $\Delta = 0$), which is an illustration of the fact that both t and ϕ are 'bad coordinates' in these regions. Interestingly, the singular behaviours of the t and ϕ coordinates 'cancel' in the expression for $\dot{r}^2$.

The above equations may in turn be used to obtain expressions relating differentials of the coordinates along the particle trajectory. In particular, we find that

$$\frac{dr}{dt} = \frac{\dot{r}}{\dot{t}} = -\Delta\left[\frac{2\mu c^2}{r}\left(1+\frac{a^2}{r^2}\right)\right]^{1/2}\left(r^2+a^2+\frac{2\mu a^2}{r}\right)^{-1}$$

$$\frac{d\phi}{dt} = \frac{\dot{\phi}}{\dot{t}} = \frac{2\mu ac}{r}\left(r^2+a^2+\frac{2\mu a^2}{r}\right)^{-1}$$

$$\frac{d\phi}{dr} = \frac{\dot{\phi}}{\dot{r}} = -\frac{2\mu a}{r\Delta}\left[\frac{2\mu}{r}\left(1+\frac{a^2}{r^2}\right)\right]^{-1/2}.$$

We note that both dt/dr and $d\phi/dr$ become infinite at the horizons (when $\Delta = 0$), but $d\phi/dt$ remains finite there. The above equations may be integrated numerically in a straightforward way to obtain the trajectory of the massive particle. In Figure 13.4, we plot such a trajectory in the (ct, r)-plane and in the (x, y)-plane (where $x = \sqrt{r^2+a^2}\cos\phi$ and $y = \sqrt{r^2+a^2}\sin\phi$) for a particle that passes through the point $(r, \phi) = (8\mu, 0)$ at $t = 0$ in a Kerr geometry with rotation parameter $a = 0.8\mu$. In particular, we note that both plotted curves have

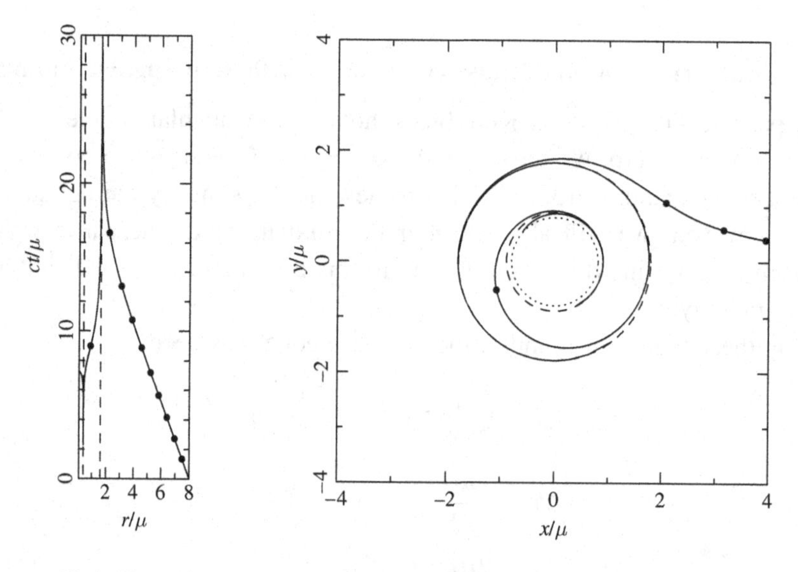

Figure 13.4 The trajectory of an initially radially moving massive particle falling from rest at infinity in a Kerr geometry with rotation parameter $a = 0.8\mu$. The trajectory (solid line) is plotted in the (ct, r)-plane (left) and the (x, y)-plane (right), where $x = \sqrt{r^2+a^2}\cos\phi$ and $y = \sqrt{r^2+a^2}\sin\phi$. The locations of the horizons (broken lines) and ring singularity (dotted line) are also indicated. The points correspond to unit intervals of $c\tau/\mu$, where τ is the proper time and we have taken $\tau = t = 0$ at $r = 8\mu$.

discontinuities at the horizons, which are shown as broken lines. This illustrates the pathology of the t- and ϕ-coordinates in these regions. The points in each plot correspond to unit intervals of $c\tau/\mu$, where τ is the proper time and we have taken $\tau = 0$ at $r = 8\mu$. The proper time increases steadily, without becoming singular; the particle reaches the ring singularity in the equatorial plane at $c\tau/\mu = 10.2$. In the right-hand plot we also note the effect of frame-dragging on the trajectory of the initially radially moving particle.

13.13 Equatorial circular motion of massive particles

For circular motion, we require that $\dot{r} = 0$ and, for the particle to remain in a circular orbit, that the radial acceleration $\ddot{r}$ must also vanish. In terms of the effective potential defined in (13.46), for a circular orbit at $r = r_c$ we thus require

$$V_{\text{eff}}(r_c; h, k) = \tfrac{1}{2}c^2(k^2 - 1) \qquad \text{and} \qquad \left.\frac{dV_{\text{eff}}}{dr}\right|_{r=r_c} = 0. \tag{13.47}$$

These two equations determine the values of the constants k and h that correspond to a circular orbit at some assigned radius $r = r_c$.

Obtaining analytic expressions for k and h is, algebraically, considerably more complicated than in the Schwarzschild case. The derivation is simplified somewhat, however, by working in terms of $u = 1/r$. Making this substitution into (13.46) and then differentiating the resulting expression with respect to u, we find that

$$\begin{aligned} -\mu c^2 u + \tfrac{1}{2}[h^2 - a^2c^2(k^2-1)]u^2 - \mu(h - ack)^2u^3 &= \tfrac{1}{2}c^2(k^2-1), \\ -\mu c^2 + [h^2 - a^2c^2(k^2-1)]u - 3\mu(h-ack)^2u^2 &= 0, \end{aligned}$$

where the second equality holds since $dV_{\text{eff}}/dr = (dV_{\text{eff}}/du)(du/dr)$, and therefore $dV_{\text{eff}}/du = 0$ implies that $dV_{\text{eff}}/dr = 0$. The algebra is further eased by introducing the variable $x = h - ack$, so that the two equations above become

$$-\mu c^2 u + \tfrac{1}{2}(x^2 + 2ackx + a^2c^2)u^2 - \mu x^2u^3 = \tfrac{1}{2}c^2(k^2-1), \tag{13.48}$$

$$-\mu c^2 + (x^2 + 2ackx + a^2c^2)u - 3\mu x^2u^2 = 0. \tag{13.49}$$

Subtracting (13.48) from u times (13.49) and performing a simple rearrangment of (13.49), we obtain

$$c^2k^2 = c^2(1 - \mu u) + \mu x^2u^3, \tag{13.50}$$

$$2xacku = x^2u(3\mu u - 1) - c^2(a^2u - \mu). \tag{13.51}$$

These are the basic equations to be used for obtaining analytic expressions for the constants k and h.

Eliminating k between (13.50) and (13.51), we quickly obtain a quadratic equation for x^2,

$$u^2\left[(3\mu u-1)^2-4a^2\mu u^3\right]x^4-2c^2u\left[(3\mu u-1)(a^2u-\mu)-2ua^2(\mu u-1)\right]x^2 + c^4(a^2u-\mu)^2=0.$$

Using the standard formula for the roots of a quadratic equation, one finds (after some straightforward but substantial algebra) that

$$x^2=\frac{c^2(a\sqrt{u}\pm\sqrt{\mu})^2}{u(1-3\mu u\mp 2a\sqrt{\mu u^3})}. \tag{13.52}$$

As we shall see below, the upper signs corresponds to the counter-rotating circular orbit and the lower signs to the co-rotating one. Furthermore, in order to obtain x we must choose either the positive or negative square root of (13.52). As we might expect from our discussion of the stability of massive particle orbits in the Schwarzschild case (see Chapter 9), the possibility exists for a circular orbit at a given coordinate radius to be either stable or unstable. It is straightforward to show that it is the negative root of (13.52) that corresponds to the stable case, in which we are most interested. We therefore consider only the solution

$$x=-\frac{c(a\sqrt{u}\pm\sqrt{\mu})}{[u(1-3\mu u\mp 2a\sqrt{\mu u^3})]^{1/2}}. \tag{13.53}$$

Inserting this solution into (13.50), for a stable circular orbit of inverse coordinate radius u we find that

$$\boxed{k=\frac{1-2\mu u\mp a\sqrt{\mu u^3}}{(1-3\mu u\mp 2a\sqrt{\mu u^3})^{1/2}},} \tag{13.54}$$

the energy of a particle of rest mass m_0 being $E=km_0c^2$. The corresponding value of the specific angular momentum for the orbit is obtained by calculating $h=x+ack$, which gives

$$\boxed{h=\mp\frac{c\sqrt{\mu}(1+a^2u^2\pm 2a\sqrt{\mu u^3})}{\sqrt{u}(1-3\mu u\mp 2a\sqrt{\mu u^3})^{1/2}}.} \tag{13.55}$$

We note that, as expected, in the limit $a\to 0$ the expressions (13.54) and (13.55) reduce to the corresponding results in the Schwarzschild case derived in Chapter 9.

13.14 Stability of equatorial massive particle circular orbits

It is worthwhile considering in some detail the stability of equatorial massive-particle orbits. Of particular astrophysical interest is the stability of the circular orbits discussed above and, especially, the coordinate radius of the innermost stable circular orbit in the co- and counter-rotating cases.

For a circular orbit of coordinate radius $r = r_c$ we require the condition (13.47). In addition, for marginal stability we require that (for $r = r_c$)

$$\frac{d^2V_{\text{eff}}}{dr^2} = \frac{d^2V_{\text{eff}}}{du^2}\left(\frac{du}{dr}\right)^2 + \frac{dV_{\text{eff}}}{du}\frac{d^2u}{dr^2} = u^3\left(u\frac{d^2V_{\text{eff}}}{du^2} + 2\frac{dV_{\text{eff}}}{du}\right) = 0,$$

where $u = 1/r$. Since $dV_{\text{eff}}/du = 0$ for a circular orbit, this additional requirement amounts to $d^2V_{\text{eff}}/du^2 = 0$. From (13.49), this reads

$$x^2 + 2ackx + a^2c^2 - 6\mu x^2 u = 0,$$

which may be more conveniently written as

$$u = \frac{x^2 + 2ackx + a^2c^2}{6\mu x^2} = \frac{h^2 - a^2c^2(k^2-1)}{6\mu x^2}.$$

Inserting the expressions (13.53–13.55) for x, k and h respectively into this equation and simplifying, one finds that

$$1 - 3a^2u^2 - 6u\mu \mp 8a\sqrt{\mu u^3} = 0.$$

Finally, using $u = 1/r$, one obtains an implicit equation for the coordinate radius r of the innermost stable circular orbit,

$$\boxed{r^2 - 6\mu r - 3a^2 \mp 8a\sqrt{\mu r} = 0,} \tag{13.56}$$

where, once again, the upper sign corresponds to the counter-rotating orbit and the lower sign to the co-rotating orbit. In the limit $a = 0$, we see that we recover $r = 6\mu$ for the innermost stable circular orbit in the Schwarzschild case. In the extreme Kerr limit $a = \mu$ we find, by inspection, that $r = 9\mu$ for the counter-rotating orbit and $r = \mu$ for the co-rotating case.

The general solution to the above quartic equation in $\sqrt{r}$ can be found analytically by standard methods, but the resulting expressions are algebraically messy. It is more instructive instead to solve the equation numerically and plot the results for a range of a/μ values, as shown in Figure 13.5 (left-hand panel). Also of particular interest is the energy E of a particle in the innermost co- and counter-rotating stable circular orbits. Using the expression (13.54), in the right-hand panel of Figure 13.5 we plot $k = E/(m_0c^2)$ for these orbits as a function of a/μ.

The difference between the energy $E = km_0c^2$ of a particle in an orbit and the energy m_0c^2 of the particle at rest at infinity is the gravitational binding energy

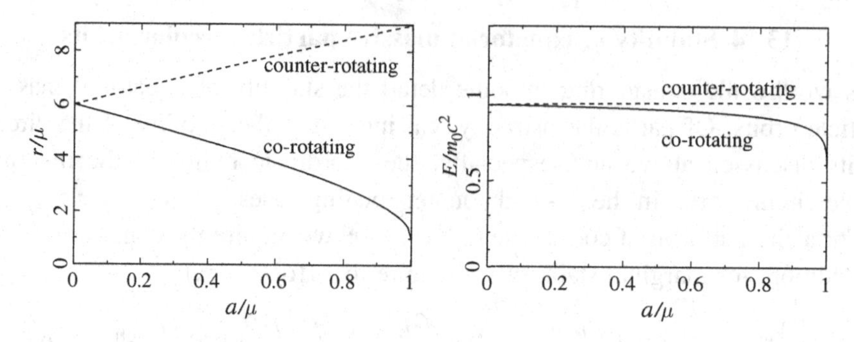

Figure 13.5 The scaled coordinate radii r/μ (left) and the constant $k = E/(m_0c^2)$ (right) for the innermost stable co-rotating and counter-rotating circular orbits in the equatorial plane of the Kerr geometry, as functions of a/μ.

of the orbit. As discussed in Chapter 9, the binding energy of a particle in an accretion disc around a compact object can be released. As the particle loses angular momentum, owing to turbulent viscosity, it gradually moves inwards, releasing gravitational energy mostly as radiation, until it reaches the innermost stable circular orbit, at which point it spirals rapidly inwards onto the compact object. The *efficiency* ε_{acc} of the accretion disc is the fraction of the rest mass energy that can be released in making the transition from rest at infinity to the innermost stable circular orbit and is given by $\varepsilon_{\text{acc}} = 1 - k$. We see from Figure 13.5 (right-hand panel) that for all values of a/μ the co-rotating orbit is the more bound, and the corresponding binding energy is greatest for an extreme Kerr black hole ($a/\mu = 1$). In this case

$$\varepsilon_{\text{acc}} = 1 - \frac{1}{\sqrt{3}} \approx 42\%,$$

and so an accretion disc around such an object could convert nearly one-half of the rest mass energy of its constituent particles into radiation. For a realistic astrophysical Kerr black hole that has been 'spun-up' by the accretion process one expects that $a/\mu \approx 0.998$, in which case $\varepsilon_{\text{acc}} \approx 32\%$, which is still substantially larger than the value of 5.7% in the Schwarzschild case.

13.15 Equatorial trajectories of photons

For photons, the null geodesics in the equatorial plane are governed by (13.40) and the 'energy' equation (13.43) with $\epsilon^2 = 0$, which reads

$$\boxed{\dot{r}^2 = c^2k^2 + \frac{a^2c^2k^2 - h^2}{r^2} + \frac{2\mu(h - ack)^2}{r^3}.} \tag{13.57}$$

As in our discussion of photon trajectories in the Schwarzschild geometry in Chapter 9, it is useful to introduce the parameter $b = h/(ck)$. By considering the limit $r \to \infty$, one again finds that b may be interpreted as an impact parameter for the trajectories that extend to infinity. We note that for $r \to \infty$ the constant k is positive and so b (or h) has the same sign as $\dot{\phi}$ in this limit.

One may rewrite the energy equation (13.57) in the form

$$\frac{\dot{r}^2}{h^2} + V_{\text{eff}}(r; b) = \frac{1}{b^2}, \qquad (13.58)$$

where we have identified the effective potential as

$$V_{\text{eff}}(r; b) = \frac{1}{r^2}\left[1 - \left(\frac{a}{b}\right)^2 - \frac{2\mu}{r}\left(1 - \frac{a}{b}\right)^2\right]. \qquad (13.59)$$

As was the case for massive particles, V_{eff} has the same r-dependence as the corresponding expression for the Schwarzschild case, derived in Chapter 9, and so the graph of V_{eff} has the same general shape. Indeed, as one would expect, in the limit $a \to 0$ equation (13.59) reduces to the Schwarzschild result. When $a \neq 0$, however, one must again be careful in interpreting (13.59) as an effective potential, since it depends on the value b (and hence k) of the particle trajectory. Nevertheless, by differentiating (13.58) with respect to τ, one finds that the radial acceleration of a particle is still given by $\ddot{r} = -h^2\, dV_{\text{eff}}/dr$. (In fact, by a rescaling $\lambda \to h\lambda$ of the affine parameter λ, the explicit h-dependence is removed from this result and (13.58).) Similarly, the stability of a circular orbit, for example, may be deduced by considering the sign of d^2V_{eff}/dr^2 in the usual manner.

13.16 Equatorial principal photon geodesics

As might be expected, radial photon geodesics do not exist in the equatorial plane of the Kerr geometry. Nevertheless, we can obtain information about the radial variation of the light-cone structure by investigating the *principal* null geodesics. These are defined by the condition $b = a$. The system of equations (13.38), (13.39), (13.57) then reduces to

$$\dot{t} = (r^2 + a^2)k/\Delta,$$
$$\dot{\phi} = ack/\Delta,$$
$$\dot{r} = \pm ck,$$

where the plus sign and the minus sign in the last equation correspond respectively to outgoing and incoming photons. We can see that such geodesics play the same role in the Kerr geometry as do the radial geodesics in the Schwarzschild case, in that the radial coordinate is described at a uniform rate with respect to the affine parameter.

Choosing $\dot{r} = +ck$ for outgoing photons, we find that

$$\frac{dt}{dr} = \frac{\dot{t}}{\dot{r}} = \frac{(r^2+a^2)}{c\Delta}, \qquad \frac{d\phi}{dr} = \frac{\dot{\phi}}{\dot{r}} = \frac{a}{\Delta}.$$

Using the fact that $\Delta > 0$ in region I, it follows that $dr/dt > 0$ in region I, thus confirming that these equations correspond to outgoing photons. If we restrict our attention to the case $a^2 < \mu^2$, the equations can be immediately integrated to give

$$ct = r + \left(\mu + \frac{\mu^2}{\sqrt{\mu^2 - a^2}}\right) \ln\left|\frac{r}{r_+} - 1\right| + \left(\mu - \frac{\mu^2}{\sqrt{\mu^2 - a^2}}\right) \ln\left|\frac{r}{r_-} - 1\right| + \text{constant}, \tag{13.60}$$

$$\phi = \frac{a}{2\sqrt{\mu^2 - a^2}} \ln\left|\frac{r - r_+}{r - r_-}\right| + \text{constant}. \tag{13.61}$$

The solution corresponding to incoming photons is obtained by choosing $\dot{r} = -ck$ and has the same form as above but with $t \to -t$ and $\phi \to -\phi$. In Figure 13.6 we plot an incoming principal null geodesic in the (ct, r)-plane and in the (x, y)-plane (with $x = \sqrt{r^2 + a^2} \cos\phi$ and $y = \sqrt{r^2 + a^2} \sin\phi$) for a photon that passes through

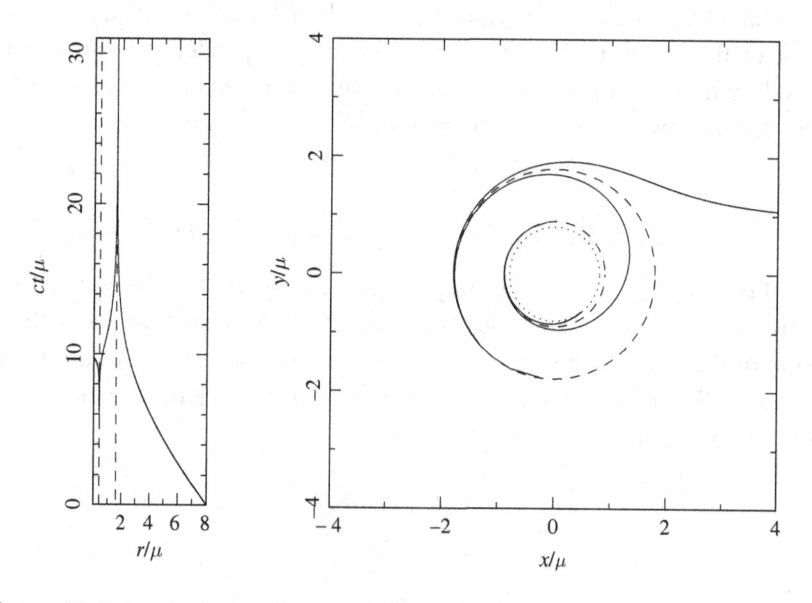

Figure 13.6 A principal null geodesic in a Kerr geometry with rotation parameter $a = 0.8\mu$. The trajectory (solid line) is plotted in the (ct, r)-plane (left) and in the (x, y)-plane (right); $x = \sqrt{r^2 + a^2} \cos\phi$ and $y = \sqrt{r^2 + a^2} \sin\phi$. The locations of the horizons (broken lines) and the ring singularity (dotted line) are also indicated.

the point $(r, \phi) = (8\mu, 0)$ at $t = 0$ in a Kerr geometry with rotation parameter $a = 0.8\mu$.

We note that, in the limit $a \to 0$, (13.60) reduces to the equation for a radial photon trajectory in the Schwarzschild geometry, as presented in Section 11.3. Indeed, the null geodesics considered above play the same role as the null radial geodesics in the Schwarzschild case, giving information about the radial variation of the light-cone structure. We can draw a spacetime diagram of the light-cone structure using these equations and we find in region I a diagram analogous to that obtained for the Schwarzschild geometry in (t, r, θ, ϕ) coordinates in Chapter 11: the light-cones narrow down as $r \to r_+$. On $r = r_+$ both t and ϕ become infinite, again indicating that this is a *coordinate singularity*.

13.17 Equatorial circular motion of photons

For circular photon motion we require $\dot{r} = 0$ and, for the photon to remain in a circular orbit, the radial acceleration $\ddot{r}$ must also vanish. In terms of the effective potential defined in (13.59), for a circular orbit at $r = r_c$ we thus require

$$V_{\text{eff}}(r_c; b) = \frac{1}{b^2} \quad \text{and} \quad \left.\frac{dV_{\text{eff}}}{dr}\right|_{r=r_c} = 0. \tag{13.62}$$

These two equations determine a single value $r = r_c$ (different for prograde and retrograde orbits) for which there exists a circular orbit, and the corresponding value of the constant b.

Using the expression (13.59), the above conditions yield respectively

$$r_c = 3\mu \frac{b-a}{b+a}, \tag{13.63}$$

$$(b+a)^3 = 27\mu^2(b-a). \tag{13.64}$$

These equations may be solved by setting $y = b + a$ in (13.64), solving the resulting cubic equation and substituting the resulting value of b into (13.63). One may easily verify that the result can be written as

$$r_c = 2\mu \left\{1 + \cos\left[\frac{2}{3}\cos^{-1}\left(\pm\frac{a}{\mu}\right)\right]\right\}, \tag{13.65}$$

$$b = 3\sqrt{\mu r_c} - a, \tag{13.66}$$

where the upper sign in (13.65) corresponds to retrograde orbits and the lower sign to prograde orbits. In the limit $a \to 0$ we recover the conditions for a circular photon orbit in the Schwarzschild case, obtained in Chapter 9, namely $r_c = 3\mu$ and $b = 3\sqrt{3}\mu$. As in the Schwarzschild case, circular photon orbits in the Kerr geometry are unstable.

13.18 Stability of equatorial photon orbits

In our discussion of the stability of photon orbits in the Schwarzschild geometry, it was useful to consider the effective potential for photon motion. As mentioned above, however, when $a \neq 0$ one must be careful in interpreting (13.59) as an effective potential since it depends on the value b (and hence k) of the particle trajectory. Nevertheless, we can still investigate the stability of the photon orbits by factorising the energy equation (13.57).[5] One finds that

$$\dot{r}^2 = \frac{(r^2+a^2)^2 - a^2\Delta}{r^4}\left[c^2k^2 - \frac{4\mu ra}{(r^2+a^2)^2-a^2\Delta}ckh - \frac{r(r-2\mu)}{(r^2+a^2)^2-a^2\Delta}h^2\right]$$

$$= \frac{(r^2+a^2)^2 - a^2\Delta}{c^2r^4}\left[c^2k - V_+(r)\right]\left[c^2k - V_-(r)\right], \qquad (13.67)$$

where $V_\pm(r)$ do not depend on k and are given by

$$V_\pm(r) = \frac{2\mu ra \pm r^2\Delta^{1/2}}{(r^2+a^2)^2-a^2\Delta}ch = \left[\omega \pm \left(\omega^2 - \frac{g^{\phi\phi}}{g^{tt}}\right)^{1/2}\right]h. \qquad (13.68)$$

The first property to notice is that if $\Delta < 0$ then the functions $V_\pm(r)$ are complex and so there are no (real) solutions to the equation $\dot{r} = 0$. This shows that the photon orbit has no turning points. Thus once a photon crosses the surface $\Delta = 0$, it cannot turn around and return back across the surface. Therefore $\Delta = 0$ defines an event horizon in the equatorial plane (in fact, as we showed earlier, that $\Delta = 0$ defines the event horizon is true in the general case).

The qualitative features of photon trajectories may be deduced by plotting the functions $V_\pm(r)$. We choose first the case $ah > 0$ (angular momentum in the same sense as the rotation of the source) and confine our attention to $r > r_+$ (i.e. outside the outer horizon). The curves are plotted in Figure 13.7. It is clear from (13.67) that photon propagation is only possible if $c^2k > V_+$ or $c^2k < V_-$, since we require $\dot{r}^2 > 0$. Thus, at any given coordinate radius r, photon propagation cannot occur if c^2k has a value lying in the region between the curves $V_-(r)$ and $V_+(r)$. However, we must also remember, from (13.38), that $c^2k = p_t$, the covariant time component of the photon's 4-momentum. This is the *energy* of the photon relative to a fixed observer at infinity. We are used to the idea of 'positive-energy' photons with $p_t > 0$. They may come in from infinity and either reach a minimum r or plunge into the black hole, depending on whether they encounter the hump in $V_+(r)$.

What about photons for which $p_t < V_-$? Some of these have $p_t > 0$ but others have $p_t < 0$. Which photons of these types, if any, can actually exist? Near the

[5] This approach is based on that presented in B. Schutz, *A First Course in General Relativity*, Cambridge University Press, 1985.

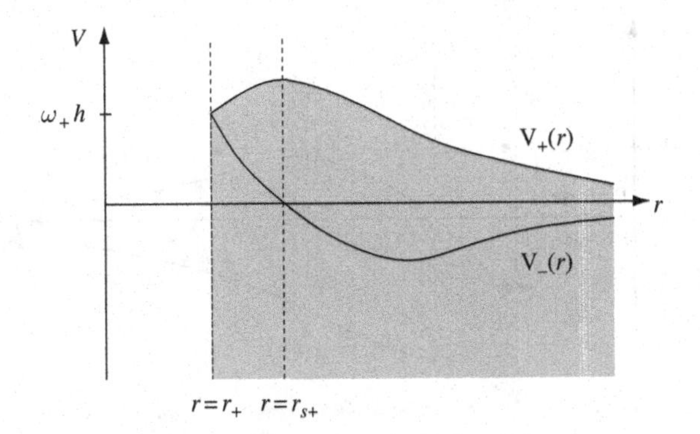

Figure 13.7 The factored effective-potential diagram for equatorial photon orbits with positive angular momentum ($ah > 0$). The quantity ω_+ is the value of ω at $r = r_+$. Photon propagation is forbidden in the shaded region.

horizon in the Kerr metric, the 'energy' p_t relative to an observer at infinity has no obvious physical meaning. The important requirement is that to an observer near the horizon the photon has a positive energy. A convenient observer, although any would suffice, is one who resides at fixed r in the equatorial plane, circling the hole with a fixed angular velocity Ω (this observer is not on a geodesic, so would need to be in a spaceship). As discussed in Section 13.8, the observer's 4-velocity $\boldsymbol{u}$ in (t, r, θ, ϕ) coordinates has components

$$[u^\mu] = u^t(1, 0, 0, \Omega).$$

Thus, he measures a photon energy

$$E = \boldsymbol{p} \cdot \boldsymbol{u} = p_t u^t + p_\phi u^\phi = u^t(p_t - \Omega h).$$

The photon must therefore have $p_t > \Omega h$. Since $p_t = c^2 k$, we thus require

$$c^2 k > \Omega h. \tag{13.69}$$

From our discussion in Section 13.8 about observers in the ergoregion, we know that Ω is restricted to lie in the range $\Omega_- < \Omega < \Omega_+$, where $\Omega_\pm$ are given by (13.28). Comparing (13.28) with (13.68) we see that, remarkably, $V_\pm = \Omega_\pm h$. Thus, any photon with $c^2 k > V_+$ also satisfies the condition (13.69) and so is allowed, while any photon with $c^2 k < V_-$ violates (13.69) and is forbidden. We conclude that here there is nothing qualitatively different from our discussion of photon orbits in the Schwarzschild geometry.

For photons moving in the opposite direction to the hole's rotation ($ah < 0$), new features do appear. If $ah < 0$ it is clear from (13.68) that the shapes of the $V_\pm(r)$ curves are simply turned upside down (see Figures 13.8 and 13.7). From (13.67) directly, we again see that in the region between the curves $V_-(r)$ and

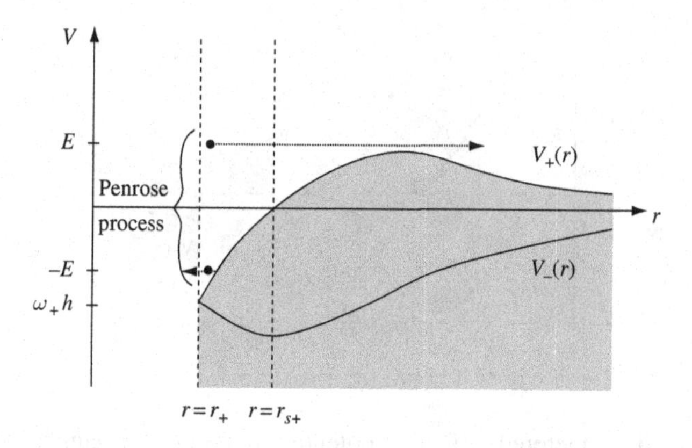

Figure 13.8 The factored effective-potential diagram for equatorial photon orbits with negative angular momentum ($ah < 0$). The quantity ω_+ is the value of ω at $r = r_+$. Photon propagation is forbidden in the shaded region. The Penrose process is also illustrated (see the text for details).

$V_+(r)$ there is no photon propagation. Moreover, the condition (13.69) means that photons must have $c^2k > V_+$ but from Figure 13.8 we see that in the region $r < r_{S+}$ (the ergoregion) some of these photons can have $c^2k < 0$. We can now understand in an alternative manner an idealised version of the Penrose process, discussed in Section 13.9. At some point between r_+ and r_{S+} it is allowable to create two photons, one having $p_t = E$ and the other having $p_t = -E$, so that their total energy is *zero*. Then the 'positive-energy' photon could be directed in such a way as to leave the hole and reach infinity, while the 'negative-energy' photon is necessarily trapped and inevitably crosses the horizon. The net effect is that the positive-energy photon will leave the hole, carrying its energy to infinity. Thus energy has been extracted.

13.19 Eddington–Finkelstein coordinates

We have seen throughout our discussion of the Kerr geometry that the Boyer–Lindquist coordinates t and ϕ are 'bad' in the region near the horizons. By analogy with our discussion of removing the coordinate singularity in the Schwarzschild geometry, we may use the equations for the principal photon geodesics, (13.60) and (13.61), to obtain a coordinate transformation that extends the solution through $r = r_+$.

Working with these equations in differential form, we have

$$c\,dt = -\frac{r^2 + a^2}{\Delta}\,dr, \tag{13.70}$$

$$d\phi = -\frac{a}{\Delta}\,dr, \tag{13.71}$$

for the ingoing photons. For advanced Eddington–Finkelstein coordinates (t', ϕ', r, θ) we want ingoing principal photon trajectories to be straight lines. Thus, for such a trajectory, we require

$$c\,dt' = -dr \qquad \text{and} \qquad d\theta = d\phi' = 0.$$

From (13.70) and (13.71), we see immediately that the required transformations are

$$c\,dt' = c\,dt + \frac{2\mu r}{\Delta}\,dr,$$

$$d\phi' = d\phi + \frac{a}{\Delta}\,dr.$$

The Kerr solution in advanced Eddington–Finkelstein coordinates then takes the form

$$\begin{aligned} ds^2 = {} & \left(1 - \frac{2\mu r}{\rho^2}\right) c^2\,dt'^2 - \frac{4\mu r}{\rho^2} c\,dt'\,dr \\ & - \left(1 + \frac{2\mu r}{\rho^2}\right) dr^2 + \frac{4\mu r a \sin^2\theta}{\rho^2} c\,dt'\,d\phi' \\ & + \frac{2(r^2 + a^2) a \sin^2\theta}{\rho^2}\,dr\,d\phi' - \rho^2\,d\theta^2 \\ & - \left[(r^2 + a^2)\sin^2\theta + \frac{2\mu r a^2 \sin^4\theta}{\rho^2}\right] d\phi'^2. \end{aligned}$$

If we define the advanced time parameter $p = ct' + r$ (such that $dp = 0$ along the photon geodesic), the Kerr solution can also be written as

$$\begin{aligned} ds^2 = {} & \left(1 - \frac{2\mu r}{\rho^2}\right) dp^2 - 2\,dp\,dr + \frac{4\mu r a \sin^2\theta}{\rho^2}\,dp\,d\phi' + 2a\sin^2\theta\,dr\,d\phi' - \rho^2\,d\theta^2 \\ & - \left[(r^2 + a^2)\sin^2\theta + \frac{2\mu r a^2 \sin^4\theta}{\rho^2}\right] d\phi'^2. \end{aligned}$$

One may alternatively straighten the outgoing photon geodesics by introducing retarded Eddington–Finkelstein coordinates (t^*, ϕ^*, r, θ) and the retarded time parameter $q = ct^* - r$, in an analogous manner.

Figure 13.9 shows a spacetime diagram along the equator of a Kerr black hole using advanced Eddington–Finkelstein coordinates. As in the Schwarzschild solution, the event horizon at r^+ marks a surface of 'no return'. Once a particle has crossed the event horizon, its future is directed towards region III, which contains the singularity – you can never return back to region I. Unlike the Schwarzschild solution, the singularity in the Kerr solution is *timelike* (the singularity in the Schwarzschild solution is *spacelike*). In theory, this means that it is possible to *avoid* the singularity by moving along a timelike path; in other words, if we

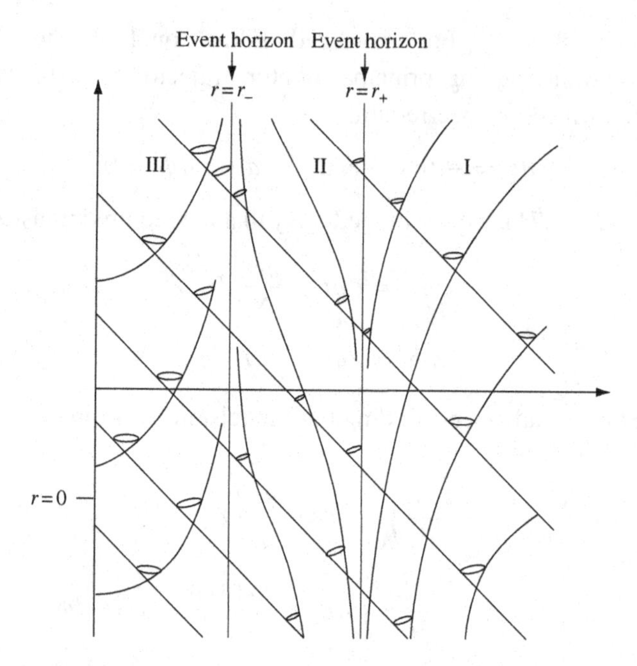

Figure 13.9 spacetime diagram of the Kerr solution in advanced Eddington–Finkelstein coordinates.

were in a spaceship (and ignoring the intense tidal forces which would make this experiment impractical) we could manoeuvre along a path to avoid the singularity. Indeed, by performing a maximal extension of the Kerr geometry in an analogous way to the Kruskal extension of the Schwarzschild geometry described in Chapter 9, one finds that a particle may re-cross the surface $r = r_-$ and eventually emerge from $r = r_+$ into a different asymptotically flat spacetime (in an analogous way to that described for the Reissner–Nordström geometry in Section 12.6). However, you should not take the internal structure of the Kerr solution too seriously. As mentioned above, region III also contains closed timelike curves (at $r < 0$), which are very bad news because they violate causality. Most theorists would hope that quantum gravity comes to the rescue and prevents causality violation. At present we do not really know what happens within region III.

Figure 13.10 shows a schematic illustration of the light-cone structure in the equatorial plane of the Kerr solution, which also illustrates the frame-dragging effect. As we approach the infinite redshift surface S^+, any particle travelling against the direction of rotation has to travel at the speed of light just to remain stationary (relative to a fixed observer at infinity). At smaller r, in the ergoregion, the light-cones are tipped over, so that photons (and massive particles) are forced to travel in the direction of rotation. At the event horizon r^+, the lightcones tip over so far that the future is directed towards region II.

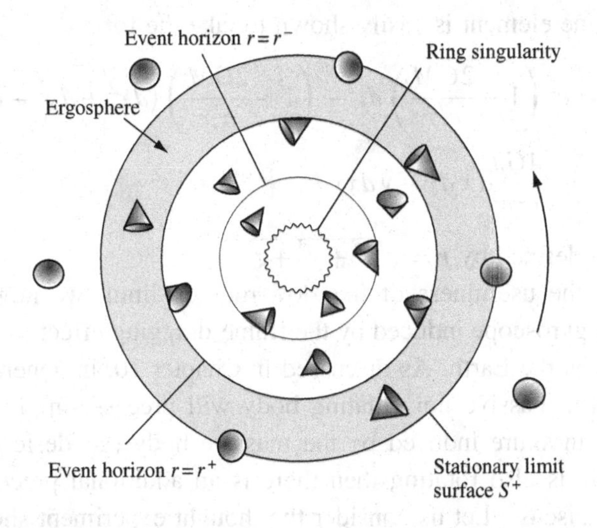

Figure 13.10 Frame dragging in the equatorial plane of the Kerr solution.

13.20 The slow-rotation limit and gyroscope precession

Since the full Kerr solution is rather complicated, it is useful to consider the simpler approximate form for the common limiting case of a *slowly rotating* body. Thus, we will only keep terms in the Kerr metric to first order in a. Writing the resulting metric in terms of the angular momentum $J = Mac$ of the rotating body, in Boyer–Lindquist coordinates we obtain

$$\boxed{ds^2 = ds^2_{\text{Schwarzschild}} + \frac{4GJ}{c^2 r}\sin^2\theta\, d\phi\, dt,} \tag{13.72}$$

where the first term on the right-hand side is the standard Schwarzschild line element. In the slow-rotation limit, Boyer–Lindquist coordinates tend to Schwarzschild coordinates. This metric is useful for performing calculations of, for example, the general-relativistic effects due to the rotation of the Earth. In fact, for terrestrial applications, and many other astrophysical situations, we may also assume the gravitational field to be weak, in which case the line element becomes

$$ds^2 = c^2\left(1 - \frac{2GM}{c^2 r}\right)dt^2 - \left(1 + \frac{2GM}{c^2 r}\right)(dr^2 + r^2\, d\theta^2 + r^2 \sin^2\theta\, d\phi^2)$$
$$+ \frac{4GJ}{c^2 r}\sin^2\theta\, d\phi\, dt. \tag{13.73}$$

It is often also convenient to work in Cartesian coordinates defined by

$$x = r\sin\theta\cos\phi, \qquad y = r\sin\theta\sin\phi, \qquad z = r\cos\theta,$$

for which the line element is easily shown to take the form

$$ds^2 = c^2\left(1-\frac{2GM}{c^2r}\right)dt^2 - \left(1+\frac{2GM}{c^2r}\right)(dx^2+dy^2+dz^2) + \frac{4GJ}{c^2r^3}(x\,dy - y\,dx)\,dt, \quad (13.74)$$

where r is now defined by $r = \sqrt{x^2+y^2+z^2}$.

To illustrate the usefulness of the slow-rotation limit, we now consider the precession of a gyroscope induced by the frame-dragging effect of a slowly rotating body, such as the Earth. As discussed in Chapter 10, in general a gyroscope in orbit around a massive non-rotating body will precess simply as a result of the spacetime curvature induced by the massive body (geodesic precession). If the central body is also rotating then there is an additional precessional effect, which we now discuss. Let us consider the thought experiment shown schematically in Figure 13.11. A gyroscope falls freely down the rotation axis of a slowly rotating body. Initially the spin axis is oriented perpendicular to the rotation axis. By symmetry, if the body were not rotating then the spin axis would remain fixed with respect to infinity (e.g. pointing constantly to one distant star), thus for this particular orbit there is no geodesic precession of the gyroscope. By this measure, the local inertial frames on the axis are not rotating with respect to infinity. However, if instead the body were rotating, even with a small angular momentum, then the gyroscope would precess, indicating that the local inertial frames are rotating with respect to infinity.

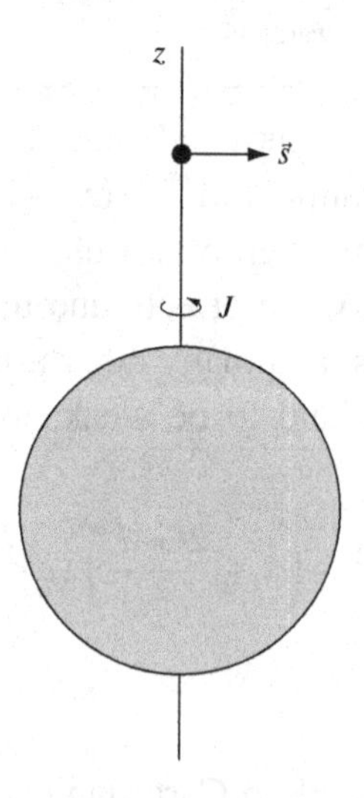

Figure 13.11 A gyroscope (solid circle) falling down the rotation axis of a spinning body.

We can use the metric (13.74) to calculate the precession rate of the gyroscope on its downward 'polar plunge' trajectory. As shown in Section 10.5, the spin 4-vector $\boldsymbol{s}(\tau)$ is parallel-transported along the geodesic trajectory. Thus, its components satisfy

$$\frac{ds^\mu}{d\tau} + \Gamma^\mu{}_{\nu\sigma} s^\nu u^\sigma = 0. \tag{13.75}$$

For the physical arrangement under consideration, the initial 4-velocity $\boldsymbol{u}$ and the spin 4-vector $\boldsymbol{s}$ of the gyroscope in Cartesian coordinates have the forms

$$[u^\mu] = (u^t, 0, 0, u^z) \qquad \text{and} \qquad [s^\mu] = (0, s^x, s^y, 0).$$

Moreover, these forms remain valid at all later times, since the trajectory is a polar plunge and $\boldsymbol{s} \cdot \boldsymbol{u}$ is conserved along it. Thus, in (13.75), the only equations we need to consider are

$$\frac{ds^x}{d\tau} + \Gamma^x{}_{xt} s^x u^t + \Gamma^x{}_{xz} s^x u^z + \Gamma^x{}_{yt} s^y u^t + \Gamma^x{}_{yz} s^y u^z = 0, \tag{13.76}$$

$$\frac{ds^y}{d\tau} + \Gamma^y{}_{xt} s^x u^t + \Gamma^y{}_{xz} s^x u^z + \Gamma^y{}_{yt} s^y u^t + \Gamma^y{}_{yz} s^y u^z = 0. \tag{13.77}$$

To continue with our calculation, we must first find the connection coefficients appearing in the above equations. This is most easily achieved using the 'Lagrangian' approach, writing down the Euler–Lagrange equation for x and remembering that on the polar axis all terms proportional to some positive power of x or y are zero and $r = z$. One finds that the only non-zero connection coefficients of the form $\Gamma^x{}_{\mu\nu}$ are

$$(\Gamma^x{}_{yt})_{z\text{-axis}} = \frac{2GJ}{c^2 z^3} \qquad \text{and} \qquad (\Gamma^x{}_{xz})_{z\text{-axis}} = -\frac{2GM}{c^2 z(z + 2GM/c^2)}.$$

By considering the symmetry properties of the metric (13.74), one immediately deduces that, on the polar axis, the only non-zero connection coefficients of the form $\Gamma^y{}_{\mu\nu}$ are

$$(\Gamma^y{}_{xt})_{z\text{-axis}} = -\frac{2GJ}{c^2 z^3} \qquad \text{and} \qquad (\Gamma^y{}_{yz})_{z\text{-axis}} = -\frac{2GM}{c^2 z(z + 2GM/c^2)}.$$

These connection coefficients can now be substituted into equations (13.76, 13.77), which can be solved once u^t and u^z have been determined from the geodesic equations. Assuming, however, that the speed of the falling gyroscope is non-relativistic, then to leading order in $1/c$ we may take $u^t \approx 1$ and $u^z \approx 0$. Thus, in this approximation, equations (13.76, 13.77) reduce to

$$\frac{ds^x}{d\tau} = -\frac{2GJ}{c^2 z^3} s^y \qquad \text{and} \qquad \frac{ds^y}{d\tau} = \frac{2GJ}{c^2 z^3} s^x.$$

Hence, as it falls, the gyroscope precesses in the same direction as the body is rotating, i.e. the local inertial frames are dragged with respect to infinity. This is called the *Lens–Thirring effect*. At a height z the rate of precession is

$$\boxed{\Omega_{\rm LT} = \frac{2GJ}{c^2 z^3}.}$$

It should be noted that we have calculated this precession rate in a Cartesian coordinate system in which the centre of the gravitating body is at rest and the gyroscope is falling. Fortunately, an observer free-falling with the gyroscope would measure the *same* precession rate since the Lorentz transformation that connects the two frames is a boost along the z-axis, which does not affect the transverse components s^x and s^y of the spin vector. Of course, the Lens–Thirring effect also results in the precession of gyroscopes following trajectories other than the polar plunge considered here, but determining the rate of precession in general requires a considerably longer calculation (see Exercise 17.26).

Exercises

13.1 Verify that the Boyer–Lindquist form of the Kerr metric satisfies the empty-space Einstein field equations.
Note: This exercise is only for the truly dedicated reader!

13.2 Show that the Boyer–Lindquist form of the Kerr metric can be written in the forms (13.12) and (13.13).

13.3 Calculate the contravariant components $g^{\mu\nu}$ of the Kerr metric in Boyer–Lindquist coordinates.

13.4 Show that, in the limit $\mu \to 0$, the Kerr metric tends to the Minkowksi metric.

13.5 Show that the Kerr–Schild form of the Kerr metric can be transformed into the Boyer–Lindquist form by the coordinate transformations (13.16–13.19).

13.6 Consider the 2-surfaces defined by $t = \text{constant}$ and $r = r_\pm$ in the Kerr geometry. Show that, for each surface, the circumference around the 'poles' is less than the circumference around the equator. Show that the same is true for the 2-surfaces defined by $t = \text{constant}$ and $r = r_{S\pm}$.

13.7 Show that the proper area of the event horizon $r = r_+$ in the Kerr geometry is given by

$$A = 4\pi\left(r_+^2 + a^2\right).$$

Hence show that, for fixed μ, the area A is a maximum for $a = 0$. Conversely, for fixed A, show that μ is a minimum for $a = 0$. Comment on your results.

13.8 An observer is at fixed (r, θ) coordinates in the ergoregion of a Kerr black hole and has angular velocity $\Omega = d\phi/dt$ with respect to a second observer at rest at infinity. Show that the allowed range for Ω is given by $\Omega_- < \Omega < \Omega_+$, where

$$\Omega_\pm = \omega \pm c\left(\frac{\Delta}{\Sigma^2}\right)^{1/2}$$

and ω, Δ and Σ^2 have their usual meanings in the Kerr metric.

13.9 Use your answer to Exercise 13.7 to show that the area of the event horizon $r = r_+$ in the Kerr geometry may be written as

$$A = \frac{8\pi G}{c^4}\left[M^2 + \sqrt{M^4 - \left(\frac{cJ}{G}\right)^2}\right],$$

where M and J are the mass and angular momentum of the black hole. Hence show that if the mass and angular momentum change by δM and δJ respectively then the corresponding change in the proper area of the horizon is given by

$$\delta A = \frac{8\pi G}{c}\frac{a}{\Omega_{\rm H}\sqrt{\mu^2 - a^2}}\left(\delta M - \Omega_{\rm H}\frac{\delta J}{c^2}\right),$$

where $\Omega_{\rm H}$ is the 'angular velocity of the horizon', defined in (13.29). Thus show that the area of the event horizon must increase in the Penrose process.

13.10 Show that, in the equatorial plane $\theta = \pi/2$ of the Kerr geometry, the contravariant metric components in Boyer–Lindquist coordinates are

$$g^{tt} = \frac{1}{c^2\Delta}\left(r^2 + a^2 + \frac{2\mu a^2}{r}\right), \qquad g^{t\phi} = \frac{2\mu a}{cr\Delta},$$

$$g^{rr} = -\frac{\Delta}{r^2}, \qquad g^{\phi\phi} = -\frac{1}{\Delta}\left(1 - \frac{2\mu}{r}\right).$$

13.11 Show that the geodesic equations for particle motion in the equatorial plane of the Kerr geometry may be written in Boyer–Lindquist coordinates as

$$\dot{t} = \frac{1}{\Delta}\left[\left(r^2 + a^2 + \frac{2\mu a^2}{r}\right)k - \frac{2\mu a}{cr}h\right],$$

$$\dot{\phi} = \frac{1}{\Delta}\left[\frac{2\mu ac}{r}k + \left(1 - \frac{2\mu}{r}\right)h\right],$$

$$\dot{r}^2 = (c^2k^2 - \epsilon^2) + \frac{2\epsilon^2\mu}{r} + \frac{a^2(c^2k^2 - \epsilon^2) - h^2}{r^2} + \frac{2\mu(h - ack)^2}{r^3},$$

where $\epsilon^2 = c^2$ for a massive particle and $\epsilon = 0$ for a photon. Verify that these equations reduce to the Schwarzschild case in the limit $a \to 0$.

13.12 The trajectory of an infalling particle of mass m in the equatorial plane of a Kerr black hole is characterised by the usual parameters k and h. If the particle eventually falls into the black hole, show that the mass and the angular momentum of the hole are changed in such a way that

$$M \to M + kmc^2, \qquad J \to J + mh.$$

Show further that the corresponding change δa in the rotation parameter of the black hole is given by

$$\delta a = \frac{m}{cM}(h - ack).$$

If the particle falls into an extreme Kerr black hole, for which $a = \mu$, show that a naked singularity would be created if

$$h > 2ck\mu.$$

However, by determining the maximum value of the effective potential $V_{\text{eff}}(r; h, k)$ defined in (13.46) for $a = \mu$, show that a particle with $h > 2ck\mu$ can never fall into the black hole.

13.13 For a Kerr black hole, using the Boyer–Lindquist coordinates show that, for a particle in circular orbit in the $\theta = \pi/2$ plane, the coordinate angular velocity $\Omega = d\phi/dt$ satisfies

$$\Omega = \frac{c\mu^{1/2}}{a\mu^{1/2} \pm r^{3/2}}.$$

This is the Kerr-metric analogue to $\Omega^2 = GM/r^3$ for the Schwarzschild metric. Here the plus sign corresponds to prograde orbits, the minus to retrograde orbits.

13.14 If a particle's motion is initially in the $\theta = \pi/2$ plane in a Kerr metric, show that the motion will remain in this plane.

13.15 Show that the values of the parameters k and h for a circular orbit of coordinate radius $r = r_c$, given in (13.54) and (13.55) respectively, satisfy the requirements

$$V_{\text{eff}}(r_c; h, k) = \tfrac{1}{2}c^2(k^2 - 1) \qquad \text{and} \qquad \left.\frac{dV_{\text{eff}}}{dr}\right|_{r=r_c} = 0.$$

Show further that for the orbit to be stable one requires

$$r_c^2 - 6\mu r_c - 3a^2 \mp 8a\sqrt{\mu r_c} = 0.$$

13.16 An observer (not necessarily free-falling) orbits a Kerr black hole in the equatorial plane in a circular orbit. His 'angular velocity with respect to a distant observer' is $\Omega = d\phi/dt$. Find the components u^t, u^ϕ, u_t and u_ϕ in terms of Ω, r, μ and a.

13.17 Suppose that the circular orbit considered in Exercise 13.16 lies outside the horizon r_+ but inside the stationary limit r_{S_+}. Show that under these circumstances Ω must be non-zero, i.e. the observer cannot remain at rest relative to a distant observer. If the orbiting observer is in the region $r_- < r < r_+$, show that the orbit cannot be circular.

13.18 Show that the effective potential for photon orbits in the equatorial plane of the Kerr geometry is given by

$$V_{\text{eff}}(r;\,b) = \frac{1}{r^2}\left[1-\left(\frac{a}{b}\right)^2 - \frac{2\mu}{r}\left(1-\frac{a}{b}\right)^2\right].$$

13.19 For a circular photon orbit of coordinate radius $r = r_c$ in the Kerr geometry, show that

$$r_c = 2\mu\left\{1+\cos\left[\frac{2}{3}\cos^{-1}\left(\pm\frac{a}{\mu}\right)\right]\right\},$$
$$b = 3\sqrt{\mu r_c} - a,$$

where the upper sign in the first equation corresponds to retrograde orbits and the lower sign to prograde orbits. Hence show that, for an extreme Kerr black hole $(a=\mu)$, $r_c = 4\mu$ for a retrograde orbit and $r_c = \mu$ for a prograde orbit.

13.20 For a photon orbit in the equatorial plane of the Kerr geometry, show that

$$\dot{r}^2 = \frac{(r^2+a^2)^2 - a^2\Delta}{c^2 r^4}[c^2k - V_+(r)][c^2k - V_-(r)],$$

where

$$V_\pm(r) = \frac{2\mu r a \pm r^2\Delta^{1/2}}{(r^2+a^2)^2 - a^2\Delta}ch = \left[\omega \pm \left(\omega^2 - \frac{g^{\phi\phi}}{g^{tt}}\right)^{1/2}\right]h.$$

13.21 The general axisymmetric stationary metric can be written in the form

$$ds^2 = A\,dt^2 - B(d\phi - \omega\,dt)^2 - C\,dr^2 - D\,d\theta^2,$$

where A, B, C, D and ω are functions only of the coordinates r and θ. Alice is an astronaut in a powered spaceship that maintains fixed (r, ϕ) coordinates in the equatorial plane $\theta = \pi/2$ (at a position for which $g_{tt} > 0$). She simultaneously emits two photons in opposite tangential directions in the equatorial plane and uses a prearranged system of mirrors to cause each photon to move along a circular (non-geodesic) path of constant r. Show that the coordinate angular velocities of the two photons are given by

$$\frac{d\phi}{dt} = \omega \pm \left(\frac{A}{B}\right)^{1/2}.$$

Hence show that the two photons do not arrive back with Alice simultaneously but are separated by a time interval

$$\Delta\tau = \frac{4\pi\omega B}{c(A - B\omega^2)^{1/2}},$$

as measured by Alice's on-board clock. Comment on the physical significance of this result.

13.22 Bob is in a powered spaceship following a circular orbit $r =$ constant in the equatorial plane of the geometry in Exercise 13.21. His angular velocity is such that the component u_ϕ of his 4-velocity is zero. Using the same arrangement of mirrors as in Exercise 13.21, he performs an experiment similar to Alice's. Show that for Bob the two photons arrive back to him simultaneously.

13.23 Which, if any, of the photons considered in Exercises 13.21 and 13.22 is redshifted from its original frequency on arriving back with Alice or Bob? Explain your reasoning.

13.24 An isolated thin rigid spherical shell has mass M and radius R. If the shell is set spinning slowly, with angular momentum J, show that inertial frames within the shell rotate with angular velocity

$$\omega = \frac{2GJ}{c^2 R^3}.$$

Comment briefly on how this result is related to Mach's principle.

14

The Friedmann–Robertson–Walker geometry

We now discuss the application of general relativity to modelling the behaviour of the universe as a whole. In order to do this, we make some far-reaching assumptions, but only those consistent with our observations of the universe. As in our derivations of the Schwarzschild and Kerr geometries, we begin by using symmetry arguments to restrict the possible forms for the metric describing the overall spacetime geometry of the universe.[1]

14.1 The cosmological principle

When we look up at the sky we see that the stars around us are grouped into a large-density concentration – the Milky Way Galaxy. On a slightly larger scale, we see that our Galaxy belongs to a small group of galaxies (called the Local Group). Our Galaxy and our nearest large neighbour, the Andromeda galaxy, dominate the mass of the Local Group. On still larger scales we see that our Local Group sits on the outskirts of a giant supercluster of galaxies centred in the constellation of Virgo. Evidently, on small scales matter is distributed in a highly irregular way but, as we look on larger and larger scales, the matter distribution looks more and more uniform. In fact, we have very good evidence (particularly from the constancy of the temperature of the cosmic microwave background in different directions on the sky) that the universe is *isotropic* on the very largest scales, to high accuracy. If the universe has no preferred centre then *isotropy* also implies *homogeneity*. We therefore have good physical reasons to study simple cosmological models in which the universe is assumed to be homogeneous and

[1] For a detailed discussion, see, for example, J. N. Islam, *An Introduction to Mathematical Cosmology*, Cambridge University Press, 1992.

isotropic[2]. We thus assume the *cosmological principle*, which states that *at any particular time, the universe looks the same from all positions in space and all directions in space at any point are equivalent.*

14.2 Slicing and threading spacetime

The intuitive statement of the cosmological principle given above needs to be made more precise. In particular, how does one define a 'particular time' in general relativity that is valid globally, when there are no global inertial frames? Also, since observers moving relative to one another will view the universe differently, according to which observers do we demand the universe to appear isotropic?

In general relativity the concept of a 'moment of time' is ambiguous and is replaced by the notion of a *three-dimensional spacelike hypersurface*. To define a 'time' parameter that is valid globally, we 'slice up' spacetime by introducing a series of non-intersecting spacelike hypersurfaces that are labelled by some parameter t. This parameter then defines a universal time in that 'a particular time' means a given spacelike hypersurface. It should be noted, however, that we may construct the hypersurfaces $t =$ constant in any number of ways. In a general spacetime there is no preferred 'slicing' and hence no preferred 'time' coordinate t.

It is useful at this point to introduce the idealised concept of *fundamental observers*, who are assumed to have no motion relative to the overall *cosmological fluid* associated with the 'smeared-out' motion of all the galaxies and other matter in the universe. A fundamental observer would, for example, measure no dipole moment in his observations of the cosmic microwave background radiation; an observer with a non-zero *peculiar velocity* would observe such a dipole as a result of the Doppler effect arising from his motion relative to the cosmological fluid. Adopting *Weyl's postulate*, the timelike worldlines of these observers are assumed to form a bundle, or *congruence*, in spacetime that diverges from a point in the (finite or infinitely distant) past or converges to such a point in the future. These worldlines are non-intersecting, except possibly at a singular point in the past or future or both. Thus, there is a unique worldline passing through each (non-singular) spacetime point. The set of worldlines is sometimes described as providing *threading* for the spacetime.

The hypersurfaces $t =$ constant may now be naturally constructed in such a way that the 4-velocity of any fundamental observer is orthogonal to the hypersurface.

[2] It is worth noting that isotropy about every point automatically implies homogeneity. However, homogeneity does *not* necessarily imply isotropy. For example, a universe with a large-scale magnetic field that pointed in one direction everywhere and had the same magnitude at every point would be homogeneous but not isotropic.

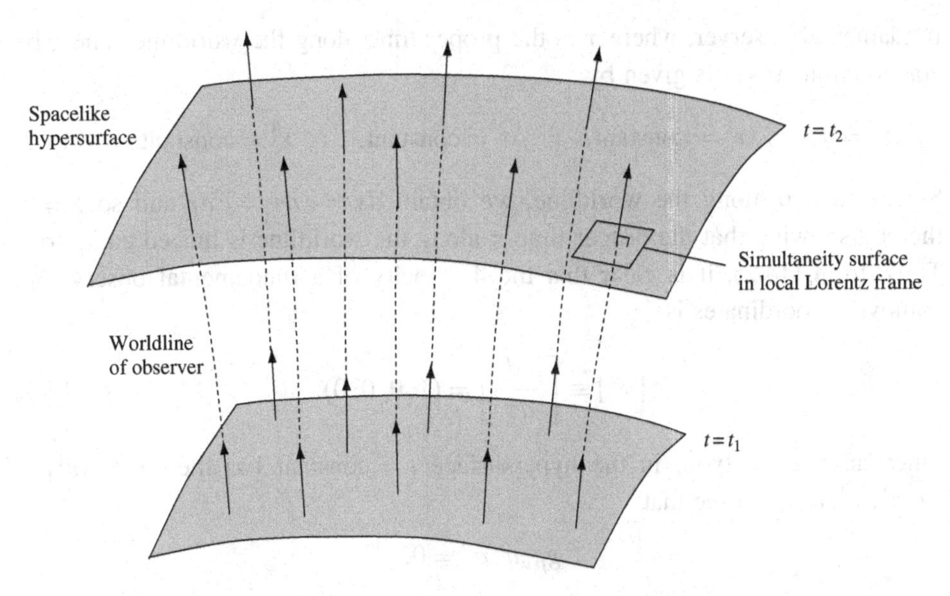

Figure 14.1 Representation (with one spatial dimension suppressed) of spacelike hypersurfaces on which fundamental observers are assumed to lie. The worldline of any fundamental observer is orthogonal to any such surface.

Thus, the surface of simultaneity of the local Lorentz frame of any such observer coincides locally with the hypersurface (see Figure 14.1). Each hypersurface may therefore be considered as the 'meshing together' of all the local Lorentz frames of the fundamental observers.

14.3 Synchronous comoving coordinates

The spacelike hypersurfaces discussed above are labelled by a parameter t, which may be taken to be the *proper time* along the worldline of any fundamental observer. The parameter t is then called the *synchronous time coordinate* or *cosmic time*. In addition, we may also introduce spatial coordinates (x^1, x^2, x^3) that are *constant* along any worldline. Thus each fundamental observer has fixed (x^1, x^2, x^3) coordinates, and so the latter are called *comoving coordinates*. Since each hypersurface $t =$ constant is orthogonal to the observer's worldline, the line element takes the form

$$ds^2 = c^2\,dt^2 - g_{ij}\,dx^i\,dx^j \qquad (\text{for } i, j = 1, 2, 3), \tag{14.1}$$

where the g_{ij} are functions of the coordinates (t, x^1, x^2, x^3).

We may verify that the metric (14.1) does indeed incorporate the properties described in the previous section, as follows. Let $x^\mu(\tau)$ be the worldline of a fundamental observer, where τ is the proper time along the worldline. Then, by construction, $x^\mu(\tau)$ is given by

$$x^0 = \tau, \qquad x^1 = \text{constant}, \qquad x^2 = \text{constant}, \qquad x^3 = \text{constant}. \qquad (14.2)$$

Since $dx^i = 0$ along the worldline, we obtain $ds = c\,d\tau = c\,dt$ and so $t = \tau$, thereby showing that the proper time τ along the worldline is indeed equal to t. Thus, from (14.2), it is clear that the 4-velocity of a fundamental observer in comoving coordinates is

$$[u^\mu] \equiv \left[\frac{dx^\mu}{d\tau}\right] = (1, 0, 0, 0). \qquad (14.3)$$

Since any vector lying in the hypersurface $t = \text{constant}$ has the form $[a^\mu] = (0, a^1, a^2, a^3)$, we see that

$$g_{\mu\nu} u^\mu a^\nu = 0,$$

because $g_{0i} = 0$ for $i = 1, 2, 3$. Hence, the observer's 4-velocity is orthogonal to the hypersurface, as we required. Finally, we may show that the worldline given by (14.2) satisfies the geodesic equation

$$\frac{d^2 x^\mu}{d\tau^2} + \Gamma^\mu{}_{\nu\sigma} \frac{dx^\nu}{d\tau} \frac{dx^\sigma}{d\tau} = 0.$$

Using (14.3), we see that we require only that $\Gamma^\mu{}_{00} = 0$. This quantity is given by

$$\Gamma^\mu{}_{00} = \tfrac{1}{2} g^{\mu\nu} (2\partial_0 g_{0\nu} - \partial_\nu g_{00}),$$

which is easily shown to be zero by using the fact that $g^{0i} = 0$ for $i = 1, 2, 3$. Thus the worldlines $x^\mu(\tau)$ are geodesics and hence can describe particles (observers) moving only under the influence of gravity.

14.4 Homogeneity and isotropy of the universe

The metric (14.1) does *not* yet incorporate the property that space is homogeneous and isotropic. Indeed this form of the metric can be used, with the help of a special coordinate system obtained by singling out a particular fundamental observer, to derive some general properties of the universe, *without the assumptions of homogeneity and isotropy*, although we will not consider such cases here.

Let us now incorporate the postulates of homogeneity and isotropy. The former demands that all points on a particular spacelike hypersurface are equivalent, whereas the latter demands that all directions on the hypersurface are equivalent for fundamental observers. The (squared) spatial separation on the same hypersurface $t =$ constant of two nearby galaxies at coordinates (x^1, x^2, x^3) and $(x^1 + \Delta x^1, x^2 + \Delta x^2, x^3 + \Delta x^3)$ is

$$d\sigma^2 = g_{ij}\Delta x^i \Delta x^j.$$

If we consider the triangle formed by three nearby galaxies at some particular time t, then isotropy requires that the triangle formed by these same galaxies at some later time must be similar to the original triangle. Moreover, homogeneity requires that the magnification factor must be *independent* of the position of the triangle in the 3-space. It thus follows that time t can enter the g_{ij} only through a common factor, so that the ratios of small distances are the same at all times. Hence the metric must take the form

$$\boxed{ds^2 = c^2\, dt^2 - S^2(t) h_{ij}\, dx^i\, dx^j,} \tag{14.4}$$

where $S(t)$ is a time-dependent *scale factor* and the h_{ij} are functions of the coordinates (x^1, x^2, x^3) only. We note that it is common practice to identify fundamental observers loosely with individual galaxies (which are assumed to be pointlike). However, since the magnification factor is independent of position, we must neglect the small peculiar velocities of real individual galaxies.

14.5 The maximally symmetric 3-space

We clearly require the 3-space spanned by the spacelike coordinates (x^1, x^2, x^3) to be homogeneous and isotropic. This leads us to study the *maximally symmetric 3-space*. In three dimensions, the curvature tensor R_{ijkl} has, in general, six independent components, each of which is a function of the coordinates. We therefore need to specify six functions to define the intrinsic geometric properties of a general three-dimensional space. Clearly, the more symmetrical the space, the fewer the functions needed to specify its properties. A *maximally symmetric space* is specified by just one number – the *curvature* K, which is independent of the coordinates. Such *constant curvature* spaces must clearly be *homogeneous and isotropic*.

The curvature tensor of a maximally symmetric space must take a particularly simple form. It must clearly depend on the constant K and on the metric tensor g_{ij}. The simplest expression that satisfies the various symmetry properties and identities of R_{ijkl} and contains just K and the metric tensor is given by

$$R_{ijkl} = K(g_{ik}g_{jl} - g_{il}g_{jk}). \tag{14.5}$$

In fact, a maximally symmetric space is *defined* as one having a curvature tensor of the form (14.5).

The Ricci tensor is given by

$$\begin{aligned} R_{jk} = g^{il} R_{ijkl} &= K g^{il} (g_{ik} g_{jl} - g_{il} g_{jk}) \\ &= K(\delta^l_k g_{jl} - \delta^l_l g_{jk}) \\ &= K(g_{jk} - 3 g_{jk}) = -2K g_{jk}. \end{aligned}$$

The curvature scalar is thus given by

$$R = R^k_k = -2K\delta^k_k = -6K.$$

As in our derivation of the general static isotropic metric in Section 9.1, the metric of an isotropic 3-space must depend only on the rotational invariants

$$\vec{x}\cdot\vec{x} \equiv r^2, \qquad d\vec{x}\cdot d\vec{x}, \qquad \vec{x}\cdot d\vec{x},$$

and in spherical polar coordinates (r, θ, ϕ) it must take the form

$$\begin{aligned} d\sigma^2 &= C(r)(\vec{x}\cdot d\vec{x})^2 + D(r)(d\vec{x}\cdot d\vec{x})^2 \\ &= C(r) r^2\, dr^2 + D(r)(dr^2 + r^2\, d\theta^2 + r^2 \sin^2\theta\, d\phi^2). \end{aligned}$$

Following our analysis in Chapter 9, we can simplify this line element by redefining the radial coordinate $\bar{r}^2 = r^2 D(r)$. Dropping the bars on the variables, the metric can thus be written as

$$d\sigma^2 = B(r)\, dr^2 + r^2 d\theta^2 + r^2 \sin^2\theta\, d\phi^2,$$

where $B(r)$ is an arbitrary function of r.

We have met this line element before – it is identical to the space part of the general static isotropic metric. In Chapter 9, we showed that the only non-zero connection coefficients are

$$\Gamma^r{}_{rr} = \frac{1}{2B(r)}\frac{dB(r)}{dr}, \qquad \Gamma^r{}_{\theta\theta} = -\frac{r}{B(r)}, \qquad \Gamma^r{}_{\phi\phi} = -\frac{r\sin^2\theta}{B(r)},$$

$$\Gamma^\theta{}_{r\theta} = \Gamma^\phi{}_{r\phi} = \frac{1}{r}, \qquad \Gamma^\theta{}_{\phi\phi} = -\sin\theta\cos\theta, \qquad \Gamma^\phi{}_{\phi\theta} = \cot\theta.$$

The Ricci tensor is given in terms of the connection coefficients by

$$R_{ij} = \partial_j \Gamma^k{}_{ik} - \partial_k \Gamma^k{}_{ij} + \Gamma^l{}_{ik}\Gamma^k{}_{lj} - \Gamma^l{}_{ij}\Gamma^k{}_{lk},$$

and, after some algebra, we find that its non-zero components are

$$R_{rr} = -\frac{1}{rB}\frac{dB}{dr},$$

$$R_{\theta\theta} = \frac{1}{B} - 1 - \frac{r}{2B^2}\frac{dB}{dr}$$

$$R_{\phi\phi} = R_{\theta\theta}\sin^2\theta$$

For our 3-space to be maximally symmetric, however, we must have

$$R_{ij} = -2Kg_{ij},$$

and so we require

$$\frac{1}{rB}\frac{dB}{dr} = 2KB(r), \tag{14.6}$$

$$1 + \frac{r}{2B^2}\frac{dB}{dr} - \frac{1}{B} = 2Kr^2. \tag{14.7}$$

Integrating (14.6) we immediately obtain

$$B(r) = \frac{1}{A - Kr^2},$$

where A is a constant of integration. Substituting this expression into (14.7) then gives

$$1 - A + Kr^2 = Kr^2,$$

from which we see that $A = 1$. Thus, we have constructed the line element for the maximally symmetric 3-space, which takes the form

$$\boxed{d\sigma^2 = \frac{dr^2}{1 - Kr^2} + r^2 d\theta^2 + r^2 \sin^2\theta \, d\phi^2} \tag{14.8}$$

and has a curvature tensor specified by one number, K, the *curvature* of the space.

Notice also that this is *exactly* the same form as the metric for a 3-sphere embedded in four-dimensional Euclidean space, which we discussed in Chapter 2. The metric contains a 'hidden symmetry', since *the origin of the radial coordinate is completely arbitrary.* We can choose any point in this space as our origin since all points are equivalent. *There is no centre in this space.* We also note that, on scales small compared with the spatial curvature, the line element (14.8) is equivalent to that of a three-dimensional Euclidean space.

14.6 The Friedmann–Robertson–Walker metric

Combining the our expression (14.8) for the maximally symmetric 3-space with the line element (14.4), which incorporates the cosmological principle and Weyl's postulate, we obtain

$$ds^2 = c^2\,dt^2 - S^2(t)\left[\frac{dr^2}{1-Kr^2} + r^2(d\theta^2 + \sin^2\theta\,d\phi^2)\right]. \tag{14.9}$$

It is usual to write this line element in an alternative form in which the arbitrariness in the magnitude of K is absorbed into the radial coordinate and the scale factor. Assuming firstly that $K \neq 0$ we define the variable $k = K/|K|$ in such a way that $k = \pm 1$ depending on whether K is positive or negative. If we introduce the rescaled coordinate

$$\bar{r} = |K|^{1/2} r,$$

then (14.9) becomes

$$ds^2 = c^2\,dt^2 - \frac{S^2(t)}{|K|}\left[\frac{d\bar{r}^2}{1-k\bar{r}^2} + \bar{r}^2(d\theta^2 + \sin^2\theta\,d\phi^2)\right].$$

Finally, we define a rescaled scale function $R(t)$ by

$$R(t) = \begin{cases} \dfrac{S(t)}{|K|^{1/2}} & \text{if } K \neq 0, \\ S(t) & \text{if } K = 0. \end{cases}$$

Then, dropping the bars on the radial coordinate, we obtain the standard form for the *Friedmann–Robertson–Walker* (FRW) line element,

$$\boxed{ds^2 = c^2\,dt^2 - R^2(t)\left[\frac{dr^2}{1-kr^2} + r^2\left(d\theta^2 + \sin^2\theta\,d\phi^2\right)\right],} \tag{14.10}$$

where k takes the values -1, 0, or 1 depending on whether the spatial section has negative, zero or positive curvature respectively. It is also clear that the coordinates (r, θ, ϕ) appearing in the FRW metric are still *comoving*, i.e. the worldline of a galaxy, ignoring any peculiar velocity, has fixed values of (r, θ, ϕ).

14.7 Geometric properties of the FRW metric

The geometric properties of the homogeneous and isotropic 3-space corresponding to the hypersurface $t =$ constant depend upon whether $k = -1$, 0 or 1. We now consider each of these cases in turn.

Positive spatial curvature: $k = 1$

In the case $k = 1$, we see that the coefficient of dr in the FRW metric becomes singular as $r \to 1$. We therefore introduce a new radial coordinate χ, defined by the relation

$$r = \sin\chi \qquad \Rightarrow \qquad dr = \cos\chi\, d\chi = (1-r^2)^{1/2}\, d\chi,$$

so that the spatial part of the FRW metric takes the form

$$d\sigma^2 = R^2\left[d\chi^2 + \sin^2\chi(d\theta^2 + \sin^2\theta\, d\phi^2)\right],$$

where R is the value of the scale factor at the particular time t defining the spacelike hypersurface of interest.

Some insight into this spatial metric may be gained by considering the 3-space as *embedded* in a four-dimensional Euclidean space with coordinates (w, x, y, z), where

$$\begin{aligned}
w &= R\cos\chi,\\
x &= R\sin\chi\sin\theta\cos\phi,\\
y &= R\sin\chi\sin\theta\sin\phi,\\
z &= R\sin\chi\cos\theta.
\end{aligned}$$

In fact we have already discussed exactly this embedding in Section 2.9. Such an embedding is possible since one can write

$$d\sigma^2 = dw^2 + dx^2 + dy^2 + dz^2 = R^2\left[d\chi^2 + \sin^2\chi(d\theta^2 + \sin^2\theta\, d\phi^2)\right],$$

where, from the transformation equations, we have the constraint

$$w^2 + x^2 + y^2 + z^2 = R^2.$$

This shows that our 3-space can be considered as a three-dimensional sphere in the four-dimensional Euclidean space. This hypersurface is defined by the coordinate ranges

$$0 \le \chi \le \pi, \qquad 0 \le \theta \le \pi, \qquad 0 \le \phi \le 2\pi.$$

The surfaces $\chi = $ constant are 2-spheres with surface area

$$A = \int_{\theta=0}^{\pi}\int_{\phi=0}^{2\pi}(R\sin\chi\, d\theta)(R\sin\chi\sin\theta\, d\phi) = 4\pi R^2\sin^2\chi,$$

and (θ, ϕ) are the standard spherical polar coordinates of these 2-spheres. Thus, as χ varies from 0 to π, the area of the 2-spheres increases from zero to a maximum value of $4\pi R^2$ at $\chi = \pi/2$, after which it decreases to zero at $\chi = \pi$. The proper radius of a 2-sphere is $R\chi$, and so the surface area is smaller than that of a sphere of radius $R\chi$ in Euclidean space.

The entire 3-space has a *finite* total volume given by

$$V = \int_{\chi=0}^{\pi}\int_{\theta=0}^{\pi}\int_{\phi=0}^{2\pi}(R\,d\chi)(R\sin\chi\,d\theta)(R\sin\chi\sin\theta\,d\phi) = 2\pi^2R^3,$$

which is the reason why, in this case, R is often referred to as the 'radius of the universe'.

Zero spatial curvature: $k = 0$

In this case, if we set $r = \chi$ (to keep our notation consistent), the 3-space line element is

$$d\sigma^2 = R^2\left[d\chi^2 + \chi^2(d\theta^2 + \sin^2\theta\,d\phi^2)\right],$$

which is simply the ordinary three-dimensional Euclidean space. As usual, under the transformation

$$x = R\chi\sin\theta\cos\phi, \qquad y = R\chi\sin\theta\sin\phi, \qquad z = R\chi\cos\theta,$$

the line element becomes

$$d\sigma^2 = dx^2 + dy^2 + dz^2.$$

Negative spatial curvature: $k = -1$

In this case, it is convenient to introduce a radial coordinate χ given by

$$r = \sinh\chi \qquad \Rightarrow \qquad dr = \cosh\chi\,d\chi = (1 + r^2)^{1/2}\,d\chi,$$

so that the spatial part of the FRW metric becomes

$$d\sigma^2 = R^2\left[d\chi^2 + \sinh^2\chi(d\theta^2 + \sin^2\theta\,d\phi^2)\right].$$

We cannot embed this 3-space in a four-dimensional Euclidean space, but it can be embedded in a four-dimensional Minkowski space with coordinates (w, x, y, z) given by

$$\begin{aligned}
w &= R\cosh\chi,\\
x &= R\sinh\chi\sin\theta\cos\phi,\\
y &= R\sinh\chi\sin\theta\sin\phi,\\
z &= R\sinh\chi\cos\theta.
\end{aligned}$$

In this case, we can write

$$d\sigma^2 = dw^2 - dx^2 - dy^2 - dz^2,$$

together with the constraint

$$w^2 - x^2 - y^2 - z^2 = R^2,$$

which shows that the 3-space can be represented as a three-dimensional hyperboloid in the four-dimensional Minkowski space. The hypersurface is defined by the coordinate ranges

$$0 \leq \chi \leq \infty, \qquad 0 \leq \theta \leq \pi, \qquad 0 \leq \phi \leq 2\pi.$$

The 2-surfaces $\chi = \text{constant}$ are 2-spheres with surface area

$$A = 4\pi R^2 \sinh^2 \chi,$$

which increases indefinitely as χ increases. The proper radius of such a 2-sphere is $R\chi$, and so the surface area is larger than the corresponding result in Euclidean space. The total volume of the space is infinite.

From the above discussion, we see that a convenient form for the FRW metric is

$$\boxed{ds^2 = c^2 dt^2 - R^2(t)\left[d\chi^2 + S^2(\chi)(d\theta^2 + \sin^2\theta\, d\phi^2)\right],} \tag{14.11}$$

where the function $r = S(\chi)$ is given by

$$\boxed{S(\chi) = \begin{cases} \sin\chi & \text{if } k = 1, \\ \chi & \text{if } k = 0, \\ \sinh\chi & \text{if } k = -1. \end{cases}} \tag{14.12}$$

Once again, it is clear that (χ, θ, ϕ) are *comoving coordinates*.

14.8 Geodesics in the FRW metric

In the comoving coordinate system(s) we have defined above, the galaxies have fixed spatial coordinates (by construction; any peculiar velocities are ignored). Thus the 'cosmological fluid' is at rest in the comoving frame we have chosen. We now consider the motion of particles travelling with respect to this comoving frame. In particular, we consider the geodesic motion of 'free' particles, i.e. those experiencing only the 'background' gravitational field of the cosmological fluid and no other forces. Examples of such particles might include a projectile shot out of a galaxy or a photon travelling through intergalactic space. We could use the 'Lagrangian' procedure to calculate the geodesic equations for the FRW metric, but instead we take advantage of the fact that the spatial part of the metric is homogeneous and isotropic to arrive at the equations rather more quickly.

It is convenient to express the FRW metric in the form (14.11) and write $[x^\mu] = (t, \chi, \theta, \phi)$, so that

$$g_{00} = c^2, \quad g_{11} = -R^2(t), \quad g_{22} = -R^2(t)S^2(\chi), \quad g_{33} = -R^2(t)S^2(\chi)\sin^2\theta.$$

The path of a particle is given by the geodesic equation

$$\dot{u}^\mu + \Gamma^\mu{}_{\nu\sigma}u^\nu u^\sigma = 0,$$

where $u^\mu = \dot{x}^\mu$ and the dot corresponds to differentation with respect to some affine parameter. For our present purposes, however, it will be more useful to rewrite the geodesic equation in the form

$$\dot{u}_\mu = \tfrac{1}{2}(\partial_\mu g_{\nu\sigma})u^\nu u^\sigma,$$

which shows, as expected, that if the metric is independent of a particular coordinate x^λ then u_λ is conserved along the geodesic.

Let us suppose that the geodesic passes through some spatial point P. Since the spatial part of the metric is spatially homogeneous and isotropic we can, without loss of generality, take the spatial origin of the coordinate system, i.e. $\chi = 0$, to be at the point P. This simplifies the analysis considerably.

Consider first the ϕ-component u^3. Since the metric is independent of ϕ, we have $\dot{u}_3 = 0$ so that u_3 is constant along the geodesic. But

$$u_3 = g_{33}u^3 = -R^2(t)S^2(\chi)\sin^2\theta\, u^3,$$

so that $u_3 = 0$ at the point P where $\chi = 0$. Thus $u_3 = 0$ along the path and so also we have $u^3 = \dot{\phi} = 0$. Hence, along the geodesic,

$$\boxed{\phi = \text{constant.}}$$

For the θ-component, we have

$$\dot{u}_2 = \tfrac{1}{2}(\partial_2 g_{\nu\sigma})u^\nu u^\sigma. \tag{14.13}$$

The only component of $g_{\mu\nu}$ that depends on $x^2 = \theta$ is g_{33}, but the contribution of the corresponding term in (14.13) vanishes since $u^3 = 0$. Thus $\dot{u}_2 = 0$ and so u_2 is constant along the geodesic. Again

$$u_2 = g_{22}u^2 = -R^2(t)S^2(\chi)u^2,$$

which vanishes at $P(\chi = 0)$, and so u_2 is zero along the geodesic, as is u^2, so that

$$\boxed{\theta = \text{constant.}}$$

For the r-component,

$$\dot{u}_1 = \tfrac{1}{2}(\partial_1 g_{\nu\sigma})u^\nu u^\sigma. \tag{14.14}$$

We have $u^2 = u^3 = 0$, while g_{00} and g_{11} are independent of χ. Thus, $\dot{u}_1 = 0$ so that u_1 is constant along the geodesic, so $u_1 = g_{11}u^1$ must be constant. Thus, we have

$$\boxed{R^2(t)\dot{\chi} = \text{constant}.} \tag{14.15}$$

Finally, u^0 can be found from the appropriate normalisation condition, $u^\mu u_\mu = c^2$ for massive particles or $u^\mu u_\mu = 0$ for photons. Thus, we have

$$\boxed{\dot{t}^2 = \begin{cases} 1 + \dfrac{R^2(t)\dot{\chi}^2}{c^2} & \text{for a massive particle,} \\ \dfrac{R^2(t)\dot{\chi}^2}{c^2} & \text{for a photon.} \end{cases}}$$

14.9 The cosmological redshift

We can use the results of the last section to derive the *cosmological redshift*. Suppose that a photon is emitted at cosmic time t_E by a comoving observer with fixed spatial coordinates $(\chi_E, \theta_E, \phi_E)$ and that the photon is received at time t_R by another observer at fixed comoving coordinates. We may take the latter observer to be at the origin of our spatial coordinate system.

For a photon one can choose an affine parameter such that the 4-momentum is $p^\mu = \dot{x}^\mu$. From our above discussion, $d\theta = d\phi = 0$ along the photon geodesic, or equivalently $p_2 = p_3 = 0$, and (14.15) shows that p_1 is constant along the geodesic. Since the photon momentum is null, we also require that $g^{\mu\nu}p_\mu p_\nu = 0$, which reduces to

$$\frac{1}{c^2}(p_0)^2 - \frac{1}{R^2(t)}(p_1)^2 = 0,$$

from which we find $p_0 = cp_1/R(t)$.

In Appendix 9A we showed that, for an emitter and receiver with fixed spatial coordinates, the frequency shift of the photon is given, in general, by

$$\frac{\nu_R}{\nu_E} = \frac{p_0(R)}{p_0(E)}\left[\frac{g_{00}(E)}{g_{00}(R)}\right]^{1/2}. \tag{14.16}$$

For the FRW metric we have $g_{00} = c^2$, and so we find immediately that

$$\boxed{1 + z \equiv \frac{\nu_E}{\nu_R} = \frac{R(t_R)}{R(t_E)}.} \tag{14.17}$$

Thus we see that if the scale factor $R(t)$ is increasing with cosmic time, so that the universe is *expanding*, then the photon is *redshifted* by an amount z. Conversely, if the universe were contracting then the photon would be blueshifted. Only if the universe were *static*, so that $R = \text{constant}$, would there be no frequency shift.

In fact, we may also arrive at this result directly from the FRW metric. Since $ds = d\theta = d\phi = 0$ along the photon path, from (14.11) we have, for an incoming photon,

$$\int_{t_E}^{t_R} \frac{c\,dt}{R(t)} = \int_0^{\chi_E} d\chi.$$

Now, if the emitter sends a second light pulse at time $t_E + \delta t_E$, which is received at time $t_R + \delta t_R$, then

$$\int_{t_E+\delta t_E}^{t_R+\delta t_R} \frac{c\,dt}{R(t)} = \int_0^{\chi_E} d\chi = \int_{t_E}^{t_R} \frac{c\,dt}{R(t)},$$

from which we see immediately that

$$\int_{t_R}^{t_R+\delta t_R} \frac{c\,dt}{R(t)} = \int_{t_E}^{t_E+\delta t_E} \frac{c\,dt}{R(t)}.$$

Assuming that δt_E and δt_R are small, so that $R(t)$ can be taken as constant in both integrals, we have

$$\frac{\delta t_R}{R(t_R)} = \frac{\delta t_E}{R(t_E)}.$$

Considering the pulses to be the successive wavecrests of an electromagnetic wave, we again find that

$$1 + z \equiv \frac{\nu_E}{\nu_R} = \frac{\delta t_R}{\delta t_E} = \frac{R(t_R)}{R(t_E)}.$$

14.10 The Hubble and deceleration parameters

In a common notation we shall write the present cosmic time, or epoch, as t_0. Thus photons received today from distant galaxies are received at t_0. If the emitting galaxy is nearby and emits a photon at cosmic time t, we can write $t = t_0 - \delta t$, where $\delta t \ll t_0$. Thus, let us expand the scale factor $R(t)$ as a power series about the present epoch t_0 to obtain

$$\begin{aligned} R(t) &= R[t_0 - (t_0 - t)] \\ &= R(t_0) - (t_0 - t)\dot{R}(t_0) + \tfrac{1}{2}(t_0 - t)^2 \ddot{R}(t_0) - \cdots \\ &= R(t_0)\left[1 - (t_0 - t)H(t_0) - \tfrac{1}{2}(t_0 - t)^2 q(t_0) H^2(t_0) - \cdots\right], \end{aligned} \quad (14.18)$$

where we have introduced the *Hubble parameter* $H(t)$ and the *deceleration parameter* $q(t)$. These are given by

$$\boxed{\begin{aligned} H(t) &\equiv \frac{\dot{R}(t)}{R(t)}, \\ q(t) &\equiv -\frac{\ddot{R}(t)R(t)}{\dot{R}^2(t)}, \end{aligned}} \tag{14.19}$$

where the dot corresponds to differentiation with respect to cosmic time t. It should be noted that these definitions are valid at any cosmic time. The present-day values of these parameters are usually denoted by $H_0 \equiv H(t_0)$ and $q_0 \equiv q(t_0)$.

Using these definitions, we can write the redshift z in terms of the 'look-back time' $t_0 - t$ as

$$z = \frac{R(t_0)}{R(t)} - 1 = \left[1 - (t_0 - t)H_0 - \tfrac{1}{2}(t_0 - t)^2 q_0 H_0^2 - \cdots\right]^{-1} - 1$$

and, assuming that $t_0 - t \ll t_0$, we have

$$z = (t_0 - t)H_0 + (t_0 - t)^2\left(1 + \tfrac{1}{2}q_0\right)H_0^2 + \cdots. \tag{14.20}$$

Since it is the redshift that is an observable quantity, it is more useful to invert the above power series to obtain the look-back time $t_0 - t$ in terms of z. Thus for $z \ll 1$ we have

$$\boxed{t_0 - t = H_0^{-1} z - H_0^{-1}\left(1 + \tfrac{1}{2}q_0\right)z^2 + \cdots.} \tag{14.21}$$

It is worth noting that, as one might expect in this approximation, the relations (14.20) and (14.21) depend only on the present-day values H_0 and q_0 of the Hubble and deceleration parameters and hence may be evaluated without knowledge of the complete expansion history $R(t)$ of the universe.

Using the Taylor expansion (14.18), we can also obtain an approximate expression for the χ-coordinate of the emitting galaxy, which is given by

$$\chi = \int_t^{t_0} \frac{c\,dt}{R(t)} = \int_t^{t_0} cR_0^{-1}[1 - (t_0 - t)H_0 - \cdots]^{-1}\,dt.$$

Assuming once more that $t_0 - t \ll t_0$, we have

$$\chi = cR_0^{-1}\left[(t_0 - t) + \tfrac{1}{2}(t_0 - t)^2 H_0 + \cdots\right]. \tag{14.22}$$

We may now substitute for the look-back time $t_0 - t$ in this result using (14.21), to obtain an expression for the χ-coordinate of the emitting galaxy in terms of its redshift (assuming $z \ll 1$), which reads

$$\boxed{\chi = \frac{c}{R_0 H_0}\left[z - \tfrac{1}{2}(1+q_0)z^2 + \cdots\right].} \tag{14.23}$$

Once again, in this approximation the results (14.22) and (14.23) only depend on the present-day values H_0 and q_0 and may be evaluated without knowing the expansion history of the universe.

From the FRW metric, we see that the proper distance d to the emitting galaxy[3] at cosmic time t_0 is $d = R_0\chi$. Thus, for very nearby galaxies, $d \approx c(t_0 - t)$. Moreover, from (14.20), in this case $z \approx (t_0 - t)H_0$. So, if we were to interpret the cosmological redshift as a Doppler shift due to a recession velocity v of the emitting galaxy, we would obtain

$$\boxed{v = cz = H_0 d,} \tag{14.24}$$

which is approximately valid for small z. The galaxies will therefore appear to recede from us with a recession speed proportional to their distance from us. This is, of course, *Hubble's law*, named after Edwin Hubble, who discovered the expansion of the universe in 1929 by comparing redshifts with distance measurements to nearby galaxies (derived from the period–luminosity relation of Cepheid variables). His results suggested a linear recession law, as in (14.24). This was an amazing result. It implies that the universe started off at high density at some finite time in the past. You will notice from (14.24) that the Hubble 'constant' has the dimensions of inverse time. As we will see later, the quantity $1/H_0$ gives the *age of the universe* to within a factor of order unity. It is clear that, in general, the Hubble parameter will vary with cosmic time t and hence with redshift z. By combining the expressions (14.18), (14.19) and (14.21), we can obtain an expression for how the Hubble parameter varies with z for small redshift,

$$\boxed{H(z) = H_0[1 + (1+q_0)z - \cdots].} \tag{14.25}$$

So far, we have been considering the low-z limit. Having introduced the Hubble parameter, however, we may use it to derive useful *general* expressions for the

[3] In order to measure the proper distance d, one would in fact have to arrange for all the 'civilisations' along the route to the galaxy to lay out measuring rods at the same cosmic time t_0. This could be synchronised by, for example, requiring the temperature of the cosmic microwave background or the mean matter density of the universe to have a given value. We will discuss more practical measures of distance shortly.

look-back time to an emitting galaxy, and for its χ-coordinate, as functions of the redshift z of the received photon. In general, we have

$$dz = d(1+z) = d\left(\frac{R_0}{R}\right) = -\frac{R_0}{R^2}\dot{R}\,dt = -(1+z)H(z)\,dt,$$

which provides a very useful relation between an interval dz in redshift and the corresponding interval dt in cosmic time. Thus, we can write the look-back time as

$$\boxed{t_0 - t = \int_t^{t_0} dt = \int_0^z \frac{dz}{(1+z)H(z)},} \tag{14.26}$$

and the galaxy's χ-coordinate is given by

$$\boxed{\chi = \int_t^{t_0} \frac{c\,dt}{R(t)} = \frac{c}{R_0}\int_0^z \frac{dz}{H(z)}.} \tag{14.27}$$

It is clear, however, that in order to evaluate either of these integrals we must know how $H(z)$ varies with z, which requires knowledge of the evolution of the scale factor $R(t)$.

14.11 Distances in the FRW geometry

Distance measures in an expanding universe can be confusing. For example, let us consider the distance to some remote galaxy. The light received from the galaxy was emitted when the universe was younger, because light travels at a finite speed c. Evidently, as we look at more distant objects, we see them as they were at an earlier time in the universe's history *when proper distances were smaller*, since the universe is expanding. What, therefore, do we mean by the 'distance' to a galaxy? In fact, interpreting and calculating distances in an expanding universe is straightforward, but one must be clear about what is meant by 'distance'.

From the FRW metric

$$ds^2 = c^2 dt^2 - R^2(t)\left[d\chi^2 + S^2(\chi)(d\theta^2 + \sin^2\theta\, d\phi^2)\right],$$

we can define a number of different measures of distance. The parameter χ is a comoving coordinate that is sometimes referred to as the *coordinate distance*, whereas the *proper distance* to an object at some cosmic time t is $d = R(t)\chi$, but this cannot be measured in practice. Thus, we must look for alternative ways of defining the distance to an object. The two most important *operationally defined* distance measures are the *luminosity distance* and the *angular diameter distance*. These distance measures form the basis for observational tests of the geometry of the universe.

Luminosity distance

In an ordinary static Euclidean universe, if a source of *absolute luminosity* L (measured in $\mathrm{W} = \mathrm{J\,s^{-1}}$) is at a distance d then the *flux* that we receive (measured in $\mathrm{W\,m^{-2}}$) is $F = L/(4\pi d^2)$. Now suppose that we are actually in an expanding FRW geometry, but we know that the source has a luminosity L and we observe a flux F. The quantity

$$\boxed{d_{\mathrm{L}} = \left(\frac{L}{4\pi F}\right)^{1/2},} \tag{14.28}$$

is called the *luminosity distance* of the source. This is an *operational definition*, and we must now investigate how to express it in terms of the FRW metric.

Consider an emitting source E with a fixed comoving coordinate χ relative to an observer O (note that, by symmetry, the emitter would assign the *same* value of χ to the observer). We assume that the absolute luminosity of E as a function of cosmic time is $L(t)$ and that the photons it emits are detected by O at cosmic time t_0. Clearly, the photons must have been emitted at an earlier time t_e. Assuming the photons to have been emitted isotropically, the radiation will be spread evenly over a sphere centred on E and passing through O (see Figure 14.2). The proper area of this sphere is

$$A = 4\pi R^2(t_0) S^2(\chi).$$

However, each photon received by O is redshifted in frequency, so that

$$\nu_0 = \frac{\nu_e}{1+z},$$

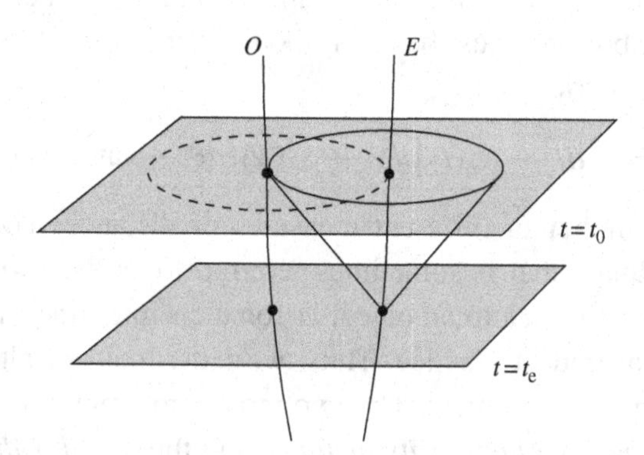

Figure 14.2 Geometry associated with the definition of luminosity distance (with one spatial dimension suppressed).

and, moreover, the arrival rate of the photons is also reduced by the same factor. Thus, the *observed* flux at O is

$$F(t_0) = \frac{L(t_e)}{4\pi[R_0 S(\chi)]^2} \frac{1}{(1+z)^2}.$$

The luminosity distance defined above is now evaluated as

$$\boxed{d_L = R_0 S(\chi)(1+z).} \tag{14.29}$$

This is an important quantity, which can be used practically, but note that it depends on the time history of the scale factor through the dependence on χ.

Angular diameter distance

Another important distance measure is based upon the notion of the existence of some standard-length 'rods', whose angular diameter we can observe. Suppose that a source has proper diameter ℓ. Then, in Euclidean space, if it were at a distance d it would subtend an angular diameter $\Delta\theta = D/d$. In an FRW geometry, we thus define the *angular diameter distance* to an object to be

$$\boxed{d_A = \frac{\ell}{\Delta\theta}.} \tag{14.30}$$

This is again an operational definition, and we now investigate how to express it in terms of the FRW metric.

Suppose we have two radial null geodesics (light paths) meeting at the observer at time t_0 with an angular separation $\Delta\theta$, having been emitted at time t_e from a source of proper diameter ℓ at a fixed comoving coordinate χ (assuming, for simplicity, that the spatial axes are oriented so that $\phi = $ constant along the photon paths); see Figure 14.3. To obtain a clearer view of the specification of the coordinates, we may look vertically down the worldline of O and define the coordinates as in Figure 14.4. From the angular part of the FRW metric we have

$$\ell = R(t_e) S(\chi) \Delta\theta,$$

so that

$$d_A = R(t_e) S(\chi) = R(t_0) \frac{R(t_e)}{R(t_0)} S(\chi) = \frac{R(t_0) S(\chi)}{1+z}.$$

Thus the angular diameter distance is given by

$$\boxed{d_A = \frac{R_0 S(\chi)}{1+z}.} \tag{14.31}$$

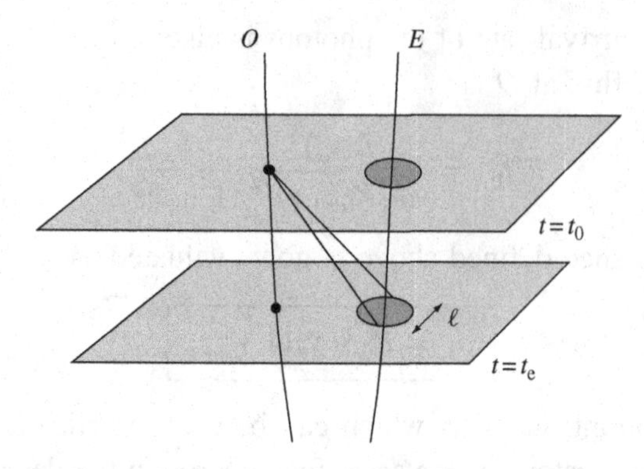

Figure 14.3 Geometry associated with the definition of angular diameter distance (with one spatial dimension suppressed).

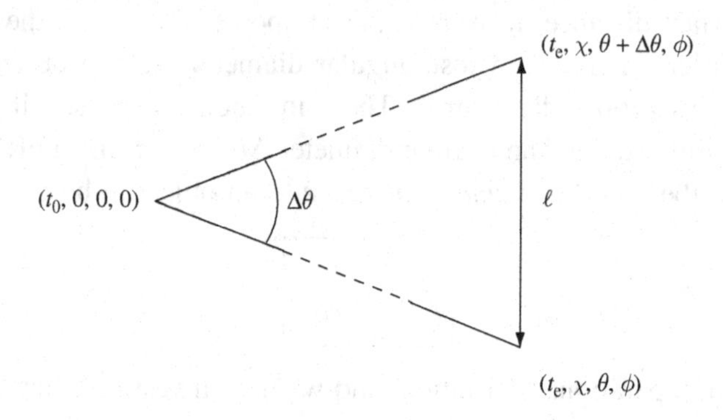

Figure 14.4 Specification of the coordinates in the definition of angular diameter distance.

This differs from the luminosity distance d_L by a factor $(1+z)^2$, emphasizing again that 'distance' depends on definition. Again, because of the χ-dependence we need to know the time history of the scale factor $R(t)$ to evaluate d_A.

14.12 Volumes and number densities in the FRW geometry

The interpretation of cosmological observations often requires one to determine the volume of some three-dimensional region of the FRW geometry. Consider a comoving cosmological observer, whom we may take to be at the origin $\chi = 0$ of our comoving coordinate system. From the FRW metric

$$ds^2 = c^2\,dt^2 - R^2(t)\left[d\chi^2 + S^2(\chi)(d\theta^2 + \sin^2\theta\,d\phi^2)\right],$$

we see that, at cosmic time t_0, the proper volume of the region of space lying in the infinitesmial coordinate range $\chi \to \chi + d\chi$ and subtending an infinitesmial solid angle $d\Omega = \sin\theta\, d\theta\, d\phi$ at the observer is

$$dV_0 = (R_0\, d\chi)\left[R_0^2 S^2(\chi)\, d\Omega\right] = R_0^3 S^2(\chi)\, d\chi\, d\Omega.$$

For the interval $\chi \to \chi + d\chi$ in the radial comoving coordinate there exists a corresponding interval $z \to z + dz$ in the redshift of objects lying in this radial range (and also a corresponding cosmic time interval $t \to t + dt$ within which the light observed by O at $t = t_0$ was emitted). We may therefore write the volume element as

$$dV_0 = R_0^3 S^2(\chi) \frac{d\chi}{dz}\, dz\, d\Omega.$$

From (14.27), however, we have

$$\frac{d\chi}{dz} = \frac{c}{R_0 H(z)},$$

and so

$$dV_0 = \frac{c R_0^2 S^2(\chi(z))}{H(z)}\, dz\, d\Omega,$$

where we have made explicit that χ is also a function of z. This volume element is illustrated in Figure 14.5. For an expanding universe, the proper volume of this

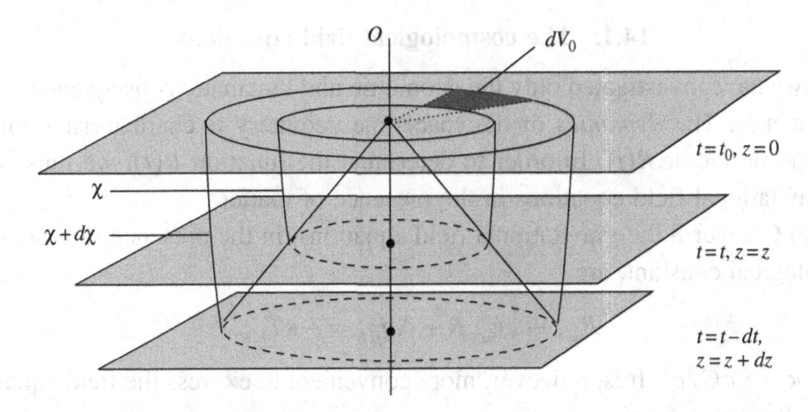

Figure 14.5 Geometry associated with the definition of a proper volume element dV_0 at cosmic time $t = t_0$ (with one spatial dimension suppressed).

comoving region will be smaller at some earlier cosmic time t (which corresponds to some redshift z). Indeed, using (14.17), we have

$$dV(z) = \frac{dV_0}{(1+z)^3} = \frac{cR_0^2 S^2(\chi(z))}{(1+z)^3 H(z)}\, dz\, d\Omega. \tag{14.32}$$

The main use of the result (14.32) is in predicting the number of galaxies (of a certain type) that one would expect to observe in a given area of sky and redshift interval, and comparing that result with observations. Suppose, for example, that the proper number density of galaxies of a certain type at a redshift z is given by $n(z)$. Using (14.32), the total number dN of such objects in the redshift interval $z \to z + dz$ and in a solid angle $d\Omega$ is

$$dN = n(z)\, dV(z) = \frac{cR_0^2 S^2(\chi(z))}{H(z)} \frac{n(z)}{(1+z)^3}\, dz\, d\Omega. \tag{14.33}$$

The above expression has been arranged to make use of the fact that, if objects are conserved (so that, once formed, galaxies are not later destroyed), we may write $n(z)/(1+z)^3 = n_0$, where n_0 is the present-day proper number density of such objects; hence the resulting expression is simplified somewhat. As an illustration, let us consider a population of galaxies which are formed instantaneously at a redshift $z = z_{\rm f}$, which are not later destroyed and which have a present-day number density n_0. From (14.33), the total number of such objects in the whole sky is

$$N = 4\pi c n_0 R_0^2 \int_0^{z_{\rm f}} \frac{S^2(\chi(z))}{H(z)}\, dz.$$

Clearly, in order to evaluate this integral one requires knowledge of the expansion history $R(t)$ of the universe.

14.13 The cosmological field equations

So far we have investigated only the geometric and kinematic consequences of the FRW metric. The *dynamics* of the spacetime geometry is characterised entirely by the scale factor $R(t)$. In order to determine the function $R(t)$, we must solve the gravitational field equations in the presence of matter.

From Chapter 8 the gravitational field equations, in the presence of a non-zero cosmological constant, are

$$R_{\mu\nu} - \tfrac{1}{2} g_{\mu\nu} R + \Lambda g_{\mu\nu} = -\kappa T_{\mu\nu},$$

where $\kappa = 8\pi G/c^4$. It is, however, more convenient to express the field equations in the alternative form

$$R_{\mu\nu} = -\kappa (T_{\mu\nu} - \tfrac{1}{2} T g_{\mu\nu}) + \Lambda g_{\mu\nu}, \tag{14.34}$$

where $T = T^{\mu}_{\mu}$. In order to solve these equations, we clearly need a model for the energy–momentum tensor of the matter that fills the universe. For simplicity, we shall grossly idealise the universe and model the matter by a simple macroscopic fluid, devoid of shear-viscous, bulk-viscous and heat-conductive properties. Thus we assume a *perfect fluid*, which is characterised at each point by its proper density ρ and the pressure p in the instantaneous rest frame. The energy–momentum tensor is given by

$$T^{\mu\nu} = \left(\rho + \frac{p}{c^2}\right) u^{\mu} u^{\nu} - p g^{\mu\nu}. \tag{14.35}$$

Since we are seeking solutions for a homogeneous and isotropic universe, the density ρ and pressure p *must be functions of cosmic time t alone.*

We may perform the calculation in any coordinate system, but the algebra is simplified slightly by adopting the comoving coordinates $[x^{\mu}] = (t, r, \theta, \phi)$, in which the FRW metric takes the form

$$ds^2 = c^2\, dt^2 - R^2(t)\left[\frac{dr^2}{1-kr^2} + r^2(d\theta^2 + \sin^2\theta\, d\phi^2)\right].$$

Thus the covariant components $g_{\mu\nu}$ of the metric are

$$g_{00} = c^2, \qquad g_{11} = -\frac{R^2(t)}{1-kr^2}, \qquad g_{22} = -R^2(t) r^2, \qquad g_{33} = -R^2(t) r^2 \sin^2\theta.$$

Since the metric is diagonal, the contravariant components $g^{\mu\nu}$ are simply the reciprocals of the covariant components.

The connection is given in terms of the metric by

$$\Gamma^{\sigma}{}_{\mu\nu} = \tfrac{1}{2} g^{\sigma\rho}(\partial_{\nu} g_{\rho\mu} + \partial_{\mu} g_{\rho\nu} - \partial_{\rho} g_{\mu\nu}),$$

from which it is straightforward to show that the only non-zero coefficients are

$$\begin{aligned}
&\Gamma^{0}{}_{11} = R\dot{R}/[c^2(1-kr^2)], && \Gamma^{0}{}_{22} = R\dot{R}r^2/c^2, && \Gamma^{0}{}_{33} = (R\dot{R}r^2\sin^2\theta)/c^2,\\
&\Gamma^{1}{}_{01} = \dot{R}/R, && \Gamma^{1}{}_{11} = kr/(1-kr^2), && \Gamma^{1}{}_{22} = -r(1-kr^2),\\
&\Gamma^{1}{}_{33} = -r(1-kr^2)\sin^2\theta, && && \\
&\Gamma^{2}{}_{02} = \dot{R}/R, && \Gamma^{2}{}_{12} = 1/r, && \Gamma^{2}{}_{33} = -\sin\theta\cos\theta,\\
&\Gamma^{3}{}_{03} = \dot{R}/R, && \Gamma^{3}{}_{13} = 1/r, && \Gamma^{3}{}_{23} = \cot\theta,
\end{aligned}$$

where the dots denote differentiation with respect to cosmic time t. We next substitute these expressions for the connection coefficients into the expression for the Ricci tensor,

$$R_{\mu\nu} = \partial_{\nu}\Gamma^{\sigma}{}_{\mu\sigma} - \partial_{\sigma}\Gamma^{\sigma}{}_{\mu\nu} + \Gamma^{\rho}{}_{\mu\sigma}\Gamma^{\sigma}{}_{\rho\nu} - \Gamma^{\rho}{}_{\mu\nu}\Gamma^{\sigma}{}_{\rho\sigma}.$$

After some tedious but straightforward algebra, we find that the off-diagonal components of the Ricci tensor are zero and the diagonal components are given by

$$\begin{aligned}
R_{00} &= 3\ddot{R}/R, \\
R_{11} &= -(R\ddot{R}+2\dot{R}^2+2c^2k)c^{-2}/(1-kr^2), \\
R_{22} &= -(R\ddot{R}+2\dot{R}^2+2c^2k)c^{-2}r^2, \\
R_{33} &= -(R\ddot{R}+2\dot{R}^2+2c^2k)c^{-2}r^2\sin^2\theta.
\end{aligned}$$

We must now turn our attention to the right-hand side of the field equations (14.34). In our comoving coordinate system (t, r, θ, ϕ), the 4-velocity of the fluid is simply

$$[u^\mu] = (1, 0, 0, 0),$$

which we can write as $u^\mu = \delta^\mu_0$. Thus the covariant components of the 4-velocity are

$$u_\mu = g_{\mu\nu}\delta^\nu_0 = g_{\mu 0} = c^2\delta^0_\mu,$$

so we can write the energy–momentum tensor (14.35) as

$$T_{\mu\nu} = (\rho c^2 + p)c^2\delta^0_\mu\delta^0_\nu - pg_{\mu\nu}.$$

Moreover, since $u^\mu u_\mu = c^2$, contraction of the energy–momentum tensor gives

$$T = T^\mu_\mu = \left(\rho + \frac{p}{c^2}\right)c^2 - p\delta^\mu_\mu = \rho c^2 - 3p.$$

Hence we can write the terms on the right-hand side of the field equations (14.34) that depend on the energy–momentum as

$$T_{\mu\nu} - \tfrac{1}{2}Tg_{\mu\nu} = (\rho c^2 + p)c^2\delta^0_\mu\delta^0_\nu - \tfrac{1}{2}(\rho c^2 - p)g_{\mu\nu}.$$

Including the cosmological-constant term, we find that the right-hand side of the field equations (14.34) vanishes for $\mu \neq \nu$. The non-zero components read

$$\begin{aligned}
-\kappa(T_{00} - \tfrac{1}{2}Tg_{00}) + \Lambda g_{00} &= -\tfrac{1}{2}\kappa(\rho c^2 + 3p)c^2 + \Lambda c^2, \\
-\kappa(T_{11} - \tfrac{1}{2}Tg_{11}) + \Lambda g_{11} &= -\left[\tfrac{1}{2}\kappa(\rho c^2 - p) + \Lambda\right]R^2/(1-kr^2), \\
-\kappa(T_{22} - \tfrac{1}{2}Tg_{22}) + \Lambda g_{22} &= -\left[\tfrac{1}{2}\kappa(\rho c^2 - p) + \Lambda\right]R^2r^2, \\
-\kappa(T_{33} - \tfrac{1}{2}Tg_{33}) + \Lambda g_{33} &= -\left[\tfrac{1}{2}\kappa(\rho c^2 - p) + \Lambda\right]R^2r^2\sin^2\theta.
\end{aligned}$$

Combining these expressions with those for the components of the Ricci tensor, we see that the three spatial field equations are equivalent, which is essentially due to the homogeneity and isotropy of the FRW metric. Thus the gravitational field equations yield just the two independent equations,

$$3\ddot{R}/R = -\tfrac{1}{2}\kappa(\rho c^2+3p)c^2+\Lambda c^2,$$
$$R\ddot{R}+2\dot{R}^2+2c^2k = \left[\tfrac{1}{2}\kappa(\rho c^2-p)+\Lambda\right]c^2R^2.$$

Eliminating $\ddot{R}$ from the second equation and remembering that $\kappa = 8\pi G/c^4$, we finally arrive at the *cosmological field equations*

$$\boxed{\begin{aligned}\ddot{R} &= -\frac{4\pi G}{3}\left(\rho+\frac{3p}{c^2}\right)R+\tfrac{1}{3}\Lambda c^2 R,\\ \dot{R}^2 &= \frac{8\pi G}{3}\rho R^2+\tfrac{1}{3}\Lambda c^2R^2-c^2k.\end{aligned}} \tag{14.36}$$

These two differential equations determine the time evolution of the scale factor $R(t)$ and are known as the *Friedmann–Lemaître equations*. In the case $\Lambda = 0$ they are often called simply the *Friedmann equations*. We will discuss the solutions to these equations in various cases in Chapter 15.

14.14 Equation of motion for the cosmological fluid

For any particular model of the universe, the two cosmological field equations (14.36) are sufficient to determine $R(t)$. Nevertheless, we can derive one further important equation (which is often useful in shortening calculations) from the fact that energy–momentum conservation requires

$$\nabla_\mu T^{\mu\nu} = 0.$$

From our discussion of a perfect fluid in Chapter 8, we know that this requirement leads to the relativistic equations of continuity and motion for the cosmological fluid. These equations read

$$\nabla_\mu(\rho u^\mu)+\frac{p}{c^2}\nabla_\mu u^\mu = 0, \tag{14.37}$$

$$\left(\rho+\frac{p}{c^2}\right)u^\mu\nabla_\mu u^\nu = \left(g^{\mu\nu}-\frac{u^\mu u^\nu}{c^2}\right)\nabla_\mu p. \tag{14.38}$$

The second equation is easily shown to be satisfied identically, since both sides equal zero. This confirms that the fluid particles (galaxies) follow geodesics, which was to be expected since p is a function of t alone, and so there is no pressure gradient to push them off geodesics. The continuity equation (14.37) can be written

$$(\partial_\mu\rho)u^\mu+\left(\rho+\frac{p}{c^2}\right)(\partial_\mu u^\mu+\Gamma^\mu{}_{\nu\mu}u^\nu) = 0.$$

Remembering that ρ is a function of t alone, and with $u^\mu = \delta^\mu_0$, this reduces to

$$\boxed{\dot{\rho}+\left(\rho+\frac{p}{c^2}\right)\frac{3\dot{R}}{R}=0,} \tag{14.39}$$

which expresses energy conservation. This equation can in fact be derived directly from the field equations (14.36) by eliminating $\ddot{R}$. Thus, only two of the three equations (14.36) and (14.39) are independent. One may use whichever two equations are most convenient in any particular calculation.

Equation (14.39) can be simply rearranged into the useful alternative form

$$\frac{d(\rho R^3)}{dt} = -\frac{3p\dot{R}R^2}{c^2}. \tag{14.40}$$

Moreover, by transforming the derivative with respect to t to a derivative with respect to R, one obtains a third useful form of the equation, namely

$$\frac{d(\rho R^3)}{dR} = -\frac{3pR^2}{c^2}. \tag{14.41}$$

Finally, we note that the density and pressure of a fluid are related by its equation of state. In cosmology, it is usual to assume that (each component of) the cosmological fluid has an equation of state of the form

$$p = w\rho c^2,$$

where the *equation-of-state parameter* w is a constant (in the more exotic cosmological models one sometimes allows w to be a function of cosmic time t, but we shall not consider such models here). The energy equation (14.41) can then be written

$$\frac{d(\rho R^3)}{dR} = -3w\rho R^2.$$

This equation has the immediate solution

$$\boxed{\rho \propto R^{-3(1+w)},} \tag{14.42}$$

which gives the evolution of the density ρ as a function of the scale factor $R(t)$. Note that in general ρc^2 is the energy density of the fluid. In particular $w=0$ for pressureless 'dust', $w=\frac{1}{3}$ for radiation and $w=-1$ for the vacuum (if the cosmological constant $\Lambda \neq 0$; see Section 8.7).

14.15 Multiple-component cosmological fluid

Suppose that the cosmological fluid in fact consists of several distinct components (for example, matter, radiation and the vacuum) that do not interact except through their mutual gravitation. Let us suppose further that each component can be modelled as a perfect fluid, as discussed above.

The energy–momentum tensor of a multiple-component fluid is given simply by

$$T^{\mu\nu} = \sum_i (T^{\mu\nu})_i,$$

where i labels the various fluid components. Since each component is modelled as a perfect fluid, we have

$$\begin{aligned} T^{\mu\nu} &= \sum_i \left[\left(\rho_i + \frac{p_i}{c^2}\right) u^\mu u^\nu - p_i g^{\mu\nu}\right] \\ &= \sum_i \left(\rho_i + \frac{p_i}{c^2}\right) u^\mu u^\nu - (\sum_i p_i) g_{\mu\nu}. \end{aligned}$$

Thus, the multicomponent fluid can itself be modelled as a single perfect fluid with

$$\boxed{\rho = \sum_i \rho_i \qquad \text{and} \qquad p = \sum_i p_i,} \tag{14.43}$$

which can be substituted directly into our cosmological field equations (14.36).[4]

Moreover, since we are assuming that the fluid components are non-interacting, conservation of energy and momentum requires that the condition

$$\nabla_\mu (T^{\mu\nu})_i = 0$$

holds separately for each component. Then each fluid will obey an energy equation of the form (14.39). Thus, if $w_i = p_i/(\rho_i c^2)$ then the density of each fluid evolves independently of the other components as

$$\boxed{\rho_i \propto R^{-3(1+w_i)}.} \tag{14.44}$$

Exercises

14.1 In an N-dimensional manifold, consider the tensor

$$R_{ijkl} = K(g_{ik} g_{jl} - g_{il} g_{jk}),$$

where K may be a function of position. Show that this tensor satisfies the symmetry properties and the cyclic identity of the curvature tensor. Show that, in order to satisfy the Bianchi identity, one requires K to be constant if $N > 2$.

[4] Unfortunately, if the individual equation-of-state parameters w_i are constants one cannot, in general, define a single effective equation-of-state parameter $w = p/(\rho c^2)$ that is also independent of cosmic time t.

14.2 For a 3-space with a line element of the form

$$d\sigma^2 = B(r)\,dr^2 + r^2 d\theta^2 + r^2 \sin^2\theta\, d\phi^2,$$

show that the non-zero components of the Ricci tensor are

$$R_{rr} = -\frac{1}{rB}\frac{dB}{dr}, \qquad R_{\theta\theta} = \frac{1}{B} - 1 - \frac{r}{2B^2}\frac{dB}{dr}, \qquad R_{\phi\phi} = R_{\theta\theta}\sin^2\theta.$$

Hence show that if the 3-space is maximally symmetric then $B(r)$ must take the form

$$B(r) = \frac{1}{A - Kr^2},$$

where A and K are constants.

14.3 In a four-dimensional Euclidean space with 'Cartesian' coordinates (w, x, y, z), a 3-sphere of radius R is defined by $w^2 + x^2 + y^2 + z^2 = R^2$. Show that the metric on the surface of the 3-sphere can be written in the form

$$d\sigma^2 = R^2\left[d\chi^2 + \sin^2\chi(d\theta^2 + \sin^2\theta\, d\phi^2)\right].$$

Show that the total volume of the 3-sphere is $V = 2\pi^2 R^3$.

14.4 In a four-dimensional Minkowski space with 'Cartesian' coordinates (w, x, y, z), a 3-hyperboloid is defined by $w^2 - x^2 - y^2 + -z^2 = R^2$. Show that the metric on the surface of the 3-hyperboloid can be written in the form

$$d\sigma^2 = R^2\left[d\chi^2 + \sinh^2\chi\,(d\theta^2 + \sin^2\theta\, d\phi^2)\right].$$

Show that the total volume of the 3-hyperboloid is infinite.

14.5 At cosmic time t_1, a massive particle is shot out into an expanding FRW universe with velocity v_1 relative to comoving cosmological observers. At a later cosmic time t_2 the particle has a velocity v_2 with respect to comoving cosmological observers. Show that, at any intermediate cosmic time t, the velocity of the particle as measured by a comoving cosmological observer is

$$v(t) = R(t)\frac{d\chi}{dt}.$$

Hence show that

$$\frac{\gamma_{v_2} v_2}{\gamma_{v_1} v_1} = \frac{R(t_1)}{R(t_2)},$$

where $\gamma_v = (1 - v^2/c^2)^{-1/2}$ and $R(t)$ is the scale factor at cosmic time t. By considering the particle momentum, show that as $v_1 \to c$ the photon redshift formula is recovered.

14.6 In the limit $z \ll 1$, show that the look-back time for a galaxy with redshift z is

$$t_0 - t = H_0^{-1} z - H_0^{-1}\left(1 + \tfrac{1}{2}q_0\right) z^2 + \cdots .$$

Show also that, in this limit, the variation of the Hubble parameter with redshift is given by

$$H(z) = H_0[1 + (1 + q_0)z - \cdots].$$

14.7 In a spatially flat FRW geometry, show that the luminosity and angular diameter distances to an object of redshift z are given, in the limit $z \ll 1$, by

$$d_L = \frac{c}{H_0}\left[z + \tfrac{1}{2}(1 - q_0)z^2 + \cdots\right],$$

$$d_A = \frac{c}{H_0}\left[z - \tfrac{1}{2}(3 + q_0)z^2 + \cdots\right].$$

Hence show that the angular diameter of a standard object can increase as z increases. Do these results still hold in a spatially curved FRW geometry?

14.8 In the FRW geometry, show that the look-back time to a nearby object at proper distance d is

$$t_0 - t = \frac{d}{c} - \frac{H_0 d^2}{2c^2} + \cdots.$$

Hence show that the redshift to the object is

$$z = \frac{H_0 d}{c} + \frac{1 + q_0}{2}\frac{H_0^2 d^2}{c^2} + \cdots.$$

14.9 The observed flux in the frequency range $[\nu_1, \nu_2]$ received from some distant comoving object is given by

$$F_{\text{obs}}(\nu_1, \nu_2) = \int_{\nu_1}^{\nu_2} f_{\text{obs}}(\nu)\, d\nu,$$

where $f_{\text{obs}}(\nu)$ is the observed flux density (in $\text{W m}^{-2}\,\text{Hz}^{-1}$) as a function of frequency. If $f_{\text{em}}(\nu)$ is the emitted (or intrinsic) flux density of the object, show that

$$f_{\text{obs}}(\nu) = \frac{f_{\text{em}}((1 + z)\nu)}{1 + z},$$

where z is the redshift of the object. If $f_{\text{em}}(\nu) \propto \nu^{\alpha}$ over a wide range of frequencies, show that

$$F_{\text{obs}}(\nu_1, \nu_2) = K_z\, F_{\text{em}}(\nu_1, \nu_2),$$

where the *K-correction* is given by $K_z = (1 + z)^{\alpha - 1}$.

14.10 The observed surface brightness Σ_{obs} of an extended object observed in the frequency range $[\nu_1, \nu_2]$ is defined as the observed flux per unit solid angle. Thus, for a (small) circular object subtending an angular diameter $\Delta\theta$ we have

$$\Sigma_{\text{obs}} = \frac{4F_{\text{obs}}(\nu_1, \nu_2)}{\pi(\Delta\theta)^2},$$

where $F_{\rm obs}(\nu_1, \nu_2)$ is defined in Exercise 14.9. Show that $\Sigma_{\rm obs}$ can be written as

$$\Sigma_{\rm obs} = \frac{4}{\pi} \frac{L_{\rm em}(\nu_1, \nu_2)}{4\pi\ell^2} \frac{K_z}{(1+z)^4},$$

where ℓ is the physical (projected) diameter of the object, $L_{\rm em}(\nu_1, \nu_2)$ is the intrinsic luminosity of the object in the frequency range $[\nu_1, \nu_2]$ and K_z is the K-correction.

Note: The above result is independent of cosmological parameters. Moreover, setting aside the K-correction, the $(1+z)^{-4}$-dependence means that the surface brightness of extended objects drops very rapidly with redshift, making the detection of high-z objects difficult.

14.11 A commonly used distance measure in cosmology is the proper-motion distance $d_{\rm M}$ defined by

$$d_{\rm M} = \frac{v}{\dot{\theta}},$$

where v is the proper transverse velocity of (some part of) the object, which is assumed known from astrophysics, and $\dot{\theta}$ is the corresponding observed angular velocity. Show that

$$d_{\rm M} = (1+z)d_{\rm A} = \frac{d_{\rm L}}{(1+z)},$$

where $d_{\rm A}$ and $d_{\rm L}$ are the angular-diameter distance and the luminosity distance to the object respectively.

14.12 A certain population of galaxies undergoes a short ultra-luminous phase at redshift $z = z_*$ that lasts for a proper time interval Δt. After this phase, such galaxies are neither created or destroyed. If $z_* \ll 1$, show that for a spatially flat universe the total number of such galaxies in the sky that are in this phase is given by

$$N = \frac{4\pi c^3 n_0 \Delta t}{H_0^2} \left[z_* + \tfrac{1}{2}(1-q_0)z_*^2 + \cdots \right],$$

where n_0 is the present-day proper number density of these galaxies.

14.13 In the comoving coordinates $[x^\mu] = (t, r, \theta, \phi)$, the FRW metric takes the form

$$ds^2 = c^2\,dt^2 - R^2(t)\left[\frac{dr^2}{1-kr^2} + r^2(d\theta^2 + \sin^2\theta\,d\phi^2)\right].$$

Using the 'Lagrangian' method, or otherwise, calculate the corresponding connection coefficients $\Gamma^\sigma{}_{\mu\nu}$. Hence calculate the non-zero elements of the Ricci tensor $R_{\mu\nu}$.

14.14 In Newtonian cosmology, the universe is modelled as an infinite gas of density $\rho(t)$ that is expanding in such a way that the relative recessional velocity of any two gas particles is $v(t) = H(t)R(t)$, where $H(t) = \dot{R}(t)/R(t)$ and $R(t)$ is the separation of the particles at time t. Use Gauss' law to determine the force on a particle of mass m on the edge of an arbitrary spherical region, and hence show that

$$\ddot{R} = -\frac{4\pi G}{3}\rho R.$$

By considering the total energy E of the above particle, show further that

$$\dot{R}^2 = \frac{8\pi G}{3}\rho R^2 - c^2 k,$$

where the constant $k = -2E/(mc^2)$. Compare these Newtonian cosmological field equations with their general-relativistic counterparts.

14.15 Show that the relativistic equation of motion for the cosmological fluid is satisfied identically and that the relativistic equation of continuity takes the form

$$\dot{\rho} + \left(\rho + \frac{p}{c^2}\right)\frac{3\dot{R}}{R} = 0.$$

Show further that this equation may also be written in the forms

$$\frac{d(\rho R^3)}{dt} = -\frac{3p\dot{R}R^2}{c^2} \quad \text{and} \quad \frac{d(\rho R^3)}{dR} = -\frac{3pR^2}{c^2}.$$

14.16 Use the cosmological field equations directly to derive the relativistic equation of continuity for the cosmological fluid given in Exercise 14.15.

14.17 Consider a spherical comoving volume of the cosmological fluid whose surface is defined by $\chi = \text{constant}$. As the universe expands show that, for the infinitesimal time interval $t \to t + dt$, the conservation of energy requires that

$$c^2\rho V = c^2(\rho + d\rho)(V + dV) + p\,dV,$$

where ρc^2 and p are the energy density and pressure of the fluid respectively. Hence show that

$$\frac{d\rho}{dR} = -3(1+w)\frac{\rho}{R},$$

where $w = p/(\rho c^2)$ and $R(t)$ is the scale factor of the universe. Show that this equation has the solution $\rho \propto R^{-3(1+w)}$.

15
Cosmological models

In the previous chapter, we considered the geometric and kinematic properties of the Friedmann–Robertson–Walker (FRW) metric and derived the *cosmological field equations* for the scale factor $R(t)$. In this chapter, we will use the cosmological field equations to determine the behaviour of the scale factor as a function of cosmic time in various *cosmological models.*

15.1 Components of the cosmological fluid

In a general cosmological model, the universe is assumed to contain both matter and radiation. In addition, the cosmological constant Λ is generally assumed to be non-zero. As discussed in Section 8.7, the modern interpretation of Λ is in terms of the energy density of the vacuum, which may also be modelled as a perfect fluid (with a peculiar equation of state). Thus, one usually adopts the viewpoint that the cosmological fluid consists of *three* components, namely matter, radiation and the vacuum, each with a different equation of state. The total equivalent mass density is simply the sum of the individual contributions,

$$\rho(t) = \rho_{\mathrm{m}}(t) + \rho_{\mathrm{r}}(t) + \rho_{\Lambda}(t), \tag{15.1}$$

where t is the cosmic time and we have adopted the commonly used cosmological notation for the equivalent mass densities of matter, radiation and the vacuum respectively. Moreover, we shall assume that these three components are non-interacting (see Section 14.15); although matter and radiation did interact in the early universe, this is a reasonable approximation for most of its history.

As mentioned in Section 14.12, each component of the cosmological fluid is modelled as a perfect fluid with an equation of state of the form

$$p_i = w_i \rho_i c^2,$$

where the equation-of-state parameter w_i is a constant (and i labels the component). In particular $w_i = 0$ for pressureless 'dust', $w_i = \frac{1}{3}$ for radiation and $w_i = -1$ for the vacuum. In general, if w_i is a constant then, requiring that the weak energy condition is satisfied (see Exercise 8.8) and that the local sound speed $(dp/d\rho)^{1/2}$ is less than c, one finds that w_i must lie in the range $-1 \leq w_i \leq 1$. We see that this is indeed the case for dust, radiation and the vacuum. We now discuss each of these components in turn and conclude with a description of their relative contributions to the total density as the universe evolves.

Matter

In general, matter in the universe may come in several different forms. In addition to the normal baryonic matter of everyday experience (such as protons and neutrons), the universe may well contain more exotic forms of matter consisting of fundamental particles that lie beyond the 'Standard Model' of particle physics. Indeed, observations of the large-scale structure in the universe suggest that most of the matter is in the form of non-baryonic *dark matter*, which interacts electromagnetically only very weakly (and is hence invisible or 'dark'). Moreover, dark matter may itself come in different forms, such as *cold dark matter* (CDM) and *hot dark matter* (HDM), the naming of which is connected to whether the typical energy of the particles is non-relativistic or relativistic. We shall not pursue this very interesting subject any further here[1] but merely note that the total matter density (at any particular cosmic time t) may be expressed as the sum of the baryonic and dark matter contributions,

$$\rho_{\mathrm{m}}(t) = \rho_{\mathrm{b}}(t) + \rho_{\mathrm{dm}}(t).$$

In the following discussion, we will not differentiate between different types of matter, since it is only the total matter density that determines how the scale factor $R(t)$ evolves with cosmic time t. We shall also make the common assumption that the matter particles (in whatever form) have a thermal energy that is much less than their rest mass energy, and so the matter can be considered to be pressureless, i.e. *dust*. In this case the equation of state parameter is simply $w = 0$. Thus, from (14.44), if the matter has a present-day proper density of $\rho_{\mathrm{m}}(t_0) \equiv \rho_{\mathrm{m},0}$, its density at some other cosmic time t is given by

$$\boxed{\rho_{\mathrm{m}}(t) = \rho_{\mathrm{m},0}\left[\frac{R_0}{R(t)}\right]^3 \qquad \text{or} \qquad \rho_{\mathrm{m}}(z) = \rho_{\mathrm{m},0}(1+z)^3,}$$

[1] For a full discussion, see (for example) T. Padmanabhan, *Structure Formation in the Universe*, Cambridge University Press, 1993.

where in the second expression we have used (14.17) to write the density in terms of the redshift z. These expressions concur with our expectation for the behaviour of the space density of dust particles in an expanding universe.

Radiation

The term radiation naturally includes photons, but also other species with very small or zero rest masses, such that they move relativistically today. An example of the latter is neutrinos (which may in fact have a small non-zero rest mass). The total equivalent mass density of radiation in the universe at some cosmic time t may then be written as the sum of the photon and neutrino contributions:

$$\rho_r(t) = \rho_\gamma(t) + \rho_\nu(t).$$

Once again, we will not differentiate between different types of radiation in our subsequent discussion, since it is only the total energy density that determines the behaviour of the scale factor. For radiation, in general, we have $w = \frac{1}{3}$. Thus, from (14.44), if the total radiation in the universe has a present-day energy density of $\rho_{r,0}c^2$ then, at other cosmic times,

$$\boxed{\rho_r(t) = \rho_{r,0}\left[\frac{R_0}{R(t)}\right]^4 \qquad \text{or} \qquad \rho_r(z) = \rho_{r,0}(1+z)^4.}$$

In this case, the variation in the space density of photons (for example) again goes as $(1+z)^3$, but there is an additional factor $1+z$ resulting from the cosmological redshift of each photon.

It is worth noting that, to a very good approximation, the dominant contribution to the radiation energy density of the universe is due to the photons of the cosmic microwave background (CMB). This radiation is (to a very high degree of accuracy) uniformly distributed throughout the universe and has a *blackbody* form. For blackbody radiation, the number density of photons with frequencies in the range $[\nu, \nu + d\nu]$ is given by

$$n(\nu, T)\, d\nu = \frac{8\pi\nu^2}{c^3\left(e^{h\nu/kT} - 1\right)}\, d\nu, \qquad (15.2)$$

where T is the 'temperature' of the radiation. Since the energy per unit frequency is simply $u(\nu, T) = n(\nu, T)h\nu$, the total equivalent mass density of the radiation is

$$\rho_r(T) = \frac{1}{c^2}\int_0^\infty u_\nu\, d\nu = \frac{aT^4}{c^2},$$

where $a = 4\pi^2 k_B^4/(60\hbar^3 c^3)$ is the reduced Stefan–Boltzmann constant. Observations show that the CMB is characterised by a present-day temperature

$T_0 = 2.726\,\text{K}$, which corresponds to a total present-day number density $n_{\gamma,0} \approx 4 \times 10^8\,\text{m}^{-3}$. It is easily shown that the CMB photon energy distribution retains its general blackbody form as the universe expands. Thus, at any given cosmic time t, the temperature of the CMB radiation in the universe is given by

$$\boxed{T(t) = T_0 \left[\frac{R_0}{R(t)}\right] \quad \text{or} \quad T(z) = T_0(1+z),} \tag{15.3}$$

from which we see that the universe must have not only been denser in the past, but also 'hotter'.

Vacuum

As mentioned above the vacuum can be modelled as a perfect fluid having an equation of state $p = -\rho c^2$, so that the fluid has a negative pressure. This corresponds to an equation of state parameter $w = -1$. Thus, from (14.44), we see that at any cosmic time t, we have

$$\boxed{\rho_\Lambda = \rho_{\Lambda,0} = \frac{\Lambda c^2}{8\pi G}.}$$

Thus, the energy density of the vacuum always has the same *constant* value.

Relative contributions of the components

On combining the above results, we find that the variation in the total equivalent mass density (15.1) may be written as

$$\rho(t) = \rho_{\text{m},0} \left[\frac{R_0}{R(t)}\right]^3 + \rho_{\text{r},0} \left[\frac{R_0}{R(t)}\right]^4 + \rho_{\Lambda,0}. \tag{15.4}$$

From this expression, we see that the relative contributions of matter, radiation and the vacuum to the total density vary as the universe evolves. The details clearly depend on the relative values of $\rho_{\text{m},0}$, $\rho_{\text{r},0}$ and $\rho_{\Lambda,0}$. Typically, however, one would expect radiation to dominate the total density when $R(t)$ is small. As the universe expands, the radiation energy density dies away the most quickly and matter becomes the dominant component. Finally, if the universe continues to expand then the matter density also dies away and the vacuum ultimately dominates the energy density. We conclude by noting that cosmologists often define the *normalised scale parameter*

$$a(t) \equiv \frac{R(t)}{R_0},$$

in terms of which the above results are more compactly written, since $a_0 = 1$ by definition. We shall make use of this parameter further in subsequent sections.

15.2 Cosmological parameters

In our very simplified model of the universe discussed above, its entire history is determined by only a handful of *cosmological parameters*. In particular, if one specifies the values of the equivalent mass densities $\rho_m(t_*)$, $\rho_r(t_*)$ and ρ_Λ at some particular cosmic time t_* then the value of each density, and hence the total density, is determined at all other cosmic times t. Indeed, specifying these quantities and the Hubble parameter $H(t_*)$ is sufficient to determine the scale factor $R(t)$ at all cosmic times using (14.36). It is most natural to take t_* to be the present-day cosmic time t_0, and so the cosmological model is entirely fixed by specifying the four quantities

$$H_0, \qquad \rho_{m,0}, \qquad \rho_{r,0}, \qquad \rho_{\Lambda,0}.$$

It is, however, both convenient and common practice in cosmology to work instead in terms of alternative *dimensionless* quantities, usually called *density parameters* or simply *densities*, which are defined by

$$\boxed{\Omega_i(t) \equiv \frac{8\pi G}{3H^2(t)}\rho_i(t)} \tag{15.5}$$

where $H(t)$ is the Hubble parameter and the label i denotes 'm', 'r' or 'Λ'. It is worth noting that $\Omega_\Lambda(t)$ is, in general, a function of cosmic time t (unlike ρ_Λ, which is a constant). In terms of these new dimensionless parameters, the cosmological model may thus be fixed by specifying the values of the four present-day quantities

$$H_0, \qquad \Omega_{m,0}, \qquad \Omega_{r,0}, \qquad \Omega_{\Lambda,0}. \tag{15.6}$$

A major goal of observational cosmology is therefore to determine these quantities for our universe. Significant advances in the last decade mean that cosmologists now know these values to an accuracy of just a few per cent.[2] We simply note here that

$$\boxed{H_0 \approx 70\,\text{km}\,\text{s}^{-1}\,\text{Mpc}^{-1}, \quad \Omega_{m,0} \approx 0.3, \quad \Omega_{r,0} \approx 5\times10^{-5}, \quad \Omega_{\Lambda,0} \approx 0.7,} \tag{15.7}$$

where the units of H_0 are those most commonly used in cosmology, in which $1\,\text{Mpc} \equiv 10^6$ parsecs $\approx 3.09\times10^{22}$ m; in SI units, $H_0 \approx 2.27\times10^{-18}\,\text{s}^{-1}$. Perhaps most astonishing is that the present-day energy density of the universe is dominated by the vacuum!

[2] How these observational advances have been achieved is discussed in, for example, J. Peacock, *Cosmological Physics*, Cambridge University Press, 1999 or P. Coles & F. Lucchin, *Cosmology: The Origin and Evolution of Cosmic Structure* (2nd edition), Wiley, 2002.

We also note, for completeness, that cosmologists define further analogous dimensionless density parameters for the individual contributions to the matter and the radiation. For example, Ω_b, Ω_dm and Ω_ν are commonly used to denote the dimensionless density of baryons, dark matter and neutrinos respectively. For our universe, cosmological observations suggest the present-day values

$$\Omega_{\text{b},0} \approx 0.05, \qquad \Omega_{\text{dm},0} \approx 0.25, \qquad \Omega_{\nu,0} \approx 0, \tag{15.8}$$

noting, in particular, that only around one-sixth of the matter density is in the form of the familiar baryonic matter. Moreover, the majority of the baryonic matter seems not to reside in ordinary (hydrogen-burning) stars; the contribution of such stars is only $\Omega_* \approx 0.008$. The values of the individual quantities (15.8) affect the astrophysical process occurring in the universe and have a profound influence on, for example, the formation of structure. For determining the overall expansion history of the universe, however, only the quantities (15.6) need be specified.

The reason for defining the densities (15.5) becomes clear when we rewrite the second of the cosmological field equations (14.36) in terms of them. Dividing this equation through by R^2 and noting that $H = \dot{R}/R$, we obtain

$$1 = \Omega_\text{m} + \Omega_\text{r} + \Omega_\Lambda - \frac{c^2 k}{H^2 R^2}, \tag{15.9}$$

where, for notational simplicity, we have dropped the explicit time dependence of the variables. Indeed, it is also common practice to define the *curvature density parameter*

$$\boxed{\Omega_k(t) = -\frac{c^2 k}{H^2(t) R^2(t)},} \tag{15.10}$$

so that, at all cosmic times t, we have the elegant relation

$$\Omega_\text{m} + \Omega_\text{r} + \Omega_\Lambda + \Omega_k = 1. \tag{15.11}$$

It should be noted that, in cosmological models with positive spatial curvature ($k = 1$), the parameter Ω_k is negative. Moreover, if the cosmological constant Λ is negative then so too is the vacuum density parameter Ω_Λ. This behaviour should be contrasted with that of Ω_m and Ω_r, which are always positive.

From (15.9), we see that the values of Ω_m, Ω_r and Ω_Λ determine the spatial curvature of the universe in a simple fashion. We have three cases:

$$\begin{aligned}
&\Omega_\text{m} + \Omega_\text{r} + \Omega_\Lambda < 1 \quad \Leftrightarrow \quad \text{negative spatial curvature } (k = -1) \quad \Leftrightarrow \quad \text{'open'},\\
&\Omega_\text{m} + \Omega_\text{r} + \Omega_\Lambda = 1 \quad \Leftrightarrow \quad \text{zero spatial curvature } (k = 0) \quad \Leftrightarrow \quad \text{'flat'},\\
&\Omega_\text{m} + \Omega_\text{r} + \Omega_\Lambda > 1 \quad \Leftrightarrow \quad \text{positive spatial curvature } (k = 1) \quad \Leftrightarrow \quad \text{'closed'}.
\end{aligned}$$

The above relations are valid at any cosmic time t but are most often applied to the present day, $t = t_0$. In particular, it is also clear from (15.9) that, although the density parameters Ω_m, Ω_r and Ω_Λ are all, in general, functions of cosmic time t, their sum cannot change sign. Thus, the universe cannot evolve from one form of the FRW geometry to another. We note that cosmologists often add to the plethora of density parameters by also defining the *total density parameter*

$$\boxed{\Omega \equiv \Omega_\text{m} + \Omega_\text{r} + \Omega_\Lambda = 1 - \Omega_k,} \tag{15.12}$$

which is related to the total equivalent mass density (15.1) by $\Omega = 8\pi G\rho/(3H^2)$. From (15.7), we see that for our universe $\Omega_0 \approx 1$ or equivalently $\Omega_{k,0} \approx 0$, and it is therefore close to being spatially flat ($k = 0$).

Finally, it is worth noting that, for any cosmological model to be spatially flat, one requires $\Omega = 1$ and it is common to describe the corresponding total equivalent mass density as the *critical density*, which is given by

$$\boxed{\rho_\text{crit} \equiv \frac{3H^2}{8\pi G}.}$$

Hence, for any given value of the Hubble parameter, this expression gives the total equivalent mass density required for the universe to be spatially flat. Since recent cosmological observations suggest that our universe is indeed close to spatially flat and, (15.7), that $H_0 \approx 70\,\text{km s}^{-1}\,\text{Mpc}^{-1}$, one finds that the present-day total equivalent mass density in our universe is

$$\rho_{\text{crit},0} = \frac{3H_0^2}{8\pi G} \approx 9.2 \times 10^{-27}\,\text{kg m}^{-3}.$$

As mentioned above, it is thought that only around 30 per cent of this equivalent mass density is in the form of matter and only around 5 per cent in the form of baryonic matter. Nevertheless, it is worth noting that $\rho_{\text{crit},0} \approx 5.5$ protons m^{-3}, and so the critical density turns out to be extremely low by laboratory standards.[3]

15.3 The cosmological field equations

Since the cosmological model can be fixed by specifying the values of the quantities listed in (15.6), it is worthwhile rewriting the cosmological field equations (14.36) in terms of these parameters. Let us begin with the second field equation. Recalling that $H = \dot{R}/R$, this may be written

$$H^2 = \frac{8\pi G}{3}\left(\sum_i \rho_i\right) - \frac{c^2 k}{R^2},$$

[3] The fact that this is a number of order unity is an accident of our choice of units!

where the label i includes matter, radiation and the vacuum. From (15.4), (15.5) and (15.10), we therefore find

$$\boxed{H^2 = H_0^2\left(\Omega_{m,0}\, a^{-3} + \Omega_{r,0}\, a^{-4} + \Omega_{\Lambda,0} + \Omega_{k,0}\, a^{-2}\right),} \qquad (15.13)$$

where we have written the result in terms of the dimensionless scale parameter $a = R/R_0$. It should be remembered that $\Omega_{k,0} = 1 - \Omega_{m,0} - \Omega_{r,0} - \Omega_{\Lambda,0}$ and may be considered merely as a convenient shorthand. It is also worth noting that, since $a = R/R_0 = (1+z)^{-1}$, equation (15.13) immediately yields an expression for the Hubble parameter $H(z)$ as a function of redshift z.

We now turn to the first cosmological field equation in (14.36). Multiplying this equation through by $R/\dot{R}^2$ and again noting that $H = \dot{R}/R$, we have

$$\frac{R\ddot{R}}{\dot{R}^2} = -\frac{4\pi G}{3H^2} \sum_i \rho_i(1+3w_i),$$

where the label i once more includes matter, radiation and the vacuum. The left-hand side is equal to minus the deceleration parameter q defined in (14.19). Thus, substituting the appropriate value of w_i for each component and using (15.5), one finds the neat relation

$$\boxed{q = \tfrac{1}{2}(\Omega_m + 2\Omega_r - 2\Omega_\Lambda).} \qquad (15.14)$$

If desired, one can easily write this equation explicitly in terms of the present-day values of the density parameters by using the result (15.13) and the relation

$$\Omega_i = \Omega_{i,0}\left(\frac{H_0}{H}\right)^2 a^{-3(1+w_i)},$$

which holds generally for matter, radiation and the vacuum.

15.4 General dynamical behaviour of the universe

The cosmological field equations (15.13) and (15.14) allow us to determine the general dynamical behaviour and the spatial geometry of the universe for any given set of values for the parameters $\Omega_{m,0}$, $\Omega_{r,0}$ and $\Omega_{\Lambda,0}$. The observations (15.7) suggest that the present-day value of the radiation density $\Omega_{r,0}$ is significantly smaller than the matter and vacuum densities. It is therefore a reasonable approximation to neglect $\Omega_{r,0}$ and parameterise a universe like our own in terms of just $\Omega_{m,0}$ and $\Omega_{\Lambda,0}$ (and H_0, which is irrelevant for our discussion in this section).

Figure 15.1 presents a summary of the properties of FRW universes dominated by matter and vacuum energy (known as Lemaitre models) as a function of

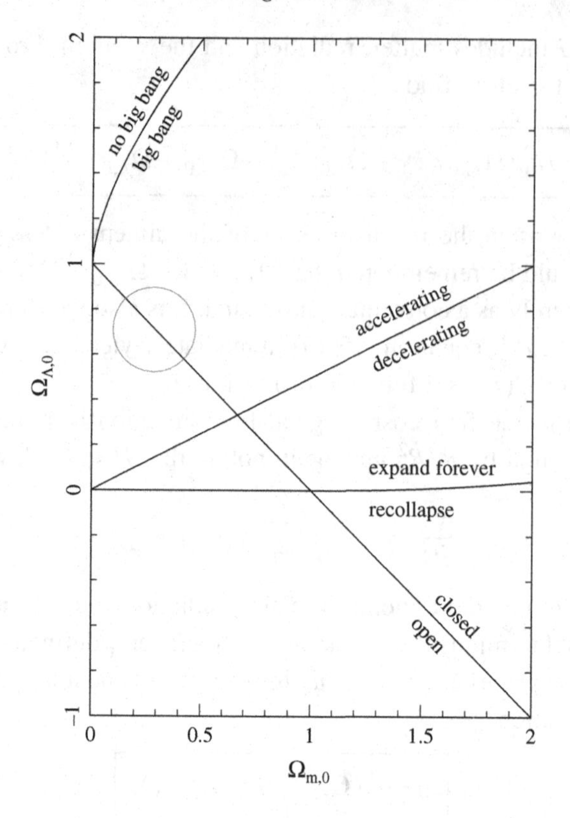

Figure 15.1 Properties of FRW universes dominated by matter and vacuum energy, as a function of the present-day density parameters $\Omega_{m,0}$ and $\Omega_{\Lambda,0}$. The circle indicates the region of the parameter space that is consistent with recent cosmological observations.

position in the $(\Omega_{m,0}, \Omega_{\Lambda,0})$ parameter space. The dividing lines between the various regions may be determined from the field equations (15.13, 15.14) and the relation (15.11). In particular, the 'open–closed' line comes directly from (15.11) evaluated at the present epoch, which gives the condition

$$\Omega_{\Lambda,0} = 1 - \Omega_{m,0}.$$

Similarly, the 'accelerating–decelerating' line is obtained immediately by setting $q_0 = 0$ in (15.14) for $t = t_0$, which gives

$$\Omega_{\Lambda,0} = \tfrac{1}{2}\Omega_{m,0}.$$

The 'expand-forever–recollapse' line and the 'big-bang–no-big-bang' line require a little more work, as we now discuss.

In fact, both these lines are determined from the expression (15.13) for the Hubble parameter. In particular, the condition for the graph of $R(t)$, or equivalently of $a(t)$, to have a turning point at some cosmic time $t = t_*$ is simply that $H(t_*) = 0$. Setting $\Omega_{k,0} = 1 - \Omega_{m,0} - \Omega_{\Lambda,0}$ in (15.13), we find that, after rearranging, this condition corresponds to

$$f(a) \equiv \Omega_{\Lambda,0}\, a^3 + (1 - \Omega_{m,0} - \Omega_{\Lambda,0})a + \Omega_{m,0} = 0. \tag{15.15}$$

This is a cubic equation for the value(s) of the scale factor $a = a_*$ at which $a(t)$ has a turning point. We are not, in fact, interested in the particular value(s) $a = a_*$ that solve (15.15) but only in whether a (real) solution exists in the region $a \geq 0$ (which is the only physically meaningful regime).

For the case $\Omega_{\Lambda,0} < 0$ we may deduce immediately from (15.14) that the universe must have started with a 'big bang', at which $a = 0$, and must eventually recollapse in a 'big crunch' as $a \to 0$ once more. In (15.14), a negative value of Ω_Λ means that the deceleration parameter q is always positive. Thus $\ddot{a}$ is always negative, and hence the $a(t)$ graph must be convex for all values of t. Since at the present epoch $\dot{a}(t_0) > 0$ (because we observe redshifts, not blueshifts), this means that $a(t)$ must have equalled zero at some point in the past, which it is usual to take as $t = 0$;[4] similar reasoning may be used to deduce that the universe must eventually recollapse, although a little more care is required in this case. As the universe expands, the vacuum energy eventually dominates and so we need only consider the Ω_Λ-term on the right-hand side of (15.14), which will not tend to zero as the scale factor increases. Thus, $\ddot{a}$ cannot tend to zero and so $a \to 0$ at some finite cosmic time in the future.

In our further analysis, we now need only consider the case in which $\Omega_{\Lambda,0} \geq 0$ in (15.15), but this still requires some care. Let us first consider the case for which $\Omega_{\Lambda,0} = 0$. Immediately, we see that equation (15.15) then has the single solution $a_* = \Omega_{m,0}/(\Omega_{m,0} - 1)$, which is negative in the range $0 \leq \Omega_{m,0} \leq 1$, indicating that there is no (physically meaningful) turning point. Therefore, over this range, the 'expand-forever–recollapse' line is simply given by $\Omega_{\Lambda,0} = 0$. We must now address the far more complicated case for which $\Omega_{\Lambda,0} > 0$. In this case $f(a) \to \pm\infty$ as $a \to \pm\infty$. Moreover $f(0) = \Omega_{m,0}$, which is positive. Thus, for $f(a)$ to have a positive root, it must have a turning point in the region $a > 0$. On evaluating the derivatives $f'(a)$ and $f''(a)$ with respect to a, it is clear that, in the limiting case of interest, $f(a)$ must have the general form illustrated in Figure 15.2. Thus, we require $f(a_*) = f'(a_*) = 0$, which quickly yields

$$a_* = \left(\frac{\Omega_{m,0}}{2\Omega_{\Lambda,0}}\right)^{1/3}. \tag{15.16}$$

[4] In fact, this reasoning is still valid in the case $\Omega_{\Lambda,0} = 0$, provided that the universe contains even an infinitesimal amount of matter (or radiation). Thus all cosmological models with $\Lambda \leq 0$ have a big-bang origin at some finite cosmic time in the past.

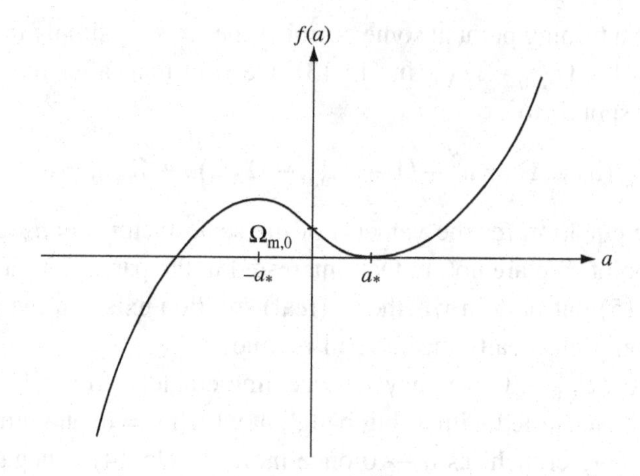

Figure 15.2 The limiting form of the cubic $f(a)$ defined in (15.15) for $\Omega_{\Lambda,0} > 0$.

On substituting this expression back into (15.15) one then obtains a separate cubic equation for $\Omega_{\Lambda,0}$, given by

$$4(1-\Omega_{m,0}-\Omega_{\Lambda,0})^3 + 27\Omega_{m,0}^2\Omega_{\Lambda,0} = 0. \tag{15.17}$$

By introducing the variable $x = [\Omega_{\Lambda,0}/(4\Omega_{m,0})]^{1/3}$, this equation quickly reduces to

$$x^3 - \frac{3x}{4} + \frac{\Omega_{m,0}-1}{4\Omega_{m,0}} = 0,$$

which is amenable to analysis using the standard formulae for finding the roots of a cubic. In particular, rewriting the resulting roots in terms of $\Omega_{\Lambda,0}$, one finds the following three cases:

- $0 < \Omega_{m,0} \leq \frac{1}{2}$, one positive root at

$$\Omega_{\Lambda,0} = 4\Omega_{m,0}\cosh^3\left[\frac{1}{3}\cosh^{-1}\left(\frac{1-\Omega_{m,0}}{\Omega_{m,0}}\right)\right]; \tag{15.18}$$

- $\frac{1}{2} < \Omega_{m,0} \leq 1$, one positive root at

$$\Omega_{\Lambda,0} = 4\Omega_{m,0}\cos^3\left[\frac{1}{3}\cos^{-1}\left(\frac{1-\Omega_{m,0}}{\Omega_{m,0}}\right)\right]; \tag{15.19}$$

- $\Omega_{m,0} > 1$, two positive roots, the larger given by (15.19) and the smaller by

$$\Omega_{\Lambda,0} = 4\Omega_{m,0}\cos^3\left[\frac{1}{3}\cos^{-1}\left(\frac{1-\Omega_{m,0}}{\Omega_{m,0}}\right) + \frac{4\pi}{3}\right]. \tag{15.20}$$

Moreover, from (15.16), one easily finds that $a_* < 1$ for (15.18, 15.19), whereas $a_* > 1$ for (15.20). Since the universe is expanding, $a_* < 1$ corresponds to a turning point in the past (i.e. no big bang), whereas $a_* > 1$ corresponds to a turning point in the future (i.e. recollapse).

The resulting lines, plotted in Figure 15.1, show some interesting features. In particular, we note that when $\Omega_{\Lambda,0} = 0$ there is a direct correspondence between the geometry of the universe and its eventual fate. In this case, open universes expand forever, whereas closed universes recollapse. This correspondence no longer holds in the presence of a non-zero cosmological constant, in which case any combination of spatial geometry and eventual fate is possible. It is also worth noting that the region of the $(\Omega_{m,0}, \Omega_{\Lambda,0})$-plane consistent with recent cosmological observations is centred on the spatially flat model (0.3,0.7) and excludes the possibility of a zero cosmological constant at high significance. These observations also show the expansion of the universe to be accelerating. They also require the universe to have started at a big bang at some finite cosmic time in the past and to expand forever in the future.

15.5 Evolution of the scale factor

So far, we have considered only the limiting behaviour of the (normalised) scale factor $a(t)$ for different values of the cosmological parameters; this was summarised in Figure 15.1. We now discuss how to find the form of the $a(t)$-curve at all cosmic times, for a given set of (present-day) cosmological parameter values. This behaviour is entirely determined by the cosmological field equation (15.13). Remembering that $H = \dot{a}/a$, this may be written as

$$\left(\frac{da}{dt}\right)^2 = H_0^2\left(\Omega_{m,0}\,a^{-1} + \Omega_{r,0}\,a^{-2} + \Omega_{\Lambda,0}a^2 + 1 - \Omega_{m,0} - \Omega_{r,0} - \Omega_{\Lambda,0}\right). \tag{15.21}$$

Instead of working directly in terms of the cosmic time t, it is more convenient to introduce the new dimensionless variable

$$\hat{t} = H_0(t - t_0), \tag{15.22}$$

which measures cosmic time relative to the present epoch in units of the 'Hubble time' H_0^{-1}. In terms of this new variable, (15.21) becomes

$$\boxed{\left(\frac{da}{d\hat{t}}\right)^2 = \Omega_{m,0}\,a^{-1} + \Omega_{r,0}\,a^{-2} + \Omega_{\Lambda,0}a^2 + 1 - \Omega_{m,0} - \Omega_{r,0} - \Omega_{\Lambda,0}.} \tag{15.23}$$

There exist some special cases, where $\Omega_{m,0}$, $\Omega_{r,0}$ and $\Omega_{\Lambda,0}$ take on particular simple values, for which equation (15.23) can be solved analytically; we will discuss some of these cosmological models in Section 15.6. In general, however, a numerical solution is necessary. Starting at the point $\hat{t} = 0$ (the present epoch), for which $a_0 = 1$, the normalised scale factor at time step $n+1$ can be approximated by the Taylor expansion

$$a_{n+1} \approx a_n + \left(\frac{da}{d\hat{t}}\right)_n \Delta\hat{t} + \frac{1}{2}\left(\frac{d^2a}{d\hat{t}^2}\right)_n (\Delta\hat{t})^2, \tag{15.24}$$

where $\Delta\hat{t}$ is the (small) step size in $\hat{t}$. The coefficient of $\Delta\hat{t}$ is given by (15.23), and the coefficient of $(\Delta\hat{t})^2$ may be obtained by differentiating (15.23). The latter is important since, without the $(\Delta\hat{t})^2$ term, equation (15.24) would not carry the integration correctly through a value of a for which $da/d\hat{t}$ is small or zero.

Figure 15.3 shows the variation in the normalised scale factor $a(\hat{t})$ as a function of $\hat{t}$ for different values of $(\Omega_{m,0}, \Omega_{\Lambda,0})$ as indicated, assuming that $\Omega_{r,0}$ is negligible, as it is for our universe. In the top panel $\Omega_{m,0} + \Omega_{\Lambda,0} = 1$ in each case, so each universe has a flat spatial geometry $(k = 0)$. The solid line corresponds to the case $(0.3, 0.7)$, which is preferred by recent cosmological observations. An interesting cosmological 'coincidence' for this model is that the present epoch, $\hat{t} = 0$, corresponds closely to the point of inflection on the $a(\hat{t})$ curve. A second such 'coincidence' is that the age of the universe in this model (i.e. the time since the big bang) is very close to one Hubble time.[5] The broken-and-dotted line in the top panel of the figure corresponds to the case $(0, 1)$, which is known as the *de Sitter model* and will be discussed further in Section 15.6. For the moment, we simply note that this model has no big-bang origin (although $a \to 0$ as $\hat{t} \to -\infty$) and will expand forever. The broken line in the top panel corresponds to the case $(1, 0)$, which is known as the *Einstein–de-Sitter model* and will also be discussed in Section 15.6. As we see from the figure, this model does have a big-bang origin. It is also on the borderline between expanding forever and recollapsing; it will in fact expand forever, but $\dot{a} \to 0$ as $\hat{t} \to \infty$.

In the bottom panel of Figure 15.3 we have $\Omega_{m,0} + \Omega_{\Lambda,0} \neq 1$ in each case, and so each universe is spatially curved; in particular the case $(0.3, 0)$ is open and the cases $(0.3, 2)$ and $(4, 0)$ are closed. We see that the case (0.3, 2) has no big-bang origin, and is, in fact, what is known as a *bounce model*, where the universe collapses from large values of the scale factor and 'bounces' at some finite minimum value of a, after which it re-expands forever. Conversely, the case (4, 0) corresponds to a cosmological model with a big-bang origin that expands to some finite maximum value of a before recollapsing to a big crunch.

Before going on to discuss cosmological models that admit an analytic solution for $a(t)$, it is worth discussing the general case in the limit $a \to 0$. Whether

[5] Whether such coincidences have some deeper significance is the subject of current cosmological research.

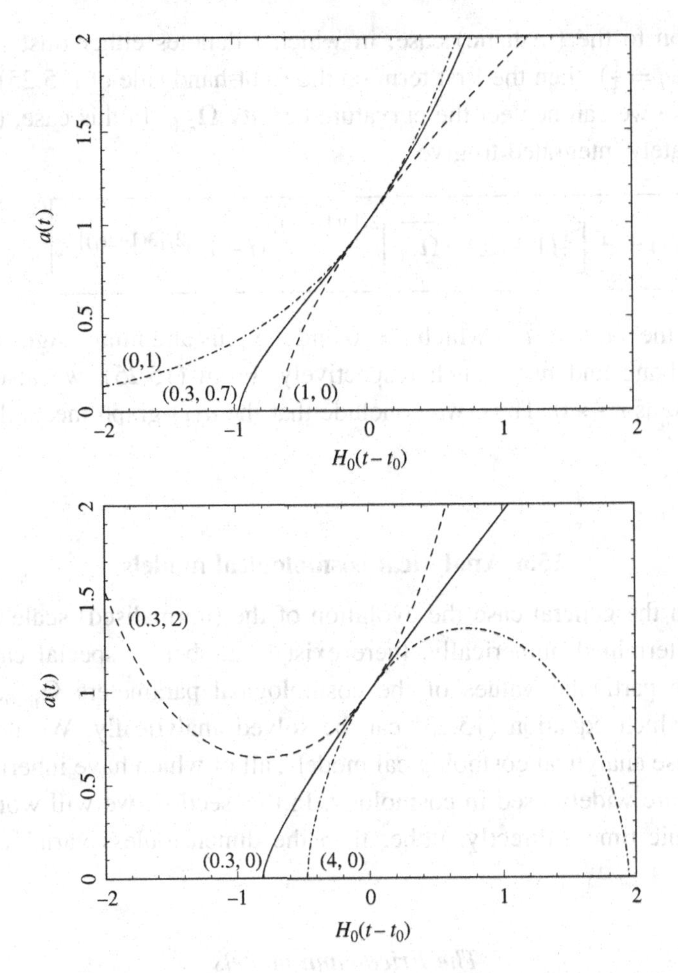

Figure 15.3 The variation in the normalised scale factor as a function of the dimensionless variable $H_0(t-t_0)$ for different values of $(\Omega_{\mathrm{m},0}, \Omega_{\Lambda,0})$ as indicated, assuming that $\Omega_{\mathrm{r},0}$ is negligible. Top panel: $\Omega_{\mathrm{m},0}+\Omega_{\Lambda,0}=1$ in each case, so the universes have a flat spatial geometry ($k=0$). Bottom panel: $\Omega_{\mathrm{m},0}+\Omega_{\Lambda,0}\neq 1$ in each case, so the universes are spatially curved; in particular, the case (0.3, 0) is open and the cases (0.3, 2) and (4, 0) are closed.

considering the big bang or the big crunch, in this limit we can assume that the energy density of the universe is dominated by one kind of source (which one will depend on the particular cosmological model under consideration). In this case, (15.23) can be written

$$\left(\frac{da}{d\hat{t}}\right)^2 = \Omega_{i,0}\, a^{-(1+3w_i)} + \Omega_{k,0}, \tag{15.25}$$

where the label i denotes the dominant form of the energy density as $a \to 0$ and w_i is the corresponding equation-of-state parameter. Moreover, if we restrict our attention to the (realistic) case, in which i denotes either dust ($w_i = 0$) or radiation ($w_i = \frac{1}{3}$), then the first term on the right-hand side of (15.25) dominates as $a \to 0$, so we can neglect the curvature density $\Omega_{k,0}$. In this case, (15.25) can be immediately integrated to give

$$a(\hat{t}) = \pm\left[\tfrac{3}{2}(1+w_i)\sqrt{\Omega_{i,0}}\right]^{2/[3(1+w_i)]}(\hat{t}-\hat{t}_*)^{2/[3(1+w_i)]}, \tag{15.26}$$

where $\hat{t}_*$ is the value of $\hat{t}$ at which $a = 0$ and the plus and minus signs correspond to the big bang and big crunch respectively. From (15.25), we also note that $da/d\hat{t} \to \infty$ as $a \to 0$. Thus, we conclude that the $a(\hat{t})$-graph meets the $\hat{t}$-axis at right angles.

15.6 Analytical cosmological models

Although in the general case the evolution of the (normalised) scale factor $a(t)$ must be determined numerically, there exist a number of special cases, corresponding to particular values of the cosmological parameters $\Omega_{\mathrm{m},0}$, $\Omega_{\mathrm{r},0}$ and $\Omega_{\Lambda,0}$, for which equation (15.23) can be solved analytically. We now discuss some of these analytical cosmological models, all of which have inherited special names that are widely used in cosmology. In this section, we will work in terms of the cosmic time t directly, rather than the dimensionless variable $\hat{t}$ defined in (15.22).

The Friedmann models

Cosmological models with a zero cosmological constant (and, strictly, a non-zero matter or radiation density) are known as the *Friedmann models*. As noted in Section 15.4, all Friedmann models have a big-bang origin at a finite cosmic time in the past. Moreover, it is possible to place a strict upper limit on the age of the universe in such models. Since the $a(t)$-curve is everywhere convex, it is clear from Figure 15.4 that it crosses the t-axis at a time that is closer to the present time $t = t_0$ than the time at which the tangent to the point (t_0, a_0) reaches the t-axis (note that $a_0 = 1$). Clearly, the point where the tangent meets the t-axis is the point at which $a(t)$ would have been zero for $\dot{a} =$ constant and $\ddot{a} = 0$. The time elapsed from that point to the present epoch is simply $\dot{a}(t_0)/a(t_0) = H_0^{-1}$. Thus, in Friedmann models, the age of the universe must be less than the Hubble time:

$$t_0 < H_0^{-1}.$$

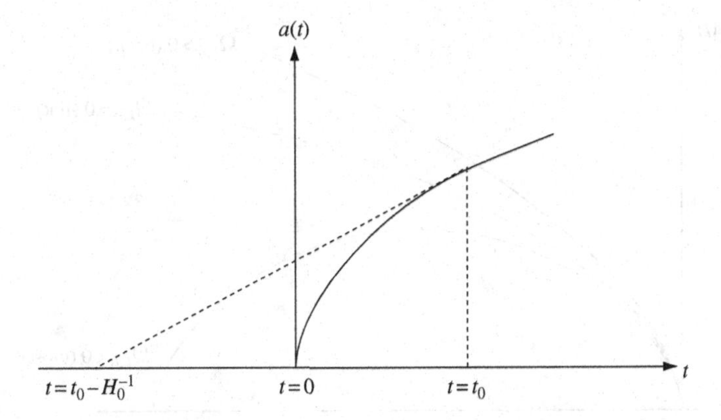

Figure 15.4 Diagram to illustrate that, for all Friedmann models, the age of the universe is less than the Hubble time $1/H_0$.

The behaviour of $a(t)$ near the big-bang origin is given by (15.26) and is independent of the curvature density parameter $\Omega_{k,0} = 1 - \Omega_{m,0} - \Omega_{r,0}$ (and hence of the sign of k). The future evolution, however, depends crucially on this constant. From (15.21) we can distinguish three possible histories, depending on the value of $\Omega_{k,0}$:

$$
\begin{aligned}
\Omega_{k,0} > 0 \quad &\Leftrightarrow \quad \text{open } (k=-1) \quad \Leftrightarrow \quad \dot{a} \to \text{non-zero constant as a} \to \infty, \\
\Omega_{k,0} = 0 \quad &\Leftrightarrow \quad \text{flat } (k=0) \quad \Leftrightarrow \quad \dot{a} \to 0 \text{ as } a \to \infty, \\
\Omega_{k,0} < 0 \quad &\Leftrightarrow \quad \text{closed } (k=1) \quad \Leftrightarrow \quad \dot{a} = 0 \text{ at some finite value } a_{\text{max}}.
\end{aligned}
$$

Thus, we see the main feature of Friedmann models, namely, that the *dynamics* of the universe is directly linked to its *geometry*. The three cases above are illustrated in Figure 15.5. We shall now find explicit analytical solutions for $a(t)$ in the special cases of a dust-only and a radiation-only Friedmann model. We will also obtain an analytic form for t as a function of a for the case of a spatially flat $(k=0)$ Friedmann model containing both matter and radiation.

Dust-only Friedmann models $(\Omega_{\Lambda,0} = 0, \Omega_{r,0} = 0)$ In this case (15.21) becomes

$$
\dot{a}^2 = H_0^2\left(\Omega_{m,0}a^{-1} + 1 - \Omega_{m,0}\right) \quad \Rightarrow \quad t = \frac{1}{H_0}\int_0^a \left[\frac{x}{\Omega_{m,0} + (1-\Omega_{m,0})x}\right]^{1/2} dx, \tag{15.27}
$$

which may be integrated straightforwardly in each of the three cases $\Omega_{m,0} = 1$, $\Omega_{m,0} > 1$ and $\Omega_{m,0} < 1$ respectively, as follows.

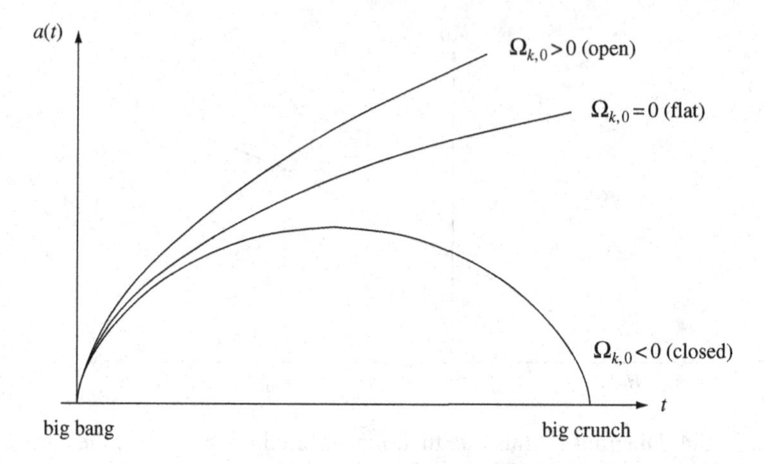

Figure 15.5 Schematic illustration of the evolution of the normalised scale factor $a(t)$ in closed, open and spatially flat Friedmann models.

- For $\Omega_{\mathrm{m},0} = 1$ ($k = 0$) the solution is immediate, and we find that

$$\boxed{a(t) = \left(\tfrac{3}{2}H_0 t\right)^{2/3}.} \tag{15.28}$$

This particular case is known as the *Einstein–de-Sitter* (or EdS) model.
- For $\Omega_{\mathrm{m},0} > 1 (k = 1)$ the integral (15.27) can be evaluated by substituting $x = [\Omega_{\mathrm{m},0}/(\Omega_{\mathrm{m},0} - 1)]\sin^2(\psi/2)$, where ψ is known as the *development angle* and varies over the range $[0, \pi]$. One then obtains

$$\boxed{a = \frac{\Omega_{\mathrm{m},0}}{2(\Omega_{\mathrm{m},0} - 1)}(1 - \cos\psi), \qquad t = \frac{\Omega_{\mathrm{m},0}}{2H_0(\Omega_{\mathrm{m},0} - 1)^{3/2}}(\psi - \sin\psi),}$$

which shows that the graph of $a(t)$ is a *cycloid*.
- For $\Omega_{\mathrm{m},0} < 1 (k = -1)$ the integral (15.27) can be evaluated by substituting $x = [\Omega_{\mathrm{m},0}/(1 - \Omega_{\mathrm{m},0})]\sinh^2(\psi/2)$, and one obtains

$$\boxed{a = \frac{\Omega_{\mathrm{m},0}}{2(1 - \Omega_{\mathrm{m},0})}(\cosh\psi - 1), \qquad t = \frac{\Omega_{\mathrm{m},0}}{2H_0(1 - \Omega_{\mathrm{m},0})^{3/2}}(\sinh\psi - \psi).}$$

In each case, one may also obtain expressions for $\rho_{\mathrm{m}}(t) = \rho_{\mathrm{m},0}a^{-3}$ and $H(t) = \dot{a}/a$, and hence for $\Omega_{\mathrm{m}}(t)$.

Radiation-only Friedmann models ($\Omega_{\Lambda,0} = 0, \Omega_{m,0} = 0$) In this case (15.21) becomes

$$\dot{a}^2 = H_0^2\left(\Omega_{r,0}a^{-2} + 1 - \Omega_{r,0}\right) \quad \Rightarrow \quad t = \frac{1}{H_0}\int_0^a \frac{x}{\sqrt{\Omega_{r,0} + (1-\Omega_{r,0})x^2}}\,dx, \tag{15.29}$$

which may again be integrated straightforwardly for $\Omega_{r,0} = 1$, $\Omega_{r,0} > 1$ and $\Omega_{r,0} < 1$ respectively.

- For $\Omega_{r,0} = 1$ ($k = 0$) the solution is again immediate, and we find that

$$\boxed{a(t) = (2H_0t)^{1/2}.}$$

- For $\Omega_{r,0} < 1 (k = -1)$ and $\Omega_{r,0} > 1 (k = 1)$ the integral (15.29) can be evaluated by inspection to give

$$\boxed{a(t) = \left(2H_0\Omega_{r,0}^{1/2}t\right)^{1/2}\left(1 + \frac{1-\Omega_{r,0}}{2\Omega_{r,0}^{1/2}}H_0t\right)^{1/2}.}$$

In each case, one may again obtain expressions for $\rho_r(t) = \rho_{r,0}a^{-4}$ and $H(t) = \dot{a}/a$, and hence for $\Omega_r(t)$.

Spatially flat Friedmann models ($\Omega_{\Lambda,0} = 0$, $\Omega_{m,0} + \Omega_{r,0} = 1$) In this case (15.21) becomes

$$\dot{a}^2 = H_0^2\left(\Omega_{m,0}a^{-1} + \Omega_{r,0}a^{-2}\right) \quad \Rightarrow \quad t = \frac{1}{H_0}\int_0^a \frac{x}{\sqrt{\Omega_{m,0}x + \Omega_{r,0}}}\,dx, \tag{15.30}$$

which may be straightforwardly integrating by substituting $y = \Omega_{m,0}x + \Omega_{r,0}$ to obtain

$$\boxed{H_0t = \frac{2}{3\Omega_{m,0}^2}\left[(\Omega_{m,0}a + \Omega_{r,0})^{1/2}(\Omega_{m,0}a - 2\Omega_{r,0}) + 2\Omega_{r,0}^{3/2}\right].}$$

Unfortunately, this expression cannot be easily inverted to give $a(t)$. Nevertheless, it is simple to show that the above expression becomes $\frac{2}{3}a^{3/2}$ for a matter-only model and $\frac{1}{2}a^2$ for a radiation-only model, and therefore agrees with our earlier results.

The Lemaitre models

The Lemaitre models are a generalisation of the Friedmann models in which the cosmological constant is non-zero. In particular, we will focus here on matter-only models ($\Omega_{r,0} = 0$), although our discussion is easily modified for radiation-only models, and can be extended to include models containing both matter and radiation. The general dynamical properties of Lemaitre models with $\Omega_{r,0} = 0$ were discussed in detail in Section 15.4, with particular focus on their limiting behaviour. We concentrate here on determining the generic form of the $a(t)$-curve for models of this type that have a big-bang origin and will expand forever. We begin by considering the general case of arbitrary spatial curvature and then specialise to the spatially flat case. A model of the latter sort appears to provide a reasonable description of our own universe, if one neglects its radiation energy density.

Matter-only Lemaitre models with arbitrary spatial curvature ($\Omega_{r,0} = 0$) In this case the cosmological field equation (15.13) reads

$$\dot{a}^2 = H_0^2 \left(\Omega_{m,0} a^{-1} + \Omega_{\Lambda,0} a^2 + \Omega_{k,0}\right), \tag{15.31}$$

where $\Omega_{k,0} = 1 - \Omega_{m,0} - \Omega_{\Lambda,0}$. Obtaining explicit formulae giving, for example, the scale factor as a function of time is in general quite complicated, since the integrals turn out to involve elliptic functions,[6] which are unfamiliar to most physicists these days. Nevertheless, we see that for small a the first term on the right-hand side dominates and the equation is easily integrated. Thus, after starting from a big-bang origin at $t = 0$, the $a(t)$-curve at first increases as

$$\boxed{a(t) = \left(\tfrac{3}{2} H_0 \sqrt{\Omega_{m,0}}\, t\right)^{2/3} \qquad \text{(for small t).}}$$

which agrees with our earlier result (15.26). As the universe expands, however, the matter energy density decreases and the vacuum energy eventually dominates. Thus, for large t (and hence large a), the second term on the right-hand side of (15.31) dominates. Once again the equation is then easily integrated to give

$$\boxed{a(t) \propto \exp\left(H_0 \sqrt{\Omega_{\Lambda,0}}\, t\right) \qquad \text{(for large t).}}$$

[6] See e.g. M. Abramowitz & I. A. Stegun, *Handbook of Mathematical Physics*, Dover, 1972.

From the above limiting behaviour at small and large t, it is clear that the universe must, at some point, make a transition from a decelerating to an accelerating phase. This occurs when $\ddot{a} = 0$, at which point the $a(t)$-curve has a point of inflection. Differentiating (15.31), we find that

$$\ddot{a} = \tfrac{1}{2} H_0^2 \left(2\Omega_{\Lambda,0} a - \Omega_{\mathrm{m},0} a^{-2}\right). \qquad (15.32)$$

From this result, we may verify immediately that at early cosmic times (when a is small) we have $\ddot{a} < 0$, and so the expansion is decelerating. As the universe expands, the deceleration gradually decreases until $\ddot{a}$ changes sign, after which the expansion accelerates ever more rapidly. We see that the value of the normalised scale factor at which the point of inflection ($\ddot{a} = 0$) occurs is given by

$$a_* = \left(\frac{\Omega_{\mathrm{m},0}}{2\Omega_{\Lambda,0}}\right)^{1/3}. \qquad (15.33)$$

It is, in fact, possible to obtain an approximate analytic expression for the normalised scale factor $a(t)$ in the vicinity of the point of inflection. To do this, we must first obtain an approximate form for the cosmological field equation (15.31) in the vicinity of this point. Denoting the cosmic time at the point of inflection by t_*, we may perform separate Taylor expansions of a and $\dot{a}^2$ about $t = t_*$ to obtain

$$a \approx a_* + \dot{a}_*(t - t_*) \qquad \text{and} \qquad \dot{a}^2 \approx \dot{a}_*^2 + \dot{a}_*\dddot{a}_*(t - t_*)^2,$$

where, for notational convenience, we have written $a_* \equiv a(t_*)$, $\dot{a}_* \equiv \dot{a}(t_*)$, etc. Using the first expression to subtitute for $(t - t_*)^2$ in the second, we obtain

$$\dot{a}^2 \approx \dot{a}_*^2 + \frac{\dddot{a}_*(a - a_*)^2}{\dot{a}_*}. \qquad (15.34)$$

Differentiating (15.32) one easily obtains an expression for $\dddot{a}$. Then, substituting (15.33) into the resulting expression, and into (15.31), one finds that (15.34) becomes

$$\dot{a}^2 \approx H_0^2 \left[\Omega_{k,0} + 3\Omega_{\Lambda,0} a_*^2 + 3\Omega_{\Lambda,0}(a - a_*)^2\right].$$

This equation can now be integrated analytically and has the solution

$$\boxed{a(t) = a_* + a_* \left[1 + \tfrac{1}{3}\Omega_{k,0}\left(\tfrac{1}{4}\Omega_{\Lambda,0}\Omega_{\mathrm{m},0}^2\right)^{-1/3}\right]^{1/2} \sinh\left[H_0(3\Omega_{\Lambda,0})^{1/2}(t - t_*)\right].} \qquad (15.35)$$

An interesting property of this type of model is that in the case of positive spatial curvature ($k = 1$), for which $\Omega_{k,0} < 0$, there is a 'coasting period' in the

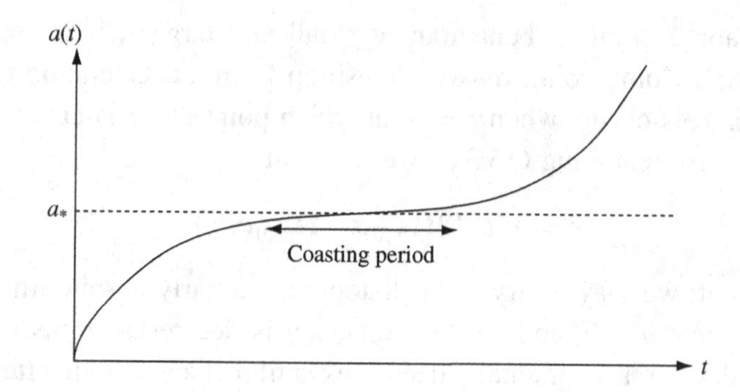

Figure 15.6 The behaviour of $a(t)$ in the Lemaitre model with $k = 1$. For $k = 0$ or $k = -1$, there is no extended coasting period.

vicinity of the point where $\ddot{a} = 0$, during which the value of $a(t)$ remains almost equal to a_* (see Figure 15.6). It is easily seen from (15.35) that, by setting the value of the quantity $\frac{1}{3}\Omega_{k,0}(\frac{1}{4}\Omega_{\Lambda,0}\Omega_{\mathrm{m},0}^2)^{-1/3}$ sufficiently close to -1, one can make the coasting period arbitrarily long. Indeed, in the limiting case, it is easy to show that one requires that $\Omega_{\mathrm{m},0}$ and $\Omega_{\Lambda,0}$ should satisfy (15.17).

Spatially flat matter-only Lemaitre models ($\Omega_{\mathrm{r},0} = 0$, $\Omega_{\mathrm{m},0} + \Omega_{\Lambda,0} = 1$) In this case one can give an explicit formula for the scale factor. Moreover, even if the universe turns out not to be exactly spatially flat, recent cosmological observations show that it is close enough to flatness for the formulae involved to act as a reasonable first approximation and so it is worthwhile to have them available.

In the spatially flat case, the cosmological field equation (15.13) may be written

$$\dot{a}^2 = H_0^2\left[(1-\Omega_{\Lambda,0})a^{-1} + \Omega_{\Lambda,0}a^2\right] \Rightarrow \quad t = \frac{1}{H_0}\int_0^a \frac{x}{\sqrt{(1-\Omega_{\Lambda,0})x + \Omega_{\Lambda,0}x^4}}\, dx.$$

This integral is a little more difficult than those considered earlier, but it can be made tractable by the substitution $y^2 = x^3|\Omega_{\Lambda,0}|/(1-\Omega_{\Lambda,0})$, which yields

$$H_0 t = \frac{2}{3\sqrt{|\Omega_{\Lambda,0}|}}\int_0^{\sqrt{a^3|\Omega_{\Lambda,0}|/(1-\Omega_{\Lambda,0})}} \frac{dy}{\sqrt{1 \pm y^2}},$$

where the plus sign in the integrand corresponds to the case $\Omega_{\Lambda,0} > 0$ and the minus sign to $\Omega_{\Lambda,0} < 0$. This may now be integrated easily to give

$$\boxed{H_0 t = \frac{2}{3\sqrt{|\Omega_{\Lambda,0}|}}\begin{cases} \sinh^{-1}\left[\sqrt{a^3|\Omega_{\Lambda,0}|/(1-\Omega_{\Lambda,0})}\right] & \text{if } \Omega_{\Lambda,0} > 0, \\ \sin^{-1}\left[\sqrt{a^3|\Omega_{\Lambda,0}|/(1-\Omega_{\Lambda,0})}\right] & \text{if } \Omega_{\Lambda,0} < 0, \end{cases}} \tag{15.36}$$

which may be inverted to give $a(t)$ in each case. One can also obtain analytic expressions for $H(t)$ and $\rho_m(t)$ (see Exercise 15.24) and thus for $\Omega_m(t)$ and $\Omega_\Lambda(t)$.

The de Sitter model

The de Sitter model is a particular special case of a Lemaitre model defined by the cosmological parameters $\Omega_{m,0} = 0$, $\Omega_{r,0} = 0$ and $\Omega_{\Lambda,0} = 1$. This model is therefore spatially flat ($k = 0$) but is not a true cosmological model in the strictest sense, since it assumes that the matter and radiation densities are zero. Nevertheless, it is interesting in its own right both for historical reasons and because of its close connection with the theory of inflation (see Section 16.1).

For the de Sitter model, the cosmological field equation (15.13) reads

$$\left(\frac{\dot{a}}{a}\right)^2 = H_0^2,$$

which immediate tells us that the Hubble parameter $H(t)$ is a constant and the normalised scale factor increases exponentially as

$$\boxed{a(t) = \exp[H_0(t - t_0)] = \exp\left[\sqrt{\Lambda/3}c(t - t_0)\right],}$$

where, in the second equality, we have expressed the solution in terms of the cosmological constant Λ. Thus, the de Sitter model has no big-bang singularity at a finite time in the past.

Einstein's static universe

All the cosmological models that we have constructed so far are *evolving* cosmologies. We know now, of course, that the universe is expanding and so there is no conflict with the field equations. Nevertheless, it is interesting historically to look at Einstein's static model of the universe. Einstein derived his field equations well before the discovery of the expansion of the universe and he was worried that he could not find static cosmological solutions. He therefore introduced the cosmological constant with the sole purpose of constructing static solutions.

For $\Lambda > 0$, we seek a solution to the field equations in which the universe is static, i.e. $\ddot{a} = \dot{a} = 0$. In this case, the Hubble parameter H is zero always, and so the dimensional densities in (15.5) are formally infinite. It is more convenient therefore to work with the field equations in their original forms (14.36). We see immediately that we require

$$\boxed{4\pi G\rho_{m,0} = \Lambda c^2 = \frac{c^2 k}{R_0^2}.}$$

In fact, the first equality can be more succinctly written as $\rho_{m,0} = 2\rho_{\Lambda,0}$. Since Λ is positive we thus require $k = 1$, and so the universe has positive spatial curvature.

How well did Einstein's static universe fit with cosmological observations of the time? The mean matter density of the universe is still a matter of great debate, but recent cosmological observations suggest that

$$\rho_{m,0} \approx 3 \times 10^{-27}\,\text{kg}\,\text{m}^{-3}.$$

In Einstein's time, this value was estimated only to within about two orders of magnitude. Nevertheless, adopting the above value of $\rho_{m,0}$ we find that the scale factor is $R_0 \approx 2 \times 10^{26}$ m ≈ 6000 Mpc, which is more than sufficient for the closed spatial geometry to be large enough to encompass the observable universe. Also $\Lambda = 1/R_0^2 = 2.5 \times 10^{-53}\,\text{m}^{-2}$, which is small enough to evade the limits on Λ from Solar System experiments ($|\Lambda| \leq 10^{-46}\,\text{m}^{-2}$). Thus the Einstein static universe was not immediately and obviously wrong.

However, aside from the fact that the model disagreed with later observations indicating an expanding universe, it has the theoretically undesirable feature of being *unstable*. The cosmological constant must be fine-tuned to match the density of the universe. Thus, if we add or subtract one proton from this universe, or convert some matter into radiation, we will disturb the finely tuned balance between gravity and the cosmological constant and the universe will begin to expand or contract.

15.7 Look-back time and the age of the universe

Since the cosmological model may be fixed by specifying the values of the four (present-day) cosmological parameters H_0, $\Omega_{m,0}$, $\Omega_{r,0}$ and $\Omega_{\Lambda,0}$, it is possible to use these quantities to determine other useful *derived* cosmological parameters. In this section we consider the look-back time and the age of the universe.

In Chapter 14, we showed that if a comoving particle (galaxy) emitted a photon at cosmic time t that is received by an observer at $t = t_0$ then the 'look-back time' $t_0 - t$ is given as a function of the photon's redshift by

$$t_0 - t = \int_0^z \frac{d\bar{z}}{(1+\bar{z})H(\bar{z})}. \tag{15.37}$$

From the cosmological field equation (15.13), on noting that $a = R/R_0 = (1+z)^{-1}$ we obtain the useful result

$$H^2(z) = H_0^2\left[\Omega_{m,0}(1+z)^3 + \Omega_{r,0}(1+z)^4 + \Omega_{\Lambda,0} + \Omega_{k,0}(1+z)^2\right]. \tag{15.38}$$

Thus, the look-back time to a comoving object with redshift z is given by

$$t_0 - t = \frac{1}{H_0}\int_0^z \frac{d\bar{z}}{(1+\bar{z})\sqrt{\Omega_{\mathrm{m},0}(1+\bar{z})^3 + \Omega_{\mathrm{r},0}(1+\bar{z})^4 + \Omega_{\Lambda,0} + \Omega_{k,0}(1+\bar{z})^2}}.$$

We note that the differential form of this relation is perhaps more useful since one is often interested simply in the cosmic time interval dt corresponding to an interval dz in redshift. In any case, a more convenient form of the integral for evaluation is obtained by making the substitution $x = (z+1)^{-1}$, which yields

$$t_0 - t = \frac{1}{H_0}\int_{(1+z)^{-1}}^1 \frac{x\,dx}{\sqrt{\Omega_{\mathrm{m},0}x + \Omega_{\mathrm{r},0} + \Omega_{\Lambda,0}x^4 + \Omega_{k,0}x^2}}. \qquad (15.39)$$

Assuming $\Omega_{\mathrm{r},0} = 0$ (which is a reasonable approximation for our universe), in Figure 15.7 we plot $H_0(t - t_0)$, the look-back time in units of the Hubble time, as a function of redshift for several values of $\Omega_{\mathrm{m},0}$ and $\Omega_{\Lambda,0}$.

In any cosmological model with a big-bang origin, an extremely important quantity is the *age of the universe*, i.e. the cosmic time interval between the point when $a(t) = 0$ and the present epoch $t = t_0$. Since $z \to \infty$ at the big bang, we may immediately obtain an expression for the age of the universe in such a model

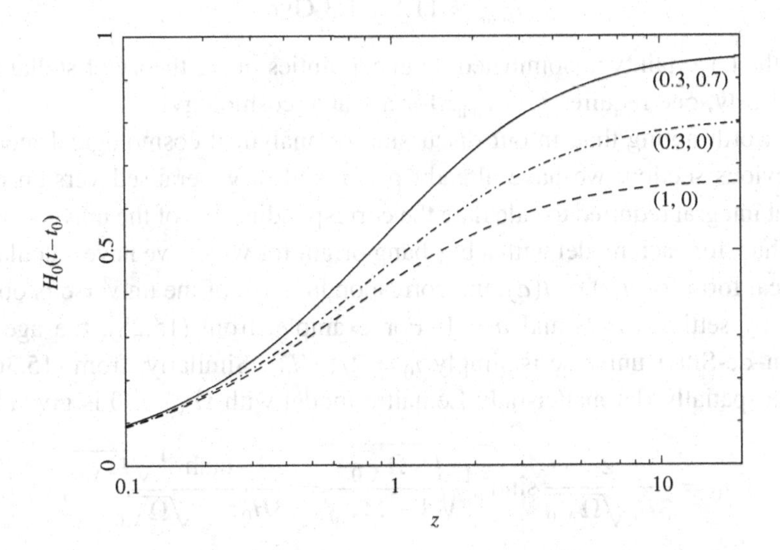

Figure 15.7 The variation in look-back time, in units of the Hubble time, as a function of redshift z for several sets of values $(\Omega_{\mathrm{m},0}, \Omega_{\Lambda,0})$ as indicated, assuming that $\Omega_{\mathrm{r},0}$ is negligible.

Table 15.1 *The age of the universe in* Gyr *for various cosmological models (with* $\Omega_{r,0} = 0$*)*

		H_0 in km s^{-1} Mpc^{-1}		
$\Omega_{m,0}$	$\Omega_{\Lambda,0}$	50	70	90
1.0	0.0	13.1	9.3	7.2
0.3	0.0	15.8	11.3	8.8
0.3	0.7	18.9	13.5	10.5

by letting $z \to \infty$ in (15.39), so that the lower limit of the integral equals zero. Since the resulting integral is dimensionless, we can write

$$t_0 = \frac{1}{H_0} f(\Omega_{m,0}, \Omega_{r,0}, \Omega_{\Lambda,0}),$$

where f is the value of the integral, which is typically a number of order unity. The age of the universe is therefore the Hubble time multiplied by a number of order unity. For general values of the density parameters $\Omega_{m,0}$, $\Omega_{r,0}$ and $\Omega_{\Lambda,0}$, it is not possible to perform the integral analytically and so one has to resort to numerical integration. Table 15.1 lists the age of the universe t_0 for the same values of $\Omega_{m,0}$ and $\Omega_{\Lambda,0}$ as considered in Figure 15.7. It is interesting to compare these values with estimates of the ages of the oldest stars in globular clusters,

$$t_{\text{stars}} \approx 11.5 \pm 1.3 \text{ Gyr},$$

where the uncertainty is dominated by uncertainties in the theory of stellar evolution. Clearly, one requires $t_0 > t_{\text{stars}}$ for a viable cosmology!

It is worth noting that, in our discussion of analytical cosmological models in the previous section, we have already performed (a generalised version of) the relevant integral required to calculate the corresponding age of the universe in each case. Thus, for each model with a big-bang origin for which we have calculated an analytical form for $a(t)$ or $t(a)$, the corresponding age of the universe is obtained simply by setting $t = t_0$ and $a = 1$. For example, from (15.28), the age of an Einstein–de-Sitter universe is simply $t_0 = 2/(3H_0)$. Similarly, from (15.36), the age of a spatially flat matter-only Lemaitre model with $\Omega_{\Lambda,0} > 0$ is given by

$$t_0 = \frac{2}{3H_0\sqrt{\Omega_{\Lambda,0}}} \sinh^{-1}\sqrt{\frac{\Omega_{\Lambda,0}}{1-\Omega_{\Lambda,0}}} = \frac{2}{3H_0} \frac{\tanh^{-1}\sqrt{\Omega_{\Lambda,0}}}{\sqrt{\Omega_{\Lambda,0}}},$$

where, in the second equality, we have rewritten the result in a more useful form involving $\Omega_{\Lambda,0}$, using standard formulae for inverse hyperbolic trigonometric functions.

15.8 The distance–redshift relation

We may also obtain a general expression for the comoving χ-coordinate of a galaxy emitting a photon at time t that is received at time t_0 with redshift z. This is given by

$$\chi = \int_t^{t_0} \frac{c\,d\bar{t}}{R(\bar{t})} = \frac{c}{R_0}\int_0^z \frac{d\bar{z}}{H(\bar{z})}.$$

We may now subsitute for $H(z)$ using the expression (15.38) derived in the previous section. Thus the χ-coordinate of a comoving object with redshift z is given by

$$\chi(z) = \frac{c}{R_0 H_0}\int_0^z \frac{d\bar{z}}{\sqrt{\Omega_{\mathrm{m},0}(1+\bar{z})^3 + \Omega_{\mathrm{r},0}(1+\bar{z})^4 + \Omega_{\Lambda,0} + \Omega_{k,0}(1+\bar{z})^2}}. \quad (15.40)$$

Once again, the differential form of this result is perhaps more useful, since one is often interested in the comoving coordinate interval $d\chi$ corresponding to an interval dz in redshift. As before, a simpler form for the integral is obtained by making the substitution $x = (1+\bar{z})^{-1}$, which yields

$$\boxed{\chi(z) = \frac{c}{R_0 H_0}\int_{(1+z)^{-1}}^{1} \frac{dx}{\sqrt{\Omega_{\mathrm{m},0}x + \Omega_{\mathrm{r},0} + \Omega_{\Lambda,0}x^4 + \Omega_{k,0}x^2}}.} \quad (15.41)$$

From (14.29) and (14.31), the corresponding luminosity distance $d_{\mathrm{L}}(z)$ and angular diameter distance $d_{\mathrm{A}}(z)$ to the object are given by

$$d_{\mathrm{L}}(z) = R_0(1+z)S(\chi(z)) \qquad \text{and} \qquad d_{\mathrm{A}}(z) = \frac{R_0}{1+z}S(\chi(z)),$$

where $S(\chi)$ is given by (14.12), whereas the proper distance to the object is simply $d(z) = R_0 S(\chi(z))$. It is useful to introduce the notation $\chi(z) = cE(z)/(R_0 H_0)$, so that $E(z)$ denotes the integral in (15.41). Using the expression (15.10) to obtain $\Omega_{k,0}$, one can then write

$$R_0 S(\chi(z)) = \frac{c}{H_0}\begin{cases} |\Omega_{k,0}|^{-1/2} S\left(\sqrt{|\Omega_{k,0}|}\,E(z)\right) & \text{for } \Omega_{k,0} \neq 0, \\ E(z) & \text{for } \Omega_{k,0} = 0, \end{cases}$$

which allows simple direct evaluation of $d_{\mathrm{L}}(z)$ and $d_{\mathrm{A}}(z)$ in each case.

As was the case in the previous section, for general values of $\Omega_{\mathrm{m},0}$, $\Omega_{\mathrm{r},0}$ and $\Omega_{\Lambda,0}$ it is not possible to perform the integral (15.41) analytically and so one has to resort to numerical integration. Figure 15.8 shows plots of dimensionless luminosity distance $(c/H_0)^{-1}d_{\mathrm{L}}(z)$ (top panel) and dimensionless angular diameter distance $(c/H_0)^{-1}d_{\mathrm{A}}(z)$ (bottom panel) for various values of $\Omega_{\mathrm{m},0}$ and $\Omega_{\Lambda,0}$, assuming that $\Omega_{\mathrm{r},0}$ is negligible; the solid, broken and dotted lines correspond to

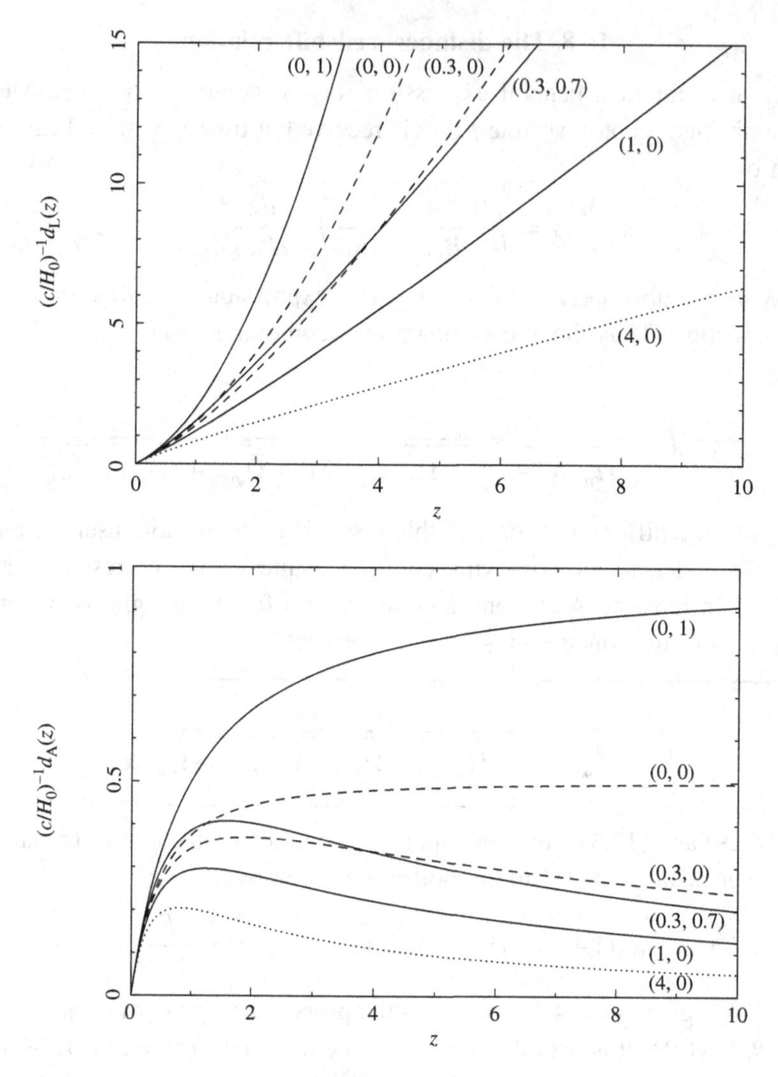

Figure 15.8 The variation in dimensionless luminosity distance (top panel) and dimensionless angular diameter distance (bottom panel) as functions of redshift, for different sets of values $(\Omega_{m,0}, \Omega_{\Lambda,0})$ as indicated, assuming that $\Omega_{r,0}$ is negligible. The solid, broken and dotted lines correspond to spatially flat, open and closed models respectively.

spatially flat, open and closed models respectively. In particular, it is worth noting that, for the models with a non-zero matter density, the angular diameter distance has a maximum at some finite value of the redshift $z = z_*$. Thus, for a source of fixed proper length ℓ, the angular diameter $\Delta\theta = \ell/d_A$ declines with redshift for $z < z_*$, as one might naively expect, but then *increases* with redshift for $z > z_*$. A very-high-redshift galaxy (if such a thing existed) would therefore cast a large, but dim, ghostly image on the sky. The physical reason for this is that the light from a distant object was emitted when the universe was much younger than it is now – the object was close to us when the light was emitted. This, coupled with gravitational focussing of the light rays by the intervening matter in the universe, means that the galaxy looks big!

The integral (15.41) can, in fact, be evaluated analytically in some simple cases. As an example, consider the Einstein–de-Sitter (EdS) model ($\Omega_{m,0} = 1$, $\Omega_{r,0} = 0$, $\Omega_{\Lambda,0} = 0$). In this case, we find that

$$\chi(z) = \frac{c}{R_0 H_0} \int_{(1+z)^{-1}}^{1} \frac{dx}{\sqrt{x}} = \frac{2c}{R_0 H_0} \left[1 - (1+z)^{-1/2}\right].$$

Thus, the luminosity distance in the EdS model is given as a function of z by

$$d_L(z) = \frac{2c}{H_0}(1+z)\left[1 - (1+z)^{-1/2}\right],$$

and the angular diameter distance by

$$d_A(z) = \frac{2c}{H_0}\frac{1}{1+z}\left[1 - (1+z)^{-1/2}\right].$$

Note that, in this case, $d_A(z)$ has a maximum at a redshift $z = 5/4$.

The relations between redshift and luminosity distance (angular diameter distance), form the basis of observational tests of the geometry of the universe. All one needs is a standard candle (for application of the luminosity-distance–redshift relation) or a standard ruler (for application of the angular-diameter-distance–redshift relation). Comparison with the predicted relations shown in Figure 15.8 can then fix the values of $\Omega_{m,0}$ and $\Omega_{\Lambda,0}$. Unfortunately, standard candles and standard rulers are hard to find in the universe! Nevertheless, in recent years there has been remarkable progress, using distant Type Ia supernovae as standard candles and anisotropies in the cosmic microwave background radiation as a standard ruler. The results of these observations suggest that we live in a spatially flat universe with $\Omega_{m,0} \approx 0.3$ and $\Omega_{\Lambda,0} \approx 0.7$.

15.9 The volume–redshift relation

In Section 14.12 we found that, at the present cosmic time t_0, the proper volume of the region of space lying in the infinitesimal coordinate range $\chi \rightarrow \chi + d\chi$ and subtending an infinitesimal solid angle $d\Omega = \sin\theta \, d\theta \, d\phi$ at the observer is

$$dV_0 = \frac{cR_0^2 S^2(\chi(z))}{H(z)} \, dz \, d\Omega, \tag{15.42}$$

the corresponding volume of this region at a redshift z being given by $dV(z) = dV_0/(1+z)^3$. We may now express dV_0 in terms of the cosmological parameters H_0, $\Omega_{\mathrm{m},0}$, $\Omega_{\mathrm{r},0}$ and $\Omega_{\Lambda,0}$. Using the expressions (15.40), (15.38) and (15.10) for $\chi(z)$, $H(z)$ and Ω_k respectively, we find immediately that

$$dV_0 = \frac{(cH_0^{-1})^3}{h(z)} \begin{cases} |\Omega_{k,0}|^{-1} S^2\left(\sqrt{|\Omega_{k,0}|}E(z)\right) & \text{for } \Omega_{k,0} \neq 0, \\ E^2(z) & \text{for } \Omega_{k,0} = 0, \end{cases} \tag{15.43}$$

where we have defined the new function

$$h(z) \equiv \frac{H(z)}{H_0} = \sqrt{\Omega_{\mathrm{m},0}(1+z)^3 + \Omega_{\mathrm{r},0}(1+z)^4 + \Omega_{\Lambda,0} + \Omega_{k,0}(1+z)^2}$$

and $E(z) \equiv \int_0^z d\bar{z}/h(\bar{z})$ is the function defined in the previous section.

For general values of $\Omega_{\mathrm{m},0}$, $\Omega_{\mathrm{r},0}$ and $\Omega_{\Lambda,0}$, one must once again resort to numerical integration to obtain dV_0. In Figure 15.9, we plot the dimensionless differential comoving volume element $(c/H_0)^{-3}\, dV_0/(dz\, d\Omega)$ as a function of redshift z for several values of $\Omega_{\mathrm{m},0}$ and $\Omega_{\Lambda,0}$, assuming that $\Omega_{\mathrm{r},0} = 0$. In particular, we note that, in the currently favoured case $(\Omega_{\mathrm{m},0}, \Omega_{\Lambda,0}) = (0.3, 0.7)$, we may explore a large comoving volume by observing objects in the redshift range $z = 2$–3.

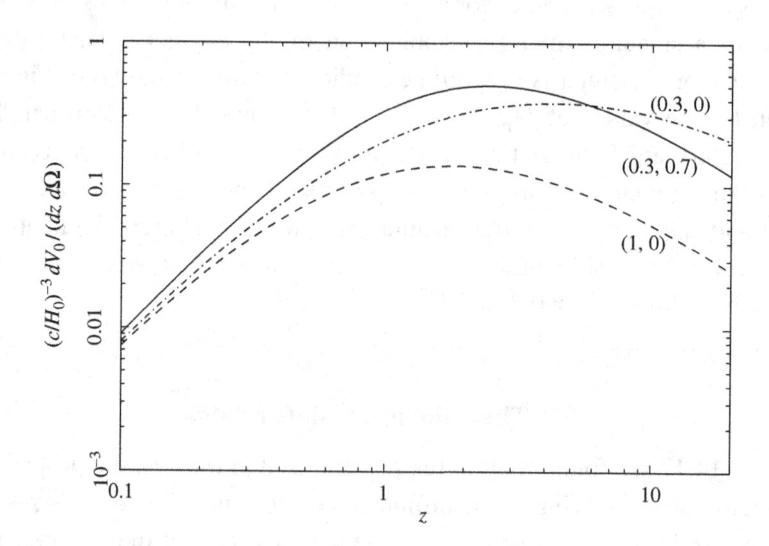

Figure 15.9 The variation in the dimensionless differential comoving volume element as a function of redshift z for several sets of values $(\Omega_{\mathrm{m},0}, \Omega_{\Lambda,0})$ as indicated, assuming that $\Omega_{\mathrm{r},0}$ is negligible.

15.10 Evolution of the density parameters

For the majority of our discussion so far, we have concentrated on exploring cosmological models with properties determined by fixing the values of the *present-day* densities $\Omega_{m,0}$, $\Omega_{r,0}$ and $\Omega_{\Lambda,0}$. From the definition (15.5), however, it is clear that each density is, in general, a function of cosmic time t. It is therefore of interest to investigate the evolution of these densities as the universe expands.

From (15.5) we have

$$\Omega_i(t) = \frac{8\pi G}{3H^2(t)}\rho_i(t) \quad \Rightarrow \quad \dot{\Omega}_i = \frac{8\pi G}{3H^2}\left(\dot{\rho}_i - \frac{2\dot{H}}{H}\rho_i\right), \tag{15.44}$$

where the label i denotes 'm', 'r' or 'Λ' and the dots denote differentiation with respect to cosmic time t. From the equation of motion (14.39) for a cosmological fluid, however, we have

$$\dot{\rho}_i = -3(1+w_i)H\rho_i,$$

where we have written $H = \dot{R}/R$, and $w_i = p_i/(\rho_i c^2)$ is the equation-of-state parameter. Thus (15.44) becomes

$$\dot{\Omega}_i = -\Omega_i H\left[3(1+w_i) + \frac{2\dot{H}}{H^2}\right], \tag{15.45}$$

where we have taken a factor of H outside the brackets for later convenience. We now need an expression for $\dot{H}$, which is given by

$$\dot{H} = \frac{d}{dt}\left(\frac{\dot{R}}{R}\right) = \frac{\ddot{R}}{R} - \left(\frac{\dot{R}}{R}\right)^2 = \frac{\ddot{R}}{R} - H^2,$$

and so we may write

$$\frac{\dot{H}}{H^2} = \frac{R\ddot{R}}{\dot{R}^2} - 1 = -(q+1),$$

where q is the deceleration parameter. Substituting this result into (15.45) and using the expression (15.14) for q, we finally obtain the neat relation

$$\dot{\Omega}_i = \Omega_i H(\Omega_m + 2\Omega_r - 2\Omega_\Lambda - 1 - 3w_i).$$

Setting $w_i = 0$, $\frac{1}{3}$ and -1 respectively for matter (dust), radiation and the vacuum, we thus obtain

$$\boxed{\begin{aligned} \dot{\Omega}_m &= \Omega_m H[(\Omega_m - 1) + 2\Omega_r - 2\Omega_\Lambda], \\ \dot{\Omega}_r &= \Omega_r H[\Omega_m + 2(\Omega_r - 1) - 2\Omega_\Lambda], \\ \dot{\Omega}_\Lambda &= \Omega_\Lambda H[\Omega_m + 2\Omega_r - 2(\Omega_\Lambda - 1)]. \end{aligned}} \tag{15.46}$$

By dividing these equations by one another, we may remove the dependence on the Hubble parameter H and the cosmic time t and hence obtain a set of coupled first-order differential equations in the variables Ω_m, Ω_r and Ω_Λ alone. Therefore, given some general point in this parameter space, these equations define a unique trajectory that passes through this point. As an illustration, let us consider the case in which $\Omega_r = 0$. Dividing the remaining two equations then gives

$$\frac{d\Omega_\Lambda}{d\Omega_m} = \frac{\Omega_\Lambda[\Omega_m - 2(\Omega_\Lambda - 1)]}{\Omega_m[(\Omega_m - 1) - 2\Omega_\Lambda]},$$

which defines a set of trajectories (or 'flow lines') in the $(\Omega_m, \Omega_\Lambda)$-plane. This equation also highlights the significance of the points $(1, 0)$ and $(0, 1)$ in this plane, which act as 'attractors' for the trajectories. This is illustrated in Figure 15.10, which shows a set of trajectories for various cosmological models. Since any general point in the plane defines a unique trajectory passing through that point, it is convenient to specify each trajectory by the present-day values $\Omega_{m,0}$ and $\Omega_{\Lambda,0}$ (although one could equally well use the values at any other cosmic time). In the left-hand panel, we plot trajectories passing through $\Omega_{m,0} = 0.3$ and $\Omega_{\Lambda,0} = 0.1, 0.2, \ldots, 1.1$, and in the right-hand panel the trajectories pass through $\Omega_{\Lambda,0} = 0.7$ and $\Omega_{m,0} = 0.1, 0.2, \ldots, 1.1$. We see that the trajectories all start at $(1, 0)$, which is an unstable fixed point, and converge on $(0, 1)$, which is a stable fixed point.

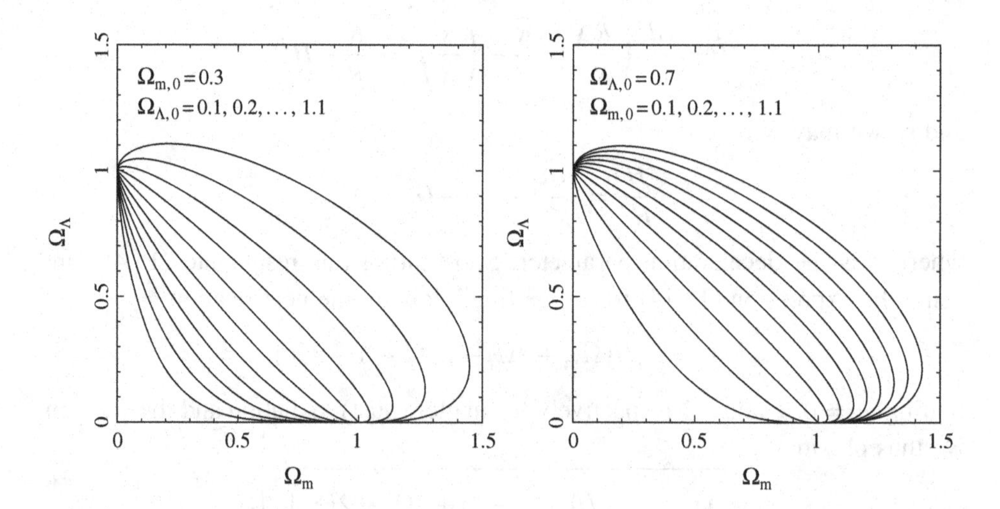

Figure 15.10 Evolution of the density parameters Ω_m and Ω_Λ for various cosmological models passing through the points $\Omega_{m,0} = 0.3$ and $\Omega_{\Lambda,0} = 0.1, 0.2, \ldots, 1.1$ (left-hand panel) and $\Omega_{\Lambda,0} = 0.7$ and $\Omega_{m,0} = 0.1, 0.2, \ldots, 1.1$ (right-hand panel).

It is worth noting the profound effect of a non-zero cosmological constant on the evolution of the density parameters. In the case $\Lambda = 0$, any slight deviation from $\Omega_m = 1$ in the early universe results in a rapid evolution away from the point $(1, 0)$ along the Ω_m-axis, tending to $(0, 0)$ for an open universe and to $(\infty, 0)$ for a closed one. If $\Lambda > 0$, however, the trajectory is 'refocussed' and tends to the spatially flat de Sitter case $(0, 1)$. Indeed, for a wide range of initial conditions, by the time the matter density has reached $\Omega_m \approx 0.3$ the universe is close to spatially flat.

15.11 Evolution of the spatial curvature

We may investigate directly the behaviour of the spatial curvature as the universe expands by determining the evolution of the curvature density parameter

$$\Omega_k = 1 - \Omega_m - \Omega_r - \Omega_\Lambda = -\frac{c^2 k}{H^2 R^2}. \tag{15.47}$$

Differentiating the final expression on the right-hand side with respect to cosmic time, or combining the derivatives (15.46), one quickly finds that

$$\boxed{\dot{\Omega}_k = 2\Omega_k H q = \Omega_k H(\Omega_m + 2\Omega_r - 2\Omega_\Lambda),} \tag{15.48}$$

where q is the deceleration parameter. We observe that if $\Omega_\Lambda = 0$ then the quantity in parentheses is always positive. Thus, in this case, if Ω_k differs slightly from zero at some early cosmic time then the spatial curvature rapidly evolves away from the spatially flat case. In particular, $\Omega_k \to 1$ in the open case and $\Omega_k \to -\infty$ in the closed case. The presence of a positive cosmological constant, however, changes this behaviour completely. In this case, at some finite cosmic time the $2\Omega_\Lambda$ term in (15.48) will dominate the matter and radiation terms, with the result that Ω_k is 'refocussed' back to $\Omega_k = 0$.

We may in fact obtain an analytic expression for the spatial curvature as a function of redshift z, in terms of the present-day values of the density parameters. Substituting for $c^2 k$ from (15.47) evaluated at $t = t_0$, and noting that $R_0/R = 1 + z$, we obtain the useful general formula

$$\Omega_k(z) = \left[\frac{H_0(1+z)}{H(z)}\right]^2 \Omega_{k,0}.$$

Using our expression (15.38) for $H(z)$ then gives

$$\boxed{\Omega_k(z) = \frac{\Omega_{k,0}}{\Omega_{m,0}(1+z) + \Omega_{r,0}(1+z)^2 + \Omega_{\Lambda,0}(1+z)^{-2} + \Omega_{k,0}}.}$$

In particular, we see that (apart from models with only vacuum energy), even if the present-day value $\Omega_{k,0}$ differs greatly from zero, at very high redshift (i.e. in the distant past) $\Omega_k(z)$ must have differed by only a tiny amount from zero. Since today we measure the value $\Omega_{k,0}$ to be (conservatively) in the range -0.5 to 0.5, this means that at very early epochs Ω_k must have been very finely tuned to near zero. This tuning of the initial conditions of the expansion is called the *flatness problem* and has no solution within standard cosmological models. From our above discussion, however, the presence of a positive cosmological constant goes some way to explaining why the universe is close to spatially flat at the present epoch.

15.12 The particle horizon, event horizon and Hubble distance

Thus far, we have considered the evolution of the entire spatial part of the FRW geometry. It is, however, interesting to consider the extent of the region 'accessible' (via light signals) to some comoving observer at a given cosmic time t.

Particle horizon

Let us consider a comoving observer O situated (without loss of generality) at $\chi = 0$. Suppose further that a second comoving observer E has coordinate χ_1 and emits a photon at cosmic time t_1, which reaches O at time t. Assuming light to be the fastest possible signal, the only signals emitted at time t_1 that O receives by the time t are from radial coordinates $\chi < \chi_1$.

The comoving coordinate χ_1 of the emitter E is determined by

$$\chi_1 = c\int_{t_1}^{t} \frac{d\bar{t}}{R(\bar{t})}. \tag{15.49}$$

If the integral on the right-hand side diverges as $t_1 \to 0$ then χ_1 can be made as large as we please by taking t_1 sufficiently small. Thus, in this case, in principle it is possible to receive signals emitted at sufficiently early epochs from *any* comoving particle (such as a typical galaxy). If, however, the integral converges as $t_1 \to 0$ then χ_1 can never exceed a certain value for a given t. In this case our vision of the universe is limited by a *particle horizon*. At any given cosmic time t, the χ-coordinate of the particle horizon is given by

$$\boxed{\chi_p(t) = c\int_0^t \frac{d\bar{t}}{R(\bar{t})} = c\int_0^{R(t)} \frac{dR}{R\dot{R}},} \tag{15.50}$$

where in the second equality we have rewritten the expression as an integral over R. The corresponding proper distance to the particle horizon is $d_p(t) = R(t)\chi_p(t)$.

We see that expression (15.50) will be *finite* if $R\dot{R} \sim R^{\alpha}$ with $\alpha < 1$, which is equivalent to the condition $\ddot{R} < 0$. Hence, *any* universe for which the expansion has been continually *decelerating* up to the cosmic time t will have a finite particle horizon at that time. Clearly, this includes all the Friedmann models that we discussed earlier, but particle horizons also occur in other cosmological models, for example in the spatially flat Lemaitre model with $\Omega_{\mathrm{m},0} \approx 0.3$ and $\Omega_{\Lambda,0} \approx 0.7$ that seems to provide a reasonable description of our universe.

On differentiating (15.50) with respect to t, we have $d\chi_{\mathrm{p}}/dt = c/R(t)$, which is always greater than zero. Thus, the particle horizon of a comoving observer grows as the cosmic time t increases, and so parts of the universe that were not in view previously must gradually come into view. This does not mean, however, that a galaxy that was not visible at one instant suddenly appears in the sky a moment later! To understand this, we note that if the universe has a big-bang origin then we have $R(t_1) \to 0$ as $t_1 \to 0$, and so $z \to \infty$. Thus, the particle horizon at any given cosmic time is the surface of *infinite redshift*, beyond which we cannot see. If the particle horizon grew to encompass a galaxy, the galaxy would therefore appear at first with an infinite redshift, which would gradually reduce as more cosmic time passed. Hence the galaxy would not simply 'pop' into view.[7]

In fact, we can obtain explicit expressions for the particle horizon in some cosmological models. For example, a matter-dominated model at early epochs obeys $R(t)/R_0 = (t/t_0)^{2/3}$, whereas a radiation-dominated model at early epochs obeys $R(t)/R_0 = (t/t_0)^{1/2}$. Substituting these expressions into (15.50) gives the proper distance to the particle horizon at cosmic time t as

$$d_{\mathrm{p}}(t) = 3ct \quad \text{(matter-dominated)}, \qquad d_{\mathrm{p}}(t) = 2ct \quad \text{(radiation-dominated)}.$$

These proper distances are larger than ct because the universe has expanded while the photon has been travelling. Alternatively, if one has an analytic expression for $\chi(z)$ for some cosmological model then the corresponding expression for χ_{p} may be obtained simply by letting $z \to \infty$.

The existence of particle horizons for the common cosmological models illustrates the *horizon problem*, i.e. how do vastly separated regions display the same physical characteristics (e.g. the nearly uniform temperature of the cosmic microwave background) when, according to standard cosmological models, these regions could never have been in causal contact? This problem, like the flatness problem, is a serious challenge to standard cosmology that can best be resolved by invoking the theory of inflation (see Chapter 16).

[7] In practice, our view of the universe is not limited by our particle horizon but by the epoch of recombination, which occurred at $z_{\mathrm{rec}} \approx 1500$ (long before the formation of any galaxies). Prior to this epoch, the universe was ionised and photons were frequently scattered by the free electrons, whereas after this point electrons and protons (and neutrons) combined to form atoms and the photons were able to propagate freely. This *surface of last scattering* is therefore the effective limit of our observable universe.

The horizon problem can be illustrated by a simple example. Consider a galaxy at a proper distance of 10^9 light years away from us. Since the age of the universe is $\sim 1.5 \times 10^{10}$ years, there has been sufficient time to exchange about 15 light signals with the galaxy. At earlier times, when the scale factor R was smaller, everything was closer together and so we might have naively expected that this would improve causal contact. In a continuously decelerating universe, however, it makes the problem *worse*. At, for example, the epoch of *recombination* (when the cosmic microwave background photons were emitted) the redshift z was approximately 1000, so $R(t_{\text{rec}})/R_0 \approx 10^{-3}$ and the proper distance to the 'galaxy' is 10^6 light years.[8] If we assume, for simplicity, that after t_{rec} the expansion followed a matter-dominated Einstein–de-Sitter universe, then

$$\left(\frac{t_{\text{rec}}}{t_0}\right)^{2/3} = \frac{R_{\text{rec}}}{R_0} = 10^{-3},$$

and so $t_{\text{rec}} = 1.5 \times 10^{5.5}$ years. However, assuming that prior to t_{rec} the expansion followed a radiation-dominated Einstein–de-Sitter model, the proper distance to the particle (causal) horizon is $2ct_{\text{rec}} = 3 \times 10^{5.5}$ light years. Thus, by t_{rec} 'we' could not have exchanged even *one* light signal with the other 'galaxy'.

Event horizon

Although our particle horizon grows as the cosmic time t increases, in some cosmological models there could be events that we may *never* see (or, conversely, never influence). Returning to our expression (15.49), we see that if the integral on the right-hand side diverges as $t \to \infty$ (or the time at which R equals zero again), then it will be possible to receive light signals from any event. However, if the integral instead converges for large t then, for light signals emitted at t_1, we will only ever receive those from events for which the χ-coordinate is less than

$$\boxed{\chi_{\text{e}}(t_1) = c\int_{t_1}^{t_{\max}} \frac{dt}{R(t)},}$$

where $t_{\max}$ is either infinity or the time of the big crunch (i.e. $R(t_{\max}) = 0$). This is called the *event horizon*. By symmetry, $\chi_{\text{e}}(t_0)$ is the maximum χ-coordinate that can be reached by a light signal sent by us today.

Hubble distance

From our discussion in Section 15.7, the elapsed cosmic time t since the big bang is, in general, of the order $H^{-1}(t)$, which is known as the Hubble time and

[8] In reality the galaxy would not yet have formed, but this does not affect the main point of the argument.

provides a characteristic time scale for the expansion of the universe. In a similar way, at a cosmic time t one can define the *Hubble distance*

$$\boxed{d_{\rm H}(t) = cH^{-1}(t),}$$

which provides a characteristic length scale for the universe. We may also define the *comoving Hubble distance*

$$\chi_{\rm H}(t) = \frac{d_{\rm H}(t)}{R(t)} = \frac{c}{H(t)R(t)} = \frac{c}{\dot{R}(t)}, \tag{15.51}$$

where in the last equality we have used the fact that $H = \dot{R}/R$. The above expression simply gives the χ-coordinate corresponding to the Hubble distance.

The Hubble distance $d_{\rm H}(t)$ corresponds to the typical length scale (at cosmic time t) over which physical processes in the universe operate coherently. It is also the length scale at which general-relativistic effects become important; indeed, on length scales much less than $d_{\rm H}(t)$, Newtonian theory is often sufficient to describe the effects of gravitation. From our discussion above, we further note that the proper distance to the particle horizon for standard cosmological models is typically

$$d_{\rm p}(t) \sim ct \sim cH^{-1}(t).$$

Thus, we see that the particle horizon in such cases is of the same order as the Hubble distance. As a result, the Hubble distance is often described simply as the 'horizon'. It should be noted, however, that the particle horizon and the Hubble distance are distinct quantities, which may differ by many orders of magnitude in *inflationary cosmologies*, which we discuss in the next chapter. In particular, we note that the particle horizon at time t depends on the entire expansion history of the universe to that point, whereas the Hubble distance is defined instantaneously at t. Moreover, once an object lies within an observer's particle horizon it remains so. On the contrary, an object can be within an observer's Hubble distance at one time, lie outside it at some later time and even come back within it at a still later epoch.

Exercises

15.1 For blackbody radiation, the number density of photons with frequencies in the range $[\nu, \nu + d\nu]$ is given by

$$n(\nu, T)\, d\nu = \frac{8\pi\nu^2}{c^3\,(e^{h\nu/kT} - 1)}\, d\nu, \tag{E15.1}$$

where T is the 'temperature' of the radiation. By conserving the total number of photons, show that the photon energy distribution of the cosmic microwave background (CMB) radiation retains its general blackbody form as the universe expands. Show further that the total number density n of photons is

$$n(T) = 0.244 \left(\frac{2\pi k_B T}{hc}\right)^3.$$

Hence show that the present-day number density of CMB photons in the universe is $n_0 \approx 4 \times 10^8\,\mathrm{m}^{-3}$, and compare this with the present-day number density of protons. How does this ratio vary with cosmic time?

Hint: $\int_0^\infty \frac{x^2}{e^x - 1}\,dx = 0.244\pi^2$.

15.2 Suppose that the present-day energy densities of radiation and matter (in the form of dust) are $\rho_r(t_0)c^2$ and $\rho_m(t_0)c^2$ respectively. Show that the energy densities of the two components were equal at a redshift z_{eq} given by

$$1 + z_{eq} = \frac{\rho_m(t_0)}{\rho_r(t_0)}.$$

What assumptions underlie this result? Hence show that

$$1 + z_{eq} = \frac{3c^2\Omega_{m,0}H_0^2}{8\pi G a T_0^4},$$

where a is the reduced Stefan–Boltzmann constant and T_0 is the present-day temperature of the cosmic microwave background. Show that for our universe $z_{eq} \approx 5000$. What was the temperature of the CMB radiation at this epoch?

15.3 Show that in the early, radiation-dominated, phase of the universe, the temperature T of the radiation satisfies the equation

$$\left(\frac{\dot{T}}{T}\right)^2 = \frac{8\pi G a T^4}{3},$$

where the dot denotes differentiation with respect to the cosmic time t and a is the reduced Stefan–Boltzmann constant. Hence show that

$$T = \left(\frac{3c^2}{32\pi G a}\right)^{1/4} t^{-1/2} \approx 1.5 \times 10^{10} \left(\frac{t}{\mathrm{s}}\right)^{-1/2}\,\mathrm{K},$$

and that the cosmic time at matter–radiation equality is $t_{eq} \approx 16\,000$ years.

15.4 The CMB radiation was emitted at the epoch of recombination at redshift $z_{rec} \approx 1500$. Show that $t_{rec} \approx 450\,000$ years.

15.5 Consider a cylindrical piston chamber of cross-sectional area A 'filled' with vacuum energy. The piston is withdrawn a linear distance dx. Show that the energy created by withdrawing the piston equals the work done by the vacuum, provided that

$$p_{vac} = -\rho_{vac}c^2.$$

Hence show that, in this case, the vacuum energy density is constant as the piston is withdrawn.

15.6 Show that the present-day value of the scale factor of the universe may be written as

$$R_0 = \frac{c}{H_0}\left(\frac{k}{\Omega_{\mathrm{k},0}}\right)^{1/2}.$$

What value does R_0 take in a spatially flat universe?

15.7 Show that, for our universe to be spatially flat, the total density must be equivalent to ≈ 5 protons m^{-3}.

15.8 In the Newtonian cosmological model discussed in Exercise 14.14, show that the total energy E of the test particle of mass m can be written as

$$E = \tfrac{1}{2}m(1-\Omega_{\mathrm{m}})R^2H^2,$$

and interpret this result physically.

15.9 Show that at all cosmic times the density parameters obey the relation

$$\Omega_{\mathrm{m}} + \Omega_{\mathrm{r}} + \Omega_{\Lambda} + \Omega_k = 1.$$

15.10 In terms of the dimensionless density parameters, show that the two cosmological field equations can be written in the forms

$$H^2 = H_0^2\left[\Omega_{\mathrm{m},0}\,a^{-3} + \Omega_{\mathrm{r},0}\,a^{-4} + \Omega_{\Lambda,0} + \Omega_{k,0}\,a^{-2}\right],$$
$$q = \tfrac{1}{2}(\Omega_{\mathrm{m}} + 2\Omega_{\mathrm{r}} - 2\Omega_{\Lambda}),$$

where H and q are the Hubble and deceleration parameters respectively, and $a = R/R_0$ is the normalised scale factor.

15.11 The *conformal time* variable is defined by $d\eta = c\,dt/R$. Hence show that the second cosmological field equation can be written as

$$\left(\frac{da}{d\eta}\right)^2 = -\frac{k}{\Omega_{k,0}}(\Omega_{\mathrm{m},0}\,a + \Omega_{\mathrm{r},0} + \Omega_{\Lambda,0}a^4 + \Omega_{k,0}\,a^2).$$

15.12 Show that the density parameter for matter, radiation or the vacuum varies with the normalised scale factor as

$$\Omega_i = \Omega_{i,0}\left(\frac{H_0}{H}\right)^2 a^{-3(1+w_i)},$$

where w_i is the appropriate equation-of-state parameter.

15.13 Show that the condition for the $a(t)$-curve to have a turning point is

$$f(a) \equiv \Omega_{\Lambda,0}a^3 + (1 - \Omega_{\mathrm{m},0} - \Omega_{\Lambda,0})a + \Omega_{\mathrm{m},0} = 0.$$

In the case $\Omega_{\Lambda,0} > 0$, show by evaluating the derivatives $f'(a)$ and $f''(a)$ that the condition for $f(a)$ to have a single positive root at $a = a_*$ is $f(a_*) = f'(a_*) = 0$. Show further that this root occurs at

$$a_* = \left(\frac{\Omega_{\mathrm{m},0}}{2\Omega_{\Lambda,0}}\right)^{1/3}.$$

Hence show that the values $\Omega_{m,0}$ and $\Omega_{\Lambda,0}$, along any dividing line in this plane that separates those models with a turning point in the $a(t)$-curve from those without, must satisfy

$$4(1-\Omega_{m,0}-\Omega_{\Lambda,0})^3+27\Omega_{m,0}^2\Omega_{\Lambda,0}=0.$$

15.14 Show that the substitution $x=[\Omega_{\Lambda,0}/(4\Omega_{m,0})]^{1/3}$ reduces the final cubic equation in Exercise 15.13 to

$$x^3-\frac{3x}{4}+\frac{\Omega_{m,0}-1}{4\Omega_{m,0}}=0.$$

By using the standard formulae for the roots of a cubic, or otherwise, verify the results (15.18–15.20).

15.15 Show that, in terms of the variable $\hat{t}=H_0(t-t_0)$, the evolution of the normalised scale factor obeys the equation

$$\left(\frac{da}{d\hat{t}}\right)^2=\Omega_{m,0}a^{-1}+\Omega_{r,0}a^{-2}+\Omega_{\Lambda,0}a^2+1-\Omega_{m,0}-\Omega_{r,0}-\Omega_{\Lambda,0}.$$

Show that, when one is integrating this equation numerically, an iterative algorithm of the form

$$a_{n+1}\approx a_n+\left(\frac{da}{d\hat{t}}\right)_n\Delta\hat{t}$$

would not be able to propagate the solution through points for which $da/d\hat{t}=0$.

15.16 For a $k=-1$ Friedmann model containing no matter or radiation, show that the line element becomes

$$ds^2=c^2dt^2-c^2t^2[d\chi^2+\sinh^2\chi\,(d\theta^2+\sin^2\theta\,d\phi^2)].$$

Show that this metric describes a Minkowski spacetime.

15.17 For a dust-only Friedmann model with $\Omega_{m,0}>1$, show that

$$a=\frac{\Omega_{m,0}}{2(\Omega_{m,0}-1)}(1-\cos\psi),\qquad t=\frac{\Omega_{m,0}}{2H_0(\Omega_{m,0}-1)^{3/2}}(\psi-\sin\psi).$$

Hence show that the $a(t)$-curve has a maximum at

$$a_{\max}=\frac{\Omega_{m,0}}{\Omega_{m,0}-1},\qquad t_{\max}=\frac{\pi}{2H_0}\frac{\Omega_{m,0}}{(\Omega_{m,0}-1)^{3/2}},$$

and that the age t_0 of such a universe is given by

$$t_0=\frac{\Omega_{m,0}}{2H_0(\Omega_{m,0}-1)^{3/2}}\left[\cos^{-1}\left(\frac{2}{\Omega_{m,0}}-1\right)-\frac{2}{\Omega_{m,0}}(\Omega_{m,0}-1)^{1/2}\right]<\frac{2}{3H_0}.$$

15.18 For the Einstein–de-Sitter model, prove the following useful results:

$$a(t)=\left(\frac{t}{t_0}\right)^{2/3},\qquad H(t)=\frac{2}{3t}=H_0(1+z)^{3/2},\qquad q_0=\frac{1}{2},\qquad \rho_m(t)=\frac{1}{6\pi Gt^2}.$$

15.19 For a radiation-only Friedmann model with $\Omega_{r,0} \neq 1$, show that

$$a(t) = (2H_0\Omega_{r,0}^{1/2}t)^{1/2}\left(1+\frac{1-\Omega_{r,0}}{2\Omega_{r,0}^{1/2}}H_0t\right)^{1/2}.$$

Hence, for $\Omega_{r,0} > 1$, show that the $a(t)$-curve has a maximum at

$$a_{\max} = \left(\frac{\Omega_{r,0}}{\Omega_{r,0}-1}\right)^{1/2}, \qquad t_{\max} = \frac{1}{H_0}\frac{\Omega_{r,0}^{1/2}}{(\Omega_{r,0}-1)},$$

and that the age t_0 of such a universe is given by

$$t_0 = \frac{1}{H_0}\frac{1}{\Omega_{r,0}^{1/2}+1} < \frac{1}{2H_0}.$$

15.20 For the spatially flat, radiation-only, Friedmann model, prove the following useful results:

$$a(t) = \left(\frac{t}{t_0}\right)^{1/2}, \qquad H(t) = \frac{1}{2t} = H_0(1+z)^2, \qquad q_0 = 1, \qquad \rho_r(t) = \frac{3}{32\pi G t^2}.$$

15.21 For a spatially flat Friedmann model containing both matter and radiation, show that

$$H_0t = \frac{2}{3\Omega_{m,0}^2}\left[(\Omega_{m,0}a+\Omega_{r,0})^{1/2}(\Omega_{m,0}a-2\Omega_{r,0})+2\Omega_{r,0}^{3/2}\right].$$

15.22 For a Lemaitre model containing no radiation, show that at the point of inflection of the $a(t)$-curve the value of the normalised scale factor is

$$a_* = \left(\frac{\Omega_{m,0}}{2\Omega_{\Lambda,0}}\right)^{1/3},$$

and calculate a_* for our universe. Show further that, in the vicinity of the point of inflection, the scale factor obeys the equation

$$\dot{a}^2 \approx H_0^2[\Omega_{k,0}+3\Omega_{\Lambda,0}a_*^2+3\Omega_{\Lambda,0}(a-a_*)^2]$$

and that this has the solution

$$a(t) = a_* + a_*\left[1+\tfrac{1}{3}\Omega_{k,0}\left(\tfrac{1}{4}\Omega_{\Lambda,0}\Omega_{m,0}^2\right)^{-1/3}\right]^{1/2}\sinh\left[H_0(3\Omega_{\Lambda,0})^{1/2}(t-t_*)\right].$$

15.23 For a spatially flat Lemaitre model containing no radiation, show that

$$H_0t = \frac{2}{3\sqrt{|\Omega_{\Lambda,0}|}}\int_0^{\sqrt{a^3|\Omega_{\Lambda,0}|/(1-\Omega_{\Lambda,0})}}\frac{dy}{\sqrt{1\pm y^2}}.$$

Hence show that

$$a(t) = \left(\frac{1-\Omega_{\Lambda,0}}{|\Omega_{\Lambda,0}|}\right)^{1/3}\begin{cases}\sinh^{2/3}\left(\tfrac{3}{2}\sqrt{\Omega_{\Lambda,0}}H_0t\right) & \text{if } \Omega_{\Lambda,0} > 0,\\ \sin^{2/3}\left(\tfrac{3}{2}\sqrt{|\Omega_{\Lambda,0}|}H_0t\right) & \text{if } \Omega_{\Lambda,0} < 0.\end{cases}$$

15.24 Show that, in general,

$$\frac{\ddot{R}}{R} = H^2 + \dot{H}.$$

Hence use the cosmological field equations to show that, for a spatially flat Lemaitre model containing no radiation, the Hubble parameter and the matter density satisfy the equations

$$2\dot{H} + 3H^2 = \Lambda c^2,$$

$$3H^2 - \Lambda c^2 = 8\pi G \rho_{\rm m}.$$

Assuming $\Lambda > 0$ and requiring $\rho_{\rm m} > 0$, thus show that

$$H(t) = \sqrt{\frac{\Lambda c^2}{3}} \coth\left(\frac{3}{2}\sqrt{\frac{\Lambda c^2}{3}}\,t\right),$$

$$\rho_{\rm m}(t) = \frac{\Lambda c^2}{8\pi G} \operatorname{cosech}^2\left(\frac{3}{2}\sqrt{\frac{\Lambda c^2}{3}}\,t\right),$$

and therefore find expressions for $\Omega_{\rm m}(t)$ and $\Omega_\Lambda(t)$. Show further that

$$t = \frac{2}{3H}\frac{\tanh^{-1}\sqrt{\Omega_\Lambda}}{\sqrt{\Omega_\Lambda}}.$$

Hint: $\int a/(a^2 - x^2)\, dx = \coth^{-1}(x/a) + \text{constant}$, for $x^2 > a^2$.

15.25 Show that for a physically reasonable perfect fluid (i.e. density > 0 and pressure ≥ 0) there is no static isotropic homogeneous solution to Einstein's equations with $\Lambda = 0$. Show that it is possible to obtain a static zero-pressure solution by the introduction of a cosmological constant Λ such that

$$\Lambda c^2 = 4\pi G \rho_{\rm m,0} = \frac{c^2 k}{R_0^2}.$$

Show that this solution is unstable, however.

15.26 Show that the comoving χ-coordinate of a galaxy emitting a photon at time t that is received at t_0 is given by

$$\chi = \frac{c}{R_0}\int_{(1+z)^{-1}}^{1} \frac{da}{a\dot{a}}.$$

Using the cosmological field equation (15.13) to substitute for $\dot{a}$, show that

$$\chi(z) = \frac{c}{R_0 H_0}\int_{(1+z)^{-1}}^{1} \frac{dx}{\sqrt{\Omega_{\rm m,0}x + \Omega_{\rm r,0} + \Omega_{\Lambda,0}x^4 + \Omega_{k,0}x^2}}.$$

15.27 For a dust-only Friedmann model, show that the luminosity–distance relation varies with redshift as

$$d_{\rm L}(z) = \frac{2c}{H_0\Omega_{\rm m,0}^2}\left[\Omega_{\rm m,0}z + (\Omega_{\rm m,0} - 2)\left(\sqrt{\Omega_{\rm m,0}z + 1} - 1\right)\right].$$

This result is known as *Mattig's formula.*

15.28 For a Friedmann model dominated by a single source of energy density, show that

$$\Omega_i^{-1}(z) - 1 = \frac{\Omega_{i,0}^{-1} - 1}{(1+z)^{1+3w_i}},$$

where w_i is the equation-of-state parameter of the source. Use this result to comment on the flatness problem.

15.29 For a general cosmological model, show that

$$\dot{\Omega}_i = \Omega_i H(\Omega_{\rm m} + 2\Omega_{\rm r} - 2\Omega_\Lambda - 1 - 3w_i),$$

where i denotes matter, radiation or the vacuum.

15.30 By differentiating the definition $\Omega_k = -kc^2/(R^2H^2)$, show that

$$\dot{\Omega}_k = 2\Omega_k H q = \Omega_k H(\Omega_{\rm m} + 2\Omega_{\rm r} - 2\Omega_\Lambda),$$

where q is the deceleration parameter.

15.31 Show that the particle horizon at cosmic time t is given

$$\chi_{\rm p}(t) = \frac{c}{R_0 H_0}\int_0^{a(t)} \frac{dx}{\sqrt{\Omega_{{\rm m},0}x + \Omega_{{\rm r},0} + \Omega_{\Lambda,0}x^4 + \Omega_{k,0}x^2}}.$$

15.32 Consider the cosmological line element

$$ds^2 = c^2dt^2 - e^{2t/b}(dr^2 + r^2d\theta^2 + r^2\sin^2\theta\, d\phi^2).$$

Light signals from a galaxy at coordinate distance r are emitted at epoch t_1 and received by an observer at epoch t_0. Show that

$$\frac{r}{bc} = e^{-t_1/b} - e^{-t_0/b}.$$

For a given r, show that there is a maximum epoch t_1 and interpret this result physically. Show that a light ray emitted by the observer asymptotically approaches the coordinate $r = bc$ but never reaches it.

16

Inflationary cosmology

In the last two sections of the previous chapter, we saw that standard cosmological models suffer, in particular, from the flatness problem and the horizon problem. To these problems, one might also add the 'expansion problem', which asks simply why the universe is expanding at all. Although this appears as an initial condition in cosmological models, one would hope to explain this phenomenon with an underlying physical mechanism. In this chapter, we therefore augment our discussion of cosmological models with a brief outline of the inflationary scenario, which seeks to solve these problems (and others) and has, over the past two decades, become a fundamental part of modern cosmological theory.[1] In particular, we will discuss the effect of inflation on the evolution of the universe as a whole and also consider how inflation gives rise to perturbations in the early universe that subsequently collapse under gravity to form all the structure we observe in the universe today. Given the general algebraic complexity of these topics (particularly the perturbation analysis), we will adopt the convention throughout this chapter that

$$\boxed{8\pi G = c = 1.}$$

This choice of units makes many of the equations far less cluttered and amounts only to a rescaling of the scalar field and its potential (see below), which we can remove at the end if desired.

16.1 Definition of inflation

As noted in Section 15.12, the horizon problem is a direct consequence of the *deceleration* in the expansion of the universe. Thus, a possible solution is to

[1] For a detailed discussion of inflationary cosmology, see, for example, A. Liddle & D. Lyth, *Cosmological Inflation and Large-scale Structure*, Cambridge University Press, 2000.

postulate an *accelerating* phase of expansion, prior to any decelerating phase. In an accelerating phase, causal contact is better at earlier times and so remotely separated parts of our present universe could have 'coordinated' their physical characteristics in the early universe. Such an accelerating phase is called a period of *inflation*. Hence the basic definition of inflation is that

$$\boxed{\ddot{R} > 0.} \tag{16.1}$$

In fact, we may recast this condition in an alternative manner that is physically more meaningful by considering the comoving Hubble distance defined in (15.51), namely $\chi_H(t) = H^{-1}(t)/R(t)$. The derivative with respect to t is given by

$$\frac{d}{dt}\left(\frac{H^{-1}}{R}\right) = \frac{d}{dt}\left(\frac{1}{\dot{R}}\right) = -\frac{\ddot{R}}{\dot{R}^2},$$

and so the condition (16.1) can be written as

$$\boxed{\frac{d}{dt}\left(\frac{H^{-1}}{R}\right) < 0.}$$

Thus, an equivalent condition for inflation is that the comoving Hubble distance *decreases* with cosmic time. Hence, when viewed in comoving coordinates, the characteristic length scale of the universe becomes *smaller* as inflation proceeds.

Let us suppose that, at some period in the early universe, the energy density is dominated by some form of matter with density ρ and pressure p. The first cosmological field equation (14.36) (with $\Lambda = 0$ and $8\pi G = c = 1$) then reads

$$\ddot{R} = -\tfrac{1}{6}(\rho + 3p)R. \tag{16.2}$$

Thus, we see that in order for the universe to accelerate, i.e. for inflation to occur, we require that

$$\boxed{p < -\tfrac{1}{3}\rho.} \tag{16.3}$$

In other words, we need the 'matter' to have an equation of state with negative pressure. In fact, the above criterion can also solve the flatness problem. The second cosmological field equation (with $\Lambda = 0$ and $8\pi G = c = 1$) reads

$$\dot{R}^2 = \tfrac{1}{3}\rho R^2 - k. \tag{16.4}$$

During a period of acceleration ($\ddot{R} > 0$), the scale factor must increase faster than $R(t) \propto t$. Provided that $p < -\frac{1}{3}\rho$, the quantity ρR^2 will increase during such a period as the universe expands and can make the curvature term on the right-hand side negligible, provided that the accelerating phase persists for sufficiently long.

We could, of course, have included the cosmological-constant terms in the two field equations, which would then be equivalent to those for a fluid with an equation of state $\rho = -p$ and so would clearly satisfy the criterion (16.3). However, we have chosen to omit such terms since, as we will see, if 'matter' in the form of a *scalar field* exists in the early universe then this can act as an *effective cosmological constant.* In order to show that the existence of such fields is likely, we must consider briefly the topic of *phase transitions* in the very early universe.

16.2 Scalar fields and phase transitions in the very early universe

The basic physical mechanism for producing a period of inflation in the very early universe relies on the existence, at such epochs, of matter in a form that can be described classically in terms of a scalar field (as opposed to a vector, tensor or spinor field, examples of which are provided by the electromagnetic field, the gravitational field and normal baryonic matter respectively). Upon quantisation, a scalar field describes a collection of spinless particles.

It may at first seem rather arbitrary to postulate the presence of such scalar fields in the very early universe. Nevertheless, their existence is suggested by our best theories for the fundamental interactions in Nature, which predict that the universe experienced a succession of *phase transitions* in its early stages as it expanded and cooled. For the purposes of illustration, let us model this expansion by assuming that the universe followed a standard radiation-dominated Friedmann model in its early stages, in which case

$$R(t) \propto t^{1/2} \propto \frac{1}{T(t)}, \qquad (16.5)$$

where the 'temperature' T is related to the typical particle energy by $T \sim E/k_B$. The basic scenario is as follows.

- $E_P \sim 10^{19}$ GeV $> E > E_{GUT} \sim 10^{15}$ GeV The earliest point at which the universe can be modelled (even approximately) as a classical system is the *Planck era*, corresponding to particle energies $E_P \sim 10^{19}$ GeV (or temperature $T_P \sim 10^{32}$ K) and time scales $t_P \sim 10^{-43}$ s (prior to this epoch, it is considered that the universe can be described only in terms of some, as yet unknown, quantum theory of gravity). At these extremely high energies, grand unified theories (GUTs) predict that the electroweak and strong forces are in fact *unified* into a single force and that these interactions bring the particles present into thermal equilibrium. Once the universe has cooled to $E_{\rm GUT} \sim 10^{15}$ GeV (corresponding to $T_{\rm GUT} \sim 10^{27}$ K), there is a spontaneous breaking of the larger symmetry group characterising the GUT into a product of smaller symmetry groups, and the electroweak and strong forces separate. From (16.5), this GUT phase transition occurs at $t_{\rm GUT} \sim 10^{-36}$ s.

- $E_{\rm GUT} \sim 10^{15}\,{\rm GeV} > E > E_{\rm EW} \sim 100\,{\rm GeV}$ During this period (which is extremely long in logarithmic terms), the electroweak and strong forces are separate and these interactions sustain thermal equilibrium. This continues until the universe has cooled to $E_{\rm EW} \sim 100\,{\rm GeV}$ (corresponding to $T_{\rm EW} \sim 10^{15}$ K), when the unified electroweak theory predicts that a second phase transition should occur in which the electromagnetic and weak forces separate. From (16.5), this electroweak phase transition occurs at $t_{\rm EW} \sim 10^{-11}$ s.
- $E_{\rm EW} \sim 100\,{\rm GeV} > E > E_{\rm QH} \sim 100\,{\rm MeV}$ During this period the electromagnetic, weak and strong forces are separate, as they are today. It is worth noting, however, that when the universe has cooled to $E_{\rm QH} \sim 100\,{\rm MeV}$ (corresponding to $T_{\rm QH} \sim 10^{12}$ K) there is a final phase transition, according to the theory of quantum chromodynamics, in which the strong force increases in strength and leads to the confinement of quarks into hadrons. From (16.5), this quark–hadron phase transition occurs at $t_{\rm QH} \sim 10^{-5}$ s.

In general, phase transitions occur via a process called *spontaneous symmetry breaking*, which can be characterised by the acquisition of certain non-zero values by *scalar* parameters known as *Higgs fields*. The symmetry is manifest when the Higgs fields have the value zero; it is spontaneously broken whenever at least one of the Higgs fields becomes non-zero. Thus, the occurrence of phase transitions in the very early universe suggests the existence of scalar fields and hence provides the motivation for considering their effect on the expansion of the universe. In the context of inflation, we will confine our attention to scalar fields present at, or before, the GUT phase transition (the most speculative of these phase transitions).

16.3 A scalar field as a cosmological fluid

For simplicity, let us consider a single scalar field ϕ present in the very early universe. The field ϕ is traditionally called the *'inflaton' field* for reasons that will become apparent shortly. The Lagrangian for a scalar field ϕ (see Section 19.6) has the usual form of a kinetic term minus a potential term:

$$L = \tfrac{1}{2} g^{\mu\nu}(\partial_\mu \phi)(\partial_\nu \phi) - V(\phi).$$

The corresponding field equation for ϕ is obtained from the Euler–Lagrange equations and reads

$$\Box^2\phi + \frac{dV}{d\phi} = 0, \tag{16.6}$$

where $\Box^2 \equiv \nabla^\mu \nabla_\mu = g^{\mu\nu}\nabla_\mu \nabla_\nu$ is the *covariant* d'Alembertian operator. A simple example is a *free* relativistic scalar field of mass m, for which the potential would be $V(\phi) = \frac{1}{2}m^2\phi^2$ and the field equation becomes the covariant Klein–Gordon equation,

$$\Box^2\phi + m^2\phi = 0.$$

For the moment, however, it is best to keep the potential function $V(\phi)$ general.

The energy–momentum tensor $T_{\mu\nu}$ for a scalar field can be derived from this variational approach (see Section 19.12), but in fact we can use our earlier experience to anticipate its form. By analogy with the forms of the energy–momentum tensor for dust and for electromagnetic radiation, we require that $T_{\mu\nu}$ is (i) symmetric and (ii) quadratic in the derivatives of the dynamical variable ϕ, and (iii) that $\nabla_\mu T^{\mu\nu} = 0$ by virtue of the field equation (16.6). It is straightforward to show that the required form must be

$$T_{\mu\nu} = (\partial_\mu\phi)(\partial_\nu\phi) - g_{\mu\nu}\left[\tfrac{1}{2}(\partial_\sigma\phi)(\partial^\sigma\phi) - V(\phi)\right]. \tag{16.7}$$

The energy–momentum tensor for a perfect fluid is

$$T_{\mu\nu} = (\rho + p)u_\mu u_\nu - p g_{\mu\nu},$$

and by comparing the two forms in a Cartesian inertial coordinate system ($g_{\mu\nu} = \eta_{\mu\nu}$) in which the fluid is at rest, we see that the scalar field acts like a perfect fluid, with an energy density and pressure given by

$$\boxed{\begin{aligned} \rho_\phi &= \tfrac{1}{2}\dot{\phi}^2 + V(\phi) + \tfrac{1}{2}(\vec{\nabla}\phi)^2, \\ p_\phi &= \tfrac{1}{2}\dot{\phi}^2 - V(\phi) - \tfrac{1}{6}(\vec{\nabla}\phi)^2. \end{aligned}} \tag{16.8}$$

In particular, we note that if the field ϕ were both temporally and spatially constant, its equation of state would be $p_\phi = -\rho_\phi$ and so the scalar field would act as a *cosmological constant* with $\Lambda = V(\phi)$ (with $8\pi G = c = 1$). In general this is not the case, but we will assume that the spatial derivatives can be neglected. This is equivalent to assuming that ϕ is a function only of t and so has no spatial variation.

16.4 An inflationary epoch

Let us suppose that the scalar field does not interact (except gravitationally) with any other matter or radiation that may be present. In this case, the scalar field will independently obey an equation of motion of the form (14.39), namely

$$\dot{\rho}+3(\rho+p)\frac{\dot{R}}{R}=0.$$

Substituting the expressions (16.8) and assuming no spatial variations, we quickly find that the equation of motion of the scalar field is

$$\boxed{\ddot{\phi}+3H\dot{\phi}+\frac{dV}{d\phi}=0.} \tag{16.9}$$

The form of this equation will be familiar to any student of classical mechanics and allows one to develop an intuitive picture of the evolution of the scalar field. If one thinks of the plot of the potential V versus ϕ as defining some curve, then the motion of the scalar field value ϕ is identical to that of a ball rolling (or, more precisely, sliding) under gravity along the curve, subject to a frictional force proportional to its speed (and to the value of the Hubble parameter).

Let us assume further that there is some period when the scalar field dominates the energy density of the universe. Moreover we will demand that the scalar field energy density is sufficiently large that we may neglect the curvature term in the cosmological field equation (16.4) although this is not strictly necessary.[2] Thus, we may write (16.4) as

$$\boxed{H^2=\tfrac{1}{3}\left[\tfrac{1}{2}\dot{\phi}^2+V(\phi)\right].} \tag{16.10}$$

This equation and (16.9) thus provide a set of coupled differential equations in ϕ and H that determine completely the evolution of the scalar field and the scale factor of the universe during the epoch of scalar-field domination. From our criterion (16.3) and the expressions (16.8), we see that inflation will occur (i.e. $\ddot{R}>0$) provided that

$$\boxed{\dot{\phi}^2<V(\phi).} \tag{16.11}$$

[2] Note that, even if the curvature term is not negligible to begin with, the initial stages of inflation will soon render it so.

16.5 The slow-roll approximation

The inflation equations (16.9) and (16.10) can easily be solved numerically, and even analytically for some special choices of $V(\phi)$. In general, however, an analytical solution is only possible in the *slow-roll approximation*, in which it is assumed that $\dot{\phi}^2 \ll V(\phi)$. On differentiating, this in turn implies that $\ddot{\phi} \ll dV/d\phi$ and so the $\ddot{\phi}$-term can be neglected in the equation of motion (16.9), to yield

$$\boxed{3H\dot{\phi} = -\frac{dV}{d\phi}.} \tag{16.12}$$

Moreover, the cosmological field equation (16.10) becomes simply

$$\boxed{H^2 = \tfrac{1}{3}V(\phi).} \tag{16.13}$$

It is worth noting that, in this approximation, the rate of change of the Hubble parameter and the scalar field can be related very easily. Differentiating (16.13) with respect to t and combining the result with (16.12), one obtains

$$\dot{H} = -\tfrac{1}{2}\dot{\phi}^2. \tag{16.14}$$

The conditions for inflation in the slow-roll approximation can be put into a useful dimensionless form. Using the two equations above and the condition $\dot{\phi}^2 \ll V(\phi)$, it is easy to show that

$$\epsilon \equiv \frac{1}{2}\left(\frac{V'}{V}\right)^2 \ll 1, \tag{16.15}$$

where $V' \equiv dV/d\phi$ and the factor $\frac{1}{2}$ is included according to the standard convention. Differentiating the above expression with respect to ϕ, one also finds that

$$\eta \equiv \frac{V''}{V} \ll 1. \tag{16.16}$$

These two conditions make good physical sense in that they require the potential $V(\phi)$ to be sufficiently 'flat' that the field ϕ 'rolls' slowly enough for inflation to occur. It is worth noting, however, that these conditions alone are necessary but not sufficient conditions for inflation, since they limit only the form of $V(\phi)$ and not that of $\dot{\phi}$, which could be chosen to violate the condition (16.11). Thus, one must also assume that (16.11) holds.

It is worth considering the special case in which the potential $V(\phi)$ is sufficiently flat that, during (some part of) the period of inflation, its value remains roughly constant. From (16.12), we see that in this case the Hubble parameter is constant and the scale factor grows *exponentially*:

$$\boxed{R(t) \propto \exp\left(\sqrt{\tfrac{1}{3}V(\phi)}\,t\right).}$$

16.6 Ending inflation

As the field value ϕ 'rolls' down the potential $V(\phi)$, the condition (16.11) will eventually no longer hold and inflation will cease. Equivalently, in the slow-roll approximation, the conditions (16.15, 16.16) will eventually no longer be satisfied. If the potential $V(\phi)$ possesses a local minimum, which is usually the case in most inflationary models, the field will no longer roll slowly downhill but will oscillate about the minimum of the potential, the oscillation being gradually damped by the $3H\dot{\phi}$ friction term in the equation of motion (16.9). Eventually, the scalar field is left stationary at the bottom of the potential. If the value of the potential at its minimum is $V_{\text{min}} > 0$ then clearly the condition (16.11) is again satisfied and the universe continues to inflate indefinitely. Moreover, in this case $p_\phi = -\rho_\phi$ and so the scalar field acts as an effective cosmological constant $\Lambda = V_{\text{min}}$. If $V_{\text{min}} = 0$, however, no further inflation occurs, the scalar field has zero energy density and the dynamics of the universe is dominated by any other fields present.

In fact, the scenario outlined above would occur only if the scalar field were not coupled to any other fields, which is almost certainly not the case. In practice, such couplings will cause the scalar field to decay during the oscillatory phase into pairs of elementary particles, into which the energy of the scalar field is thus converted. The universe will therefore contain roughly the same energy density as it did at the start of inflation. The process of decay of the scalar field into other particles is therefore termed *reheating*. These particles will interact with each other and subsequently decay themselves, leaving the universe filled with normal matter and radiation in thermal equilibrium and thereby providing the initial conditions for a standard cosmological model.

16.7 The amount of inflation

Although the motivation for the introduction of the inflationary scenario was (in part) to solve the flatness and horizon problems, we have not yet considered the amount of inflation required to achieve this goal. From our present understanding of particle physics, it is thought that inflation occurs at around the era of the GUT phase transition, or earlier. For illustration, let us assume that the universe has followed a standard radiation-dominated Friedmann model for (the majority of) its history since the epoch of inflation at $t \sim t_*$. From (16.5), we thus have

$$\frac{R_*}{R_0} \sim \left(\frac{t_*}{t_0}\right)^{1/2} \sim \frac{T_0}{T_*}, \qquad (16.17)$$

where $T_0 \sim 3\,\mathrm{K}$ is the present-day temperature of the cosmic microwave background radiation and $t_0 \sim 1/H_0 \sim 10^{18}$ s is the present age of the universe.

Let us first consider the flatness problem. From (15.47), the ratio of the spatial curvature density at the inflationary epoch to that at the present epoch is given by

$$\frac{\Omega_{k,*}}{\Omega_{k,0}} = \left(\frac{H_0}{H_*}\right)^2 \left(\frac{R_0}{R_*}\right)^2 \sim \frac{t_*}{t_0}, \qquad (16.18)$$

where we have used the fact that $H_0/H_* \sim t_*/t_0$. Assuming inflation to occur at some time between the Planck era and the GUT phase transition, so that $t_\mathrm{P} < t_* < t_\mathrm{GUT}$, from Section 16.2 we find that the ratio (16.18) lies in the range $\sim 10^{-60}$–10^{-54}. Thus, if the present-day value $\Omega_{k,0}$ is of order unity then the required degree of fine-tuning of $\Omega_{k,*}$ is extreme, in a standard cosmological model. Since the ratio above depends on $1/R_*^2$ we thus find that, to solve the flatness problem (in order that Ω_k can also be of order unity prior to inflation), we require the scale factor to grow during inflation by a factor $\sim 10^{27}$–10^{30}. In terms of the required number N of e-foldings of the scale factor, we thus have

$$\boxed{N \gtrsim 60\text{–}70 \qquad \text{(flatness problem).}}$$

We now turn to the horizon problem. If the universe followed a standard radiation-dominated Friedmann model in its earliest stages, then (reinstating c for the moment) the particle horizon at the inflationary epoch is

$$d_{\mathrm{p},*} = 2ct_*,$$

which, taking $t_\mathrm{P} < t_* < t_\mathrm{GUT}$, gives the size of a causally connected region at this time as $\sim 10^{-34}$–10^{-27} m. From (16.17), we see that the size of such a region today would be only $\sim 10^{-3}$–1 m. The current size of the observable universe, however, is given approximately by the present-day Hubble distance,

$$d_{\mathrm{H},0} = cH_0^{-1} \sim 10^{26}\,\mathrm{m} \sim 3000\,\mathrm{Mpc}.$$

To solve the horizon problem, we thus require the scale factor to grow by a factor of $\sim 10^{26}$–10^{29} during the period of inflation. Expressing this result in terms of the required number N of e-foldings, we once again find

$$\boxed{N \gtrsim 60\text{–}70 \qquad \text{(horizon problem).}}$$

We have thus found that *both* the flatness and the horizon problems can be solved by a period of inflation, provided that the scale factor undergoes more than around 60–70 e-foldings during this period. We may now consider the constraints placed by this condition on the form of the scalar field potential $V(\phi)$. In the slow-roll approximation, the number of e-foldings that occur while the scalar field 'rolls' from ϕ_1 to ϕ_2 is given by

$$\boxed{N = \int_{t_1}^{t_2} H\,dt = \int_{\phi_1}^{\phi_2} \frac{H}{\dot{\phi}}\,d\phi = -\int_{\phi_1}^{\phi_2} \frac{V}{V'}\,d\phi.}$$

If the potential is reasonably smooth then $V' \sim V/\phi$. Thus, if $\Delta\phi = |\phi_{\text{start}} - \phi_{\text{end}}|$ is the range of ϕ-values over which inflation occurs, one finds $N \sim (\Delta\phi)^2$. In order to solve the flatness and horizon problems, one hence requires $\Delta\phi \gg 1$.

16.8 Starting inflation

The observant reader will have noticed that so far we have not discussed how inflation may start. During the inflationary epoch, the scalar field rolls downhill from ϕ_{start} to ϕ_{end}, but we have not yet considered how the universe can arrive at an appropriate starting state. The details will depend, in fact, on the precise inflationary cosmology under consideration, but there are generally two main classes of model. In early models of inflation, the inflationary epoch is an 'interlude' in the evolution of a standard cosmological model. In such models, the inflaton field ϕ is usually identified with a scalar Higgs field operating during the GUT phase transition. It is thus assumed that the universe was in a state of thermal equilibrium from the very beginning and that this state was relatively homogeneous and large enough to survive until the beginning of inflation at the GUT era; an example of this sort is provided by the *'new' inflation* model discussed below in Section 16.9. In more recent models of inflation, the scalar field ϕ is not identified with the Higgs field in the GUT phase transition but is some generic scalar field present in the very early universe. In particular, in these models the universe may inflate soon after it exits the Planck era, thereby avoiding the above assumptions regarding the state of the universe prior to the inflationary epoch; an example of such a model is the *chaotic inflation* scenario discussed in Section 16.10. We will also discuss briefly the natural extension of the chaotic inflation model, called *stochastic inflation* (or *eternal inflation*) in Section 16.11.

16.9 'New' inflation

In the *'new' inflation* model,[3] the inflationary epoch occurs when the universe goes through the GUT phase transition. As we will see, models of this general type typically require a rather special form for the potential $V(\phi)$ in order to produce an effective period of inflation. In particular, identifying the inflaton field ϕ with the scalar Higgs field operating during the GUT spontaneous-symmetry-breaking phase transition, considerations from quantum field theory suggest a form for the potential $V(\phi, T)$ which is actually also a function of temperature T. The typical form for $V(\phi, T)$ is shown in Figure 16.1 for several values of T. At very high temperatures the potential is parabolic with a minimum at $\phi = 0$, which is the *true vacuum* state (i.e. the state of lowest energy). Thus at very high temperatures we would expect the scalar field to have the value $\phi = 0$. However, for lower temperatures the form of the potential changes until at the critical temperature

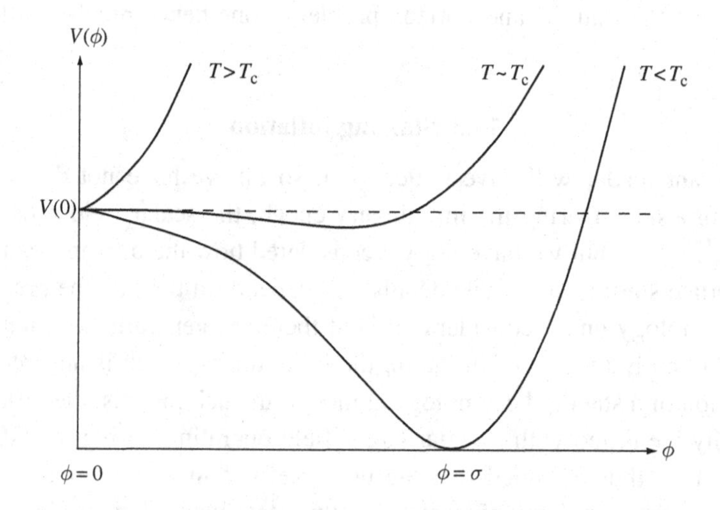

Figure 16.1 The temperature-dependent potential function for a Higgs-like scalar field ϕ.

[3] The 'new' inflationary model is so called in order to distinguish it from the original 'old' inflation model of Guth, in which the scalar Higgs field executed quantum mechanical tunnelling at $T \sim T_c$, where T_c is the critical temperature, from the metastable false ground state at $\phi = 0$ through a potential barrier to the true ground state with $\phi > 0$. Although this model provided the genesis for the inflationary idea, it was quickly shown to predict a universe very different to the one we observe. In short, the tunnelling process produces bubble nucleation and it turns out that these bubbles are too small to be identified with the observable universe and are carried apart too quickly by the intervening inflating space for them to coalesce, hence resulting in a highly inhomogeneous universe, contrary to observations.

$T = T_c$ the potential develops a lower energy state than that at $\phi = 0$. Thus this new non-zero value of ϕ is now the true vacuum state, and $\phi = 0$ is now a *false vacuum* state. For even lower values the new true vacuum state becomes more pronounced until a final form is reached for 'low' temperatures.

Let us now consider the evolution of the scale factor $R(t)$, the radiation energy density ρ_r and the scalar field ϕ.

Phase 1 When the temperature is very high, i.e. far above the GUT phase transition scale of $T_c \sim 10^{27}$ K, from Figure 16.1 we would expect the scalar field to have the value $\phi = 0$ (i.e. at the true vacuum state for these temperatures), and Figure 16.1 shows that it will remain at $\phi = 0$. Since $\rho_r \propto R^{-4}$, however, we would expect the radiation to dominate over the scalar field at very early epochs. Thus we have the standard early-time radiation-dominated Friedmann model, in which we can neglect the curvature constant k. Thus, for $T \gg T_c$,

$$\boxed{R \propto t^{1/2}, \qquad \rho_r \propto t^{-2}, \qquad \phi = 0.}$$

Phase 2 It is clear from the above equations that there will come a time when the scalar-field energy density dominates over that of the radiation. Provided that this occurs for $T > T_c$ the scalar field remains at $\phi = 0$, in which case it acts as an effective cosmological constant of value $\Lambda = V(0)$. Thus, in this phase, the scale factor undergoes an *exponential expansion*:

$$\boxed{R(t) \propto \exp\left(\sqrt{\frac{1}{3}V(0)}\,t\right).}$$

As a result of the exponential expansion, however, there is a corresponding exponential decrease in the temperature T, which results in a rapid change of the potential function. Thus $T \sim T_c$ is reached very quickly, and so this phase is *extremely short-lived*, and very little expansion is actually achieved. Indeed, if $T \sim T_c$ is reached before the scalar-field energy density dominates over that of the radiation then phase 2 does not occur at all.

Phase 3 Once $T \sim T_c$, we see from Figure 16.1 that the scalar field is now able to roll downhill away from $\phi = 0$ and so the GUT phase transition occurs. Provided that the potential is sufficiently flat, the slow-roll approximation holds and the universe inflates, the evolution of the scalar field being determined by (16.12) and the Hubble parameter by (16.13). If the potential is roughly constant then the *exponential* expansion continues. The rapid growth of the scale factor once again causes the evolution of the potential function as the temperature drops. The duration of this period of inflation depends critically on the flatness and length of the plateau of the $V(\phi)$ function for $T < T_c$. For certain 'reasonable' potentials the universe can easily inflate in such a way that the number of *e*-foldings $N \gtrsim 60$, and can be considerably larger. This is therefore the main *inflationary phase*. According to detailed calculations, phase 3 occurs between $t_1 \sim 10^{-36}$ s and $t_2 \sim 10^{-34}$ s and the scale factor increases by a factor of around 10^{50}.

Phase 4 Eventually, the slow-roll approximation fails and inflation ends. The scalar field then rolls rapidly down towards the true vacuum state at $\phi = \sigma$, oscillating about the minimum point, and follows the behaviour outlined in Section 16.6. In particular, if $V(\sigma) = 0$ then the universe will revert to the standard radiation-dominated Friedmann model with

$$\boxed{R(t) \propto t^{1/2}.}$$

Hence, at $t \sim 10^{-34}$ s, the universe starts a standard Friedmann expansion, albeit with the desired 'initial' conditions. Thus, the inflationary model incorporates all the observationally verified predictions of the standard cosmological models.

Although the 'new' inflation model still has its advocates, it suffers from undesirable features. In particular, the scenario only provides an effective period of inflation if $V(\phi, T)$ has a very flat plateau near $\phi = 0$, which is somewhat artificial. Moreover, the period of thermal equilibrium prior to the inflationary phase (so one can speak sensibly of the universe having a particular temperature) requires many particles to interact with one another, and so already one requires the universe to be very large and contain many particles. Finally, the universe could easily recollapse before inflation starts. As a result of these difficulties, new inflation may not be a viable model, and so there are strong theoretical reasons to believe that the inflaton field ϕ *cannot* be identified with the GUT symmetry-breaking Higgs field. Thus, the hope that GUTs could provide the mechanism for the homogeneity and flatness of the universe may have to be abandoned.

16.10 Chaotic inflation

In more recent models of inflation, the scalar field ϕ is not identified with the Higgs field in the GUT phase transition but is regarded as a generic scalar field present in the very early universe. In particular, these models invoke the idea of *chaotic inflation*. In this scenario, as the universe exits the Planck era at $t \sim 10^{-43}$ s the initial value of the scalar field ϕ_{start} is set chaotically, i.e. it acquires different random values in different regions of the universe. In some regions, ϕ_{start} is somewhat displaced from the minimum of the potential and

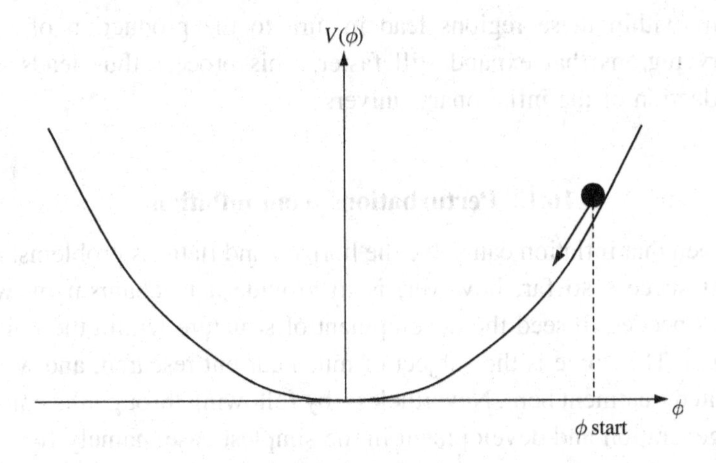

Figure 16.2 The potential $V(\phi) = \frac{1}{2}m^2\phi^2$ for a free scalar field. The field is initially displaced from the minimum of the potential due to chaotic initial conditions as the universe comes out of the Planck era.

so the field subsequently rolls downhill. If the potential is sufficiently flat, the field is more likely to be displaced a greater distance from its minimum and will roll slowly enough, and for a sufficiently prolonged period, for the region to undergo an effective period of inflation. Conversely, in other regions $\phi_{\rm start}$ may not be displaced sufficiently from the minimum of the potential for the region to inflate. Thus, on the largest scales the universe is highly inhomogeneous, but our observable universe lies (well) within a region that underwent a period of inflation.

According to this scenario, inflation may occur even in theories with very simple potentials, such as $V(\phi) \sim \phi^n$, and is thus a very generic process that can take place under a broad range of conditions. Indeed, the potential function need not depend on the temperature T. A very simple example is a free scalar field, for which $V(\phi) = \frac{1}{2}m^2\phi^2$ (see Figure 16.2). Moreover, in the chaotic scenario, inflation may begin even if there is no thermal equilibrium in the early universe, and it may even start just after the Planck epoch.

16.11 Stochastic inflation

A natural extension to the chaotic inflation model is the mechanism of *stochastic* (or *eternal*) inflation. The main idea in this scenario is to take account of quantum fluctuations in the evolution of the scalar field, which we have thus far ignored by modelling the field entirely classically. If, in the chaotic assignment of initial values of the scalar field, some regions have a large value of $\phi_{\rm start}$ then quantum fluctuations can cause ϕ to move further *uphill* in the potential $V(\phi)$. These regions inflate at a greater rate than the surrounding ones, and the fraction of the total volume of the universe containing the growing ϕ-field increases. Quantum fluctuations within these regions lead in turn to the production of some new inflationary regions that expand still faster. This process thus leads to eternal self-reproduction of the inflationary universe.

16.12 Perturbations from inflation

We have seen that inflation can solve the horizon and flatness problems. Arguably its greatest success so far, however, is to provide a mechanism by which the fluctuations needed to seed the development of structure within the universe can be generated. This topic is the subject of much current research, and we can give only a limited treatment here. Nevertheless, by following through the equations for structure generation and development in the simplest case, namely for a spatially flat universe with a simple 'gauge choice' (see below), we hope that the reader will be able to get a flavour of the physics involved.

The current opinion of how structure in the universe originated is that it was via amplification, during a period of inflation, of initial *quantum* irregularities of the scalar field that drives inflation. Thus what we need to do can be divided into two broad categories. First, we need to work out the equations of motion for spatial perturbations in the scalar field. This can be done classically, i.e. taking the scalar field as a classical source linked self-consistently to the gravitational field via a classical energy–momentum tensor. Second, we need to derive initial conditions for these perturbations, and this demands that we understand the quantum field theory of the perturbations themselves. This sounds formidable but actually turns out to be no more complicated than considering the quantum physics of a mass on a spring, albeit one in which the mass changes as a function of time. These topics are discussed in detail in the remainder of this chapter.

16.13 Classical evolution of scalar-field perturbations

We assume that the scalar field ϕ, which hitherto has been a function of cosmic time t only, now has perturbations that are functions of space and time. We can thus write

$$\boxed{\phi(t) \rightarrow \phi_0(t) + \delta\phi(t, \vec{x}).} \qquad (16.19)$$

These perturbations will lead to a perturbed energy–momentum tensor, which we shall derive shortly. The Einstein field equations then imply that the Einstein tensor is also perturbed away from its background value. In turn, therefore, we must have a metric different from the Friedmann–Robertson–Walker one assumed so far. We thus need to assume a form for this metric in order to calculate the new Einstein tensor. It is at this point we must make the choice of 'gauge' (i.e.

coordinate system) referred to above. Once perturbations are present there is no preferred way to define a spacetime slicing of the universe. The details of this are quite subtle but amount simply to the fact that by choosing different coordinate systems we can change the apparent character of the perturbations considerably. For example, suppose that we choose, as a new time coordinate, one for which surfaces of constant time have a constant value of the new perturbed scalar field on them. This is always possible and, in such a gauge, the spatial fluctuations of ϕ have apparently totally vanished!

To meet such problems, methods that deal only with gauge-invariant quantities have been developed. We will make contact with such methods below, when we introduce the so-called 'curvature' perturbations. These are gauge invariant and therefore represent physical quantities. To reach this point, however, we first work with a specific simple form of gauge known as the as the *longitudinal* or *Newtonian* gauge, and indeed with a restricted form of this – one where only one extra function (known here as a 'potential function') is introduced. The justification for using such a restricted form is that it leads to an Einstein tensor with the correct extra degrees of freedom to match the extra terms in the scalar-field energy–momentum tensor arising from the field perturbations.

For a spatially flat ($k = 0$) background FRW model, which is what we will assume, we adopt Cartesian comoving coordinates and write the perturbed metric as

$$\boxed{ds^2 = (1+2\Phi)\,dt^2 - (1-2\Phi)\,R^2(t)\left(dx^2+dy^2+dz^2\right),} \tag{16.20}$$

where Φ is a general infinitesimal function of all four coordinates (and should not to be confused with the scalar field ϕ). Its assumed smallness means that we will only need to consider quantities to first order in Φ. A general discussion of this *linearising* process is presented in the next chapter, but for the time being we simply note that one can consider Φ as representing the *Newtonian potential* of the perturbations. For instance, for a spherically symmetric perturbation of mass M and radius r, if we put $\Phi = -GM/rc^2$ then the first term of (16.20) recovers the tt-term of the Schwarzschild metric.

The perturbed Einstein field equations

We now need to find both the new energy–momentum tensor of the scalar field ϕ and the new Einstein tensor corresponding to our perturbed metric. Equating them will link our two perturbation variables $\delta\phi$ and Φ and provide us with the equations of evolution that we need. The first step is to calculate the connection coefficients corresponding to the perturbed metric (16.20) to first order in Φ. These are easily shown to take the form $\Gamma^\sigma{}_{\mu\nu} = (\Gamma_0)^\sigma{}_{\mu\nu} + \delta\Gamma^\sigma{}_{\mu\nu}$, where the first term corresponds to the connection coefficients of the unperturbed metric (i.e. with $\Phi = 0$) and the perturbation terms are given by

$$\begin{aligned}
\delta\Gamma^0{}_{0\mu} &= \partial_\mu\Phi, \\
\delta\Gamma^0{}_{ii} &= -R^2(\dot{\Phi}+4H\Phi), \\
\delta\Gamma^j{}_{ii} &= \delta^{jk}\partial_k\Phi \text{ (for } i \neq j), \\
\delta\Gamma^i{}_{00} &= \frac{1}{R^2}\delta^{ik}\partial_k\Phi, \\
\delta\Gamma^i{}_{i\mu} &= -\partial_\mu\Phi.
\end{aligned}$$

In these expressions, $H = \dot{R}/R$ is the Hubble parameter of the unperturbed background, and no sum over repeated i indices is implied. The remaining perturbed connection coefficients either follow from symmetry or are zero.

These connection coefficients yield a Riemann and hence an Einstein tensor. Again working to first order in Φ, the perturbed part of the Einstein tensor is found to be

$$\boxed{\begin{aligned}
\delta G^0_i &= -2\partial_i(\dot{\Phi}+H\Phi), \\
\delta G^0_0 &= -2(\vec{\nabla}^2\Phi - 3H\dot{\Phi} - 3H^2\Phi), \\
\delta G^i_i &= 2[\ddot{\Phi}+4\dot{\Phi}H+(2\dot{H}+3H^2)\Phi],
\end{aligned}} \tag{16.21}$$

where again no sum over repeated i indices is implied and the remaining entries either follow from symmetry or are zero. The symbol $\vec{\nabla}^2$ here denotes the spatial Laplacian, which in this simple flat case is given by

$$\vec{\nabla}^2 = \frac{1}{R^2}\left(\frac{\partial^2}{\partial x^2}+\frac{\partial^2}{\partial y^2}+\frac{\partial^2}{\partial z^2}\right). \tag{16.22}$$

It is worth noting that, in the entries of (16.21), the time derivative of the Hubble parameter appears. From (16.14), this can be rewritten as

$$\dot{H} = -\tfrac{1}{2}\dot{\phi}_0^2, \tag{16.23}$$

remembering that this equation now applies to the background FRW spacetime.

We also need to evaluate the perturbed part of the scalar-field energy–momentum tensor. Substituting (16.19) into (16.7) and working to first order in ϕ, one quickly finds that

$$\boxed{\begin{aligned}
\delta T^0_i &= \dot{\phi}_0\partial_i(\delta\phi), \\
\delta T^0_0 &= -\dot{\phi}_0^2\Phi+\dot{\phi}_0\delta\dot{\phi}+V'\delta\phi, \\
\delta T^i_i &= \dot{\phi}_0^2\Phi-\dot{\phi}_0\delta\dot{\phi}+V'\delta\phi,
\end{aligned}} \tag{16.24}$$

where $V' = dV/d\phi_0$ and the remaining components either follow from symmetry or are zero.

We may now use the Einstein field equations to relate the Einstein tensor and the scalar-field energy–momentum tensor. Since the unperturbed part of the field equations is automatically satisfied, one simply requires that $\delta G^{\nu}_{\mu} = -\delta T^{\nu}_{\mu}$ (since $\kappa = 8\pi G/c^4$ equals unity in our chosen system of units). We may thus equate, with the inclusion of a minus sign, the components shown in equations (16.21) and (16.24). At first sight, it is by no means obvious that we have allowed ourselves enough freedom in including only one extra function, Φ, in the metric. Nevertheless, as we now show, everything in fact works out. Let us start with the $\binom{0}{i}$-components, for which we have the equation

$$2\partial_i(\dot{\Phi} + H\Phi) = \dot{\phi}_0 \partial_i(\delta\phi). \tag{16.25}$$

Remembering that H and ϕ_0 have no spatial dependence, we can integrate this immediately to obtain

$$\boxed{\dot{\Phi} + H\Phi = \tfrac{1}{2}\dot{\phi}_0 \delta\phi.} \tag{16.26}$$

One next equates the $\binom{i}{i}$-components, which gives

$$-2[\ddot{\Phi} + 4\dot{\Phi}H + (2\dot{H} + 3H^2)\Phi] = \dot{\phi}_0^2\Phi - \dot{\phi}_0\delta\dot{\phi} + V'\delta\phi, \tag{16.27}$$

but we may show that this contains no information beyond that already obtained from the $\binom{0}{i}$-components. In particular, differentiating (16.26) with respect to time gives

$$\ddot{\Phi} + \dot{H}\Phi + H\dot{\Phi} = \tfrac{1}{2}\ddot{\phi}_0\delta\phi + \tfrac{1}{2}\dot{\phi}_0\delta\dot{\phi}; \tag{16.28}$$

then, using equations (16.9) and (16.23) to substitute for $\ddot{\phi}_0$ and $\dot{H}$ respectively, one finds that (16.27) is satisfied if (16.26) holds, thus establishing consistency. The only new information must therefore come from equating the $\binom{0}{0}$-components. Using (16.26) and eliminating V' again then yields

$$\boxed{\left(\dot{\phi}_0^2 + 2\vec{\nabla}^2\right)\Phi = \dot{\phi}_0^2 \frac{d}{dt}\left(\frac{\delta\phi}{\dot{\phi}_0}\right).} \tag{16.29}$$

Perturbation equations in Fourier space

The results (16.26) and (16.29) are the basic equations relating Φ and $\delta\phi$. To make further progress, however, it is convenient to work instead in terms of the Fourier decomposition of these quantities and analyse what happens to a perturbation corresponding to a given comoving spatial scale. Thus, we assume that Φ and $\delta\phi$ are decomposed into a superposition of plane-wave states with comoving wavevector $\vec{k}$, so that

$$\Phi(\vec{x}) = \frac{1}{(2\pi)^{3/2}} \int \Phi_{\vec{k}} \exp(i\vec{k}\cdot\vec{x})\, d^3\vec{k},$$

where (with a slight abuse of notation) $\vec{x} = (x, y, z)$ and a similar expression holds for $\delta\phi$. The evolution of a mode amplitude $\Phi_{\vec{k}}$ depends only on the comoving wavenumber $k = |\vec{k}|$; the corresponding actual physical wavenumber is $k/R(t)$. We thus work simply in terms of Φ_k and $\delta\phi_k$. In terms of these variables, the action of $\vec{\nabla}^2$ will be just to multiply Φ_k by $-k^2/R^2(t)$, whereas the time derivatives remain unchanged. Equations (16.26) and (16.29) therefore become

$$\boxed{\begin{aligned} \dot{\Phi}_k + H\Phi_k &= \tfrac{1}{2}\dot{\phi}_0\delta\phi_k, \\ \left(1 - \frac{2k^2}{R^2\dot{\phi}_0^2}\right)\Phi_k &= \frac{d}{dt}\left(\frac{\delta\phi_k}{\dot{\phi}_0}\right). \end{aligned}} \tag{16.30}$$

Thus, we see that we have obtained two coupled first-order differential equations for the quantities Φ_k and $\delta\phi_k$, which are the amplitudes of the plane-wave perturbations of comoving wavenumber k in the metric and in the scalar field respectively. Clearly, what we could do next is to eliminate one quantity in terms of derivatives of the other and then obtain a single second-order equation in terms of just one of them (plus the background quantities, of course, but the evolution of these is assumed known). In fact, this leads to rather messy equations and, moreover, in terms of the discussion given above the results are not *gauge invariant*, since neither Φ_k nor $\delta\phi_k$ is gauge invariant on its own.

16.14 Gauge invariance and curvature perturbations

As mentioned above, gauge invariance is related to how we define spatial 'slices' of the perturbed spacetime. By transforming to a new time coordinate, one can apparently make the perturbations in the scalar field come and go at will. There are two ways to take care of this difficulty. First, one can choose variables that are insensitive to such changes and therefore definitely describe something physical. These are called gauge-invariant variables. Second, one can use variables which would change if one altered the slicing but which are defined relative to a particular slicing that can itself be defined physically. These are then also physical variables and are, perhaps confusingly, also sometimes called gauge invariant, although this is not really a good description. Note that changing spatial coordinates within a particular slicing also induces changes, but these are not relevant to our discussion here and we concentrate just on changes in time coordinate.

Let us start our discussion by taking the first of the two routes outlined above, namely describing the perturbations in terms of truly gauge-invariant quantities. For any scalar function f in spacetime, consider the effects upon it of the change in time coordinate $t \to t' = t + \Delta t$. We may define a new, perturbed, function by

$$f'(t') = f(t), \tag{16.31}$$

where, as just stated, we suppress the $\vec{x}$-dependences in what follows. Thus, to first order in Δt, we may write

$$f'(t) = f'(t' - \Delta t) \approx f(t) - \dot{f}\,\Delta t \tag{16.32}$$

where we do not have to specify whether it is f or f' that is being differentiated with respect to time to obtain $\dot{f}$ or whether the latter is evaluated at t or t', since these would be second-order differences. Hence the perturbation in the scalar function due to the 'gauge transformation' $t \to t + \Delta t$ is given by[4]

$$\boxed{\Delta f = -\dot{f}\,\Delta t.} \tag{16.33}$$

We may now evaluate the change in the perturbed spacetime metric corresponding to the gauge transformation $t \to t + \Delta t$. To do this, however, one must distinguish between the two occurrences of the Φ-variable in (16.20). For an *arbitrary* scalar perturbation, the general form of the perturbed metric in fact takes the form

$$ds^2 = (1 + 2\Psi)\,dt^2 - (1 - 2\Phi)\,R^2(t)\left(dx^2 + dy^2 + dz^2\right), \tag{16.34}$$

in which Ψ and Φ are *different* functions. Nevertheless, for matter with no 'anisotropic stress' (so that all the off-diagonal components of the space part of the stress–energy tensor are zero), the two functions may be taken as equal; this is the case for a perfect fluid or a scalar field and hence leads to (16.20). Even in this case, however, the two functions behave differently under the gauge transformation. We need consider only the Φ-function above, which clearly takes the role of a *spatial curvature* term since it modifies the space part of the metric by a multiplicative factor. Under $t \to t + \Delta t$ we find that

$$R^2(1 - 2\Phi) \to R^2(1 - 2\Phi) + \left[2R\dot{R}(1 - 2\Phi) - 2R^2\dot{\Phi}\right]\Delta t. \tag{16.35}$$

[4] This is the simplest version of the 'Lie derivative', which describes the change in a (possibly tensor) function when 'dragged back' along 'flow lines' in parameter space; see, for example, B. Schutz, *Geometrical Methods of Mathematical Physics*, Cambridge University Press, 1980.

Since both Φ and Δt are infinitesimal, we may employ the same arguments that led to (16.33). Then, to first order, we have

$$\Delta\Phi = H\Delta t, \tag{16.36}$$

where we have also used the fact that $R\dot{R} = R^2 H$. Thus, for *any* scalar function f with perturbations δf, we see that the combination

$$\zeta_f = \Phi + \frac{H\,\delta f}{\dot{f}} \tag{16.37}$$

is *gauge invariant* under the gauge transformation $t \to t + \Delta t$ that we are considering, since to first order we have

$$\zeta_f \to \zeta_f' = \Phi + H\Delta t + \frac{H(\delta f - \dot{f}\Delta t)}{\dot{f}} = \zeta_f. \tag{16.38}$$

Thus, for the specific example of our scalar-field perturbation $\delta\phi$, we may identify the corresponding gauge-invariant quantity as

$$\boxed{\zeta = \Phi + \frac{H\,\delta\phi}{\dot{\phi}_0}.} \tag{16.39}$$

We will therefore use this variable (or its Fourier transform) in our subsequent discussion in later sections. In the literature this quantity is called the *curvature perturbation*, for reasons that will become clear shortly.

Before going on to consider the evolution of these curvature perturbations, let us first discuss briefly the second route outlined at the start of this section for defining a physically meaningful perturbation variable. This route can be illustrated directly with the Φ-function, and one begins by defining the quantity

$$\boxed{\mathcal{R} \equiv -\Phi|_{\text{co}},} \tag{16.40}$$

where the subscript indicates that Φ is to be evaluated on comoving slices. By 'comoving' we mean a time-slicing that is orthogonal to the worldlines of the 'fluid' that makes up the matter. For an ordinary fluid, this would amount to choosing frames in which, at each instant and position, the fluid appears to be at rest. The same applies here and, because the frame involved is physically defined, the variable $\mathcal{R}$, which measures the spatial curvature in the given frame, is itself physically well defined. Thus the quantity $\mathcal{R}$ is also called the 'curvature perturbation' in the literature. As we now show, it is in fact equal to minus the variable ζ defined in (16.39), and so both may be described as such.

For any scalar density perturbation $\delta\rho$, one can write the spatial curvature in the comoving slice as

$$\mathcal{R} = -\zeta_\rho + \frac{H\delta\rho_{\rm co}}{\dot{\rho}}. \tag{16.41}$$

Let us therefore consider what happens for the particular case of a perturbation in a scalar field. As shown in (16.24), the $\binom{0}{i}$-components of the perturbed stress–energy tensor read

$$\delta T_i^0 = \dot{\phi}_0\, \partial_i(\delta\phi). \tag{16.42}$$

In the comoving frame, this momentum density must vanish, by definition, and so the scalar-field perturbation cannot depend on the spatial coordinates and thus vanishes. Hence, for a scalar field, we have

$$\boxed{\mathcal{R} = -\zeta.} \tag{16.43}$$

16.15 Classical evolution of curvature perturbations

We now consider the evolution of the Fourier transform of the gauge-invariant perturbation (16.39), namely

$$\boxed{\zeta_k \equiv \Phi_k + H\frac{\delta\phi_k}{\dot{\phi}_0},} \tag{16.44}$$

which is clearly itself gauge invariant. Using (16.30), the second-order differential equation satisfied by this quantity is quite simply shown to be

$$\boxed{\ddot{\zeta}_k + \left(\frac{\dot{\phi}_0^2}{H} + \frac{2\ddot{\phi}_0}{\dot{\phi}_0} + 3H\right)\dot{\zeta}_k + \frac{k^2}{R^2}\zeta_k = 0.} \tag{16.45}$$

Given a potential $V(\phi_0)$ and some initial conditions for H and ϕ_0, we can integrate the background evolution equations numerically and obtain H and ϕ_0 as functions of cosmic time t. If we simultaneously integrate ζ_k using (16.45), we can thereby trace the evolution of the curvature perturbation over the time period of interest. An example of the results of this procedure is shown in Figures 16.3 and 16.4,

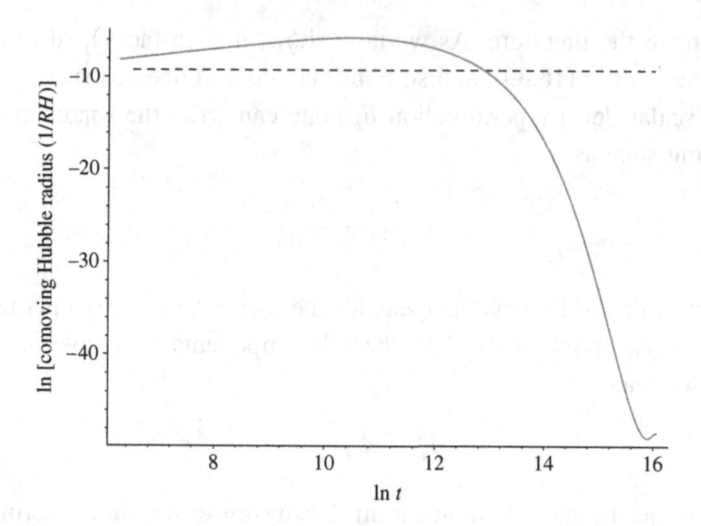

Figure 16.3 Evolution of the logarithm of the comoving Hubble distance $\ln[1/(RH)]$ versus $\ln t$ (solid line) in a chaotic inflation model driven by a free scalar field of mass $m \sim 2 \times 10^{-6}$, the initial values of H and ϕ_0 being chosen in such a way that there is a period of inflation lasting approximately for the period $\ln t \approx 11$–16. Also shown (broken line) is the fixed comoving scale $1/k$, where $k = 10^4$ is the comoving wavenumber of the perturbation shown in Figure 16.4. Note that all quantities are in Planck units.

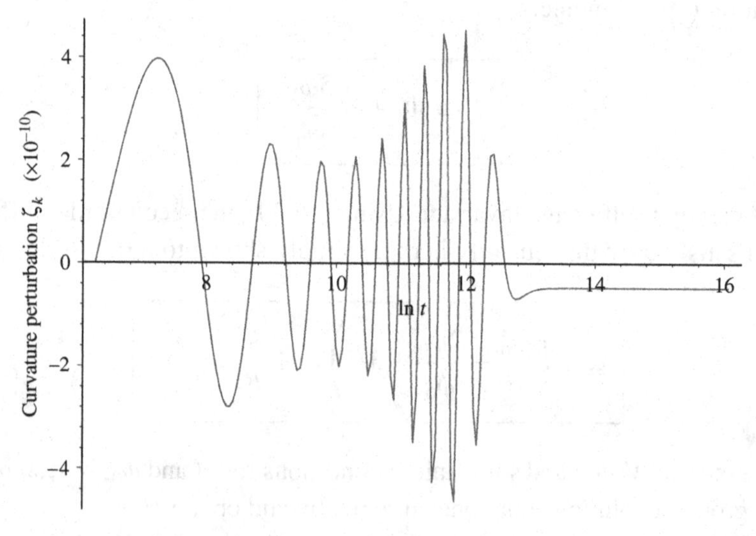

Figure 16.4 Evolution of the curvature perturbation ζ_k versus $\ln t$ for $k = 10^4$ in a chaotic inflation model driven by a free scalar field of mass $m \sim 2 \times 10^{-6}$. Note that all quantities are in Planck units.

for the particular choice of potential $V(\phi_0) = \frac{1}{2}m^2\phi_0^2$ (chaotic inflation) with $m \sim 2 \times 10^{-6}$ (a typical value in such theories). The initial conditions for H and ϕ_0 were chosen to give inflation over the period $\ln t \approx 11$–16, and the comoving wavenumber of the perturbation[5] was chosen as $k = 10^4$.

From Figure 16.3, one can verify that the universe is indeed inflating during the period $\ln t \approx 11$–16, since the comoving Hubble distance $1/(RH)$ is *decreasing* with cosmic time (see Section 16.1). In inflationary theory, this quantity is loosely called the 'horizon' but must be distinguished from the 'particle horizon', as discussed in Section 15.12. The broken line in Figure 16.3 is the natural logarithm of the reciprocal of the comoving wavenumber k, which is of course constant for a given perturbation. This reciprocal, $1/k$, gives another dimensionless scale and (ignoring possible factors of 2π that, one could argue, should be introduced) can be thought of as the comoving wavelength scale of the perturbation itself.

The behaviour of the curvature perturbation ζ_k is shown in Figure 16.4 for $k = 10^4$ and can be understood from the behaviour of the comoving Hubble distance (or horizon) in Figure 16.3.[6] Whilst the perturbation scale $1/k$ is less than the horizon radius $1/(RH)$ the curvature perturbation ζ_k just oscillates. Once the comoving horizon radius has dropped below $1/k$, however, we see that (at $\ln t \sim 13$) the perturbation suddenly 'freezes' and no longer oscillates. We speak of this moment, when $1/k$ becomes greater than $1/(RH)$, as the perturbation 'leaving the horizon' and, in intuitive terms, we can understand that beyond this point the perturbation is no longer able to feel its own self-gravity, since it is larger than the characteristic scale over which physical processes in the universe operate coherently. The curvature perturbation thereafter remains frozen at whatever value it has reached at this point until much later in the history of the universe, when the comoving horizon scale eventually catches up with $1/k$ again. At this point, the perturbation is said to 're-enter the horizon', and oscillations will begin again (though at this stage it is not expected that these will be in the scalar field itself, since the latter is thought to decay into other particles, via the process of *reheating*, shortly after inflation ends – see Section 16.6).

The key point to note is that, via inflation, one has produced 'super-horizon' scale fluctuations in the early universe. These fluctuations later go on to provide the seeds for galaxy formation and the perturbations in the cosmic microwave

[5] Note that all quantities here are measured in Planck units, e.g. the masses are in inverse Planck lengths and the times in Planck times.

[6] The initial conditions used for examining the classical behaviour of ζ_k can of course be chosen arbitrarily. The starting values of ζ_k and its time derivative used in Figure 16.4 in fact correspond to 'quantum' conditions, where field-theoretic values for the initial fluctuation are set. This is discussed in Section 16.6 below, where a new variable ξ_k, related to ζ_k, is introduced. The specific values used correspond to evaluating the imaginary part of equation (16.51) and its time derivative, followed by a global phase shift such that the initial phase is zero.

background radiation that we observe today. By studying the distribution of galaxies and CMB fluctuations as a function of scale, it is possible to obtain an idea of the underlying primordial spectrum of perturbations that produced them. Thus, by predicting this primordial spectrum, we can perform a test of the whole inflationary picture for the origin of fluctuations. This is obviously an area of great current interest. We can give only a simplified treatment, but the basic equations are within our reach, as we now discuss.

16.16 Initial conditions and normalisation of curvature perturbations

The key concept we need for predicting the primordial spectrum of perturbations produced during inflation can be stated in the following question: what sets the initial conditions for the perturbation ζ_k itself? If we knew this for each k, then, since the evolution of ζ_k through to the point where it freezes would be known, given the evolution of the background model we could compute a spectrum of curvature perturbations as a function of k.

The basic idea for setting the initial conditions for the perturbations is that they come from quantum-field-theoretic fluctuations in the value of the scalar field ϕ. Thus the 'classical' perturbations discussed above need to be quantized, in a field theory sense, and this will allow their initial values to be set. A rigorous way of performing this quantisation has been developed[7] and, although the process is complicated, the final result in our case is very simple. To apply the result, we must first make two changes of variable in our discussion above.

- Convert from cosmic time t to a new dimensionless time variable η known as 'conformal time' and defined by $d\eta/dt = c/R$.
- Convert from the curvature perturbation ζ_k to a new variable ξ_k given by $\xi_k = \alpha\zeta_k$, where $\alpha = R\dot{\phi}_0/H$.

The formal procedure then shows that the correct quantisation may be achieved simply by treating ξ_k as a free complex scalar field and quantising it in the standard fashion. The evolution equation for the quantum perturbations turns out to be identical to the 'classical' equation for ξ_k. Thus, having fixed initial conditions for ξ_k quantum mechanically, one may follow the classical evolution.

Let us first derive the classical evolution equation for ξ_k. Making the transformation of variables noted above, equation (16.45) becomes even simpler. In particular, the intermediate variable α was chosen in order to remove the first-derivative term in (16.45), so as to make it more like a simple harmonic oscillator

[7] See, for example, V. F. Mukhanov, H. A. Feldman & R. H. Brandenberger, Theory of cosmological perturbations, *Physics Reports* **215**, 203–333, 1992.

equation. Using a prime to denote a derivative with respect to conformal time, we obtain

$$\boxed{\xi_k'' + \left(k^2 - \frac{\alpha''}{\alpha}\right)\xi_k = 0.} \tag{16.46}$$

It is now clear that we are dealing with the equation for the kth mode of a scalar field with a time-variable mass given by $m_\xi^2 = -\alpha''/\alpha$. The explicit expression for this effective mass is given in terms of the background quantities by

$$m_\xi^2 = \frac{\phi_0'^2}{2} - \frac{2\phi_0'\phi_0''}{RH} - \frac{\phi_0'^4}{2R^2H^2} - \frac{\phi_0'''}{\phi_0'}, \tag{16.47}$$

$$= -2R^2H^2\left(1 + \frac{\dot{\phi}_0^2}{2H^2} + \frac{\dot{\phi}_0^4}{4H^4} + \frac{\dot{\phi}_0\ddot{\phi}_0}{H^3} + \frac{3\ddot{\phi}_0}{2H\dot{\phi}_0} + \frac{\dddot{\phi}_0}{2H^2\dot{\phi}_0}\right), \tag{16.48}$$

where, in the last line, we have re-expressed the result in terms of derivatives with respect to cosmic time t rather than conformal time, which we will find useful momentarily. Perhaps surprisingly, it is the ϕ_0'''/ϕ_0' term in (16.47) that gives rise to the leading-order term $2R^2H^2$ in (16.48)!

To set the initial conditions for ξ_k, we will study the variable-mass term m_ξ^2 in the form (16.48). In the 'slow-roll' approximation, $\ddot{\phi}_0$ and higher derivatives were neglected. Furthermore, here we shall assume that $\dot{\phi}_0 \ll H$ during the periods of interest. In this case $m_\xi^2 \sim -2R^2H^2$ and (16.46) becomes

$$\boxed{\xi_k'' + (k^2 - 2R^2H^2)\,\xi_k = 0.} \tag{16.49}$$

In this form, we can see the origin of the behaviour discussed above in terms of a perturbation 'leaving the horizon'. When $k \gg RH$ the perturbation length scale is within the horizon (since $1/k \ll 1/(RH)$) and we have oscillatory behaviour. When $k \ll RH$, however, the perturbation length scale exceeds the horizon and we have exponential growth in ξ_k. Moreover, in the latter case we see directly from (16.46) that, if k can be neglected, we may immediately deduce the solution $\xi_k \propto \alpha$. Since $\xi_k = \alpha\zeta_k$, this means that the curvature perturbation ζ_k is constant, which is exactly the behaviour seen in Figure 16.4.

Let us now consider further the regime $k \gg RH$, when the perturbations are well inside the horizon, which is where the initial conditions for ξ_k can be set. In this regime, (16.49) becomes simply the harmonic oscillator equation $\xi_k'' + k^2\xi_k = 0$, the quantisation of which is well understood. This quantisation demands that the norm of any state evaluates to unity in Planck units, or equivalently that the conserved current of the field χ is unity, so that

$$-i(\chi\chi'^* - \chi'\chi^*) = 1. \tag{16.50}$$

It is this condition that sets the absolute scale of the perturbations. Hence, the properly normalised positive-energy solution in the regime $k \gg RH$ is given (up to a constant phase factor) by

$$\xi_k = \frac{1}{\sqrt{2k}} \exp(-ik\eta), \tag{16.51}$$

which is therefore the form to which any solution of (16.49) must tend well within the horizon.

We may now attempt to obtain a full solution to (16.49) and can in fact achieve this quite simply. Consider the following series of manipulations concerning the conformal time η, in which we carry out an integration by parts:

$$\begin{aligned}\eta = \int \frac{dt}{R} = \int \frac{dR}{R^2 H} &= \left[-\frac{1}{RH} \right] - \int \frac{dH}{RH^2} \\ &= \left[-\frac{1}{RH} \right] - \int \frac{\dot{H}}{H^2} \frac{dR}{R^2 H} \\ &= \left[-\frac{1}{RH} \right] + \int \frac{\dot{\phi}_0^2}{2H^2} \frac{dt}{R}.\end{aligned}$$

Again ignoring a term in $\dot{\phi}_0^2/H^2$, we can thus write

$$\eta = \eta_{\text{end}} - \frac{1}{RH}, \tag{16.52}$$

where η_{end} is the value at which the conformal time saturates at the end of inflation (that it does indeed saturate is obvious from the facts that $d\eta/dt = 1/R$ and that R is increasing exponentially during inflation). Figure 16.5 shows that (16.52) is indeed a good approximation during inflation in our current numerical example. Equation (16.49) now becomes

$$\xi_k'' + \left[k^2 - \frac{2}{(\eta_{\text{end}} - \eta)^2} \right] \xi_k = 0, \tag{16.53}$$

which finally is exactly soluble. There is a unique solution (up to a constant phase factor) that tends to (16.51) for small η; it is given by

$$\boxed{\xi_k = \frac{1}{\sqrt{2k^3}} \frac{i + k(\eta_{\text{end}} - \eta)}{\eta_{\text{end}} - \eta} e^{-ik\eta}.} \tag{16.54}$$

By inspection this has the correct property for $\eta \ll \eta_{\text{end}}$ provided that $k\eta_{\text{end}} \gg 1$. Comparison with Figure 16.5 shows that this is indeed the case for k-values of interest (for the figure, $k = 10^4$ and $\eta_{\text{end}} \approx 0.64$).

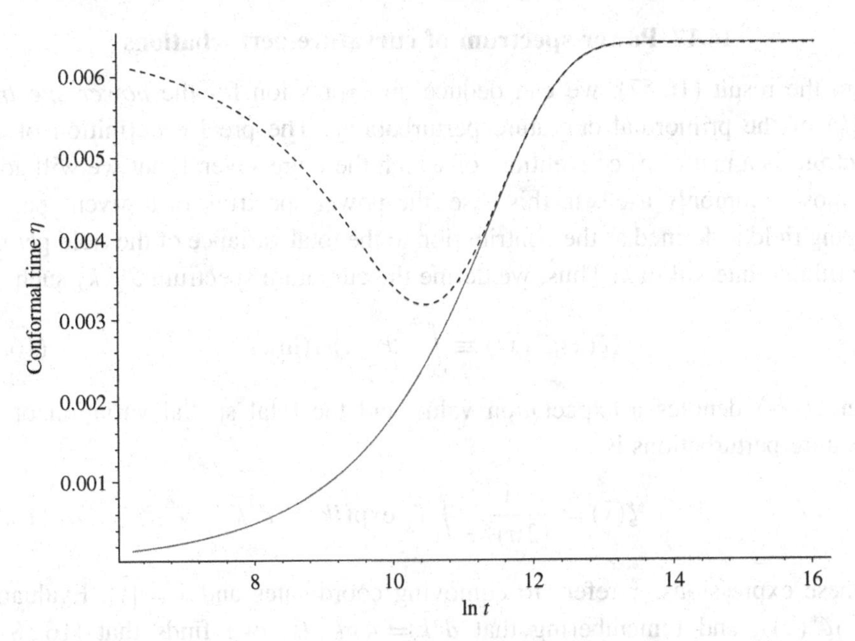

Figure 16.5 Evolution of conformal time η for the same numerical case as that illustrated in Figures 16.3 and 16.4 (solid curve). The broken curve shows the approximation given in equation (16.52), which is seen to be very good once inflation starts, around $\ln t \sim 11$.

Now that we have a correctly normalised general solution for ξ_k, let us consider the regime $k \ll RH$ at which the perturbation length scale exceeds the horizon. We use (16.52) to rewrite the solution just found as

$$\xi_k = \frac{1}{\sqrt{2k^3}}(k + iRH)e^{ik/RH} \approx \frac{iRH}{\sqrt{2k^3}}, \tag{16.55}$$

where the final expression is valid for $k \ll RH$. Thus, for such modes,

$$\zeta_k = \frac{\xi_k}{\alpha} \approx \frac{i}{\sqrt{2k^3}} \frac{H^2}{\dot{\phi}_0}. \tag{16.56}$$

Since we have demonstrated that ζ_k is constant after the mode has left the horizon, this means we are free to evaluate the right-hand side at the horizon exit itself. We therefore write schematically

$$\boxed{\zeta_k \approx \frac{i}{\sqrt{2k^3}} \left(\frac{H^2}{\dot{\phi}_0} \right) \bigg|_{k=RH}} \tag{16.57}$$

This is a famous and important result in inflationary theory; it gives the (constant) value of the amplitude of the plane-wave curvature perturbation having comoving wavenumber k for modes whose length scale exceeds the horizon.

16.17 Power spectrum of curvature perturbations

From the result (16.57), we can deduce an expression for the *power spectrum* $\mathcal{P}_\zeta(k)$ of the primordial curvature perturbations. The precise definition of this spectrum is a matter of convention, of which there are several, but we will adopt the most commonly used. In this case, the power spectrum of a given spatially varying field is defined as the contribution to the total variance of the field per unit logarithmic interval in k. Thus, we define the curvature spectrum $\mathcal{P}_\zeta(k)$ such that

$$\langle \zeta(\vec{x})\zeta^*(\vec{x})\rangle \equiv \int_0^\infty \mathcal{P}_\zeta(k)\, d(\ln k), \tag{16.58}$$

where $\langle\cdots\rangle$ denotes a expectation value and the total spatial variation of the curvature perturbations is

$$\zeta(\vec{x}) = \frac{1}{(2\pi)^{3/2}} \int \zeta_{\vec{k}} \exp(i\vec{k}\cdot\vec{x})\, d^3\vec{k}. \tag{16.59}$$

In these expressions, $\vec{x}$ refers to comoving coordinates and $k = |\vec{k}|$. Evaluating $\langle \zeta(\vec{x})\zeta^*(\vec{x})\rangle$, and remembering that $d^3\vec{k} = 4\pi k^2 dk$, one finds that (16.58) is satisfied providing that

$$\langle \zeta_{\vec{k}}\, \zeta^*_{\vec{k}'}\rangle = \frac{2\pi^2}{k^3}\mathcal{P}_\zeta(k)\, \delta^{(3)}(\vec{k}-\vec{k}'),$$

where $\delta^{(3)}(\vec{k}-\vec{k}')$ is the three-dimensional delta function. We may therefore write $\mathcal{P}_\zeta(k) = k^3\langle|\zeta_k|^2\rangle/(2\pi^2)$ and, using (16.57), we finally obtain[8]

$$\boxed{\mathcal{P}_\zeta(k) = \left(\frac{H^2}{2\pi\dot{\phi}_0}\right)^2_{k=RH}}. \tag{16.60}$$

In the slow-roll approximation, we know that H is only slowly decreasing whilst $\dot{\phi}_0$ is approximately constant. To a first approximation, therefore, the power spectrum of the perturbations expected from inflation, as measured by the contribution to the total fluctuation per unit logarithmic interval, is constant. Such a spectrum is called *scale invariant* and was proposed in the late 1960s as being the most likely to lead to structure appropriately distributed over the scales we see today. It is also known as a *Harrison–Zel'dovich* spectrum, after its two co-proposers. Here we can see it emerging as a *prediction* of inflation. We can go further, however, by noting that, during inflation, H is slowly declining, $\dot{\phi}_0$ is approximately constant and R is increasing exponentially. Thus modes with

[8] We note that the quantity $\mathcal{P}_\zeta(k)$ is often written using the alternative notation $\Delta^2_\zeta(k)$. In addition, it is common to define the quantity $P_\zeta(k) \equiv \langle|\zeta_k|^2\rangle$, which is also often called the power spectrum and is related to $\mathcal{P}_\zeta(k)$ by $\mathcal{P}_\zeta(k) = k^3 P_\zeta(k)/(2\pi^2)$.

higher k, which leave the horizon later in time, have a slightly lower value of $\mathcal{P}_\zeta(k)$ since H is lower there. As a result, the spectrum is predicted to be not exactly constant, but slightly declining, as a function of k. The details of this depend on the details of the potential $V(\phi_0)$, but we can see from the analysis here that this is a generic prediction of inflation (assuming that slow-roll is an accurate model).

Before going on to discuss in the next section the comparison of the prediction (16.60) with cosmological observations, it is worthwhile re-deriving this result in a more heuristic (and perhaps enlightening) manner. For a scalar field in an ordinary Minkowski spacetime, the zero-point uncertainty fluctuation is given by

$$\delta\phi_{k_\mathrm{p}} \approx \frac{1}{V^{1/2}} \frac{e^{-ik_\mathrm{p}t}}{\sqrt{2k_\mathrm{p}}} \tag{16.61}$$

for a mode with physical wavenumber k_p, where V is a normalising volume. Here, instead of k_p, we wish to use the comoving wavenumber k, which is related to the physical wavenumber by $k = Rk_\mathrm{p}$. Moreover, an obvious length scale for the normalising volume is the scale factor R. Thus, in our expanding FRW spacetime, we assume that

$$\delta\phi_k \approx \frac{e^{-ikt/R}}{R\sqrt{2k}}. \tag{16.62}$$

As explained above, the corresponding power spectrum of the fluctuations $\delta\phi_k$ is obtained by multiplying its squared norm by $4\pi k^3/(2\pi)^3$, which gives

$$\mathcal{P}_\phi(k) = \left(\frac{k}{2\pi R}\right)^2_{k=RH} = \left(\frac{H}{2\pi}\right)^2. \tag{16.63}$$

As above, we have evaluated the second expression at the 'horizon crossing' value of k, RH, since fluctuations on larger length scales are 'frozen in' at the value they reached at this point. To translate this result into the power spectrum of curvature perturbations $\mathcal{R}$, we need to link $\mathcal{R}$ and $\delta\phi$. Consider the change Δt in time coordinate that would be needed to move between the 'comoving slicing', in which $\delta\phi$ vanishes, to a 'flat slicing', in which Φ vanishes. Since ζ, as defined in (16.39), remains constant in this process, we see that, in this case,

$$\Delta\Phi = \Phi = -\mathcal{R} = H\Delta t \quad \text{and} \quad \Delta\phi = \delta\phi = -\dot{\phi}_0\Delta t. \tag{16.64}$$

Eliminating Δt we find that $\mathcal{R} = H\delta\phi/\dot{\phi}_0$ and hence we recover the result

$$\boxed{\mathcal{P}_\mathcal{R}(k) = \mathcal{P}_\zeta(k) = \left(\frac{H^2}{2\pi\dot{\phi}_0}\right)^2_{k=RH}.} \tag{16.65}$$

16.18 Power spectrum of matter-density perturbations

As discussed in Section 16.6, at the end of inflation the scalar field decays into other particles. Thus, one is left with a spectrum of curvature perturbations, which from (16.40) is equivalent to the spectrum of fluctuations in the gravitational potential Φ in a comoving slicing. These in turn may be related to the corresponding fluctuations $\delta\rho$ in the matter density. The full general-relativistic equations describing the evolution under gravity of these density fluctuations may be obtained by repeating all the above discussion for a perfect fluid rather than a scalar field. We will not pursue this calculation here but merely note that the resulting equations are the same as those obtained using Newtonian theory, except for a term that is important only on super-horizon scales. Therefore, on sub-horizon scales, to a good approximation we may take these potential fluctuations as obeying the perturbed Poisson's equation in Newtonian gravity,

$$\vec{\nabla}^2(\Phi) = 4\pi G(\delta\rho),$$

where $\delta\rho$ is the fluctuation in the matter density corresponding to that in the potential. Indeed, we might have expected the Newtonian theory to be a good approximation on sub-horizon scales since the gravitational field associated with the perturbations is weak.

It is more common to work instead in terms of the fractional-density fluctuation $\delta \equiv \delta\rho/\rho_0$, where ρ_0 is the background matter density. Thus, working in Fourier space, we have

$$\Phi_k = -\frac{4\pi G\rho_0 R^2}{k^2}\delta_k.$$

Using $\rho_0 = 3H^2/(8\pi G)$, for the simple spatially flat case that we are considering we see that

$$\delta_k = -\frac{2}{3}\left(\frac{k}{RH}\right)^2 \Phi_k, \qquad (16.66)$$

from which we deduce that $\langle|\delta_k|^2\rangle \propto k^4\langle|\Phi_k|^2\rangle$. Therefore, defining the *matter power spectrum* by $P_\delta(k) \equiv \langle|\delta_k|^2\rangle$ (note that this differs from definition of $\mathcal{P}_\mathcal{R}(k)$ by a factor $k^3/(2\pi^2)$, as mentioned earlier), we find that $P_\delta(k) \propto k\mathcal{P}_\mathcal{R}(k)$. Since $\mathcal{P}_\mathcal{R}(k)$ is roughly constant for slow-roll inflation, we thus obtain

$$\boxed{P_\delta(k) \propto k.} \qquad (16.67)$$

In general, the matter power spectrum is parameterised as $P_\delta(k) \propto k^n$, where n is known as the *primordial spectral index*. We therefore see that inflation naturally predicts $n = 1$, which is also known as the Harrison–Zel'dovich spectrum.

An alternative way of characterising this spectrum is to note from (16.66) that if we do not define the perturbation spectrum at a single instant of cosmic time but evaluate it when a given scale re-enters the horizon ($k = RH$) then

$$\boxed{\delta_k \propto \Phi_k.}$$

Since the spectrum $\mathcal{P}_\phi(k)$, defined as the contribution to the total variance per unit logarithm interval of k at a single instant of time, is roughly constant then so too is the matter power spectrum defined in the same way but evaluated at horizon entry. This is why the Harrison–Zel'dovich spectrum is also known as the *scale-invariant* spectrum. The fractional-density perturbations, as they enter the horizon, make a constant contribution to the total variance per unit logarithmic interval of k.

Finally, we note from (16.66) that, at a given k, the time evolution of the fractional-density perturbation δ_k is given by

$$\delta_k \propto \frac{1}{(RH)^2}.$$

For a radiation-dominated model we have $R \propto t^{1/2}$ and $H = 1/(2t)$, whereas for a matter-dominated model $R \propto t^{2/3}$ and $H = 2/(3t)$. Thus, we find

$$\delta_k(t) \propto \begin{cases} t & \text{(radiation-dominated)}, \\ t^{2/3} & \text{(matter-dominated)}, \end{cases}$$

which provides a quick derivation of the time dependence of what is known as the *growing mode* of the matter-density perturbations. In particular, we note that the time dependence of this mode is the same as that of the scale factor R in the matter-dominated case.

16.19 Comparison of theory and observation

The details of the comparison of the inflationary prediction for the perturbation spectrum with cosmological observations would take us too far afield here. We thus content ourselves with two brief illustrations. Figure 16.6 shows the prediction for the power spectrum of anisotropies in the cosmic microwave background radiation, assuming an early-universe perturbation spectrum that is exactly scale invariant. The anisotropies in the temperature of the CMB radiation provide a 'snapshot' of the (projected) density perturbations at the epoch of recombination

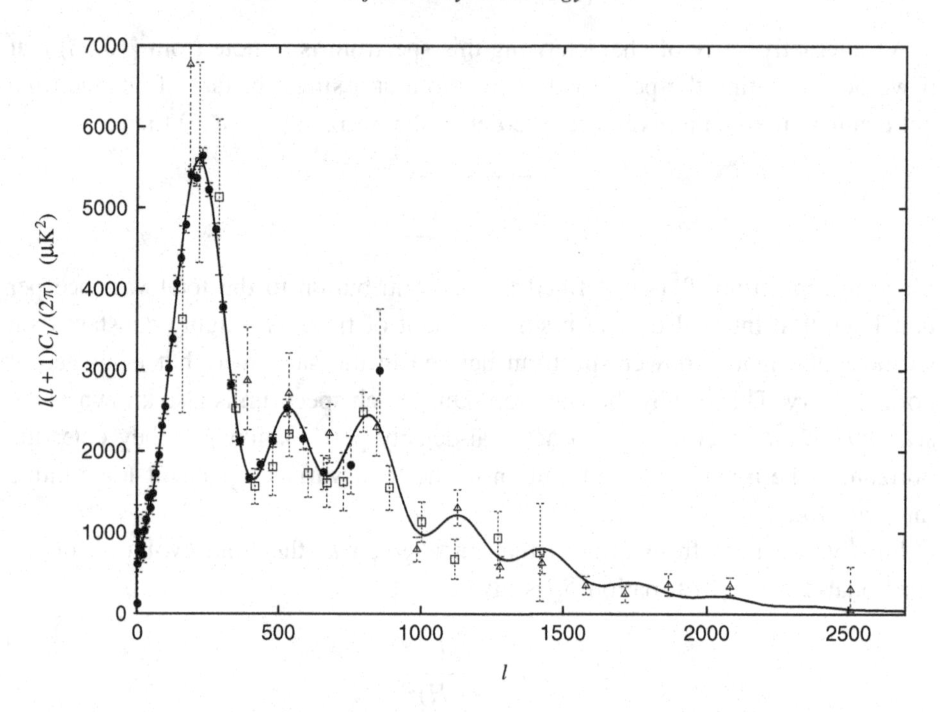

Figure 16.6 The predicted power spectrum of CMB temperature anisotropies (solid line), assuming an early-universe perturbation spectrum that is exactly scale invariant. The points show the results of recent observations of the CMB anisotropies by the Wilkinson Microwave Anisotropy Probe (WMAP, circles), Very Small Array (VSA, squares) and Arcminute Cosmology Bolometer Array Receiver (ACBAR, triangles) experiments. The vertical error bars indicate the 68 per cent uncertainty in the measured value.

($z_{\rm rec} \approx 1500$). The CMB anisotropies over the sky are usually decomposed in terms of spherical harmonics as

$$\Delta T(\theta, \phi) = \sum_{\ell=2}^{\infty} \sum_{m=-\ell}^{\ell} a_{\ell m} Y_{\ell m}(\theta, \phi),$$

where the $\ell = 0$ (constant) and $\ell = 1$ (dipole) terms are usually ignored, since the former is unrelated to the anisotropies and the latter is due to the peculiar velocity of the Earth with respect to the comoving frame of the CMB. The power in the fluctuations as a function of angular scale is therefore characterised by the spectrum

$$C_\ell = \frac{1}{2\ell+1} \sum_{m=-\ell}^{\ell} |a_{\ell m}|^2.$$

The characteristic peaks in the predicted CMB power spectrum (solid line) are a consequence of another feature of inflation that we have already seen in our equations, namely that all modes outside the horizon are frozen and can only start to oscillate once they re-enter the horizon later in the universe's evolution. This means that a 'phasing up' is able to occur, in which all modes of interest start from effectively a 'zero velocity' state when they begin the oscillations, during the epoch of recombination, that lead to the CMB imprints. This is what enables peaks to be visible in the power spectrum, with modes on different scales able to complete a different number of oscillations before the end of recombination. Coherence, leading to peaks, is maintained since each mode has the same starting conditions. This is only possible if the modes of interest are indeed on super-horizon scales prior to recombination, and the only known way of achieving this is via inflation. Thus the peaks visible in the predictions of Figure 16.6 are a powerful means of testing for inflation. The points shown in the figure are the results of recent observations of the CMB anisotropies by the WMAP (circles), VSA (squares) and ACBAR (triangles) experiments, which yield a very impressive confirmation of the peak structure and thereby a direct confirmation that inflation occurred.

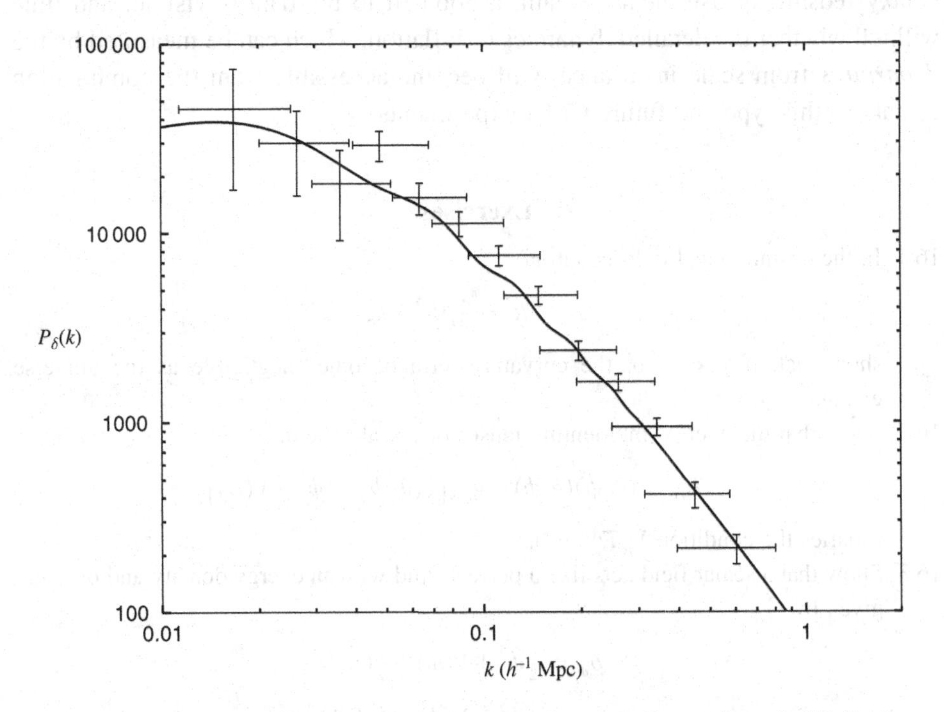

Figure 16.7 The predicted power spectrum of matter fluctuations (solid line) assuming an early-universe perturbation spectrum that is exactly scale invariant. The points show the results derived from the 2dF sample of galaxy redshift measurements. The horizontal error bars indicate the width of the bin in k-space over which the measurement is made and the vertical error bars indicate the 68 per cent uncertainty in the measured value.

From the current data it is, however, not possible to tell whether the primordial spectrum is exactly scale invariant, as assumed in generating the prediction, or whether it has the slight decrease at larger k, and therefore smaller scales, that we said was also a generic prediction of inflation. This question should be resolved by future experimental results, particularly from the CMB on smaller angular scales and from measurements of the matter distribution on a range of scales. In the latter case, one may compare the observed distribution of galaxies (both over the sky and in redshift) with the predicted power spectrum for matter fluctuations in the universe. The primordial matter power spectrum $P_\delta(k)$ in (16.67) is modified by the evolution under gravity of the perturbations once they re-enter the horizon. This effect may be calculated, and the predicted matter power spectrum resulting from an exactly scale-invariant primordial spectrum from inflation is shown as the solid line in Figure 16.7. Once again, we see that the predicted spectrum has oscillations resulting from an mechanism analogous to that which produces the oscillations in the CMB power spectrum discussed above. The points in the figure show the measurements derived from the 2dF (2 degree field) sample of galaxy redshift measurments. Again, a good fit to the data is visible, and time will tell whether the detailed dynamics of inflation, which can be measured by the *departures* from scale invariance, will become accessible from the combination of data of this type and future CMB experiments.

Exercises

16.1 In the cosmological field equation

$$\dot{R}^2 = \tfrac{1}{3}\rho R^2 - k,$$

show that, if $p < -\frac{1}{3}\rho$, the curvature term becomes negligible as the universe expands.

16.2 Show that the energy–momentum tensor of a scalar field,

$$T_{\mu\nu} = (\partial_\mu \phi)(\partial_\nu \phi) - g_{\mu\nu}\left[\tfrac{1}{2}(\partial_\sigma \phi)(\partial^\sigma \phi) - V(\phi)\right],$$

satisfies the condition $\nabla_\mu T^{\mu\nu} = 0$.

16.3 Show that a scalar field acts like a perfect fluid with an energy density and pressure given by

$$\rho_\phi = \tfrac{1}{2}\dot{\phi}^2 + V(\phi) + \tfrac{1}{2}(\vec{\nabla}\phi)^2,$$

$$p_\phi = \tfrac{1}{2}\dot{\phi}^2 - V(\phi) - \tfrac{1}{6}(\vec{\nabla}\phi)^2.$$

Show further that if the field ϕ is spatially constant then inflation will occur, provided that $\dot{\phi}^2 < V(\phi)$. If, in addition, the scalar field does not change with time, show that its equation of state is $p_\phi = -\rho_\phi$ and that it thus acts as an effective cosmological constant.

16.4 Show that the equation of motion for a scalar field with potential $V(\phi)$ is

$$\ddot{\phi} + 3H\dot{\phi} + \frac{dV}{d\phi} = 0.$$

Hence find the general solution for a free scalar field ϕ, for which $V = \frac{1}{2}m^2\phi^2$, in the case where H is approximately constant.

16.5 For a potential of the form

$$V(\phi) = V_0 \exp(-\lambda\phi),$$

where λ is a positive constant, show that the inflation equations can be solved exactly to give

$$R(t) = R_0\left(\frac{t}{t_0}\right)^{2/\lambda^2}, \qquad \phi(t) = \phi_0 + \frac{2}{\lambda}\ln\left(\sqrt{\frac{V_0}{2(6-\lambda^2)}}\lambda^2 t\right).$$

Hence show that, provided $\lambda < \sqrt{2}$, the solution corresponds to a period of inflation. Show further that the slow-roll parameters for this model are $\epsilon = \frac{1}{2}\eta = \frac{1}{2}\lambda^2$, and so the inflationary epoch never ends. This model is known as *power-law inflation*.

16.6 Show that, in general,

$$\ddot{R} = R(\dot{H} + H^2).$$

Show that $\dot{H} > 0$ only if $p < -\rho$, which is forbidden by the weak energy condition (see Exercise 8.8). Hence show that, for inflation to occur, one requires

$$-\frac{\dot{H}}{H^2} < 1,$$

and thus that the first slow-roll parameter must obey $\epsilon < 1$.

16.7 In the slow-roll approximation, show that

$$\dot{H} = -\tfrac{1}{2}\dot{\phi}^2.$$

Assuming that $\dot{\phi}$ varies monotonically with t throughout the period of inflation, show that

$$\dot{\phi} = -2H'(\phi),$$

where H is now considered as a function of ϕ, and hence that we may write the cosmological field equation as

$$[H'(\phi)]^2 - \tfrac{3}{2}H^2(\phi) = -\tfrac{1}{2}V(\phi).$$

This is known as the *Hamilton–Jacobi formalism* for inflation.

16.8 Repeat Exercise 16.5 using the Hamilton–Jacobi formalism developed in Exercise 16.7.

16.9 In the Hamilton–Jacobi formalism developed in Exercise 16.7, show that the condition for inflation to occur is

$$2\left[\frac{H'(\phi)}{H(\phi)}\right]^2 < 1.$$

16.10 Show that, during an exponential expansion phase of the universe, the proper distance between any two comoving objects separated by more than H^{-1} grows at a speed exceeding the speed of light. Hence show that an observer in such a universe can only see processes occurring inside the 'horizon' radius H^{-1}, and so the process of inflation in any spatial domain of radius H^{-1} (or 'mini-universe') occurs independently of any events outside it.

16.11 A fluctuation $\delta\phi$ in the scalar inflation field leads to a local delay of the end of inflation by $\delta t \sim \delta\phi/\dot{\phi}$. Assuming that the density of the universe after inflation decreases as t^{-2}, show that the fluctuation in the scalar field leads to a relative density contrast at the end of inflation given by

$$\frac{\delta\rho}{\rho} \sim \frac{H\delta\phi}{\dot{\phi}}.$$

Assuming the root mean square (rms) scalar field perturbation to be $\delta\phi_{\text{rms}} \sim H/(2\pi)$, show that

$$\left(\frac{\delta\rho}{\rho}\right)_{\text{rms}} \sim \frac{H^2}{2\pi\dot{\phi}}.$$

16.12 Consider an inflationary domain (or mini-universe in the context of Exercise 16.10) of initial radius H^{-1}, in which the value of the scalar field $\phi \gg 1$. In a time interval $\Delta t = H^{-1}$, show that classically, in the slow-roll approximation, the value of ϕ will change by

$$\Delta\phi \approx -\frac{2}{\phi}.$$

Assuming that the typical amplitude of quantum fluctuations in the scalar field is $\delta\phi \approx H/(2\pi)$, show that

$$\delta\phi \approx \frac{1}{2\pi}\sqrt{\frac{V(\phi)}{3}}.$$

Hence, for the case $V(\phi) = \frac{1}{2}m^2\phi^2$, show that the decrease in the value of the scalar field due to its classical motion is less than changes due to quantum fluctuations generated in the same time interval, provided that

$$\phi \gg \frac{6}{\sqrt{m}}.$$

Assuming that the typical wavelength of the quantum fluctuation is $\delta\phi$ is H^{-1}, show that, after a time interval $\Delta t = H^{-1}$, the original domain becomes effectively divided into $e^3 \sim 20$ domains of radius H^{-1}, each containing a roughly homogeneous scalar field $\phi + \Delta\phi + \delta\phi$. Thus, on average, the volume of the universe containing a growing ϕ-field increases by a factor ~ 10 after every time interval $\Delta t = H^{-1}$.

Note: This is the mechanism underlying stochastic inflation.

16.13 For the line element

$$ds^2 = (1+2\Phi)\, dt^2 - (1-2\Phi)R^2(t)(dx^2+dy^2+dz^2),$$

show that, to first order in Φ, the perturbed parts of the connection coefficients take the form

$$\begin{aligned}
\delta\Gamma^0{}_{0\mu} &= \partial_\mu \Phi, \\
\delta\Gamma^0{}_{ii} &= -R^2(\dot{\Phi}+4H\Phi), \\
\delta\Gamma^j{}_{ii} &= \delta^{jk}\partial_k \Phi \text{ (for } i \neq j) \\
\delta\Gamma^i{}_{00} &= \frac{1}{R^2}\delta^{ijk}\partial_k \Phi, \\
\delta\Gamma^i{}_{i\mu} &= -\partial_\mu \Phi,
\end{aligned}$$

where no sum over repeated i indices is implied and and the remaining perturbed coefficients either follow from symmetry or are zero. Hence show that the perturbed part of the Einstein tensor is given by

$$\begin{aligned}
\delta G^0_i &= -2\partial_i(\dot{\Phi}+H\Phi), \\
\delta G^0_0 &= -2(\vec{\nabla}^2\Phi - 3H\dot{\Phi} - 3H^2\Phi), \\
\delta G^i_i &= 2[\ddot{\Phi}+4\dot{\Phi}H+(2\dot{H}+3H^2)\Phi],
\end{aligned}$$

where again no sum over repeated i indices is implied and the remaining entries either follow from symmetry or are zero.

16.14 For the scalar-field perturbation

$$\phi(t) \to \phi_0(t) + \delta\phi(t, \vec{x}),$$

show that, to first order in $\delta\phi$, the perturbed parts of the scalar-field energy–momentum tensor are given by

$$\begin{aligned}
\delta T^0_i &= \dot{\phi}_0 \partial_i(\delta\phi), \\
\delta T^0_0 &= -\dot{\phi}_0^2\Phi + \dot{\phi}_0\delta\dot{\phi} + V'\delta\phi, \\
\delta T^i_i &= \dot{\phi}_0^2\Phi - \dot{\phi}_0\delta\dot{\phi} + V'\delta\phi,
\end{aligned}$$

where $V' = dV/d\phi_0$ and the remaining components either follow from symmetry or are zero.

16.15 Use your answers to Exercises 16.13 and 16.14 to show that the perturbed Einstein field equations yield only the two equations

$$\dot{\Phi} + H\Phi = \frac{1}{2}\dot{\phi}_0\delta\phi,$$

$$\left(\dot{\phi}_0^2 + 2\vec{\nabla}^2\right)\Phi = \dot{\phi}_0^2 \frac{d}{dt}\left(\frac{\delta\phi}{\dot{\phi}_0}\right).$$

16.16 Show that the gauge-invariant Fourier curvature perturbation

$$\zeta_k \equiv \Phi_k + H\frac{\delta\phi_k}{\dot{\phi}_0}$$

satisfies the equation of motion

$$\ddot{\zeta}_k + \left(\frac{\dot{\phi}_0^2}{H} + \frac{2\ddot{\phi}_0}{\dot{\phi}_0} + 3H\right)\dot{\zeta}_k + \frac{k^2}{R^2}\zeta_k = 0.$$

Defining the new variables $d\eta = c\,dt/R$ and $\xi_k = \alpha\zeta_k$, where $\alpha = R\dot{\phi}_0/H$, show further that

$$\xi_k'' + \left(k^2 - \frac{\alpha''}{\alpha}\right)\xi_k = 0,$$

where a prime denotes $d/d\eta$ and the 'effective mass' $m_\xi^2 = -\alpha''/\alpha$ is given by

$$\begin{aligned} m_\xi^2 &= \frac{\phi_0'^2}{2} - \frac{2\phi_0'\phi_0''}{RH} - \frac{\phi_0'^4}{2R^2H^2} - \frac{\phi_0'''}{\phi_0'}, \\ &= -2R^2H^2\left(1 + \frac{\dot{\phi}_0^2}{2H^2} + \frac{\dot{\phi}_0^4}{4H^4} + \frac{\dot{\phi}_0\ddot{\phi}_0}{H^3} + \frac{3\ddot{\phi}_0}{2H\dot{\phi}_0} + \frac{\dddot{\phi}_0}{2H^2\dot{\phi}_0}\right). \end{aligned}$$

16.17 Consider the equation of motion

$$\xi_k'' + \left[k^2 - \frac{2}{(\eta_{\text{end}} - \eta)^2}\right]\xi_k = 0.$$

Show that the unique solution (up to a phase factor) that tends to (16.51) for small η is given by

$$\xi_k = \frac{1}{\sqrt{2k^3}}\frac{i + k(\eta_{\text{end}} - \eta)}{\eta_{\text{end}} - \eta}e^{-ik\eta}.$$

17

Linearised general relativity

The gravitational field equations give a quantitative description of how the curvature of spacetime at any event is related to the energy–momentum distribution at that event. The high degree of *non-linearity* in these field equations means that a general solution for an arbitrary matter distribution is analytically intractable. Consequently, thus far we have concentrated primarily on investigating a number of special solutions that represent spacetimes with particular symmetries (aside from our discussion of perturbations in the previous chapter). In this chapter, we return to a more general investigation of the gravitational field equations and their solutions. To enable such a study, however, one must make the physical assumption that the gravitational fields are *weak*. Mathematically, this assumption corresponds to *linearising* the gravitational field equations.

17.1 The weak-field metric

As discussed in Sections 7.6 and 8.6, a weak gravitational field corresponds to a region of spacetime that is only 'slightly' curved. Thus, throughout such a region, there exist coordinate systems x^μ in which the spacetime metric takes the form

$$g_{\mu\nu} = \eta_{\mu\nu} + h_{\mu\nu} \qquad \text{where } |h_{\mu\nu}| \ll 1, \tag{17.1}$$

and the first and higher partial derivatives of $h_{\mu\nu}$ are also small.[1] Such coordinates are often termed quasi-Minkowskian coordinates, since they allow the metric to be written in a close-to-Minkowski form. Clearly, $h_{\mu\nu}$ must be symmetric with respect to the swapping of its indices. We also note that, when previously

[1] We note that one could equally well consider small perturbations about some other background metric, such that $g_{\mu\nu} = g^{(0)}_{\mu\nu} + h_{\mu\nu}$. This was the case in our discussion of inflationary perturbations in the previous chapter, in which $g^{(0)}_{\mu\nu}$ was the metric for the background Friedmann–Robertson–Walker spacetime in comoving Cartesian coordinates.

considering the weak-field limit, we further assumed that the metric was stationary, so that $\partial_0 g_{\mu\nu} = \partial_0 h_{\mu\nu} = 0$ where x^0 is the timelike coordinate. In our present discussion, however, we wish to retain the possibility of describing time-varying weak gravitational fields, and so we shall *not* make this additional assumption here.

As we have stressed many times, coordinates are arbitrary and, in principle, one could develop the description of weak gravitational fields in any coordinate system. Nevertheless, by adopting quasi-Minkowsian coordinates the mathematical labour of pursuing our analysis is greatly simplified, as is the interpretation of the resulting expressions. If one coordinate system exists in which (17.1) holds, however, then there must be many such coordinate systems. Indeed, two different types of coordinate transformation connect quasi-Minkowskian systems to each other: global Lorentz transformations and infinitesimal general coordinate transformations, both of which we now discuss.

Global Lorentz transformations

Global Lorentz transformations are of the form

$$x'^{\mu} = \Lambda^{\mu}{}_{\nu} x^{\nu}, \qquad \text{where } \eta_{\mu\nu} = \Lambda^{\rho}{}_{\mu} \Lambda^{\sigma}{}_{\nu} \eta_{\rho\sigma}$$

and the quantities $\Lambda^{\mu}{}_{\nu}$ are constant everywhere. These transform the metric coefficients as follows:

$$g'_{\mu\nu} = \frac{\partial x^{\rho}}{\partial x'^{\mu}} \frac{\partial x^{\sigma}}{\partial x'^{\nu}} g_{\rho\sigma} = \Lambda_{\mu}{}^{\rho} \Lambda_{\nu}{}^{\sigma} (\eta_{\rho\sigma} + h_{\rho\sigma}) = \eta_{\mu\nu} + \Lambda_{\mu}{}^{\rho} \Lambda_{\nu}{}^{\sigma} h_{\rho\sigma}.$$

Thus, $g'_{\mu\nu}$ is also of the form (17.1), with

$$\boxed{h'_{\mu\nu} = \Lambda_{\mu}{}^{\rho} \Lambda_{\nu}{}^{\sigma} h_{\rho\sigma}.}$$

Moreover, we see from this expression that, under a Lorentz transformation, $h_{\mu\nu}$ *itself* transforms like the components of a tensor in Minkowski spacetime.

The above property suggests a convenient alternative viewpoint when describing weak gravitational fields. Instead of considering a slightly curved spacetime representing the general-relativistic weak field, we can consider $h_{\mu\nu}$ simply as a symmetric rank-2 tensor field defined on the flat Minkowski background spacetime in Cartesian inertial coordinates. In other words, $h_{\mu\nu}$ is considered as a special-relativistic gravitational field, in an analogous way to that in which the 4-potential A_{μ} describes the electromagnetic field in Minkowski spacetime, as discussed in Chapter 6; we return to this point below. We note, however, that $h_{\mu\nu}$ does not transform as a tensor under a general coordinate transformation but only under the restricted class of global Lorentz transformations; for this reason $h_{\mu\nu}$ and tensors derived from it are sometimes called pseudotensors, although we will not use this terminology.

Infinitesimal general coordinate transformations

Infinitesimal general coordinate transformations take the form

$$x'^{\mu} = x^{\mu} + \xi^{\mu}(x), \tag{17.2}$$

where the $\xi^{\mu}(x)$ are four arbitrary functions of position of the same order of smallness as the $h_{\mu\nu}$. Infinitesimal transformations of this sort make tiny changes in the forms of all scalar, vector and tensor fields, but these can be ignored in all quantities except the metric, where tiny deviations from $\eta_{\mu\nu}$ contain all the information about gravity. From (17.2), we have

$$\frac{\partial x'^{\mu}}{\partial x^{\nu}} = \delta^{\mu}_{\nu} + \partial_{\nu}\xi^{\mu},$$

and, working to first order in small quantities, it is straightforward to show that the inverse transformation is given by[2]

$$\frac{\partial x^{\mu}}{\partial x'^{\nu}} = \delta^{\mu}_{\nu} - \partial_{\nu}\xi^{\mu}. \tag{17.3}$$

Thus, again working to first order in small quantities, the metric transforms as follows:

$$\begin{aligned} g'_{\mu\nu} = \frac{\partial x^{\rho}}{\partial x'^{\mu}}\frac{\partial x^{\sigma}}{\partial x'^{\nu}} g_{\rho\sigma} &= (\delta^{\rho}_{\mu} - \partial_{\mu}\xi^{\rho})(\delta^{\sigma}_{\nu} - \partial_{\nu}\xi^{\sigma})(\eta_{\rho\sigma} + h_{\rho\sigma}) \\ &= \eta_{\mu\nu} + h_{\mu\nu} - \partial_{\mu}\xi_{\nu} - \partial_{\nu}\xi_{\mu}, \end{aligned}$$

where we have defined $\xi_{\mu} = \eta_{\mu\nu}\xi^{\nu}$. Hence, we see that $g'_{\mu\nu}$ is also of the form (17.1), the new metric perturbation functions being related to the old ones via

$$\boxed{h'_{\mu\nu} = h_{\mu\nu} - \partial_{\mu}\xi_{\nu} - \partial_{\nu}\xi_{\mu}.} \tag{17.4}$$

If we adopt the viewpoint in which $h_{\mu\nu}$ is considered as a tensor field defined on the flat Minkowski background spacetime, then (17.4) can be considered as analogous to a gauge transformation in electromagnetism. As discussed in Chapter 6, if A_{μ} is a solution of the electromagnetic field equations then another solution that describes precisely the same physical situation is given by

$$A^{(\text{new})}_{\mu} = A_{\mu} + \partial_{\mu}\psi,$$

where ψ is any scalar field. An analogous situation holds in the case of the gravitational field. From (17.4), it is clear that if $h_{\mu\nu}$ is a solution to the linearised

[2] Note that, for the remainder of this chapter, the normal symbol for equality will be used to indicate equality up to first order in small quantities as well as exact equality.

gravitational field equations (see below) then the same physical situation is also described by

$$h^{(\text{new})}_{\mu\nu} = h_{\mu\nu} - \partial_\mu \xi_\nu - \partial_\nu \xi_\mu. \qquad (17.5)$$

In this interpretation, however, (17.5) is viewed as a gauge transformation rather than a coordinate transformation. In other words, we are still working in the same set of coordinates x^μ and have defined a new tensor $h^{(\text{new})}_{\mu\nu}$ whose components in this basis are given by (17.5).

Now that we have considered the coordinate transformations that preserve the form of the metric $g_{\mu\nu}$ in (17.1), it is useful to obtain the corresponding form for the contravariant metric coefficients $g^{\mu\nu}$. By demanding that $g^{\mu\nu}g_{\nu\sigma} = \delta^\mu_\sigma$, it is straightforward to verify that, to first order in small quantities, we must have

$$\boxed{g^{\mu\nu} = \eta^{\mu\nu} - h^{\mu\nu},}$$

where $h^{\mu\nu} = \eta^{\mu\rho}\eta^{\nu\sigma}h_{\rho\sigma}$. Moreover, it follows that indices on small quantities may be respectively raised and lowered using $\eta^{\mu\nu}$ and $\eta_{\mu\nu}$ rather than $g^{\mu\nu}$ and $g_{\mu\nu}$. For example, to first order in small quantities, we may write

$$h^\mu_\nu = g^{\mu\sigma}h_{\sigma\nu} = (\eta^{\mu\sigma} - h^{\mu\sigma})h_{\sigma\nu} = \eta^{\mu\sigma}h_{\sigma\nu}.$$

17.2 The linearised gravitational field equations

In the weak-field approximation to general relativity, one expands the gravitational field equations in powers of $h_{\mu\nu}$, using a coordinate system where (17.1) holds. On keeping only the linear terms, we thus arrive at the linearised version of general relativity. The Einstein gravitational field equations were derived in Section 8.4 and read

$$R_{\mu\nu} - \tfrac{1}{2}g_{\mu\nu}R = -\kappa T_{\mu\nu}.$$

To obtain the linearised form of these equations, we thus need to find the linearised expression for the Riemann tensor $R^\sigma{}_{\mu\nu\rho}$; the corresponding expressions for the Ricci tensor $R_{\mu\nu}$ and the Ricci scalar R then follow by the contraction of indices.

To perform this task, we first need the linearised form of the connection coefficients $\Gamma^\sigma{}_{\mu\nu}$. To first order in small quantities we have

$$\Gamma^\sigma{}_{\mu\nu} = \tfrac{1}{2}\eta^{\sigma\rho}(\partial_\nu h_{\rho\mu} + \partial_\mu h_{\rho\nu} - \partial_\rho h_{\mu\nu}) = \tfrac{1}{2}(\partial_\nu h^\sigma_\mu + \partial_\mu h^\sigma_\nu - \partial^\sigma h_{\mu\nu}), \qquad (17.6)$$

where we have defined $\partial^\sigma \equiv \eta^{\sigma\rho}\partial_\rho$. We may now substitute (17.6) directly into the expression (7.13) for the Riemann tensor, namely

$$R^\sigma{}_{\mu\nu\rho} = \partial_\nu\Gamma^\sigma{}_{\mu\rho} - \partial_\rho\Gamma^\sigma{}_{\mu\nu} + \Gamma^\tau{}_{\mu\rho}\Gamma^\sigma{}_{\tau\nu} - \Gamma^\tau{}_{\mu\nu}\Gamma^\sigma{}_{\tau\rho}. \qquad (17.7)$$

The last two terms on the right-hand side are products of connection coefficients and so will clearly be second order in $h_{\mu\nu}$; they will therefore be ignored. Hence, to first order, we obtain

$$\begin{aligned} R^{\sigma}{}_{\mu\nu\rho} &= \tfrac{1}{2}\partial_{\nu}(\partial_{\rho}h^{\sigma}_{\mu} + \partial_{\mu}h^{\sigma}_{\rho} - \partial^{\sigma}h_{\mu\rho}) - \tfrac{1}{2}\partial_{\rho}(\partial_{\nu}h^{\sigma}_{\mu} + \partial_{\mu}h^{\sigma}_{\nu} - \partial^{\sigma}h_{\mu\nu}) \\ &= \tfrac{1}{2}(\partial_{\nu}\partial_{\mu}h^{\sigma}_{\rho} + \partial_{\rho}\partial^{\sigma}h_{\mu\nu} - \partial_{\nu}\partial^{\sigma}h_{\mu\rho} - \partial_{\rho}\partial_{\mu}h^{\sigma}_{\nu}), \end{aligned}$$

which is easily shown to be invariant to a gauge transformation of the form (17.5). The linearised Ricci tensor is obtained by contracting the above expression for $R^{\sigma}{}_{\mu\nu\rho}$ on its first and last indices. This yields

$$R_{\mu\nu} = \tfrac{1}{2}(\partial_{\nu}\partial_{\mu}h + \Box^2 h_{\mu\nu} - \partial_{\nu}\partial_{\rho}h^{\rho}_{\mu} - \partial_{\rho}\partial_{\mu}h^{\rho}_{\nu}), \tag{17.8}$$

where we have defined the trace $h \equiv h^{\sigma}_{\sigma}$ and the d'Alembertian operator $\Box^2 \equiv \partial_{\sigma}\partial^{\sigma}$. The Ricci scalar is obtained by a further contraction, giving

$$R = R^{\mu}_{\mu} = \eta^{\mu\nu}R_{\mu\nu} = \Box^2 h - \partial_{\rho}\partial_{\mu}h^{\mu\rho}. \tag{17.9}$$

Substituting the expressions (17.8) and (17.9) into the gravitational field equations we obtain the linearised form

$$\partial_{\nu}\partial_{\mu}h + \Box^2 h_{\mu\nu} - \partial_{\nu}\partial_{\rho}h^{\rho}_{\mu} - \partial_{\rho}\partial_{\mu}h^{\rho}_{\nu} - \eta_{\mu\nu}(\Box^2 h - \partial_{\rho}\partial_{\sigma}h^{\sigma\rho}) = -2\kappa T_{\mu\nu}. \tag{17.10}$$

The number of terms on the left-hand side of the field equations has clearly increased in the linearisation process. This can be simplified somewhat by defining the 'trace reverse' of $h_{\mu\nu}$, which is given by

$$\bar{h}_{\mu\nu} \equiv h_{\mu\nu} - \tfrac{1}{2}\eta_{\mu\nu}h.$$

On contracting indices we find that $\bar{h} = -h$. It is also straightforward to show that $\bar{\bar{h}}_{\mu\nu} = h_{\mu\nu}$, i.e. $h_{\mu\nu} = \bar{h}_{\mu\nu} - \tfrac{1}{2}\eta_{\mu\nu}\bar{h}$. On substituting these expressions into (17.10), the field equations become

$$\boxed{\Box^2 \bar{h}_{\mu\nu} + \eta_{\mu\nu}\partial_{\rho}\partial_{\sigma}\bar{h}^{\rho\sigma} - \partial_{\nu}\partial_{\rho}\bar{h}^{\rho}_{\mu} - \partial_{\mu}\partial_{\rho}\bar{h}^{\rho}_{\nu} = -2\kappa T_{\mu\nu}.} \tag{17.11}$$

These are the basic field equations of linearised general relativity and are valid whenever the metric takes the form (17.1). Unless otherwise stated, for the remainder of this chapter we will adopt the viewpoint that $h_{\mu\nu}$ is simply a symmetric tensor field (under global Lorentz transformations) defined in quasi-Cartesian coordinates on a flat Minkowski background spacetime.

17.3 Linearised gravity in the Lorenz gauge

The field equations (17.11) can be simplified further by making use of the gauge transformation (17.5). Denoting the gauge-transformed field by $h'_{\mu\nu}$ for convenience, the components of its trace-reverse transform as

$$\begin{aligned}\bar{h}'^{\mu\rho} &= h'^{\mu\rho} - \tfrac{1}{2}\eta^{\mu\rho}h' \\ &= h^{\mu\rho} - \partial^{\mu}\xi^{\rho} - \partial^{\rho}\xi^{\mu} - \tfrac{1}{2}\eta^{\mu\rho}(h - 2\partial_{\sigma}\xi^{\sigma}) \\ &= \bar{h}^{\mu\rho} - \partial^{\mu}\xi^{\rho} - \partial^{\rho}\xi^{\mu} + \eta^{\mu\rho}\partial_{\sigma}\xi^{\sigma},\end{aligned} \qquad (17.12)$$

and hence we find that

$$\partial_{\rho}\bar{h}'^{\mu\rho} = \partial_{\rho}\bar{h}^{\mu\rho} - \Box^2\xi^{\mu}.$$

Therefore, if we choose the functions $\xi^{\mu}(x)$ so that they satisfy

$$\Box^2\xi^{\mu} = \partial_{\rho}\bar{h}^{\mu\rho}$$

then we have $\partial_{\rho}\bar{h}'^{\mu\rho} = 0$. The importance of this result is that, in this new gauge, each of the last three terms on the left-hand side of (17.11) vanishes. Thus, the field equations in the new gauge become

$$\Box^2\bar{h}'_{\mu\nu} = -2\kappa T'_{\mu\nu}.$$

Let us take stock of the simplification we have just achieved. Dropping primes and raising indices for convenience, we have found that the linearised field equations may be written in the simplified form

$$\boxed{\Box^2\bar{h}^{\mu\nu} = -2\kappa T^{\mu\nu},} \qquad (17.13)$$

provided that the $\bar{h}^{\mu\nu}$ satisfy the *gauge condition*

$$\boxed{\partial_{\mu}\bar{h}^{\mu\nu} = 0.} \qquad (17.14)$$

Moreover, we note that this gauge condition is preserved by any further gauge transformation of the form (17.5) provided that the functions ξ^{μ} satisfy $\Box^2\xi^{\mu} = 0$.

The above simplification is entirely analogous to that introduced in electromagnetism in Chapter 6. In that case, the electromagnetic field equations were reduced to the simple form $\Box^2 A^{\mu} = \mu_0 j^{\mu}$ by adoption of the Lorenz gauge condition $\partial_{\mu}A^{\mu} = 0$. This condition is preserved by any further gauge transformation $A_{\mu} \to A_{\mu} + \partial_{\mu}\psi$ if and only if $\Box^2\psi = 0$. As a result of the similarities between the electromagnetic and gravitational cases, (17.14) is often also referred to as the Lorenz gauge.

17.4 General properties of the linearised field equations

Now that we have arrived at the form of the field equations for linearised general relativity, it is instructive to consider some general consequences of our linearisation process for the resulting physical theory. The non-linearity of the original Einstein equations is a direct result of the fact that 'gravity gravitates'. In other words, any form of energy–momentum acts as a source for the gravitational field, *including* the energy–momentum associated with the gravitational field itself. By linearising the field equations we have ignored this effect.

One may straightforwardly take steps to address this shortcoming by 'bootstrapping' the theory as follows: (i) the energy–momentum carried by the linearised gravitational field $h_{\mu\nu}$ is calculated; (ii) this energy–momentum acts as a source for corrections $h^{(1)}_{\mu\nu}$ to the field; (iii) the energy–momentum carried by the corrections $h^{(1)}_{\mu\nu}$ is calculated ; (iv) this energy–momentum acts as a source for corrections $h^{(2)}_{\mu\nu}$ to the corrections $h^{(1)}_{\mu\nu}$; and so on. It is widely stated in the literature[3] that, on completing this bootstrapping process, one arrives back at the original non-linear field equations of general relativity, although this claim has recently been brought into question.[4] In either case, it is worth noting that this approach allows the resulting equations to be interpreted simply as a (fully self-consistent) relativistic theory of gravity in a fixed *Minkowski* spacetime. This viewpoint brings gravitation closer in spirit to the field theories describing the other fundamental forces. Indeed, the remarkable point is that only the field theory of gravitation has the elegant geometrical interpretation that we have spent so long exploring.

Returning to the linearised theory, one result of ignoring the energy–momentum carried by the gravitational field is an inconsistency between the linearised field equations (17.11) and the equations of motion for matter in a gravitational field. Raising the indices μ and ν on (17.11) and operating on both sides of the resulting equation with ∂_μ, one quickly finds that

$$\partial_\mu T^{\mu\nu} = 0. \tag{17.15}$$

This should be contrasted with the requirement, derived from the full non-linear field equations, that $\nabla_\mu T^{\mu\nu} = 0$. As was shown in Section 8.8, the latter requirement leads directly to the geodesic equation of motion for the worldline $x^\mu(\tau)$ of a test particle, namely

$$\ddot{x}^\mu + \Gamma^\mu{}_{\nu\sigma}\dot{x}^\nu\dot{x}^\sigma = 0, \tag{17.16}$$

where the dots denote differentiation with respect to the proper time τ. Performing a similar calculation for the condition (17.15), however, leads to the equation

[3] See, for example, R. P. Feynman, F. B. Morinigo & W. G. Wagner, *Feynman Lectures on Gravitation*, Addison–Wesley, 1995.

[4] See T. Padmanabhan, *From Gravitons to Gravity: Myths and Reality*, http://arxiv.org/abs/gr-qc/0409089.

of motion $\ddot{x}^\mu = 0$, which means that the gravitational field has *no effect* on the motion of the particle. In general, this clearly contradicts the geodesic postulate.

An alternative way to uncover this inconsistency is to note that an immediate consequence of having linearised the field equations is that solutions can be added. In other words, if the pairs of tensors $(h_{\mu\nu})_i$ and $(T_{\mu\nu})_i$ individually satisfy (17.11) for $i = 1, 2, \ldots$ then the quantity $\sum_i (h_{\mu\nu})_i$ is also a solution, corresponding to the energy–momentum tensor $\sum_i (T_{\mu\nu})_i$. Thus, for example, two point masses could remain at a fixed separation from one another indefinitely, the resulting gravitational field being simply the superposition of their individual radial fields.

Despite this inconsistency, linearised general relativity is still a useful approximation, provided that we are interested only in the far field of sources whose motion we know *a priori* and that we are willing to neglect the 'gravity of gravity'. In such cases, the effect of weak gravitational fields on test particles can be computed by inserting the form (17.6) for the connection coefficients into the geodesic equations (17.16). To calculate how the sources themselves move under their own gravity, however, one would need to re-insert into the field equations the non-linear terms that the linear theory discards.

17.5 Solution of the linearised field equations *in vacuo*

In empty space, the linearised field equations in the Lorenz gauge reduce to the wave equation

$$\boxed{\Box^2 \bar{h}^{\mu\nu} = 0,} \tag{17.17}$$

with the attendant gauge condition

$$\boxed{\partial_\mu \bar{h}^{\mu\nu} = 0.} \tag{17.18}$$

It is straightforward to show that the field equations have plane-wave solutions of the form

$$\bar{h}^{\mu\nu} = A^{\mu\nu} \exp(ik_\rho x^\rho), \tag{17.19}$$

where the $A^{\mu\nu}$ are constant (and, in general, complex) components of a symmetric tensor, and the k_μ are the constant (real) components of a vector. Substituting the expression (17.19) into the wave equation (17.17) and using the fact that $\partial_\rho \bar{h}^{\mu\nu} = k_\rho \bar{h}^{\mu\nu}$, we find that

$$\Box^2 \bar{h}^{\mu\nu} = \eta^{\rho\sigma} \partial_\rho \partial_\sigma \bar{h}^{\mu\nu} = \eta^{\rho\sigma} k_\rho k_\sigma \bar{h}^{\mu\nu} = 0.$$

This can only be satisfied if

$$\eta^{\rho\sigma} k_\rho k_\sigma = k^\sigma k_\sigma = 0, \tag{17.20}$$

and hence the vector $\mathbf{k}$ must be *null*. Since the linearised Einstein equations only take the simple form (17.17) in the Lorenz gauge, we must also take into account the gauge condition (17.18). On substituting into the latter the plane-wave form (17.19), we immediately find that the gauge condition is satisfied provided that one obeys the additional constraint

$$A^{\mu\nu}k_{\nu} = 0. \tag{17.21}$$

Thus any plane wave of the form (17.19) is a valid solution of the linearised vacuum field equations in the Lorenz gauge, provided that the vector k^{μ} satisfies (17.20) and (17.21). We will discuss plane gravitational waves in detail in the next chapter.

Since the vacuum field equations are linear (by design), *any* solution of them may be written as a superposition of such plane-wave solutions of the form

$$\bar{h}^{\mu\nu}(x) = \int A^{\mu\nu}(\vec{k})\exp(ik_{\rho}x^{\rho})\, d^3\vec{k}, \tag{17.22}$$

where $[k^{\mu}] = (k^0, \vec{k})$ and the integral is taken over all values of $\vec{k}$. Physical solutions are obtained by taking the real part of (17.22).

17.6 General solution of the linearised field equations

We now consider the general form of the solution to the linearised field equations in the presence of some non-zero energy–momentum tensor $T^{\mu\nu}$. In this case, the field equations take the form of an inhomogeneous wave equation for each component,

$$\Box^2\bar{h}^{\mu\nu} = -2\kappa T^{\mu\nu}, \tag{17.23}$$

together with the attendant gauge condition $\partial_{\mu}\bar{h}^{\mu\nu} = 0$. The general solution to (17.23) is most easily obtained by using a Green's function, in a similar manner to that employed for solving the analogous problem in electromagnetism. We will now outline this approach.

One begins by considering the solution to the inhomogeneous wave equation when the source is a δ-function, i.e. it is located at a definite event in spacetime. If this event has coordinates y^{σ}, one is therefore interested in solving an equation of the form

$$\Box^2_x \mathcal{G}(x^{\sigma} - y^{\sigma}) = \delta^{(4)}(x^{\sigma} - y^{\sigma}), \tag{17.24}$$

where the subscript on $\Box^2_x$ makes explicit that the d'Alembertian operator is with respect to the coordinates x^{σ} and $\mathcal{G}(x^{\sigma} - y^{\sigma})$ is the Green's function for our problem, which in the absence of boundaries must be a function only of the difference $x^{\sigma} - y^{\sigma}$. Since the field equations (17.23) are linear, sources that are more general can be built up by adding further δ-function sources located at different events. Thus, the general solution to the linearised field equations can be written[5]

$$\bar{h}^{\mu\nu}(x^\sigma) = \bar{h}^{\mu\nu}_{(0)}(x^\sigma) - 2\kappa \int \mathcal{G}(x^\sigma - y^\sigma) T^{\mu\nu}(y^\sigma)\, d^4y, \tag{17.25}$$

where, for completeness, we have made use of the freedom to add any solution $\bar{h}^{\mu\nu}_{(0)}(x)$ of the *homogeneous* field equations (i.e. the *in vacuo* field equations). It may be verified immediately by direct substitution that (17.25) does indeed solve (17.23). For the discussions in this chapter, however, we will take $\bar{h}^{\mu\nu}_{(0)}(x) = 0$ without loss of generality.

The problem of obtaining a general solution of the linearised field equations has thus been reduced to solving (17.24) to obtain the appropriate Green's function. This may be achieved in a number of ways, and here we shall take a physically motivated approach. For convenience, we begin by placing the δ-function source at the origin of our coordinate system. We will also make the identifications $[x^\mu] = (ct, \vec{x})$ and $r = |\vec{x}|$. With the source at the origin, we may write (17.24) as

$$\partial_\mu \partial^\mu \mathcal{G}(x^\sigma) = \delta^{(4)}(x^\sigma). \tag{17.26}$$

We first integrate this equation over a four-dimensional hypervolume V. Since the spatial spherical symmetry of the problem suggests that the Green's function should only depend on ct and r, we choose the hypervolume to be a sphere of radius r in its spatial dimensions and we integrate in t from $-\infty$ to ∞. The geometry of the bounding surface S of the hypervolume is illustrated by the vertical cylinder in Figure 17.1, in which the third spatial dimension x^3 has been suppressed. Performing the integration of (17.26) over V we obtain

$$\int_V \partial_\mu \partial^\mu \mathcal{G}(x^\sigma)\, d^4x = \int_S [\partial_\mu \mathcal{G}(x^\sigma)] n^\mu\, dS = 1, \tag{17.27}$$

where in the first equality we have used the divergence theorem to rewrite the volume integral as an integral over the bounding surface S with unit normal n^μ. Since we are working with a metric of signature $(+, -, -, -)$, it should be noted that n^μ is chosen to be outward pointing if it is timelike and inward pointing if spacelike.

Let us now consider the contributions to this surface integral over S. Since gravitational field variations travel at speed c, the only points in spacetime that can be influenced by a δ-function source at the origin are those lying on the

[5] Note that there is no need to include $\sqrt{-g}$ factors in our integral or delta-function definition, since we are considering the problem simply as a tensor field $\bar{h}^{\mu\nu}(x)$ defined on a Minkowski spacetime background in a Cartesian coordinate system.

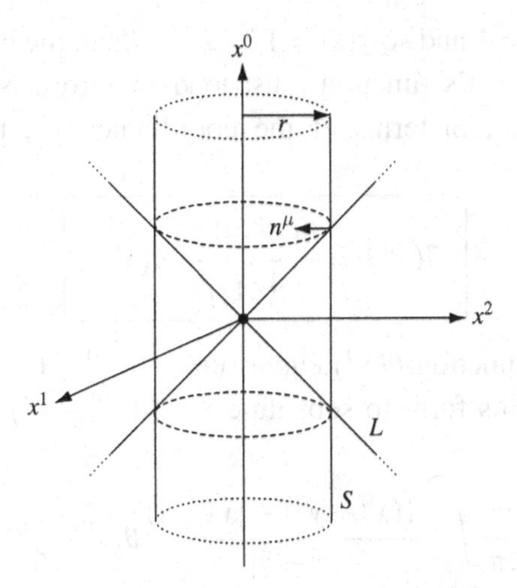

Figure 17.1 The geometry of the surface S in spacetime used to evaluate the Green's function for the wave equation. The lightcone L emanating from the origin is also shown. The x^3-direction has been suppressed.

future-pointing part of the lightcone L. Thus $\mathcal{G}(x^\sigma)$ must be zero at all points in spacetime except those lying on the future lightcone, and so must be of the form

$$\mathcal{G}(x^\sigma) = \begin{cases} f(r)\delta(ct-r) & \text{for } ct \geq 0, \\ 0 & \text{for } ct < 0, \end{cases} \tag{17.28}$$

where f is an arbitrary function of r. The intersection of the future lightcone with the surface S is a sphere (corresponding to a circle in Figure 17.1) of radius r lying in the spatial hypersurface $ct = r$. Thus, the only contribution to the surface integral in (17.27) is from this sphere (a circle in the figure), for which the (spacelike) unit normal n^μ points in the inward spatial radial direction (as illustrated). Rewriting the surface integral using $dS = c\,dt\,d\Omega$ (where $d\Omega$ is an element of solid angle) and $n^\mu \partial_\mu = -\partial_r$, and performing the integral over the spatial sphere, we thus have

$$-4\pi r^2 \int_{-\infty}^{\infty} \frac{\partial \mathcal{G}(x^\sigma)}{\partial r} c\,dt = 1, \tag{17.29}$$

where the only contribution to the integral over t occurs at $ct = r$. Substituting (17.28) into (17.29), we find that

$$4\pi r^2 f(r) \int_0^{\infty} \delta'(ct-r)c\,dt - 4\pi r^2 \frac{df(r)}{dr} \int_0^{\infty} \delta(ct-r)c\,dt = 1, \tag{17.30}$$

where the prime on the δ-function denotes differentiation with respect to its argument. Integration by parts quickly shows the first integral on the left-hand side of (17.30) to be zero, whereas the second integral equals unity. We therefore require $-4\pi r^2 df/dr = 1$ and so $f(r) = 1/(4\pi r)$, where the constant of integration vanishes since the Green's function must tend to zero at spatial infinity. Thus, re-expressing the result in terms of the coordinates x^σ, the required Green's function is

$$\boxed{\mathcal{G}(x^\sigma) = \frac{\delta(x^0 - |\vec{x}|)}{4\pi|\vec{x}|}\theta(x^0),}$$

where the Heaviside function $\theta(x^0)$ equals unity if $x^0 \geq 0$ and zero if $x^0 < 0$.

We may now use this form to substitute for $\mathcal{G}(x^\sigma - y^\sigma)$ in (17.25), with $\bar{h}^{\mu\nu}_{(0)}$ set to zero, to obtain

$$\bar{h}^{\mu\nu}(x^\sigma) = -\frac{\kappa}{2\pi}\int \frac{\delta\left((x^0 - y^0) - |\vec{x} - \vec{y}|\right)}{|\vec{x} - \vec{y}|}\theta(x^0 - y^0)T^{\mu\nu}(y^\sigma)\, d^4y.$$

Using the delta function to perform the integral over y^0, we finally find that the general solution to the linearised field equations (17.23) is given by

$$\boxed{\bar{h}^{\mu\nu}(ct, \vec{x}) = -\frac{4G}{c^4}\int \frac{T^{\mu\nu}(ct - |\vec{x} - \vec{y}|, \vec{y})}{|\vec{x} - \vec{y}|}\, d^3\vec{y}.} \qquad (17.31)$$

The interpretation of (17.31) requires some words of explanation. Here $\vec{x}$ represents the spatial coordinates of the field point at which $\bar{h}^{\mu\nu}$ is determined, $\vec{y}$ represents the spatial coordinates of a point in the source and $|\vec{x} - \vec{y}|$ is the spatial distance between them. We see that the disturbance in the gravitational field at the event $(ct, \vec{x})$ is the integral over the region of spacetime occupied by the points of the source at the *retarded times* t_r given by

$$ct_r = ct - |\vec{x} - \vec{y}|.$$

This region is the intersection of the past lightcone of the field point with the world tube of the source. An illustration of the geometric meaning of the retarded time is shown in Figure 17.2.

Although we have shown that (17.31) satisfies the linearised field equations (17.23), this form of the field equations is only valid in the Lorenz gauge. We must therefore verify that (17.31) also satisfies the Lorenz gauge condition $\partial_\mu \bar{h}^{\mu\nu} = 0$. Before embarking on this we first remind ourselves how to differentiate a function of retarded time. Setting $x^0_r \equiv ct_r = x^0 - |\vec{x} - \vec{y}|$, for any function f we have

$$\frac{\partial f(x^0_r, \vec{y})}{\partial x^\mu} = \left[\frac{\partial f(y^0, \vec{y})}{\partial y^0}\right]_r \frac{\partial x^0_r}{\partial x^\mu}, \qquad (17.32)$$

$$\frac{\partial f(x^0_r, \vec{y})}{\partial y^i} = \left[\frac{\partial f(y^0, \vec{y})}{\partial y^i}\right]_r + \left[\frac{\partial f(y^0, \vec{y})}{\partial y^0}\right]_r \frac{\partial x^0_r}{\partial y^i}, \qquad (17.33)$$

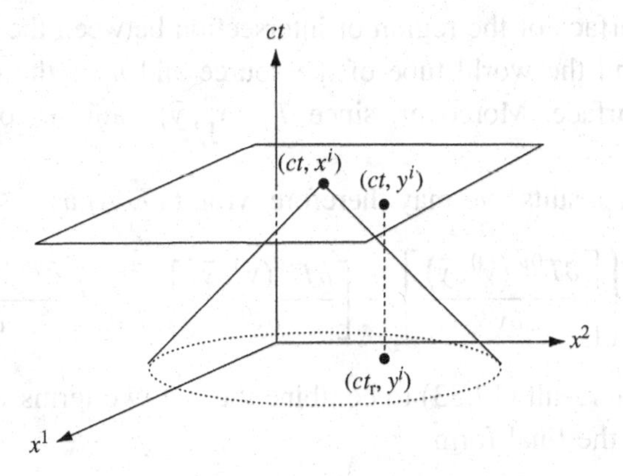

Figure 17.2 The disturbance in the gravitational field at the event (ct, x^i) is the sum of the influences of the energy and momentum sources at the points (ct_r, y^i) on the past lightcone.

where $[\ ]_r$ denotes that the expression contained within the brackets is evaluated at $y^0 = x_r^0$ and where $i = 1, 2, 3$. In addition to (17.33), we note that $\partial f(x_r^0, \vec{y})/\partial y^0 = 0$.

Let us now verify that the solution (17.31) does indeed satisfy the Lorenz gauge condition. Differentiating, we obtain

$$\begin{aligned}
\frac{\partial \bar{h}^{\mu\nu}(x^0, \vec{x})}{\partial x^\mu} &= -\frac{4G}{c^4} \int \left[\frac{1}{|\vec{x}-\vec{y}|} \frac{\partial T^{0\nu}(x_r^0, \vec{y})}{\partial x^0} + \frac{\partial}{\partial x^i} \left(\frac{T^{i\nu}(x_r^0, \vec{y})}{|\vec{x}-\vec{y}|} \right) \right] d^3\vec{y}, \\
&= -\frac{4G}{c^4} \int \left[\frac{1}{|\vec{x}-\vec{y}|} \frac{\partial T^{\mu\nu}(x_r^0, \vec{y})}{\partial x^\mu} + T^{i\nu}(x_r^0, \vec{y}) \frac{\partial}{\partial x^i} \left(\frac{1}{|\vec{x}-\vec{y}|} \right) \right] d^3\vec{y},
\end{aligned} \tag{17.34}$$

where we show explicitly that the partial derivatives are with respect to the coordinates x^μ. Using (17.32), the derivative in the first term of the integrand can be rewritten as follows:

$$\frac{\partial T^{\mu\nu}(x_r^0, \vec{y})}{\partial x^\mu} = \left[\frac{\partial T^{\mu\nu}(y^0, \vec{y})}{\partial y^0} \right]_r \frac{\partial x_r^0}{\partial x^\mu} = \left[\frac{\partial T^{0\nu}(y^0, \vec{y})}{\partial y^0} \right]_r - \left[\frac{\partial T^{i\nu}(y^0, \vec{y})}{\partial y^0} \right]_r \frac{\partial x_r^0}{\partial y^i},$$

where in the second equality we have used the fact that $\partial x_r^0/\partial x^i = -\partial x_r^0/\partial y^i$. Returning to (17.34), in a similar manner we may replace $\partial/\partial x^i$ by $-\partial/\partial y^i$ in the second term of the integrand, which then allows this term to be integrated by parts, since

$$\int T^{i\nu}(x_r^0, \vec{y}) \frac{\partial}{\partial y^i} \left(\frac{1}{|\vec{x}-\vec{y}|} \right) d^3\vec{y} = \int_S \frac{T^{i\nu}(x_r^0, \vec{y})}{|\vec{x}-\vec{y}|} n_i \, dS - \int \frac{1}{|\vec{x}-\vec{y}|} \frac{\partial T^{i\nu}(x_r^0, \vec{y})}{\partial y^i} d^3\vec{y},$$

where S is the surface of the region of intersection between the past lightcone of the field point and the world tube of the source and n_i is the outward-pointing normal to the surface. Moreover, since $T^{i\nu}(x^0_{\rm r}, \vec{y})$ vanishes on S, the surface integral is zero.

Combining our results, we may therefore write (17.34) as

$$\frac{\partial \bar{h}^{\mu\nu}}{\partial x^\mu} = -\frac{4G}{c^4}\int \left\{ \left[\frac{\partial T^{0\nu}(y^0, \vec{y})}{\partial y^0}\right]_{\rm r} - \left[\frac{\partial T^{i\nu}(y^0, \vec{y})}{\partial y^0}\right]_{\rm r} \frac{\partial x^0_{\rm r}}{\partial y^i} + \frac{\partial T^{i\nu}(x^0_{\rm r}, \vec{y})}{\partial y^i} \right\} \frac{d^3\vec{y}}{|\vec{x}-\vec{y}|}.$$

Making use of the result (17.33) to combine the last two terms within the braces, we thus arrive at the final form

$$\frac{\partial \bar{h}^{\mu\nu}}{\partial x^\mu} = -\frac{4G}{c^4}\int \frac{1}{|\vec{x}-\vec{y}|}\left[\frac{\partial T^{\mu\nu}(y^0, \vec{y})}{\partial y^\mu}\right]_{\rm r} d^3\vec{y}. \qquad (17.35)$$

As shown in Section 17.4, however, in the linearised theory the energy–momentum tensor obeys $\partial_\mu T^{\mu\nu} = 0$. Thus the integrand in (17.35) vanishes, and so we have verified that the solution (17.31) satisfies the Lorenz gauge condition $\partial_\mu \bar{h}^{\mu\nu} = 0$.

17.7 Multipole expansion of the general solution

In general, the source of the gravitational field may be dynamic and have a spatial extent that is not small compared with the distance to the point at which one wishes to calculate the field. In such cases, obtaining a simple expression for the solution (17.31) is often analytically intractable. In an analogous manner to that used in electromagnetism, it is often convenient to perform a *multipole expansion* of (17.31), which lends itself to the calculation of successive approximations to the solution. One begins by writing down the Taylor expansion

$$\begin{aligned}\frac{1}{|\vec{x}-\vec{y}|} &= \frac{1}{r} + (-y^i)\partial_i\left(\frac{1}{r}\right) + \frac{1}{2!}(-y^i)(-y^j)\partial_i\partial_j\left(\frac{1}{r}\right) + \cdots, \\ &= \frac{1}{r} + y^i\frac{x_i}{r^3} + y^i y^j\left(\frac{3x_i x_j - r^2\delta_{ij}}{r^5}\right) + \cdots,\end{aligned}$$

where $r \equiv |\vec{x}|$ is the spatial distance from the origin to the field point and $\partial_i \equiv \partial/\partial x^i$. One may then write the solution (17.31) as

$$\begin{aligned}\bar{h}^{\mu\nu}(ct, \vec{x}) = -\frac{4G}{c^4}\bigg[\frac{1}{r}\int T^{\mu\nu}(ct_{\rm r}, \vec{y})\, d^3\vec{y} + \frac{x_i}{r^3}\int T^{\mu\nu}(ct_{\rm r}, \vec{y})y^i\, d^3\vec{y} \\ + \frac{3x_i x_j - r^2\delta_{ij}}{r^5}\int T^{\mu\nu}(ct_{\rm r}, \vec{y})y^i y^j\, d^3\vec{y} + \cdots\bigg],\end{aligned} \qquad (17.36)$$

where $ct_r = ct - |\vec{x} - \vec{y}|$. This multipole expansion may be written in a particularly compact form:

$$\bar{h}^{\mu\nu}(ct, \vec{x}) = -\frac{4G}{c^4} \sum_{\ell=0}^{\infty} \frac{(-1)^\ell}{\ell!} M^{\mu\nu i_1 i_2 \cdots i_\ell}(ct_r) \partial_{i_1} \partial_{i_2} \cdots \partial_{i_\ell} \left(\frac{1}{r}\right),$$

where the *multipole moments* of the source distribution at any time t are given by

$$M^{\mu\nu i_1 i_2 \cdots i_\ell}(ct) = \int T^{\mu\nu}(ct, \vec{y}) y^{i_1} y^{i_2} \ldots y^{i_\ell} \, d^3\vec{y}.$$

Since the fall-off with distance of the term associated with the ℓth multipole moment goes as $1/r^{\ell+1}$, the gravitational field at large distances from the source is well approximated by only the first few terms of the multipole expansion.

17.8 The compact-source approximation

Let us suppose that the source is some matter distribution localised near the origin O of our coordinate system. If we take our field point $\vec{x}$ to be a distance r from O that is large compared with the spatial extent of the source, we need consider only the first term in the multipole expansion (17.36). Moreover, we assume that the source particles have speeds that are sufficiently small compared with c for us to take $ct_r = ct - r$ in the argument of the stress–energy tensor. Thus, the solution in the *compact-source approximation* is given by

$$\boxed{\bar{h}^{\mu\nu}(ct, \vec{x}) = -\frac{4G}{c^4 r} \int T^{\mu\nu}(ct - r, \vec{y}) \, d^3\vec{y}.} \qquad (17.37)$$

In this approximation, we are thus considering only the *far-field* solution to the linearised gravitational equations, which varies as $1/r$.

From (17.37), we see that calculating the gravitational field has been reduced to integrating $T^{\mu\nu}$ over the source at a fixed retarded time $ct - r$. The physical interpretation of the various components of this integral is as follows:

$\int T^{00} \, d^3\vec{y}$, total energy of source particles (including rest mass energy) $\equiv Mc^2$;
$\int T^{0i} \, d^3\vec{y}$, $c\times$ total momentum of source particles in the x^i-direction $\equiv P^i c$;
$\int T^{ij} \, d^3\vec{y}$, integrated internal stresses in the source.

For an isolated source, the quantities M and P^i are constants in the linear theory (this is easily proved directly from the conservation equation $\partial_\mu T^{\mu\nu} = 0$).[6] Moreover, without loss of generality, we may take our spatial coordinates x^i to

[6] We shall see later that a source does in fact lose energy via the emission of gravitational radiation, but the energy–momentum carried away by the gravitational field is quadratic in $h_{\mu\nu}$ and hence neglected in the linear theory.

correspond to the 'centre-of-momentum' frame of the source particles, in which case $P^i = 0$. Thus, from (17.37), in centre-of-momentum coordinates we have

$$\bar{h}^{00} = -\frac{4GM}{c^2 r}, \qquad \bar{h}^{i0} = \bar{h}^{0i} = 0. \tag{17.38}$$

The remaining components of the gravitational field are then given by the integrated stress within the source,

$$\bar{h}^{ij}(ct, \vec{x}) = -\frac{4G}{c^4 r}\left[\int T^{ij}(ct', \vec{y})\, d^3\vec{y}\right]_{\rm r}, \tag{17.39}$$

where $[\ ]_{\rm r}$ denotes that the expression in the brackets is evaluated at $ct' = ct - r$.

The integral in (17.39) is surprisingly troublesome to evaluate directly. Fortunately, there exists an alternative route that leads to a very neat expression for this quantity. We first recall that $\partial_\mu T^{\mu\nu} = 0$ (where, for consistency with (17.39), we are considering $T^{\mu\nu}$ as a function of the coordinates $(ct', \vec{y})$ and so $\partial_0 = \partial/\partial(ct')$ and $\partial_k = \partial/\partial y^k$). From this result, we may write

$$\partial_0 T^{00} + \partial_k T^{0k} = 0, \tag{17.40}$$

$$\partial_0 T^{i0} + \partial_k T^{ik} = 0. \tag{17.41}$$

Let us now consider the integral

$$\int \partial_k (T^{ik} y^j)\, d^3\vec{y} = \int (\partial_k T^{ik})\, y^j\, d^3\vec{y} + \int T^{ij}\, d^3\vec{y},$$

where the integral is taken over a region of space enclosing the source, so that $T^{\mu\nu} = 0$ on the boundary surface S of the region. Using Gauss' theorem to convert the integral on the left-hand side to an integral over the surface S, we find that its value is zero. Hence, on using (17.41), we can write

$$\int T^{ij}\, d^3\vec{y} = -\int (\partial_k T^{ik})\, y^j\, d^3\vec{y} = \int (\partial_0 T^{i0})\, y^j\, d^3\vec{y} = \frac{1}{c}\frac{d}{dt'}\int T^{i0} y^j\, d^3\vec{y}.$$

For later convenience, interchanging i and j and adding gives

$$\int T^{ij}\, d^3\vec{y} = \frac{1}{2c}\frac{d}{dy^0}\int (T^{i0} y^j + T^{j0} y^i)\, d^3\vec{y}. \tag{17.42}$$

We must now consider the integral

$$\int \partial_k (T^{0k} y^i y^j)\, d^3\vec{y} = \int (\partial_k T^{0k})\, y^i y^j\, d^3\vec{y} + \int (T^{0i} y^j + T^{0j} y^i)\, d^3\vec{y},$$

where, once again, we may use Gauss' theorem to show that the left-hand side is zero. Using (17.40), we thus have

$$\int (T^{0i} y^j + T^{0j} y^i)\, d^3\vec{y} = \frac{1}{c}\frac{d}{dt'}\int T^{00} y^i y^j\, d^3\vec{y}. \tag{17.43}$$

Combining (17.42) and (17.43) yields

$$\int T^{ij}\, d^3\vec{y} = \frac{1}{2c^2}\frac{d^2}{dt'^2}\int T^{00} y^i y^j\, d^3\vec{y}.$$

Inserting this expression into (17.39), we finally obtain the *quadrupole formula*

$$\bar{h}^{ij}(ct, \vec{x}) = -\frac{2G}{c^6 r}\left[\frac{d^2 I^{ij}(ct')}{dt'^2}\right]_{\mathrm{r}}, \tag{17.44}$$

where we have defined the *quadrupole-moment tensor* of the energy density of the source,

$$I^{ij}(ct) = \int T^{00}(ct, \vec{y})\, y^i y^j\, d^3\vec{y}, \tag{17.45}$$

which is a constant tensor on each hypersurface of constant time. In the next chapter, we will use this formula to determine the far-field gravitational radiation generated by a time-varying matter source.

17.9 Stationary sources

Let us return to the general solution (17.31) to the linearised field equation. In the previous section we confined our attention to the far-field solution for a compact source. This behaves like $1/r$ as a function of distance and depends only on the mass and inertia tensor of the source. As shown in the multipole expansion (17.36), other properties of the source generate a field that falls off more rapidly with distance. In general, it is often impossible to obtain a simple expression for the solution (17.31). Nevertheless, the solution simplifies somewhat when the source is *stationary*.

A stationary source has $\partial_0 T^{\mu\nu} = 0$, i.e. the energy–momentum tensor is constant in time. Note that this does not necessarily imply that the source is static (so that its constituent particles are not moving), which would additionally require the form of $T^{\mu\nu}$ to be invariant to the transformation $t \to -t$. A typical example of a stationary, but non-static, source is a uniform rigid sphere rotating with constant angular velocity. The main advantage of the stationary-source limit is that the time dependence vanishes and thus retardation is irrelevant. Hence, the general solution (17.31) to the linearised field equations reduces to

$$\bar{h}^{\mu\nu}(\vec{x}) = -\frac{4G}{c^4}\int \frac{T^{\mu\nu}(\vec{y})}{|\vec{x}-\vec{y}|}\, d^3\vec{y}. \tag{17.46}$$

One can perform a multipole expansion of this solution identical to that given in (17.36) but for which all time dependence is omitted. Indeed, in this case, it becomes somewhat simpler to interpret the various multipole moments physically.

A particularly interesting special case is the *non-relativistic* stationary source. Consider a source having a well-defined spatial velocity field $u^i(\vec{x})$, where the speed u of any constituent particle is small enough compared with c that we can neglect terms of order u^2/c^2 and higher in its energy–momentum tensor. In particular, we will take $\gamma_u = (1-u^2/c^2)^{-1/2} \approx 1$. Moreover, the pressure p within the source is everywhere much smaller than the energy density and may thus be neglected. From the discussion of energy–momentum tensors in Section 8.1 we see that, for such a source,

$$T^{00} = \rho c^2, \qquad T^{0i} = c\rho u^i, \qquad T^{ij} = \rho u^i u^j,$$

where $\rho(\vec{x})$ is the proper-density distribution of the source. We see that $|T^{ij}|/|T^{00}| \sim u^2/c^2$ and so we should take $T^{ij} \approx 0$ to the order of our approximation. The corresponding solution (17.46) to the linearised field equations can then be written as

$$\bar{h}^{00} = \frac{4\Phi}{c^2}, \qquad \bar{h}^{0i} = \frac{A^i}{c}, \qquad \bar{h}^{ij} = 0, \tag{17.47}$$

where we have defined the gravitational scalar potential Φ and gravitational spatial vector potential A^i by

$$\Phi(\vec{x}) \equiv -G\int \frac{\rho(\vec{y})}{|\vec{x}-\vec{y}|}\, d^3\vec{y}, \tag{17.48}$$

$$A^i(\vec{x}) \equiv -\frac{4G}{c^2}\int \frac{\rho(\vec{y})u^i(\vec{y})}{|\vec{x}-\vec{y}|}\, d^3\vec{y}. \tag{17.49}$$

The corresponding components of $h^{\mu\nu}$ are given by $h^{\mu\nu} = \bar{h}^{\mu\nu} - \frac{1}{2}\eta^{\mu\nu}\bar{h}$. The result (17.47) implies that $\bar{h} = \bar{h}^{00}$ and, on lowering indices, we find that the non-zero components are

$$\boxed{h_{00} = h_{11} = h_{22} = h_{33} = \frac{2\Phi}{c^2}, \qquad h_{0i} = \frac{A_i}{c}.} \tag{17.50}$$

It should be remembered that raising or lowering a spatial (roman) index introduces a minus sign. Thus the numerical value of A_i is minus that of A^i, the latter being the ith component of the spatial vector $\vec{A}$. The obvious analogy between the equations (17.48, 17.49) and their counterparts in the theory of electromagnetism will be discussed in detail in Appendix 17A.

For the most part, in this chapter we adopt the viewpoint that $h_{\mu\nu}$ is simply a rank-2 tensor field defined in Cartesian coordinates on a background Minkowski spacetime. At this point, however, it is useful to revert to the viewpoint in which $g_{\mu\nu} = \eta_{\mu\nu} + h_{\mu\nu}$ defines the metric of a (slightly) curved spacetime. From (17.50), we may therefore write the line element, in the limit of the non-relativistic source considered here and in quasi-Minkowski coordinates, as

$$ds^2 = \left(1 + \frac{2\Phi}{c^2}\right) c^2\, dt^2 + \frac{2A_i}{c} c\, dt\, dx^i - \left(1 - \frac{2\Phi}{c^2}\right) \sum (dx^i)^2, \qquad (17.51)$$

in which it is worth noting that $A_i\, dx^i = -\delta_{ij} A^i\, dx^j = -\vec{A} \cdot d\vec{x}$. Determining the geodesics of this line element provides a straightforward means of calculating the trajectories of test particles in the gravitational field of a non-relativistic source (in the weak-field limit). In particular, we note that we need not assume that the test particles are slow-moving, and so the trajectories of photons in this limit may also be found by determining the null geodesics of the line element (17.51).

17.10 Static sources and the Newtonian limit

A special case of stationary sources are *static* sources, for which the constituent particles are not moving. In this case the only non-zero component of the source energy–momentum tensor is the rest energy $T^{00} = \rho c^2$, where $\rho(\vec{x})$ is the proper density distribution of the source. Indeed, this *Newtonian source limit* is clearly equivalent to a stationary source with a vanishing velocity field $u^i(\vec{x}) = 0$. Thus, from (17.50), we immediately find that in this case the non-zero elements of $h_{\mu\nu}$ are

$$h_{00} = h_{11} = h_{22} = h_{33} = \frac{2\Phi}{c^2}. \qquad (17.52)$$

In fact, the above solution remains valid to a good approximation even if the source particles are moving, provided that the source energy–momentum tensor is still dominated by the rest energy of the matter distribution, so that $|T^{00}| \gg |T^{0i}|$ and $|T^{00}| \gg |T^{ij}|$.

The line element corresponding to (17.52) is given by

$$ds^2 = \left(1 + \frac{2\Phi}{c^2}\right) c^2\, dt^2 - \left(1 - \frac{2\Phi}{c^2}\right) d\sigma^2, \qquad (17.53)$$

where $d\sigma^2 = dx^2 + dy^2 + dz^2$; (17.53) is often referred to as the line element in the *Newtonian limit*. Moreover, this line element is easily adapted to allow for arbitrary spatial coordinate transformations, since $d\sigma^2$ is simply the line element of three-dimensional Euclidean space. Thus if, for example, we adopt spatial spherical polar coordinates then one need only rewrite the spatial line element as $d\sigma^2 = dr^2 + r^2\, d\theta^2 + r^2 \sin^2\theta\, d\phi^2$.

It is interesting to compare (17.53) with our discussion of the Newtonian limit in Chapter 7, where we considered weak gravitational fields, static sources and slowly moving test particles. Under these assumptions, we found that we recovered the Newtonian equation of motion for a test particle provided that we made the identification $h_{00} = 2\Phi/c^2$, where Φ is the Newtonian gravitational potential. In the solution (17.53), we have arrived at the Newtonian limit without making any restriction on the velocity of the test particle. This generalisation is important, as previously we needed to consider only the effects of the g_{00}-component of the metric, but, as the above solution shows, the trajectories of relativistic test particles and photons also depend on the metric spatial components.

As an example of the line element (17.53), let us consider the simple case of a static spherical object of mass M, so that the Newtonian gravitational potential is given by $\Phi = -GM/r$, where r is a radial coordinate. In this case, adopting spherical polar spatial coordinates, the line element in the Newtonian limit is given by

$$ds^2 = c^2\left(1 - \frac{2GM}{c^2 r}\right) dt^2 - \left(1 + \frac{2GM}{c^2 r}\right)(dr^2 + r^2\, d\theta^2 + r^2 \sin^2\theta\, d\phi^2),$$

which is straightforwardly shown to be identical to the Schwarzschild solution, to first order in M. In the Solar System, this approximation is sufficiently accurate to determine correctly the bending of light and gravitational redshifts (the Shapiro effect) induced by the Sun, giving identical results to those discussed in Chapter 10. The accuracy of the above approximation is, however, insufficient to predict perihelion shifts correctly. This is not surprising, since perihelion shift is a cumulative effect.

17.11 The energy–momentum of the gravitational field

Physically, one would expect the gravitational field to carry energy–momentum just as, for example, the electromagnetic field does. Unfortunately, the task of assigning an energy density to a gravitational field is famously difficult, both technically and in principle. From our discussion of the equivalence principle in Chapter 7, we know that transforming coordinates to a freely falling frame can always eliminate gravitational effects at any one event. As a result, there is no local notion of gravitational energy density in general relativity. Moreover, in a general spacetime there is no reason why energy and momentum should be conserved. In electromagnetism, for example, the conservation of energy and momentum for the field is a direct consequence of the symmetries of the Minkowski spacetime assumed in the theory. In a general spacetime, however, there are no such symmetries. Even in the linearised gravitational theory developed in this chapter, the field $h_{\mu\nu}$ represents a weak distortion of Minkowski space and so the Lorentz symmetry properties are lost.

Nevertheless, as we have remarked several times, one can also regard the linearised theory as describing a simple rank-2 tensor field $h_{\mu\nu}$ in Cartesian inertial coordinates propagating in a fixed Minkowski spacetime background. We might therefore hope to assign an energy–momentum tensor to this field just as we do for electromagnetism, or any other field theory in Minkowski spacetime. As was discussed in Section 17.4, the linearised gravitational theory ignores the energy–momentum associated with the gravitational field itself (i.e. the 'gravity of gravity'). To include this contribution, and thereby go beyond the linearised theory, one must modify the linearised field equations to read

$$G^{(1)}_{\mu\nu} = -\frac{8\pi G}{c^4}(T_{\mu\nu} + t_{\mu\nu}),$$

where $G^{(1)}_{\mu\nu}$ is the linearised Einstein tensor, $T^{\mu\nu}$ is the energy–momentum tensor of any matter present and $t_{\mu\nu}$ is the energy–momentum tensor of the gravitational field itself. Trivially rearranging this equation gives

$$G^{(1)}_{\mu\nu} + \frac{8\pi G}{c^4} t_{\mu\nu} = -\frac{8\pi G}{c^4} T_{\mu\nu}.$$

Returning to the *exact* Einstein equations, however, we may expand beyond first order to obtain

$$G_{\mu\nu} \equiv G^{(1)}_{\mu\nu} + G^{(2)}_{\mu\nu} + \cdots = -\frac{8\pi G}{c^4} T_{\mu\nu}, \tag{17.54}$$

where superscripts in parentheses indicate the order of the expansion in $h_{\mu\nu}$. This suggests that, to a good approximation, we should make the identification

$$t_{\mu\nu} \equiv \frac{c^4}{8\pi G} G^{(2)}_{\mu\nu}. \tag{17.55}$$

This is also in keeping with our experience of other field theories in Minkowski spacetime, such as electromagnetism, in which the energy–momentum tensor is quadratic in the field variable. One should not, however, be too firmly guided by the analogy with electromagnetism. The reason why the electromagnetic energy–momentum tensor is quadratic in the field variable is that the electromagnetic field (constituted by photons in the quantum description) does not carry charge and so cannot act as its own source. Indeed, this is the physical reason why electromagnetism is a linear theory. In the gravitational case, however, one could in fact include the higher-order terms in (17.54) in the definition of $t_{\mu\nu}$; these terms correspond to the contribution to the total energy–momentum arising from the gravitational interaction of the gravitational field with itself. Nevertheless, when the gravitational field is weak these higher-order terms may be neglected.

As one might expect from such an heuristic approach, however, there are some shortcomings of the identification (17.55), which we now outline. The terms in the Einstein tensor that are second-order in $h_{\mu\nu}$ are given by

$$G^{(2)}_{\mu\nu} = R^{(2)}_{\mu\nu} - \tfrac{1}{2}\eta_{\mu\nu}R^{(2)} - \tfrac{1}{2}h_{\mu\nu}R^{(1)} + \tfrac{1}{2}\eta_{\mu\nu}h^{\rho\sigma}R^{(1)}_{\rho\sigma}, \tag{17.56}$$

where $R^{(2)}_{\mu\nu}$ denotes the terms in the Ricci tensor that are second-order in $h_{\mu\nu}$ and $R^{(1)}$ and $R^{(2)}$ denote the terms in the Ricci scalar that are first- and second-order in $h_{\mu\nu}$ respectively. Although (17.56), and hence $t_{\mu\nu}$, is covariant under global Lorentz transformations (although not under general coordinate transformations, as one might expect), it may be may shown, after considerable algebra, that it is *not* invariant under the gauge transformation (17.5) (or equivalently the infinitesimal coordinate transformation (17.4)). One way of circumventing this problem is to take seriously the fact that the energy–momentum of a gravitational field at a point in spacetime has no real meaning in general relativity, since at any particular event one can always transform to a free-falling frame in which gravitational effects disappear. This suggests that, at each point in spacetime, one should average $G^{(2)}_{\mu\nu}$ over a small region in order to probe the physical curvature of the spacetime, which gives a gauge-invariant measure of the gravitational field strength. Denoting this averaging process by $\langle\cdots\rangle$, one should thus replace (17.55) by

$$\boxed{t_{\mu\nu} \equiv \frac{c^4}{8\pi G}\left\langle G^{(2)}_{\mu\nu}\right\rangle.} \tag{17.57}$$

Having made this identification, our task is now an algebraic one of determining the form of $\left\langle G^{(2)}_{\mu\nu}\right\rangle$ as a function of $h_{\mu\nu}$. This is rather a cumbersome calculation, but the job is made somewhat easier by averaging over small spacetime regions. Since we are averaging over all directions at each point, first derivatives average to zero. Thus, for any function of position $a(x)$, we have $\langle\partial_\mu a\rangle = 0$. This has the important consequence that $\langle\partial_\mu(ab)\rangle = \langle(\partial_\mu a)b\rangle + \langle a(\partial_\mu b)\rangle = 0$, and hence we may swap derivatives in products and inherit only a minus sign, i.e.

$$\langle(\partial_\mu a)b\rangle = -\langle a(\partial_\mu b)\rangle. \tag{17.58}$$

Let us begin by considering the last two terms on the right-hand side of (17.56), which depend on the first-order Ricci tensor and Ricci scalar. It will prove most convenient to express these in terms of the energy–momentum tensor $T_{\mu\nu}$ of any matter present. The first-order (linearised) field equation (17.11) can be written as

$$R^{(1)}_{\mu\nu} = -\kappa(T_{\mu\nu} - \tfrac{1}{2}\eta_{\mu\nu}T),$$

where $T \equiv T^{\mu}_{\mu}$ and $\kappa \equiv 8\pi G/c^4$. We also note from this equation that $R^{(1)} = \kappa T$. Thus, we may write (17.56) as

$$G^{(2)}_{\mu\nu} = R^{(2)}_{\mu\nu} - \tfrac{1}{2}\eta_{\mu\nu}R^{(2)} - \tfrac{1}{2}\kappa(\bar{h}_{\mu\nu}T + \eta_{\mu\nu}h^{\rho\sigma}T_{\rho\sigma}). \tag{17.59}$$

It therefore remains only to find the form of $R^{(2)}_{\mu\nu}$, from which $R^{(2)}$ may be obtained by contraction. The standard expression for the full Ricci tensor is obtained by contracting (17.7) on its first and last indices. Thus, the terms second-order in $h_{\mu\nu}$ are given by

$$R^{(2)}_{\mu\nu} = \partial_\nu \Gamma^{(2)\sigma}{}_{\mu\sigma} - \partial_\sigma \Gamma^{(2)\sigma}{}_{\mu\nu} + \Gamma^{(1)\rho}{}_{\mu\sigma}\Gamma^{(1)\sigma}{}_{\rho\nu} - \Gamma^{(1)\rho}{}_{\mu\nu}\Gamma^{(1)\sigma}{}_{\rho\sigma}, \tag{17.60}$$

where, on the right-hand side, the superscripts in parentheses denote the order of expansion in $h_{\mu\nu}$ for the connection coefficients. The connection coefficients to first order were calculated in (17.6), and now including the second-order terms we have

$$\begin{aligned}\Gamma^{\sigma}{}_{\mu\nu} &= \Gamma^{(1)\sigma}{}_{\mu\nu} + \Gamma^{(2)\sigma}{}_{\mu\nu} + \cdots \\ &= \tfrac{1}{2}(\partial_\nu h^\sigma_\mu + \partial_\mu h^\sigma_\nu - \partial^\sigma h_{\mu\nu}) - \tfrac{1}{2}h^{\sigma\tau}(\partial_\nu h_{\tau\mu} + \partial_\mu h_{\tau\nu} - \partial_\tau h_{\mu\nu}) + \cdots .\end{aligned}$$

Inserting these expressions into (17.60) and simplifying, one finds after a little algebra

$$\begin{aligned}R^{(2)}_{\mu\nu} = &-\tfrac{1}{4}(\partial_\mu h^{\rho\sigma})\partial_\nu h_{\rho\sigma} + \tfrac{1}{2}h^{\rho\sigma}(\partial_\mu\partial_\sigma h_{\nu\rho} + \partial_\nu\partial_\sigma h_{\mu\rho} - \partial_\mu\partial_\nu h_{\rho\sigma} - \partial_\rho\partial_\sigma h_{\mu\nu}) \\ &+ \tfrac{1}{2}(\partial^\sigma h^\rho_\nu)(\partial_\rho h_{\sigma\mu} - \partial_\sigma h_{\rho\mu}) + \tfrac{1}{2}(\partial_\sigma h^{\rho\sigma} - \tfrac{1}{2}\partial^\rho h)(\partial_\mu h_{\nu\rho} + \partial_\nu h_{\mu\rho} - \partial_\rho h_{\mu\nu}).\end{aligned} \tag{17.61}$$

Although the third group of terms on the right-hand side is not manifestly symmetric in μ and ν, this symmetry is easy to verify. In fact, in subsequent calculations it is convenient to maintain manifest symmetry by writing out this term again with μ and ν reversed and multiplying both terms by one-half.

To evaluate the averaged expression (17.57), we must now calculate $\left\langle R^{(2)}_{\mu\nu}\right\rangle$. One first makes use of the result (17.58) to rewrite products of first derivatives in (17.61) in terms of second derivatives. Using the first-order field equation (17.10) to substitute for terms of the form $\Box^2 h_{\mu\nu}$, and then applying (17.58) once more to rewrite terms containing second derivatives as products of first derivatives, one finally obtains

$$\left\langle R^{(2)}_{\mu\nu}\right\rangle = \tfrac{1}{4}\Big\langle(\partial_\mu h_{\rho\sigma})\partial_\nu h^{\rho\sigma} - 2(\partial_\sigma h^{\rho\sigma})\partial_{(\mu}h_{\nu)\rho} + 2(\partial_\rho h)\partial_{(\mu}h^{\rho}_{\nu)} - (\partial_\mu h)\partial_\nu h$$
$$+\kappa\left(2h_{\mu\nu}T + 2hT_{\mu\nu} - \eta_{\mu\nu}hT - 4h_{\rho(\mu}T^{\rho}_{\nu)}\right)\Big\rangle, \qquad (17.62)$$

where we have made use of the symmetrisation notation discussed in Chapter 4. Contracting this expression, and once again making use of the result (17.58) and the first-order field equation (17.10), one quickly finds that

$$\langle R^{(2)}\rangle = -\tfrac{1}{2}\kappa\langle h^{\rho\sigma}T_{\rho\sigma}\rangle. \qquad (17.63)$$

Combining the expressions (17.59), (17.62) and (17.63) and writing the result (mostly) in terms of the trace reverse field $\bar{h}_{\mu\nu} = h_{\mu\nu} - \frac{1}{2}\eta_{\mu\nu}h$, we thus find that the energy–momentum tensor (17.57) of the gravitational field is given by

$$\boxed{\begin{aligned} t_{\mu\nu} = \tfrac{1}{4}\kappa^{-1}\Big\langle&(\partial_\mu \bar{h}_{\rho\sigma})\partial_\nu \bar{h}^{\rho\sigma} - 2(\partial_\sigma \bar{h}^{\rho\sigma})\partial_{(\mu}\bar{h}_{\nu)\rho} - \tfrac{1}{2}(\partial_\mu \bar{h})\partial_\nu \bar{h}\\ &-\kappa\left(4\bar{h}_{\rho(\mu}T^{\rho}_{\nu)} + \eta_{\mu\nu}h^{\rho\sigma}T_{\rho\sigma}\right)\Big\rangle. \end{aligned}} \qquad (17.64)$$

It may be verified by direct substitution that this expression is indeed invariant under the gauge transformation (17.5), as required. We shall use this tensor in the next chapter to determine the energy carried by gravitational waves.

Appendix 17A: The Einstein–Maxwell formulation of linearised gravity

In our discussion of non-relativistic stationary sources in Section 17.9, we found that the expressions for the gravitational field exhibited a remarkable similarity to the corresponding results in electromagnetism. We now pursue further the analogy between linearised general relativity and electromagnetism for non-relativistic stationary sources.

As discussed in Section 17.9, for such a source we may write

$$h^{00} = h^{11} = h^{22} = h^{33} = \frac{2\Phi_g}{c^2}, \qquad h^{0i} = \frac{A^i_g}{c}, \qquad h^{ij} = 0 \quad (i \neq j); \qquad (17.65)$$

here we denote the gravitational scalar and vector potentials by Φ_g and $\vec{A}_g$ respectively. The linearised field equations may then be written as

$$\nabla^2\Phi_g = 4\pi G\rho \qquad \text{and} \qquad \nabla^2\vec{A}_g = \frac{16\pi G}{c^2}\vec{j}, \qquad (17.66)$$

where we have defined the momentum density (or matter current density) $\vec{j} \equiv \rho\vec{v}$. These equations have the solutions (17.48, 17.49), which we write as

$$\Phi_g(\vec{x}) = -G\int \frac{\rho(\vec{y})}{|\vec{x}-\vec{y}|}\, d^3\vec{y} \qquad \text{and} \qquad \vec{A}_g(\vec{x}) = -\frac{4G}{c^2}\int \frac{\vec{j}(\vec{y})}{|\vec{x}-\vec{y}|}\, d^3\vec{y}.$$

Comparing the above results with the corresponding equations in electromagnetism for the electric potential and the magnetic vector potential in the absence of time-varying fields, there is a direct analogy on making the identifications

$$\epsilon_0 \leftrightarrow -\frac{1}{4\pi G} \qquad \text{and} \qquad \mu_0 \leftrightarrow -\frac{16\pi G}{c^2}.$$

The minus signs in these relations are a result of the fact that the electric force repels like charges, whereas the gravitational force attracts (like) masses. Clearly, in the electromagnetic case, ρ and $\vec{j}$ correspond to the charge and current densities respectively, rather than the matter and momentum densities. We can take the analogy further by defining the *gravitoelectric* and *gravitomagnetic* fields

$$\vec{E}_g = -\vec{\nabla}\Phi_g \qquad \text{and} \qquad \vec{B}_g = \vec{\nabla}\times\vec{A}_g. \tag{17.67}$$

Using the equations (17.66), it is straightforward to verify that the fields $\vec{E}_g$ and $\vec{B}_g$ satisfy the *gravitational Maxwell equations*

$$\boxed{\begin{aligned} &\vec{\nabla}\cdot\vec{E}_g = -4\pi G\rho, && \vec{\nabla}\cdot\vec{B}_g = 0, \\ &\vec{\nabla}\times\vec{E}_g = 0, && \vec{\nabla}\times\vec{B}_g = -\frac{16\pi G}{c^2}\vec{j}. \end{aligned}}$$

The equations for $\vec{E}_g$ describe the standard gravitational field produced by a static mass distribution, whereas the equations for $\vec{B}_g$ provide a notationally familiar means of determining the 'extra' gravitational field produced by *moving* masses in a stationary non-relativistic source.

Although the gravitational Maxwell equations completely determine the gravitational fields produced by a non-relativistic stationary source, they do not determine the effect of such fields on the motion of a test particle. In electromagnetism one must, in addition, postulate the Lorentz force law. From our discussion in Section 8.8, however, one might suspect that in the case of gravitation the corresponding force law could be derived rather than postulated. The equation of motion for a test particle in a gravitation field is the geodesic equation

$$\ddot{x}^\sigma + \Gamma^\sigma{}_{\mu\nu}\dot{x}^\mu\dot{x}^\nu = 0, \tag{17.68}$$

where the dots denote differentiation with respect to the proper time τ of the particle. Let us assume that the test particle is *slow-moving*, i.e. its speed v is sufficiently small compared with c that we may neglect terms in v^2/c^2 and higher. Hence we may take $\gamma_v = (1 - v^2/c^2)^{-1/2} \approx 1$. Writing $[x^\mu] = (ct, \vec{x})$, the 4-velocity of the particle may thus be written

$$[\dot{x}^\mu] = \gamma_v(c, \vec{v}) \approx (c, \vec{v}).$$

This immediately implies that $\ddot{x}^0 = 0$ and, moreover, that $dt/d\tau = 1$, so we may replace dots with derivatives with respect to t. Thus, the spatial components of (17.68) may be written as

$$\frac{d^2x^i}{dt^2} \approx -\left(c^2\Gamma^i{}_{00} + 2c\Gamma^i{}_{0j}v^j + \Gamma^i{}_{ij}v^iv^j\right) \approx -\left(c^2\Gamma^i{}_{00} + 2c\Gamma^i{}_{0j}v^j\right), \qquad (17.69)$$

where in the first approximate equality we have expanded the summation in (17.68) into terms containing respectively two time components, one time and one spatial component, and two spatial components. In the second approximation, we have neglected the purely spatial terms since their ratio with respect to the purely temporal term $c^2\Gamma^i{}_{00}$ is of order v^2/c^2. To first order in the gravitational field $h_{\mu\nu}$, the connection coefficients are given by (17.6). Inserting this expression into (17.69) and remembering that for a stationary field $\partial_0 h_{\mu\nu} = 0$, one obtains

$$\frac{d^2x^i}{dt^2} \approx \tfrac{1}{2}c^2\partial^i h_{00} + c\left(\partial^i h_{0j} - \partial_j h^i_0\right) v^j = -\tfrac{1}{2}c^2\delta^{ij}\partial_j h_{00} - c\delta^{ik}(\partial_k h_{0j} - \partial_j h_{0k})v^j.$$

Substituting the expressions (17.65) and remembering that one inherits a minus sign on raising or lower a spatial (roman) index, the equation of motion may be written as

$$\frac{d^2\vec{x}}{dt^2} \approx -\vec{\nabla}\Phi_g + \vec{v} \times (\vec{\nabla} \times \vec{A}_g).$$

Thus, using (17.67), one obtains the *gravitational Lorenz force law*

$$\boxed{\frac{d^2\vec{x}}{dt^2} \approx \vec{E}_g + \vec{v} \times \vec{B}_g,}$$

for slow-moving particles in the gravitational field of a stationary non-relativistic source. The first term on the right-hand side gives the standard Newtonian result for the motion of a test particle in the field of a static non-relativistic source, whereas the second term gives a notationally familiar result for the 'extra' force felt by a *moving* test particle in the presence of the 'extra' field produced by *moving* masses in a stationary non-relativistic source.

Exercises

17.1 In a region of spacetime with a weak gravitational field, there exist coordinates in which the metric takes the form $g_{\mu\nu} = \eta_{\mu\nu} + h_{\mu\nu}$. Show that $h_{\mu\nu}$ is not a tensor under a general coordinate transformation. Show further that, to first order in $h_{\mu\nu}$,

$$g^{\mu\nu} = \eta^{\mu\nu} - h^{\mu\nu}$$

where $h^{\mu\nu} = \eta^{\mu\rho}\eta^{\nu\sigma}h_{\rho\sigma}$.

17.2 For an infinitesimal general coordinate transformation $x'^{\mu} = x^{\mu} + \xi^{\mu}(x)$, show that to first order in ξ^{μ} the inverse transformation is given by

$$\frac{\partial x^{\mu}}{\partial x'^{\nu}} = \delta^{\mu}_{\nu} - \partial_{\nu}\xi^{\mu}.$$

17.3 If $g_{\mu\nu} = \eta_{\mu\nu} + h_{\mu\nu}$ with $|h_{\mu\nu}| \ll 1$, verify that, to first order in $h_{\mu\nu}$,

$$\begin{aligned} R^{\sigma}{}_{\mu\nu\rho} &= \tfrac{1}{2}(\partial_{\nu}\partial_{\mu}h^{\sigma}_{\rho} + \partial_{\rho}\partial^{\sigma}h_{\mu\nu} - \partial_{\nu}\partial^{\sigma}h_{\mu\rho} - \partial_{\rho}\partial_{\mu}h^{\sigma}_{\nu}), \\ R_{\mu\nu} &= \tfrac{1}{2}(\partial_{\nu}\partial_{\mu}h + \Box^2 h_{\mu\nu} - \partial_{\nu}\partial_{\rho}h^{\rho}_{\mu} - \partial_{\rho}\partial_{\mu}h^{\rho}_{\nu}), \\ R &= \Box^2 h - \partial_{\rho}\partial_{\mu}h^{\mu\rho}. \end{aligned}$$

Hence show that the linearised Einstein field equations are given by

$$\partial_{\nu}\partial_{\mu}h + \Box^2 h_{\mu\nu} - \partial_{\nu}\partial_{\rho}h^{\rho}_{\mu} - \partial_{\rho}\partial_{\mu}h^{\rho}_{\nu} - \eta_{\mu\nu}(\Box^2 h - \partial_{\rho}\partial_{\sigma}h^{\sigma\rho}) = -2\kappa T_{\mu\nu}.$$

17.4 The trace reverse of $h_{\mu\nu}$ is defined by

$$\bar{h}_{\mu\nu} \equiv h_{\mu\nu} - \tfrac{1}{2}\eta_{\mu\nu}h.$$

Show that $\bar{h} = -h$ and $\bar{\bar{h}}_{\mu\nu} = h_{\mu\nu}$. Hence show that the linearised Einstein field equations in Exercise 17.3 can be written as

$$\Box^2 \bar{h}_{\mu\nu} + \eta_{\mu\nu}\partial_{\rho}\partial_{\sigma}\bar{h}^{\rho\sigma} - \partial_{\nu}\partial_{\rho}\bar{h}^{\rho}_{\mu} - \partial_{\mu}\partial_{\rho}\bar{h}^{\rho}_{\nu} = -2\kappa T_{\mu\nu}.$$

17.5 Obtain an expression for the covariant components $R_{\sigma\mu\nu\rho}$ of the linearised Riemann tensor in Exercise 17.3 and show that it is invariant under a gauge transformation of the form (17.5). Hence show that the linearised Einstein field equations are also invariant under such a gauge transformation.

17.6 From the linearised Einstein field equations, show that $\partial_{\mu}T^{\mu\nu} = 0$.

17.7 For a plane gravity wave of the form $h_{\mu\nu} = A_{\mu\nu}\exp(ik_{\lambda}x^{\lambda})$, show that the linearised Riemann tensor is given by

$$R_{\sigma\mu\nu\rho} = \tfrac{1}{2}(k_{\nu}k_{\sigma}h_{\mu\rho} + k_{\rho}k_{\mu}h_{\sigma\nu} - k_{\nu}k_{\mu}h_{\sigma\rho} - k_{\rho}k_{\sigma}h_{\mu\nu}).$$

Hence show that the linearised Ricci tensor is given by

$$R_{\mu\nu} = \tfrac{1}{2}(k_{\nu}w_{\mu} + k_{\mu}w_{\nu} - k^2 h_{\mu\nu}),$$

where $k^2 = k_\rho k^\rho$ and $w_\mu = k^\rho \bar{h}_{\mu\rho}$. Hence show that the linearised Einstein field equations require that

$$k^2 h_{\mu\nu} = k_\nu w_\mu + k_\mu w_\nu.$$

17.8 From your answer to Exercise 17.7, show that for $k^2 \neq 0$ one requires $R_{\sigma\mu\nu\rho} = 0$. Hence show that this case does not correspond to a physical wave but merely a periodic oscillation of the coordinate system.

17.9 From your answer to Exercise 17.7, show that for $k^2 = 0$ one requires $k^\rho \bar{h}_{\mu\rho} = 0$. Hence show that the wavevector k^ρ is an eigenvector of the Riemann tensor in the sense that $R_{\sigma\mu\nu\rho} k^\rho = 0$.

17.10 Show explicitly that

$$\bar{h}^{\mu\nu}(x^\sigma) = \bar{h}^{\mu\nu}_{(0)}(x^\sigma) - 2\kappa \int \mathcal{G}(x^\sigma - y^\sigma) T^{\mu\nu}(y^\sigma)\, d^4y,$$

is a solution of the linearised Einstein field equations in the Lorenz gauge if $\bar{h}^{\mu\nu}_{(0)}(x^\sigma)$ is any solution of the linearised field equations *in vacuo* and $\mathcal{G}(x^\sigma - y^\sigma)$ satisfies

$$\Box^2_x \mathcal{G}(x^\sigma - y^\sigma) = \delta^{(4)}(x^\sigma - y^\sigma).$$

17.11 The Green's function $\mathcal{G}(x^\sigma - y^\sigma)$ satisfies the equation

$$\Box^2_x \mathcal{G}(x^\sigma - y^\sigma) = \delta^{(4)}(x^\sigma - y^\sigma).$$

Show that the four-dimensional Dirac delta function can be written as

$$\delta^{(4)}(x^\sigma - y^\sigma) = \frac{1}{(2\pi)^4} \int \exp\left[ik_\lambda (x^\lambda - y^\lambda)\right] d^4k.$$

Hence, by writing the Green's function in terms of its Fourier transform $\widetilde{\mathcal{G}}(k^\sigma)$, show that

$$\mathcal{G}(x^\sigma - y^\sigma) = -\frac{1}{(2\pi)^4} \int \frac{1}{k^2} \exp\left[ik_\lambda (x^\lambda - y^\lambda)\right] d^4k,$$

where $k^2 = k_\lambda k^\lambda$.

17.12 Verify that the solution in Exercise 17.10 satisfies the Lorenz gauge condition.

17.13 Prove the results (17.32, 17.33) for the derivatives of a function of retarded time.

17.14 By writing $r \equiv |\vec{x}| = (\delta_{ij} x^i x^j)^{1/2}$, show that

$$\partial_i \left(\frac{1}{r}\right) = -\frac{x_i}{r}.$$

Hence show that

$$\partial_i \partial_j \left(\frac{1}{r}\right) = \frac{3x_i x_j - r^2 \delta_{ij}}{r^5}.$$

17.15 In Newtonian gravity, the gravitational potential $\Phi(\vec{x})$ produced by some density distribution $\rho(\vec{x})$ is given by

$$\Phi(\vec{x}) = -G\int_V \frac{\rho(\vec{y})}{|\vec{x}-\vec{y}|}\,d^3\vec{y},$$

where the integral extends over the volume of the distribution. Show that

$$\frac{1}{|\vec{x}-\vec{y}|} = \frac{1}{|\vec{x}|} + \frac{\vec{x}\cdot\vec{y}}{|\vec{x}|^3} + \mathcal{O}\left(\frac{1}{|\vec{x}|^3}\right).$$

Hence show that the gravitational potential can be written as

$$\Phi(\vec{x}) = -\frac{GM}{|\vec{x}|} - \frac{G\vec{d}\cdot\vec{x}}{|\vec{x}|^3} + \mathcal{O}\left(\frac{1}{|\vec{x}|^3}\right),$$

where

$$M = \int_V \rho(\vec{y})\,d^3\vec{y} \qquad \text{and} \qquad \vec{d} = \int_V \rho(\vec{y})\vec{y}\,d^3\vec{y}.$$

17.16 From the conservation equation $\partial_\mu T^{\mu\nu} = 0$, show that

$$\partial_0 T^{00} + \partial_k T^{0k} = 0 \qquad \text{and} \qquad \partial_0 T^{i0} + \partial_k T^{ik} = 0.$$

By integrating each equation over a spatial volume V whose bounding surface S encloses the energy–momentum source and using the three-dimensional divergence theorem, show that the quantities

$$Mc^2 \equiv \int_V T^{00}\,d^3\vec{y} \qquad \text{and} \qquad P^i \equiv \int T^{i0}\,d^3\vec{y}$$

are constants and give a physical interpretation of them.

17.17 For a stationary source, show that $\partial_0 T^{0i} = 0$. Hence show that

$$\int_V (T^{0i}y^j + T^{0j}y^i)\,d^3\vec{y} = 0,$$

where the spatial volume V encloses the source.

17.18 For a non-relativistic stationary source, show that, in centre-of-momentum coordinates,

$$\bar{h}^{00}(\vec{x}) = -\frac{4GM}{c^2|\vec{x}|} + \mathcal{O}\left(\frac{1}{|\vec{x}|^2}\right),$$

$$\bar{h}^{0i}(\vec{x}) = \frac{2G}{c^3|\vec{x}|^3}x_j J^{ij} + \mathcal{O}\left(\frac{1}{|\vec{x}|^3}\right),$$

$$\bar{h}^{ij}(\vec{x}) = 0,$$

where the quantities M and J^{ij} are given by

$$M = \int_V \rho(\vec{y})\,d^3\vec{y} \qquad \text{and} \qquad J^{ij} = \int_V \left[y^i p^j(\vec{y}) - y^j p^i(\vec{y})\right] d^3\vec{y},$$

in which $\rho(\vec{y})$ is the proper density distribution of the source and $p^i(\vec{y}) = \rho(\vec{y})u^i(\vec{y})$ is the momentum density distribution of the source. Give a physical interpretation of J^{ij}.

Hint: You will find your answer to Exercise 17.17 useful.

17.19 Use your answer to Exercise 17.18 to show that, for a stationary non-relativistic source, the gravitational scalar and vector potentials respectively are given to leading order in $1/|\vec{x}|$ by

$$\Phi_g(\vec{x}) = -\frac{GM}{|\vec{x}|} \qquad \text{and} \qquad \vec{A}_g(\vec{x}) = -\frac{2G}{c^2|\vec{x}|^3}\vec{J}\times\vec{x},$$

where $\vec{J} = \int(\vec{y}\times\vec{p})\,d^3\vec{y}$ is the total angular momentum vector of the source. Show further that these expressions are exact in the linear theory for a spherically symmetric source.

17.20 Use your answer to Exercise 17.19 to show that, in the linear theory, the line element outside a spherically symmetric matter distribution rotating about the z-axis at a steady rate is given by

$$ds^2 = c^2\left(1-\frac{2GM}{c^2r}\right)dt^2 + \frac{4GJ}{c^2r^3}(x\,dy - y\,dx)\,dt - \left(1+\frac{2GM}{c^2r}\right)(dx^2+dy^2+dz^2),$$

where $r = |\vec{x}|$. Show that this is equal to the Kerr line element to first order in M and J.

Hint: $A_i\,dx^i = -\delta_{ij}A^j\,dx^i = -\vec{A}\cdot d\vec{x}$ *and* $(\vec{J}\times\vec{x})\cdot d\vec{x} = \vec{J}\cdot(\vec{x}\times d\vec{x})$.

17.21 If $g_{\mu\nu} = \eta_{\mu\nu} + h_{\mu\nu}$, show that the terms in the Einstein tensor that are second order in $h_{\mu\nu}$ are given by

$$G^{(2)}_{\mu\nu} = R^{(2)}_{\mu\nu} - \tfrac{1}{2}\eta_{\mu\nu}R^{(2)} - \tfrac{1}{2}h_{\mu\nu}R^{(1)} + \tfrac{1}{2}\eta_{\mu\nu}h^{\rho\sigma}R^{(1)}_{\rho\sigma},$$

where $R^{(2)}_{\mu\nu}$ denotes the terms in the Ricci tensor that are second order in $h_{\mu\nu}$, and $R^{(1)}$ and $R^{(2)}$ denote the terms in the Ricci scalar that are first and second order in $h_{\mu\nu}$ respectively. Show further that this quantity is not invariant under a gauge transformation of the form (17.5).

17.22 Verify that the energy–momentum tensor of the linearised gravitational field is given by (17.64). Show further that this tensor is invariant under a gauge transformation of the form (17.5).

17.23 Use your answer to Exercise 17.19 to show that, in the linear theory, a spherically symmetric body of mass M rotating steadily with angular momentum $\vec{J}$ produces gravitoelectric and gravitomagnetic fields given respectively by

$$E_g(\vec{x}) = -\frac{GM}{|\vec{x}|^2}\hat{\vec{x}} \qquad \text{and} \qquad B_g(\vec{x}) = \frac{2G}{c^2|\vec{x}|^3}\left[\vec{J} - 3\left(\vec{J}\cdot\hat{\vec{x}}\right)\hat{\vec{x}}\right],$$

where $\hat{\vec{x}}$ is a unit vector in the $\vec{x}$- direction.

Hint: For any scalar field ϕ and spatial vector fields $\vec{a}$ and $\vec{b}$ one has $\vec{\nabla}(\phi\vec{a}) = \vec{\nabla}\phi\times\vec{a} + \phi(\vec{\nabla}\times\vec{a})$ and $\vec{\nabla}\times(\vec{a}\times\vec{b}) = \vec{a}(\vec{\nabla}\cdot\vec{b}) - \vec{b}(\vec{\nabla}\cdot\vec{a}) + (\vec{b}\cdot\vec{\nabla})\vec{a} - (\vec{a}\cdot\vec{\nabla})\vec{b}$. Also, $\vec{\nabla}(1/|\vec{x}|^3) = -3\vec{x}/|\vec{x}|^5$.

17.24 Consider a particle moving under gravity at speed v in a circular orbit of radius r in the equatorial plane of the body in Exercise 17.23. Show that the vector acceleration of the particle is given by

$$\vec{a} = -\frac{GM}{r^2}\hat{\vec{r}} \pm \frac{2GJv}{c^2 r^3}\hat{\vec{r}},$$

where $\vec{r}$ is the position vector of the orbiting particle and the plus and minus signs corresponding to prograde and retrograde orbits respectively. Hence show that the angular velocity ω of the particle is given to first order in J by

$$\omega^2 = \frac{GM}{r^3} \mp \frac{2GJ}{c^2 r^4}\sqrt{\frac{GM}{r}},$$

where the minus and plus signs now correspond to prograde and retrograde orbits respectively. Thus show that the retrograde orbit has a shorter period than the prograde orbit.

17.25 In electromagnetism, the magnetic dipole moment of a current density distribution $\vec{j}(\vec{y})$ is defined by $\vec{m} = \frac{1}{2}\int (\vec{y} \times \vec{j})\, d^3\vec{y}$, and the force and torque on the dipole in a magnetic field $\vec{B}$ are given by $\vec{F} = (\vec{m} \cdot \vec{\nabla})\vec{B}$ and $\vec{T} = \vec{m} \times \vec{B}$ respectively. Hence deduce that, in linearised gravity, the force and torque exerted by the gravitomagnetic field $\vec{B}_g$ on a spinning body with spin angular momentum $\vec{s}$ are given respectively by

$$\vec{F}_g = \tfrac{1}{2}(\vec{s} \cdot \vec{\nabla})\vec{B}_g \qquad \text{and} \qquad \vec{T}_g = \tfrac{1}{2}(\vec{s} \times \vec{B}_g).$$

Thus show that the spin angular momentum of the body will evolve as

$$\frac{d\vec{s}}{dt} = \tfrac{1}{2}(\vec{s} \times \vec{B}_g)$$

and therefore that $\vec{s}$ precesses about $\vec{B}_g$ with angular velocity $\Omega = -\frac{1}{2}|\vec{B}_g|$ (i.e. in the negative sense). This is called the *Lens–Thirring precession.*

17.26 A gyroscope is in orbit about the massive rotating body in Exercise 17.23. Use your answer to Exercise 17.25 to show that the precessional angular velocity vector of the gyroscope is given by

$$\vec{\Omega} = \frac{G}{c^2|\vec{x}|^3}\left[3(\vec{J} \cdot \hat{\vec{x}})\hat{\vec{x}} - \vec{J}\right],$$

where $\vec{x}$ is the position vector of the gyroscope relative to the centre of the massive body. Show that this result agrees with that derived in Section 13.20 when $\vec{x}$ points along $\vec{J}$.

18

Gravitational waves

In the previous chapter, we saw that the linearised field equations of general relativity could be written in the form of a wave equation

$$\Box^2 \bar{h}^{\mu\nu} = -2\kappa T^{\mu\nu}, \tag{18.1}$$

provided that the $\bar{h}^{\mu\nu}$ satisfy the Lorenz gauge condition

$$\partial_\mu \bar{h}^{\mu\nu} = 0. \tag{18.2}$$

This suggests the existence of gravitational waves in an analogous manner to that in which Maxwell's equations predict electromagnetic waves. In this chapter, we discuss in detail the propagation, generation and detection of such gravitational radiation. As in the previous chapter, we will adopt the viewpoint that $h_{\mu\nu}$ is simply a symmetric tensor field (under global Lorentz transformations) defined on a flat Minkowski background spacetime.

18.1 Plane gravitational waves and polarisation states

In Section 17.5, we showed that the general solution of the linearised field equations *in vacuo* may be written as the superposition of plane-wave solutions of the form

$$\bar{h}^{\mu\nu} = A^{\mu\nu} \exp(ik_\rho x^\rho), \tag{18.3}$$

where the $A^{\mu\nu}$ are constant (and, in general, complex) components of a symmetric tensor and k_μ are the constant (real) components of a vector. The Lorenz gauge condition is satisfied provided that the additional constraint

$$A^{\mu\nu} k_\nu = 0 \tag{18.4}$$

is obeyed. Physical solutions corresponding to propagating plane gravitational waves in empty space may be obtained by taking the real part of (18.3):

$$\begin{aligned}\bar{h}^{\mu\nu} &= \Re[A^{\mu\nu}\exp(ik_\rho x^\rho)]\\ &= \tfrac{1}{2}A^{\mu\nu}\exp(ik_\rho x^\rho) + \tfrac{1}{2}(A^{\mu\nu})^*\exp(-ik_\rho x^\rho),\end{aligned}$$

which is clearly just a superposition of two plane waves of the form (18.3).

The constants $A^{\mu\nu}$ are the components of the *amplitude tensor*, and the $k^\mu \equiv \eta^{\mu\nu}k_\nu$ are the components of the *4-wavevector*. It is conventional to denote the components of the 4-wavevector by $[k^\mu] = (\omega/c, \vec{k})$, where $\vec{k}$ is the spatial 3-wavevector in the direction of propagation and ω is the angular frequency of the wave. The nullity of $\boldsymbol{k}$ implies that $\omega^2 = c^2|\vec{k}|^2$, and so both the group and phase velocity of a gravitational wave are equal to the speed of light.

Since $A^{\mu\nu} = A^{\nu\mu}$, the amplitude tensor has 10 different (complex) components, but the four Lorenz gauge conditions (18.4) reduce the number of independent components to six. Moreover, we still have the freedom to make a further gauge transformation of the form (17.5), which will preserve the Lorenz gauge provided that we choose the four functions $\xi^\mu(x)$ so that they satisfy $\Box^2\xi^\mu = 0$. As we show below, this may be used to reduce the number of independent components in the amplitude matrix from six to just two. This results in two possible *polarisations* for plane gravitational waves.

It is convenient to consider the concrete example of a plane gravitational wave propagating in the x^3-direction, in which case the components of the 4-wavevector are

$$[k^\mu] = (k, 0, 0, k), \tag{18.5}$$

where $k = \omega/c$. The Lorenz gauge condition (18.4) then immediately gives $A^{\mu 3} = A^{\mu 0}$. Together with the symmetry of the amplitude tensor, this implies that all the components $A^{\mu\nu}$ can be expressed in terms of the six quantities $A^{00}, A^{01}, A^{02}, A^{11}, A^{12}, A^{22}$:

$$[A^{\mu\nu}] = \begin{pmatrix} A^{00} & A^{01} & A^{02} & A^{00}\\ A^{01} & A^{11} & A^{12} & A^{01}\\ A^{02} & A^{12} & A^{22} & A^{02}\\ A^{00} & A^{01} & A^{02} & A^{00}\end{pmatrix}.$$

We may now perform a gauge transformation of the form (17.5) to simplify the amplitude tensor still further. To preserve the Lorenz gauge condition we must ensure that $\Box^2\xi^\mu = 0$. A suitable transformation, which satisfies this condition, is given by

$$\xi^\mu = \epsilon^\mu \exp(ik_\rho x^\rho),$$

where the ϵ^μ are constants. Substituting this expression into the transformation law (17.12) for the trace reverse tensor $\bar{h}^{\mu\nu}$, which we assume to be of the form (18.3), one quickly finds that the amplitude tensor transforms as

$$A'^{\mu\nu} = A^{\mu\nu} - i\epsilon^\mu k^\nu - i\epsilon^\nu k^\mu + i\eta^{\mu\nu}\epsilon^\rho k_\rho. \tag{18.6}$$

Using the expression (18.5) for the 4-wavevector and the result (18.6), we obtain

$$\begin{aligned} A'^{00} &= A^{00} - ik(\epsilon^0 + \epsilon^3), & A'^{11} &= A^{11} - ik(\epsilon^0 - \epsilon^3), \\ A'^{01} &= A^{01} - ik\epsilon^1, & A'^{12} &= A^{12}, \\ A'^{02} &= A^{02} - ik\epsilon^2, & A'^{22} &= A^{22} - ik(\epsilon^0 - \epsilon^3). \end{aligned}$$

Now, by choosing the constants ϵ^μ as follows,

$$\begin{aligned} \epsilon^0 &= -i(2A^{00} + A^{11} + A^{22})/(4k), & \epsilon^1 &= -iA^{01}/k, \\ \epsilon^2 &= -iA^{02}/k, & \epsilon^3 &= -i(2A^{00} - A^{11} - A^{22})/(4k), \end{aligned}$$

we obtain

$$A'^{00} = A'^{01} = A'^{02} = 0 \quad \text{and} \quad A'^{11} = -A'^{22}.$$

On dropping primes, the first condition means that only A^{11}, A^{12} and A^{22} are non-zero. Moreover, the second condition means that only two of these can be specified independently. Choosing $A^{11} \equiv a$ and $A^{12} \equiv b$ as the two independent (in general, complex) components in our new gauge, we thus have

$$[A^{\mu\nu}_{\text{TT}}] = \begin{pmatrix} 0 & 0 & 0 & 0 \\ 0 & a & b & 0 \\ 0 & b & -a & 0 \\ 0 & 0 & 0 & 0 \end{pmatrix}. \tag{18.7}$$

for a wave travelling in the x^3-direction. As indicated, the new gauge we have adopted is known as the *transverse-traceless gauge* (or *TT gauge*), which we will discuss in more detail in Section 18.3. For now we simply note that (18.7) implies $\bar{h}_{\text{TT}} = 0 = h_{\text{TT}}$ (hence the term traceless) and $\bar{h}^{\mu\nu}_{\text{TT}} = h^{\mu\nu}_{\text{TT}}$ for our plane wave.

It is also convenient to introduce the two *linear polarisation tensors* $e_1^{\mu\nu}$ and $e_2^{\mu\nu}$, the components of which are obtained by setting $a = 1, b = 0$ and $a = 0, b = 1$ respectively in (18.7). The general amplitude tensor in the TT gauge for a wave travelling in the x^3-direction can then be written as

$$A^{\mu\nu}_{\text{TT}} = ae_1^{\mu\nu} + be_2^{\mu\nu}.$$

It follows that all possible polarisations of the gravitational wave may be obtained by superposing just two polarisations, with arbitrary amplitudes and relative phases.

18.2 Analogy between gravitational and electromagnetic waves

Before going on to discuss gravitational waves in more detail, it is instructive to illustrate the close analogy with electromagnetic waves. By adopting the Lorenz gauge condition $\partial_\mu A^\mu = 0$, the electromagnetic field equations in free space take the form $\Box^2 A^\mu = 0$. These admit plane-wave solutions of the form

$$A^\mu = \Re[Q^\mu \exp(ik_\rho x^\rho)],$$

where the Q^μ are the constant components of the amplitude vector. The field equations again imply that the 4-wavevector $\boldsymbol{k}$ is null and the Lorenz gauge condition requires that $Q^\mu k_\mu = 0$, thereby reducing the number of independent components in the amplitude vector to three. In particular, if we again consider a wave propagating in the x^3-direction then $[k^\mu] = (k, 0, 0, k)$ and the Lorenz gauge condition implies that $Q^0 = Q^3$, so that

$$[Q^\mu] = (Q^0, Q^1, Q^2, Q^0).$$

The Lorenz gauge condition is preserved by any further gauge transformation of the form $A_\mu \to A_\mu + \partial_\mu \psi$, provided that $\Box^2 \psi = 0$. An appropriate gauge transformation that satisfies this condition is

$$\psi = \epsilon \exp(ik_\rho x^\rho),$$

where ϵ is a constant. This yields $Q'^\mu = Q^\mu + i\epsilon k^\mu$, and so

$$Q'^0 = Q^0 + i\epsilon k, \qquad Q'^1 = Q^1, \qquad Q'^2 = Q^2.$$

By choosing $\epsilon = iQ^0/k$, on dropping primes we have $Q^0 = 0$. In the new gauge, the amplitude vector has just two independent components, Q^1 and Q^2, and the electromagnetic fields are transverse to the direction of propagation. By introducing the two linear polarisation vectors

$$e_1^\mu = (0, 1, 0, 0) \qquad \text{and} \qquad e_2^\mu = (0, 0, 1, 0),$$

we may write the general amplitude vector as

$$Q^\mu = a e_1^\mu + b e_2^\mu,$$

where a and b are arbitrary (in general, complex) constants.

If $b = 0$ then as the electromagnetic wave passes a free positive test charge this will oscillate in the x^1-direction with a magnitude that varies sinusoidally with time. Similarly, if $a = 0$ then the test charge will oscillate in the x^2-direction. The particular combinations of linear polarisations given by $b = \pm ia$ give circularly polarised waves, in which the mutually orthogonal linear oscillations combine in such a way that the test charge moves in a circle.

18.3 Transforming to the transverse-traceless gauge

In Section 18.1, we considered only the transformation into the TT gauge of a plane gravitational wave travelling in the x^3-direction. We now consider a general gravitational perturbation $\bar{h}^{\mu\nu}$ satisfying the empty-space linearised field equation and the Lorenz gauge condition. As discussed previously, a gauge transformation of the form (17.5) will preserve the Lorenz gauge condition provided that the four functions $\xi^\mu(x)$ satisfy $\Box^2\xi^\mu = 0$. From (17.12), the trace-reverse field tensor transforms as

$$\bar{h}'^{\mu\rho} = \bar{h}^{\mu\rho} - \partial^\mu\xi^\rho - \partial^\rho\xi^\mu + \eta^{\mu\rho}\partial_\sigma\xi^\sigma.$$

Since the components $\bar{h}_{\mu\nu}$ also satisfy the *in vacuo* wave equation $\Box^2\bar{h}_{\mu\nu} = 0$, this gauge transformation may be used to set any four linear combinations of the $\bar{h}'_{\mu\nu}$ to zero. The TT gauge is *defined* by choosing

$$\boxed{\bar{h}^{0i}_{\mathrm{TT}} \equiv 0 \qquad \text{and} \qquad \bar{h}_{\mathrm{TT}} \equiv 0.} \tag{18.8}$$

This last condition means that $\bar{h}^{\mu\nu}_{\mathrm{TT}} = h^{\mu\nu}_{\mathrm{TT}}$, and these quantities may therefore be used interchangeably. Moreover, setting $\nu = 0$ and $\nu = j$ respectively in the Lorenz gauge condition $\partial_\mu\bar{h}^{\mu\nu}_{\mathrm{TT}} = 0$, and using (18.8), gives the constraints

$$\boxed{\partial_0\bar{h}^{00}_{\mathrm{TT}} = 0 \qquad \text{and} \qquad \partial_i\bar{h}^{ij}_{\mathrm{TT}} = 0.} \tag{18.9}$$

We note that, if the gravitational field perturbation is non-stationary (i.e. it depends on t), as for a general gravitational wave disturbance, the first constraint in (18.9) implies that h^{00}_{TT} also vanishes and so $h^{\mu 0}_{\mathrm{TT}} = 0$ for all μ. In other words, in this case only the spatial components h^{ij}_{TT} are non-zero.

Let us now consider the particular case of an arbitrary plane gravitational wave of the form (18.3) and satisfying the Lorenz gauge condition. The conditions (18.4) immediately imply that

$$A^{0i}_{\mathrm{TT}} = 0 \qquad \text{and} \qquad (A_{\mathrm{TT}})^\mu_{\ \mu} = 0.$$

Moreover, the conditions (18.9) also require that

$$A^{00}_{\mathrm{TT}} = 0 \qquad \text{and} \qquad A^{ij}_{\mathrm{TT}}k_j = 0.$$

These last conditions ensure that, quite generally, a plane gravitational wave is transverse, like electromagnetic waves.

The above conditions tell us the constraints on the form of $A^{\mu\nu}_{\mathrm{TT}}$. We must now consider how to construct this tensor for a plane wave with a given spatial wavevector $\vec{k}$ and amplitude matrix $A^{\mu\nu}$. First, it is clear that we need consider only the spatial components A^{ij}_{TT}, since the remaining components are all zero. Moreover, from the above conditions, this spatial tensor must be orthogonal to $\vec{k}$ and traceless. We therefore introduce the spatial projection tensor

$$P_{ij} \equiv \delta_{ij} - n_i n_j,$$

which projects spatial tensor components onto the surface orthogonal to the unit spatial vector with components n^i. The action of the projection tensor is easily illustrated by applying it to an arbitrary spatial vector v^i. One quickly finds that $n_i P^i_{\ j} v^j = 0$ and $P^i_{\ k} P^k_{\ j} v^j = P^i_{\ j} v^j$, as required. In the case of our plane gravitational wave, we choose n^i to lie in the direction of the spatial wavevector, so that $n^i = \hat{k}^i$, and thus obtain the components of the spatial amplitude tensor that are transverse to the direction of propagation, namely

$$A^{ij}_{\text{T}} = P^i_{\ k} P^j_{\ l} A^{kl}.$$

The trace of this tensor is given by $(A_{\text{T}})^i_{\ i} = P_{kl} A^{kl}$, which in general does not vanish. Using the fact that $P^i_{\ i} = 3 - 1 = 2$, we may however construct a traceless tensor that still remains transverse to $\vec{k}$; this is given by

$$\boxed{A^{ij}_{\text{TT}} = (P^i_{\ k} P^j_{\ l} - \tfrac{1}{2} P^{ij} P_{kl}) A^{kl}.} \tag{18.10}$$

For a plane gravitational wave travelling in the x^3-direction, so that $[k^\mu] = (k, 0, 0, k)$, it is a simple matter to verify that (18.10) produces an amplitude matrix of the form

$$[A^{\mu\nu}_{\text{TT}}] = \begin{pmatrix} 0 & 0 & 0 & 0 \\ 0 & \frac{1}{2}(A^{11} - A^{22}) & A^{12} & 0 \\ 0 & A^{12} & \frac{1}{2}(A^{22} - A^{11}) & 0 \\ 0 & 0 & 0 & 0 \end{pmatrix}, \tag{18.11}$$

which agrees with that given in (18.7). In fact this result illustrates that there is a quick and simple algorithm for transforming a plane wave travelling along one of the coordinate directions into the TT gauge. We see that the transformation (18.11) corresponds to *setting to zero all components that are not transverse to the direction of wave propagation and subtracting one-half the resulting trace from the remaining diagonal elements, to make the final tensor traceless.* There is, however, nothing special about our choice of x^3-direction and so the above prescription must be true for a plane wave travelling in *any* of the three coordinate directions.

18.4 The effect of a gravitational wave on free particles

Let us now consider the motion of a set of test particles, initially at rest, in the presence of a gravitational wave. In fact, in the latter case it is not enough to consider the trajectory of just a single test particle, as we discuss below. To obtain a coordinate-independent measure of the effects of the wave, it is necessary to consider the relative motion of a set of nearby particles.

First consider a single free test particle, whose 4-velocity u^σ must satisfy the geodesic equation

$$\frac{du^\sigma}{d\tau} + \Gamma^\sigma{}_{\mu\nu} u^\mu u^\nu = 0.$$

Suppose that the particle is initially at rest in our chosen coordinate system, so that $[u^\mu] = c(1, 0, 0, 0)$. The geodesic equation then reads

$$\frac{du^\sigma}{d\tau} = -c^2 \Gamma^\sigma{}_{00} = -\tfrac{1}{2} c^2 \eta^{\sigma\rho} (\partial_0 h_{\rho 0} + \partial_0 h_{0\rho} - \partial_\rho h_{00}),$$

where in the last equality we have used (17.6) to obtain the connection coefficients to first order in terms of the derivatives of $h_{\mu\nu}$. Let us now adopt the TT gauge, which we may do for any general gravitational wave disturbance *in vacuo*. From the discussion in Section 18.3, we know that $h^{\mathrm{TT}}_{\rho 0} = 0$ for all values of ρ. Thus, initially, $du^\sigma/d\tau = 0$ and so the particle will still be at rest a moment later. The argument may then be repeated, showing that the particle remains at rest forever, regardless of the passing of the gravitational wave. In other words $[u^\sigma] = c(1, 0, 0, 0)$ is a solution of the geodesic equation in this case, as may readily be verified by direct substitution.

What has gone wrong here? The key point is that 'at rest' in this context means simply that the particle has constant spatial coordinates. What we have uncovered is that by choosing the TT gauge we have found a coordinate system that stays attached to individual particles. This has no coordinate-invariant physical meaning. To obtain a proper physical interpretation of the effect of a passing gravitational wave, we must consider a set of nearby particles.

Let us therefore consider a cloud of non-interacting free test particles. From the above discussion, the worldlines of the particles are curves having constant spatial coordinates. Thus the small spacelike vector $[\xi^\mu] = (0, \xi^1, \xi^2, \xi^3)$ giving the coordinate separation between any two nearby particles is constant (this may also be shown explicitly by demonstrating that the equation of geodesic deviation (7.24) has $\xi^\mu = $ constant as a solution in this case). Although the coordinate separation of the particles is constant, this does not mean that their *physical* spatial separation l is constant. The latter is given by

$$l^2 = -g_{ij} \xi^i \xi^j = (\delta_{ij} - h_{ij}) \xi^i \xi^j,$$

where not all the h_{ij} are constant (in any gauge) and $i, j = 1, 2, 3$. Thus we see that the passing of a gravitational wave will indeed cause the physical separation of nearby particles to vary. It is convenient at this point to introduce the quantities

$$\zeta^i = \xi^i + \tfrac{1}{2} h^i_k \xi^k. \tag{18.12}$$

One then finds straightforwardly that, in terms of these new variables (to first order in $h_{\mu\nu}$),

$$l^2 = \delta_{ij} \zeta^i \zeta^j,$$

which is again valid in any gauge. Thus, the ζ^i may be regarded as the components of a position vector giving the correct physical spatial separation when contracted with the Euclidean metric tensor δ_{ij}.

Let us now discuss the particular case of a plane gravitational wave propagating in the x^3-direction and consider a set of particles initially at rest in the (x^1, x^2)-plane, i.e. the plane perpendicular to the direction of wave propagation. Thus, the coordinate separation vector between any two particles has $\xi^3 = 0$. In the TT gauge, however, we see from (18.7) that $(h_{\text{TT}})^3_k = 0$, and so (18.12) implies that $\zeta^3 = 0$ throughout the passage of the wave. Hence the particles remain in the plane perpendicular to the wave propagation direction; it is only the physical separations in the transverse directions that vary. Thus the gravitational wave is transverse not only in its mathematical description $(h^{\mu\nu}_{\text{TT}})$ but also in its physical effects.

We first consider the effect of the passage of a gravitational wave with $A^{\mu\nu} = a e_1^{\mu\nu}$ (i.e. a single polarisation), where we take a to be real and positive for convenience, and $e_1^{\mu\nu}$ was introduced at the end of Section 18.1. Remembering that $\bar{h}^{\mu\nu}_{\text{TT}} = h^{\mu\nu}_{\text{TT}}$, we thus have

$$h^{\mu\nu}_{\text{TT}} = a e_1^{\mu\nu} \cos k_\mu x^\mu = a e_1^{\mu\nu} \cos k(x^0 - x^3)$$

where $k = \omega/c$, and using (18.12) we quickly find that

$$[\zeta^i] = (\xi^1, \xi^2, 0) - \tfrac{1}{2} a \cos k(x^0 - x^3)(\xi^1, -\xi^2, 0).$$

Thus, for two particles initially separated in the x^1-direction ($\xi^1 \neq 0$) the physical separation in the x^1-direction will oscillate, and likewise for two particles with an initial x^2 separation. Let us consider a set of particles that, when $\cos k(x^0 - x^3) = 0$, form a circle in the (x^1, x^2)-plane with a reference particle at the centre, with respect to which we refer to the other particles, using the ζ^i-vector components. Then, as the wave passes, the particles remain coplanar and at other times have spatial separations as illustrated in Figure 18.1.

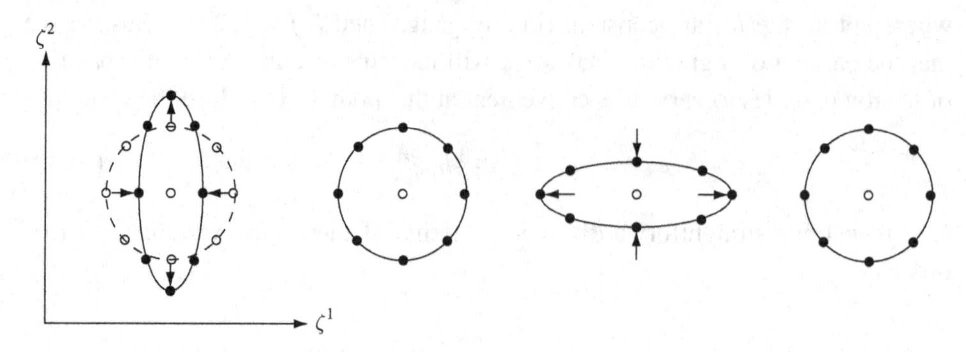

Figure 18.1 The solid dots show the effect of a plane gravitational wave with $A^{\mu\nu} = ae_1^{\mu\nu}$ on a transverse circle of particles. The initial configuration of particles is shown by the open dots. From left to right, $k(x^0 - x^3)$ is equal to $2n\pi$, $\left(2n+\frac{1}{2}\right)\pi$, $(2n+1)\pi$, $\left(2n+\frac{3}{2}\right)\pi$ respectively.

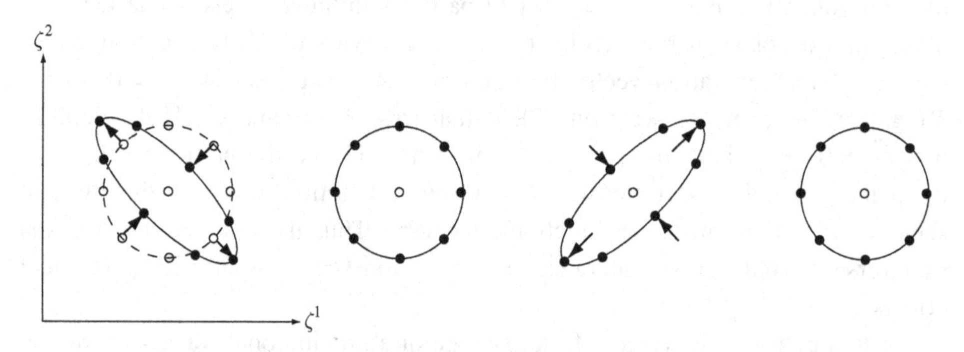

Figure 18.2 The solid dots show the effect of a plane gravitational wave with $A^{\mu\nu} = be_2^{\mu\nu}$ on a transverse circle of particles. The initial configuration of particles is shown by the open dots. From left to right, $k(x^0 - x^3)$ is equal to $2n\pi$, $\left(2n+\frac{1}{2}\right)\pi$, $(2n+1)\pi$, $\left(2n+\frac{3}{2}\right)\pi$ respectively.

We may straightforwardly repeat our analysis for a gravitational wave with the other polarisation, i.e. $A^{\mu\nu} = be_2^{\mu\nu}$ again with real and positive b. In this case one finds that

$$[\zeta^i] = (\xi^1, \xi^2, 0) - \tfrac{1}{2}b\cos k(x^0 - x^3)(\xi^2, \xi^1, 0),$$

and this results in our initial circle of particles having spatial separations as illustrated in Figure 18.2, which may be obtained from Figure 18.1 by a 45° rotation.

Having determined the relative displacements of test particles induced by the two separate polarisations of a plane gravitational wave, it is straightforward to find the effect in the general case in which $A^{\mu\nu} = ae_1^{\mu\nu} + be_1^{\mu\nu}$, where a and b

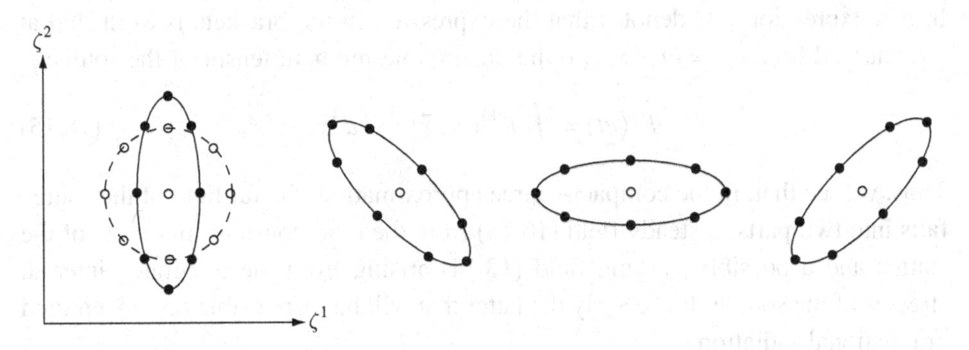

Figure 18.3 The solid dots show the effect of a plane gravitational wave with $A^{\mu\nu} = a\left(e_1^{\mu\nu} + ie_2^{\mu\nu}\right)$ (i.e. right-handed circular polarisation) on a transverse circle of particles. The initial configuration of particles is shown by the open dots. From left to right, $k(x^0 - x^3)$ is equal to $2n\pi$, $\left(2n+\frac{1}{2}\right)\pi$, $(2n+1)\pi$, $\left(2n+\frac{3}{2}\right)\pi$ respectively.

may, in general, be complex. Of particular interest are the left- and right-handed *circularly polarised* modes, for which $b = -ia$ and $b = ia$ respectively. The effect of, for example, a right-handed circularly polarised wave would be to distort our initial circle of particles into an ellipse and to rotate the ellipse in a right-handed sense, as illustrated in Figure 18.3. Note that the individual particles do not move around the ring but instead execute small circular 'epicycles'.

18.5 The generation of gravitational waves

Let us suppose that we have a matter distribution (the source) localised near the origin O of our coordinate system that we and take our field point $\vec{x}$ to be a distance r from O that is large compared with the spatial extent of the source. We may therefore use the *compact-source approximation* discussed in Section 17.8. Without loss of generality, we may take our spatial coordinates x^i to correspond to the 'centre-of-momentum' frame of the source particles, in which case from (17.38) we have

$$\bar{h}^{00} = -\frac{4GM}{c^2 r}, \qquad \bar{h}^{i0} = \bar{h}^{0i} = 0. \tag{18.13}$$

The remaining (spatial) components of the gravitational field are given by the integrated stress within the source, which may be written in terms of the *quadrupole formula* (17.44) as

$$\bar{h}^{ij}(ct, \vec{x}) = -\frac{2G}{c^6 r}\left[\frac{d^2 I^{ij}(ct')}{dt'^2}\right]_r. \tag{18.14}$$

In this expression $[\ ]_r$ denotes that the expression in the brackets is evaluated at the retarded time $ct' = ct - r$, and the quadrupole-moment tensor of the source is

$$I^{ij}(ct) = \int T^{00}(ct, \vec{y}) y^i y^j \, d^3\vec{y}. \qquad (18.15)$$

Thus, we see that, in the compact-source approximation, the far field of the source falls into two parts: a steady field (18.13) from the total constant 'mass' M of the source and a possibly varying field (18.14) arising from the integrated internal stresses of the source. It is clearly the latter that will be responsible for any emitted gravitational radiation.

For slowly moving source particles we have $T^{00} \approx \rho c^2$, where ρ is the proper density of the source, and so the integral (18.15) may be written as

$$\boxed{I^{ij}(ct) = c^2 \int \rho(ct, \vec{x}) x^i x^j \, d^3\vec{x}.} \qquad (18.16)$$

Thus, the gravitational wave produced by an isolated non-relativistic source is proportional to the second derivative of the *quadrupole moment* of the matter-density distribution. By contrast, the leading contribution to electromagnetic radiation is the first derivative of the *dipole* moment of the charge density distribution. This fundamental difference between the two theories may be easily understood from elementary considerations. Using ρ to denote either the proper mass density or the proper charge density, the volume integral $\int \rho \, dV$ over the source is constant in time for both electromagnetism and linearised gravitation and so generates no radiation. Now consider the next moment $\int \rho x^i \, dV$, i.e. the dipole moment. For electromagnetism, this gives the position of the centre of charge of the source, which can move with time and hence have a non-zero time derivative; this provides the dominant contribution in the generation of electromagnetic radiation. For gravitation, however, $\int \rho x^i \, dV$ gives the centre of mass of the source and, for an isolated system, conservation of momentum means that it cannot change with time and so cannot contribute to the generation of gravitational waves. Thus, it is the generally much smaller quadrupole moment, which measures the shape of the source, that is dominant in generating gravitational waves. This fact, and the weak coupling of gravitation to matter, means that gravitational radiation is much weaker than electromagnetic radiation. As a corollary, we note that a spherically symmetric system has a zero quadrupole moment and thus cannot emit gravitational radiation.

As an illustration of the generation of gravitational waves, let us consider two particles A and B of equal mass M moving (non-relativistically) in circular orbits of radius a about their common centre of mass with an angular speed Ω (see Figure 18.4). This might represent a simple model of a binary star system, in

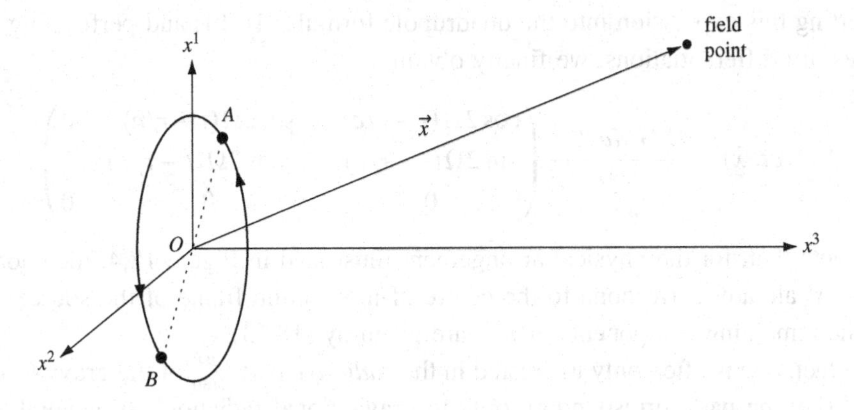

Figure 18.4 Two particles A, B of equal mass M rotating at angular speed Ω in circular orbits of radius a about their common centre of mass.

which mutual gravitational attraction keeps the particles (stars) in orbit. In this case, treating the motion in the Newtonian limit, we require that

$$\Omega = \left(\frac{GM}{4a^3}\right)^{1/2}. \tag{18.17}$$

Alternatively, in a more terrestrial setting, one might imagine the particles to be connected by a light rod of length $2a$ that is spun with constant angular velocity about its centre point, in which case Ω need not be related to M and a. For simplicity, we shall assume the particle orbits to lie in the plane $x^3 = 0$, as illustrated in Figure 18.4.

At any time t, the coordinates of particles A and B may be written

$$[x^i_A] = (a\cos\Omega t, a\sin\Omega t, 0), \qquad [x^i_B] = -(a\cos\Omega t, a\sin\Omega t, 0).$$

Thus, the proper density of the source is given by

$$\begin{aligned}\rho(ct, \vec{x}) = M\big[&\delta(x^1 - a\cos\Omega t)\,\delta(x^2 - a\sin\Omega t)\\ &+ \delta(x^1 + a\cos\Omega t)\,\delta(x^2 + a\sin\Omega t)\big]\,\delta(x^3).\end{aligned}$$

On substituting into (18.16) and making use of the standard trigonometric identities $2\cos^2\Omega t = 1+\cos 2\Omega t$, $2\sin^2\Omega t = 1-\cos 2\Omega t$ and $2\sin\Omega t\cos\Omega t = \sin 2\Omega t$, one quickly finds the quadrupole-moment tensor,

$$[I^{ij}(ct)] = Mc^2a^2\begin{pmatrix}1+\cos 2\Omega t & \sin 2\Omega t & 0\\ \sin 2\Omega t & 1-\cos 2\Omega t & 0\\ 0 & 0 & 0\end{pmatrix}. \tag{18.18}$$

Inserting this expression into the quadrupole formula (18.14) and performing the necessary differentiations, we finally obtain

$$[\bar{h}^{ij}(ct,\vec{x})] = \frac{8GMa^2\Omega^2}{c^4 r}\begin{pmatrix} \cos 2\Omega(t-r/c) & \sin 2\Omega(t-r/c) & 0 \\ \sin 2\Omega(t-r/c) & -\cos 2\Omega(t-r/c) & 0 \\ 0 & 0 & 0 \end{pmatrix}.$$

We note that, for the physical arrangement illustrated in Figure 18.4, the coordinates x^i already correspond to the centre-of-momentum frame of the source and so the remaining components of $\bar{h}^{\mu\nu}$ are given by (18.13).

In fact, one is often only interested in the *radiative* part $\bar{h}^{\mu\nu}_{\text{rad}}$ of the gravitational field (i.e. the part corresponding only to gravitational radiation). In general, the remaining components of $\bar{h}^{\mu 0}_{\text{rad}}$ may be found from the spatial components $\bar{h}^{ij}_{\text{rad}}$ using the Lorenz gauge condition. For the two-particle system discussed above, we see from (18.13) that all the remaining components $\bar{h}^{\mu 0}_{\text{rad}}$ are zero, and so

$$\left[\bar{h}^{\mu\nu}_{\text{rad}}(ct,\vec{x})\right] = \frac{8GMa^2\Omega^2}{c^4 r}\begin{pmatrix} 0 & 0 & 0 & 0 \\ 0 & \cos 2\Omega(t-r/c) & \sin 2\Omega(t-r/c) & 0 \\ 0 & \sin 2\Omega(t-r/c) & -\cos 2\Omega(t-r/c) & 0 \\ 0 & 0 & 0 & 0 \end{pmatrix}. \tag{18.19}$$

Since the amplitude goes as $1/r$, the gravitational perturbation has the form of a spherical wave rather than a plane wave. Nevertheless, for large r the wave is well approximated by a plane wave in a small range of angles about any particular direction. We also note that the angular frequency of the wave is twice the rotational angular frequency of the two particles.

It is of interest to determine the polarisation of the gravitational waves received by observers located in different directions relative to the orbiting particles. To do this, one must transform to the TT gauge appropriate to each observer. Let us first consider an observer located on the x^3-axis (at some large distance from O). By comparing with (18.7), we see that (18.19) is already in transverse-traceless form for a wave travelling in the x^3-direction. Remembering that $\bar{h}^{\mu\nu}_{\text{TT}} = h^{\mu\nu}_{\text{TT}}$ and using the fact that $r = x^3$, it is straightforward to show that

$$\left(h^{TT}_{\text{rad}}\right)^{\mu\nu} = \frac{8GMa^2\Omega^2}{c^4 r}\Re\left[\left(e_1^{\mu\nu} - ie_2^{\mu\nu}\right)\exp 2i\Omega(t - x^3/c)\right], \tag{18.20}$$

where $e_1^{\mu\nu}$ and $e_2^{\mu\nu}$ are the linear polarisation tensors introduced at the end of Section 18.1. Since the amplitude tensor has the form $A^{\mu\nu} = ae_1^{\mu\nu} + be_2^{\mu\nu}$ with $b = -ia$, this corresponds to right-handed circularly polarised radiation, as one might expect.

Let us now consider an observer located on the x^1-axis. The form (18.19) is *not* in the transverse-traceless gauge for a wave travelling in the x^1-direction. To transform to the TT gauge, we follow the prescription outlined in Section 18.3. We first set to zero all non- transverse components, i.e. all entries except those with $(i, j) = (2,2)$, $(2, 3)$, $(3, 2)$ and $(3, 3)$. We then subtract one-half of the resulting trace from the remaining diagonal elements $(2, 2)$ and $(3, 3)$ to make the final tensor traceless. Remembering that $r = x^1$ in this case, and that $\bar{h}^{\mu\nu}_{\text{TT}} = h^{\mu\nu}_{\text{TT}}$, we obtain

$$[(h^{\text{TT}}_{\text{rad}})^{\mu\nu}(ct, \vec{x})] = \frac{4GMa^2\Omega^2}{c^4 r}\begin{pmatrix} 0 & 0 & 0 & 0 \\ 0 & 0 & 0 & 0 \\ 0 & 0 & -\cos 2\Omega(t - r/c) & 0 \\ 0 & 0 & 0 & \cos 2\Omega(t - r/c) \end{pmatrix}$$

$$= \frac{4GMa^2\Omega^2}{c^4 r}\Re\left[-\tilde{e}_1^{\mu\nu}\exp 2i\Omega(t - x^1/c)\right], \tag{18.21}$$

where $\tilde{e}_1^{\mu\nu}$ is a linear polarisation tensor analogous to those used above, but for propagation in the x^1- direction. Thus, the gravitational waves received by the observer are linearly polarised in the '+' orientation illustrated in Figure 18.1 – again as one might have expected.

18.6 Energy flow in gravitational waves

Physically, one would expect gravitational waves to carry energy away from a radiating source. As discussed in Section 17.11, however, the task of assigning an energy density to a gravitational field is notoriously difficult. Nevertheless, bearing in mind the caveats made in Section 17.11, from (17.64) an appropriate expression for the energy–momentum tensor of the gravitational field *in vacuo* is

$$t_{\mu\nu} = \frac{c^4}{32\pi G}\langle(\partial_\mu \bar{h}_{\rho\sigma})\partial_\nu \bar{h}^{\rho\sigma} - 2(\partial_\sigma \bar{h}^{\rho\sigma})\partial_{(\mu}\bar{h}_{\nu)\rho} - \tfrac{1}{2}(\partial_\mu \bar{h})\partial_\nu \bar{h}\rangle,$$

where $\langle\cdots\rangle$ denotes an average over a small region at each point in spacetime. If we adopt the TT gauge, however, the Lorenz gauge condition $\partial_\mu \bar{h}^{\mu\nu}_{\text{TT}} = 0$ is automatically satisfied, and also $\bar{h}_{\text{TT}} = 0$ and $\bar{h}^{\mu\nu}_{\text{TT}} = h^{\mu\nu}_{\text{TT}}$. Thus in this gauge the energy–momentum tensor *in vacuo* reduces to

$$t_{\mu\nu} = \frac{c^4}{32\pi G}\langle(\partial_\mu h^{\text{TT}}_{\rho\sigma})\,\partial_\nu h^{\rho\sigma}_{\text{TT}}\rangle.$$

We will assume further that we are considering only the radiative part of the gravitational field, in which case we know from the discussion in Section 18.3 that $h^{\mu 0}_{\rm TT} = 0$, and so

$$\boxed{t_{\mu\nu} = \frac{c^4}{32\pi G}\left\langle (\partial_\mu h^{\rm TT}_{ij})\, \partial_\nu h^{ij}_{\rm TT} \right\rangle.} \tag{18.22}$$

In particular, from our discussion of energy–momentum tensors in Section 8.1, at any given time and spatial position the energy flux (i.e. the energy crossing unit area per unit time) of the gravitational radiation in the unit spatial direction n^i is

$$\boxed{F(\vec{n}) = -ct^{0k}n_k,} \tag{18.23}$$

where the minus sign appears as a result of our choice of metric signature, since then $F(\vec{n}) = -c\eta_{kj}t^{0k}n^j = \delta_{kj}t^{0k}n^j$, as required.

As an illustration of these general results, let us calculate the energy flux in the direction of propagation for a plane gravitational wave of the form

$$h^{ij}_{\rm TT} = A^{ij}_{\rm TT} \cos k_\lambda x^\lambda,$$

where $A^{ij}_{\rm TT}$ are constants and, for convenience, we have chosen the arbitrary phase of the wave in such a way that the amplitude matrix is real. Substituting this expression into (18.22), and using the fact that $\langle \sin^2(k_\lambda x^\lambda)\rangle = \frac{1}{2}$ when averaged over several wavelengths, the energy–momentum tensor reads

$$t_{\mu\nu} = \frac{c^4}{64\pi G} k_\mu k_\nu A^{ij}_{\rm TT} A^{\rm TT}_{ij}. \tag{18.24}$$

Thus, the flux F in the $\vec{k}$-direction is given by

$$F = -ct^{0l}\hat{k}_l = -\frac{c^5}{64\pi G}k^0 k^l \hat{k}_l A^{ij}_{\rm TT} A^{\rm TT}_{ij} = \frac{c^5}{64\pi G} k^0 k^0 A^{ij}_{\rm TT} A^{\rm TT}_{ij} = ct^{00}, \tag{18.25}$$

where in the third equality we have used the fact that $k^0 = |\vec{k}| = -k^l\hat{k}_l$, since the wavevector is null. The final expression is simply the energy density associated with the plane wave multiplied by its speed, and hence makes good physical sense as the energy flux carried by the wave in its direction of propagation.

Specialising still further, we may calculate the forms of the expressions (18.24) and (18.25) explicitly for a wave travelling in the x^3-direction, in which case

$[k_\mu] = (k, 0, 0, -k)$ where $k = \omega/c$ and $A^{\mu\nu}_{\text{TT}}$ is given by (18.7). Thus, in this case, the energy–momentum tensor (18.24) can be written as

$$[t_{\mu\nu}] = \frac{c^2}{32\pi G}\omega^2(a^2+b^2)\begin{pmatrix} 1 & 0 & 0 & -1 \\ 0 & 0 & 0 & 0 \\ 0 & 0 & 0 & 0 \\ -1 & 0 & 0 & 1 \end{pmatrix}.$$

and the flux in the direction of propagation is

$$F = \frac{c^3}{32\pi G}\omega^2(a^2+b^2). \tag{18.26}$$

Clearly, similar results hold for a plane gravitational wave travelling along any of the coordinate axes.

Using the result (18.26) and the expressions (18.20) and (18.21), we find that, for the two-particle rotating system considered in the previous section, the gravitational-wave energy flux at a (large) distance r in the x^1- and x^3- directions respectively is

$$F_1 = \frac{c^3}{32\pi G}(2\Omega)^2\left(\frac{4GMa^2\Omega^3}{c^4 r}\right)^2 = \frac{2G}{\pi c^5}\left(\frac{Ma^2\Omega^3}{r}\right)^2,$$

$$F_3 = 2\frac{c^3}{32\pi G}(2\Omega)^2\left(\frac{8GMa^2\Omega^3}{c^4 r}\right)^2 = \frac{16G}{\pi c^5}\left(\frac{Ma^2\Omega^3}{r}\right)^2.$$

Thus, we see that the energy flux in the x^3-direction is eight times that in the x^1-direction (or, by symmetry, in any direction in the $x^3 = 0$ plane). Hence the energy flux due to the gravitational radiation emitted from this system is highly anisotropic.

18.7 Energy loss due to gravitational-wave emission

Since gravitational waves carry away energy, we expect energy to be lost at a corresponding rate by the physical system generating the gravitational radiation. Let us suppose that the source matter distribution is localised near the origin O of our coordinates. To calculate the rate at which the physical system loses energy, we equate it to the energy flux of the emitted gravitational radiation evaluated over a sphere S of large radius r centered on O. Thus, if E is the energy of the physical system, we have

$$\frac{dE}{dt} = -L_{\text{GW}} = -r^2\int_{4\pi} F(\vec{e}_r)\, d\Omega, \tag{18.27}$$

where L_{GW} is the total gravitational-wave luminosity, $F(\vec{e}_r)$ is the gravitational-wave energy flux at a radius r in the (unit) radial direction $\vec{e}_r$ and $d\Omega$ is an element of solid angle.

In general, using (18.22) and (18.23) we may write the gravitational-wave flux in a unit spatial direction $\vec{n}$ as

$$F(\vec{n}) = -\frac{c^4}{32\pi G}\left\langle (\partial_t h^{\rm TT}_{ij})\,(\partial_k h^{ij}_{\rm TT})\right\rangle n^k = -\frac{c^4}{32\pi G}\left\langle (\partial_t h^{\rm TT}_{ij})\,(\vec{n}\cdot\vec{\nabla}) h^{ij}_{\rm TT}\right\rangle,$$

where we have made the identification $x^0 \equiv ct$ and where $\partial_t \equiv \partial/\partial t$. In the second equality the operator $\vec{n}\cdot\vec{\nabla}$ returns simply the rate of change of its argument in the direction $\vec{n}$. Thus, taking $\vec{n}$ to lie in the radial direction and writing $\partial_r \equiv \partial/\partial r$, we have

$$F(\vec{e}_{\rm r}) = -\frac{c^4}{32\pi G}\left\langle (\partial_t h^{\rm TT}_{ij})\,(\partial_r h^{ij}_{\rm TT})\right\rangle. \tag{18.28}$$

To obtain a general formula for (18.27), we must calculate the above energy flux in terms of properties of the source distribution. From the quadrupole formula (18.14) we have

$$\bar{h}^{ij} = -\frac{2G}{c^6 r}\left[\ddot{I}^{ij}\right]_{\rm r},$$

where I^{ij} is the quadrupole-moment tensor of the source distribution defined in (18.16), the dots denote d/dt and $[\;]_{\rm r}$ denotes that the expression should be evaluated at the retarded time $ct_{\rm r} = ct - r$. It will, in fact, be more convenient to work in terms of the *reduced quadrupole-moment tensor* of the source distribution, which is defined by

$$J_{ij} = I_{ij} - \tfrac{1}{3}\delta_{ij} I, \tag{18.29}$$

where $I = I^j_j$ is the trace of the original tensor. One immediately sees that J_{ij} is simply the traceless version of I_{ij}. As a result, we may write the transverse-traceless part of the gravitational field tensor as

$$h^{ij}_{\rm TT} = \bar{h}^{ij}_{\rm TT} = -\frac{2G}{c^6 r}\left[\ddot{I}^{ij}_{\rm TT}\right]_{\rm r} = -\frac{2G}{c^6 r}\left[\ddot{J}^{ij}_{\rm TT}\right]_{\rm r}, \tag{18.30}$$

where $J^{ij}_{\rm TT}$ is the transverse-traceless part of (18.29). Since at any point on the sphere S the direction of gravitational-wave propagation is radial, from (18.10) we have

$$J^{ij}_{\rm TT} = \left(P^i_k P^j_l - \tfrac{1}{2}P^{ij}P_{kl}\right) J^{kl}, \tag{18.31}$$

where $P^{ij} = \delta^{ij} - e^i_r e^j_r$ is the spatial projection tensor, which projects tensor components onto the spatial surface orthogonal to the radial direction at any point.

Using (18.30) and the expressions (17.32, 17.33) for the derivatives of time-retarded quantities, the derivatives in the expression (18.28) for the energy flux are given by

$$\partial_t h_{ij}^{\mathrm{TT}} = -\frac{2G}{c^6 r}\left[\dddot{J}_{\mathrm{TT}}^{ij}\right]_{\mathrm{r}},$$

$$\partial_r h_{ij}^{\mathrm{TT}} = \frac{2G}{c^6 r^2}\left[\ddot{J}_{\mathrm{TT}}^{ij}\right]_{\mathrm{r}} + \frac{2G}{c^7 r}\left[\dddot{J}_{\mathrm{TT}}^{ij}\right]_{\mathrm{r}} \approx \frac{2G}{c^7 r}\left[\dddot{J}_{\mathrm{TT}}^{ij}\right]_{\mathrm{r}},$$

where, in the second equation, we have retained only the term in $1/r$, which dominates for large r. Substituting these expressions into (18.28) we obtain

$$F(\vec{e}_r) = \frac{G}{8\pi r^2 c^9}\left\langle\left[\dddot{J}_{ij}^{\mathrm{TT}}\dddot{J}_{\mathrm{TT}}^{ij}\right]_{\mathrm{r}}\right\rangle.$$

For convenience, we now use (18.31) to rewrite the product of transverse-traceless quadrupole moments in terms of products of reduced moments. Denoting the components e_r^i of the unit radial vector by $\hat{x}^i$, this yields

$$J_{ij}^{\mathrm{TT}} J_{\mathrm{TT}}^{ij} = J_{ij}J^{ij} - 2J_i^j J^{ik}\hat{x}_j\hat{x}_k + \tfrac{1}{2}J^{ij}J^{kl}\hat{x}_i\hat{x}_j\hat{x}_k\hat{x}_l,$$

where we have made use of the fact that J_{ij} is traceless. Thus, the total gravitational-wave luminosity is given by

$$L_{\mathrm{GW}} = \frac{G}{8\pi c^9}\int_{4\pi}\left\langle\left[\dddot{J}_{ij}\dddot{J}^{ij} - 2\dddot{J}_i^j\dddot{J}^{ik}\hat{x}_j\hat{x}_k + \tfrac{1}{2}\dddot{J}^{ij}\dddot{J}^{kl}\hat{x}_i\hat{x}_j\hat{x}_k\hat{x}_l\right]_{\mathrm{r}}\right\rangle d\Omega.$$

Since the reduced quadrupole moment J_{ij} is defined as an integral over all space, it does not depend on the angular coordinates and so may be taken outside the integral. The three remaining integrals are easily evaluated to give

$$\int_{4\pi} d\Omega = 4\pi, \quad \int_{4\pi}\hat{x}_i\hat{x}_j\, d\Omega = \frac{4\pi}{3}\delta_{ij},$$

$$\int_{4\pi}\hat{x}_i\hat{x}_j\hat{x}_k\hat{x}_l\, d\Omega = \frac{4\pi}{15}(\delta_{ij}\delta_{kl} + \delta_{ik}\delta_{jl} + \delta_{il}\delta_{jk}).$$

The first result is trivial. The second result may be obtained by noting that integration over all angles yields zero for $i \neq j$, whereas on raising one index and setting $i = j$ the integrand becomes $\hat{x}_i\hat{x}^i = 1$ and so the integral equals 4π. Similar reasoning leads to the third result. Substituting these three results into (18.27) and simplifying, one finally obtains

$$\boxed{\frac{dE}{dt} = -L_{\mathrm{GW}} = -\frac{G}{5c^9}\left\langle\left[\dddot{J}_{ij}\dddot{J}^{ij}\right]_{\mathrm{r}}\right\rangle.} \qquad (18.32)$$

As an illustration, let us apply the general formula (18.32) to the specific example of the two-particle rotating system discussed in Section 18.5. The quadrupole-moment tensor I^{ij} for this system is given in (18.18), from which we quickly find that the reduced quadrupole-moment tensor (18.29) is given by

$$[J^{ij}] = Mc^2a^2 \begin{pmatrix} \frac{1}{3}+\cos 2\Omega t & \sin 2\Omega t & 0 \\ \sin 2\Omega t & \frac{1}{3}-\cos 2\Omega t & 0 \\ 0 & 0 & -\frac{2}{3} \end{pmatrix}.$$

The corresponding third time derivative reads

$$[\dddot{J}^{ij}] = 8Mc^2a^2\Omega^3 \begin{pmatrix} \sin 2\Omega t & -\cos 2\Omega t & 0 \\ -\cos 2\Omega t & -\sin 2\Omega t & 0 \\ 0 & 0 & 0 \end{pmatrix},$$

and so (18.32) becomes

$$\begin{aligned} \frac{dE}{dt} = -L_{\rm GW} &= -\frac{G}{5c^9}(8Mc^2a^2\Omega^3)^2\langle 2\sin^2 2\Omega(t-r/c)+2\cos^2 2\Omega(t-r/c)\rangle \\ &= -\frac{G}{5c^5}(128M^2a^4\Omega^6). \end{aligned} \tag{18.33}$$

18.8 Spin-up of binary systems: the binary pulsar PSR B1913+16

As discussed in Section 18.5, our simple two-particle rotating system can be used to model an equal-mass astrophysical binary system, in which case Ω is given by (18.17). Inserting this expression into (18.33), we find that the total energy E of the binary system obeys

$$\frac{dE}{dt} = -\frac{2}{5}\frac{G^4M^5}{c^5a^5}. \tag{18.34}$$

Treating the binary in the Newtonian limit, the total energy is simply

$$E = \frac{1}{2}(2Mv^2) - \frac{GM^2}{2a},$$

where v is the orbital speed of either object. Using the radial equation of motion $Mv^2/a = GM^2/(2a)^2$, we may write

$$E = -\frac{GM^2}{4a} = -Mv^2,$$

from which we see that the total energy is negative, since the binary system is gravitationally bound. Moreover, we note that as E decreases (i.e. becomes more negative), according to (18.34) the radius a of the orbit must *decrease* whereas the orbital speed v must *increase*. Thus, the emission of gravitational radiation causes the binary system to 'spin-up', ending ultimately in the coalescence of the two objects.

For comparison with observations of binary systems, the most useful way of characterising the spin-up is by the rate of change of the orbital period P. For our simple system $P = 2\pi a/v$, and so we may write the total energy as

$$E = -\left(\frac{\pi^2 G^2 M^5}{4}\right)^{1/3} P^{-2/3}. \tag{18.35}$$

Differentiating this expression with respect to t and inverting, we find that the rate of change of the orbital period is related to the rate of change of energy by

$$\frac{dP}{dt} = -\frac{3P}{2E}\frac{dE}{dt}. \tag{18.36}$$

Substituting $a = -GM^2/4E$ into (18.34) and then substituting for E using (18.35), we find that (18.36) can be written as follows:

$$\frac{dP}{dt} = -\frac{96}{5} 4^{1/3} \frac{\pi}{c^5} \left(\frac{2\pi GM}{P}\right)^{5/3}.$$

This expression gives the rate of change of the orbital period solely in terms of some constants and P itself, which can be determined straightforwardly from observations.

The spin-up of a binary system resulting from the emission of gravitational waves has already been observed in the binary pulsar PSR B1913 + 16. This system was discovered in 1974 by Hulse and Taylor and consists of a pulsar and an unseen companion, each with a mass of about $1.4M_\odot$; the orbital period is 7.75 hours. The pulsar provides a very accurate clock, so that the change in the orbital period as the system loses energy can be measured. In practice, our results above have to be modified slightly to allow for the considerable eccentricity of the orbit ($e = 0.617$), but this is relatively straightforward. Timing measurements made by Taylor and colleagues over several decades show that the decrease in orbital period as a function of time is in agreement with that predicted from the emission of gravitational radiation, to within one-third of one per cent. This constitutes an additional, and highly accurate, experimental verification of general relativity (albeit in the weak-field regime), for which Hulse and Taylor received the Nobel Prize in Physics in 1993.

18.9 The detection of gravitational waves

Although the measurement of the spin-up of the binary pulsar PSR B1913 + 16 provides indirect evidence of the existence of gravitational radiation, a major goal of modern experimental astrophysics is to make a *direct* detection of gravitational waves by measuring their influence on some test bodies.

There are two distinct approaches to gravitational-wave detection, 'free-particle' and 'resonant' detection. In our discussion in Section 18.4, we found that the effect of a gravitational wave on a cloud of free test particles is a variation in their relative separations. Thus one may attempt to detect gravitational waves by measuring the separations of a set of free test particles as a function of time, which is the basis of free-particle detection experiments. Alternatively, if the particles are not free, but are instead the constituent particles of some elastic body, then tidal forces on the particles induced by a gravitational wave will give rise to vibrations in the body, which one can attempt to measure. In particular, if the incident gravitational radiation were in the form of a plane wave of a given frequency then the amplitude of the induced vibrations would be enhanced if the elastic body were designed to have a resonant frequency close to that of the incident wave. This is the basis of resonant detection.

Resonant detectors are the older type of realistic gravitational-wave detector, having been pioneered by Weber in the early 1960s and refined by him and others over several decades. We will concentrate our discussion, however, on free-particle gravitational-wave detectors, which have gained in popularity over recent years and are also very much easier to analyse. In our discussion of the motion of free test particles in the presence of a passing gravitational wave, we showed in Section 18.4 that the relative physical separation l of two free particles varies as

$$l^2 = (\delta_{ij} - h_{ij})\xi^i \xi^j,$$

where ξ^i is the separation vector between the two particles. In the absence of a gravitational wave, the undisturbed distance l_0 between the particles is given by $l_0 = \delta_{ij}\xi^i\xi^j$. To first order in h_{ij}, the fractional change in the physical separation of the particles is therefore given by

$$\boxed{\frac{\delta l}{l_0} = -\tfrac{1}{2} h_{ij} n^i n^j,}$$

where n^i is a unit vector in the direction of separation of the two particles. Thus, we see that the passing of a gravitational wave produces a *linear strain*, i.e. the change in the relative separation of the particles is proportional to their original undisturbed separation. For typical astrophysical sources, the largest strain one might reasonably expect to receive at the Earth is of order

$$\frac{\delta l}{l} \sim 10^{-21}.$$

Thus, even if the two test masses were separated by a distance $l_0 = 1\,\text{km}$, the change δl in this distance is of order $10^{-16}\,\text{cm}$, which corresponds to $\sim 10^{-8}$ of the size of the atoms that comprise the test masses!

Fortunately, laser Michelson interferometers provide a means of measuring such tiny changes in the separation of the test masses. The principle of operation of such an experiment is quite straightforward and is illustrated in Figure 18.5. The basic system of made up of three test masses. Two have mirrors M attached to them, and to the third is attached a beamsplitter B. Each mass is suspended from a support that isolates the mass from external vibrations but allows it to swing freely in the horizontal direction. A laser L (with typical wavelength $\lambda \sim 10^{-4}\,\text{cm}$) is aimed at B, which splits the laser light into two beams directed down the arms of the interferometer. The beams are reflected by the mirrors at the end of each arm and then recombined in B before being detected in the detector D. When the beams are recombined they will interfere constructively if the lengths of the two arms L_1 and L_2 differ by an amount $\Delta L = n\lambda$ and will interfere destructively if $\Delta L = (n + \frac{1}{2})\lambda$, where n is an integer. The system is arranged so that the beams interfere destructively if all three masses are perfectly stationary. In practice, the experimental set-up is more sophisticated than the simple Michelson

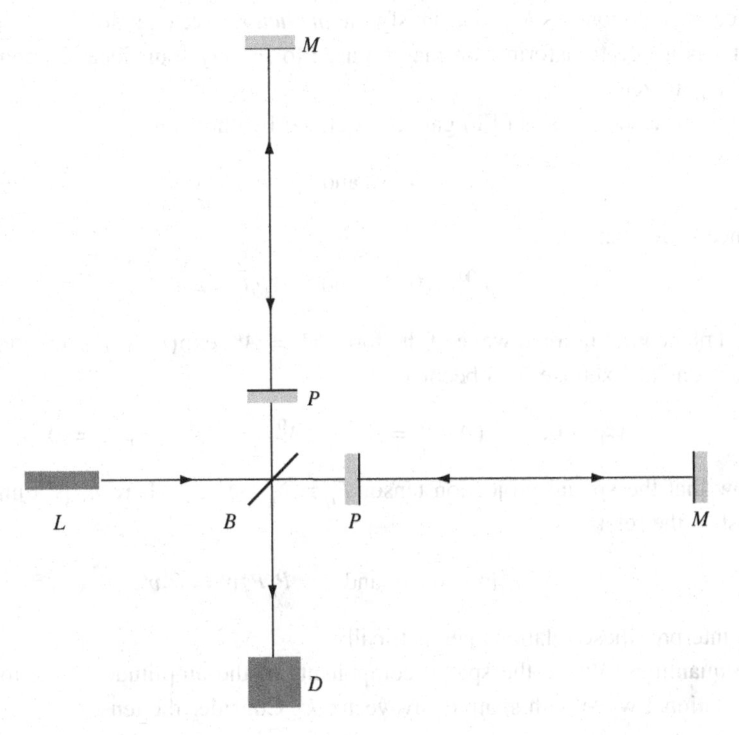

Figure 18.5 A schematic representation of a laser Michelson interferometer designed to detect gravitational waves (see the main text for details).

interferometer we have discussed. The most important improvement is the introduction of an additional test mass with a partially reflecting mirror P in each arm of the interferometer, thereby forming a 'cavity', as illustrated in Figure 18.5. A typical photon may travel up and down this cavity many times before eventually arriving at the beamsplitter, thereby greatly increasing the effective arm length of the interferometer. The use of large laser Michelson interferometers as a means for attempting to detect gravitational waves is currently being actively pursued by a number of laboratories around the world.

Exercises

18.1 For a plane gravitational wave of the form $\bar{h}^{\mu\nu} = A^{\mu\nu} \exp(ik_\rho x^\rho)$, show that, under the gauge transformation (17.5) with $\xi^\mu = \epsilon^\mu \exp(ik_\rho x^\rho)$, the amplitude tensor transforms as

$$A'^{\mu\nu} = A^{\mu\nu} - i\epsilon^\mu k^\nu - i\epsilon^\nu k^\mu + i\eta^{\mu\nu}\epsilon^\rho k_\rho.$$

18.2 The trace-reverse gravitational-field tensor transforms as

$$\bar{h}'^{\mu\rho} = \bar{h}^{\mu\rho} - \partial^\mu \xi^\rho - \partial^\rho \xi^\mu + \eta^{\mu\rho}\partial_\sigma \xi^\sigma.$$

Since the components $\bar{h}_{\mu\nu}$ also satisfy the *in vacuo* wave equation $\Box^2 \bar{h}_{\mu\nu} = 0$, show that this gauge transformation may be used to set any four linear combinations of the $\bar{h}'_{\mu\nu}$ to zero.

18.3 The transverse-traceless (TT) gauge is defined by choosing

$$\bar{h}^{0i}_{\mathrm{TT}} = 0 \qquad \text{and} \qquad \bar{h}_{\mathrm{TT}} = 0.$$

Hence show that

$$\partial_0 \bar{h}^{00}_{\mathrm{TT}} = 0 \qquad \text{and} \qquad \partial_i \bar{h}^{ij}_{\mathrm{TT}} = 0.$$

18.4 For a plane gravitational wave of the form $\bar{h}^{\mu\nu} = A^{\mu\nu} \exp(ik_\rho x^\rho)$, show that the four conditions in Exercise 18.3 become

$$A^{0i}_{\mathrm{TT}} = 0, \qquad (A_{\mathrm{TT}})^\mu_\mu = 0, \qquad A^{00}_{\mathrm{TT}} = 0, \qquad A^{ij}_{\mathrm{TT}} k_j = 0.$$

18.5 Show that the spatial projection tensor $P_{ij} \equiv \delta_{ij} - n_i n_j$, where n_i is a unit vector, satisfies the relations

$$n_i P^i_j v^j = 0 \qquad \text{and} \qquad P^i_k P^k_j v^j = P^i_j v^j,$$

and interpret these relations geometrically.

18.6 The quantities A^{ij} are the spatial components of the amplitude tensor for a plane gravitational wave with spatial wavevector k^i. Consider the tensor

$$A^{ij}_{\mathrm{TT}} = \left(P^i_k P^j_l - \tfrac{1}{2} P^{ij} P_{kl}\right) A^{kl},$$

where $P_{ij} = \delta_{ij} - \hat{k}^i\hat{k}^j$. Show that A^{ij}_{TT} is both transverse, so that $A^{ij}_{\mathrm{TT}}k_j = 0$, and traceless.

18.7 Use your answer to Exercise 18.6 to show that, for a plane gravitational wave propagating in the x^1-direction,

$$[A^{\mu\nu}_{\mathrm{TT}}] = \begin{pmatrix} 0 & 0 & 0 & 0 \\ 0 & 0 & 0 & 0 \\ 0 & 0 & \frac{1}{2}(A^{22} - A^{33}) & A^{23} \\ 0 & 0 & A^{23} & \frac{1}{2}(A^{33} - A^{22}) \end{pmatrix}.$$

18.8 In the TT gauge show that, to first order in $h_{\mu\nu}$,

$$\Gamma^{\mu}{}_{00} = 0 \qquad \text{and} \qquad \Gamma^{\mu}{}_{0\nu} = \tfrac{1}{2}\partial_0 (h_{\mathrm{TT}})^{\mu}_{\nu}.$$

18.9 Consider two nearby particles, initially at rest in our chosen coordinate system x^μ, which have a coordinate separation given by a small spacelike connecting vector $[\xi^\mu] = (0, \xi^1, \xi^2, \xi^3)$. During the passage of a gravitational wave show that, to first order in $h_{\mu\nu}$ in the TT gauge, the equation of geodesic deviation may be written as

$$\frac{D^2\xi^\mu}{D\tau^2} = c^2 R^{\mu}{}_{00\nu}\xi^\nu = \tfrac{1}{2}c^2\,(\partial_0\partial_0 h^{\mu}_{\nu})\,\xi^\nu.$$

Show further that, to the same order of approximation, in the TT gauge one has

$$\frac{D^2\xi^\mu}{D\tau^2} = \frac{d^2\xi^\mu}{d\tau^2} + \tfrac{1}{2}c^2\,(\partial_0\partial_0 h^{\mu}_{\nu})\,\xi^\nu.$$

Hence show that $\xi^\mu = \text{constant}$ is a solution of the geodesic equation, and so the coordinate separation of the two particles remains unaltered during the passage of the gravitational wave.

18.10 If ξ^i is the spatial coordinate separation vector of two nearby particles, show that the square of their physical separation is given by

$$l^2 = \delta_{ij}\zeta^i\zeta^j,$$

where $\zeta^i = \xi^i + \frac{1}{2}h^i_k\xi^k$. Show that, during the passage of a gravitational wave with $A^{\mu\nu} = b e_2^{\mu\nu}$ that is travelling in the x^3-direction,

$$[\zeta^i] = (\xi^1, \xi^2, 0) - \tfrac{1}{2}b\cos k(x^0 - x^3)(\xi^2, \xi^1, 0).$$

18.11 For two test particles reacting to the passage of a circularly polarised gravitational wave, show that one particle moves in a circle with respect to the other.

18.12 For the two-particle system considered in Section 18.5, verify that

$$[\bar{h}^{\mu\nu}_{\mathrm{rad}}(ct, \vec{x})] = \frac{8GMa^2\Omega^2}{c^4 r}\begin{pmatrix} 0 & 0 & 0 & 0 \\ 0 & \cos 2\Omega(t - r/c) & \sin 2\Omega(t - r/c) & 0 \\ 0 & \sin 2\Omega(t - r/c) & -\cos 2\Omega(t - r/c) & 0 \\ 0 & 0 & 0 & 0 \end{pmatrix},$$

and hence show that an observer on the x^3-axis measures a right-handed circularly polarised gravitational wave of the form

$$\left(h_{\text{rad}}^{TT}\right)^{\mu\nu} = \frac{8GMa^2\Omega^2}{c^4 r}\Re\left[\left(e_1^{\mu\nu} - ie_2^{\mu\nu}\right)\exp 2i\Omega(t - x^3/c)\right].$$

18.13 Consider a system of four equal masses attached to the ends of a cross formed from massless rods of equal length, set at 90°. If the system rotates freely about an axis through the centre of the cross and perpendicular to its plane, show that in the far field there is no quadrupole gravitational radiation.
Hint: Consider the system as the superposition of two systems, each like that in Exercise 18.12 but 90° out of phase.

18.14 For a plane gravitational wave of the form

$$h_{\text{TT}}^{ij} = A_{\text{TT}}^{ij} \cos k_\lambda x^\lambda.$$

travelling in the x^3-direction, verify that the energy–momentum tensor of the linearised gravitational field is given by

$$[t_{\mu\nu}] = \frac{c^2}{32\pi G}\omega^2(a^2 + b^2)\begin{pmatrix} 1 & 0 & 0 & -1 \\ 0 & 0 & 0 & 0 \\ 0 & 0 & 0 & 0 \\ -1 & 0 & 0 & 1 \end{pmatrix}$$

and that the flux in the direction of propagation is

$$F = \frac{c^3}{32\pi G}\omega^2(a^2 + b^2).$$

18.15 For the two-particle system considered in Section 18.5, verify that the gravitational-wave energy flux at a (large) distance r is, in the x^1- and x^3-directions respectively,

$$F_1 = \frac{c^3}{32\pi G}(2\Omega)^2\left(\frac{4GMa^2\Omega^3}{c^4 r}\right)^2 = \frac{2G}{\pi c^5}\left(\frac{Ma^2\Omega^3}{r}\right)^2,$$

$$F_3 = 2\frac{c^3}{32\pi G}(2\Omega)^2\left(\frac{8GMa^2\Omega^3}{c^4 r}\right)^2 = \frac{16G}{\pi c^5}\left(\frac{Ma^2\Omega^3}{r}\right)^2.$$

18.16 If $J_{\text{TT}}^{ij} = \left(P_k^i P_l^j - \frac{1}{2}P^{ij}P_{kl}\right)J^{kl}$ and $P^{ij} = \delta^{ij} - \hat{x}^i\hat{x}^j$, show that

$$J_{ij}^{\text{TT}} J_{\text{TT}}^{ij} = J_{ij}J^{ij} - 2J_i^j J^{ik}\hat{x}_j\hat{x}_k + \frac{1}{2}J^{ij}J^{kl}\hat{x}_i\hat{x}_j\hat{x}_k\hat{x}_l.$$

18.17 If $\hat{x}^i$ is a unit radial vector, show that

$$\int_{4\pi}\hat{x}_i\hat{x}_j\,d\Omega = \frac{4\pi}{3}\delta_{ij}, \qquad \int_{4\pi}\hat{x}_i\hat{x}_j\hat{x}_k\hat{x}_l\,d\Omega = \frac{4\pi}{15}(\delta_{ij}\delta_{kl} + \delta_{ik}\delta_{jl} + \delta_{il}\delta_{jk}).$$

18.18 For the two-particle system considered in Section 18.5, verify that gravitational-wave emission causes the the total energy E of the system to decrease according to

$$\frac{dE}{dt} = -\frac{G}{5c^5}(128M^2a^4\Omega^6).$$

18.19 For a binary star system containing two stars of mass M and separation $2a$, show that the orbital angular speed is

$$\Omega = \left(\frac{GM}{4a^3}\right)^{1/2}.$$

Hence show that gravitational-wave emission causes the the total energy E of the system to decrease according to

$$\frac{dE}{dt} = -\frac{2}{5}\frac{G^4M^5}{c^5a^5}.$$

Thus show that the orbital period P decreases according to

$$\frac{dP}{dt} = -\frac{96}{5}4^{1/3}\frac{\pi}{c^5}\left(\frac{2\pi GM}{P}\right)^{5/3}.$$

18.20 Show that, to first order in h_{ij}, the fractional change in the physical separation of the particles during the passage of a gravitational wave is

$$\frac{\delta l}{l_0} = -\tfrac{1}{2}h_{ij}n^i n^j,$$

where n^i is a unit vector in the direction of separation of the two particles.

18.21 Consider a line element of the form

$$ds^2 = c^2\,dt^2 - dx^2 - f^2(u)\,dy^2 - g^2(u)\,dz^2,$$

where $f(u)$ and $g(u)$ are functions of $u = ct - x$. Calculate the connection coefficients and hence the Ricci tensor for this line element. Hence show that the line element is a solution to the full empty-space field equations $R_{\mu\nu} = 0$, provided that

$$\frac{f''}{f} + \frac{g''}{g} = 0,$$

where a prime denotes d/du. Show that this solution may be interpreted, with no approximation, as a linearly polarised plane gravitational wave travelling in the x-direction.

19

A variational approach to general relativity

Most of classical and quantum physics can be expressed in terms of variational principles, and it is often when written in this form that the physical meaning is most clearly understood. Moreover, once a physical theory has been written as a variational principle it is usually straightforward to identify conserved quantities, or symmetries of the system of interest, that otherwise might have been found only with considerable effort. Conversely, by demanding that the variational principle be invariant under some symmetry, one ensures that the equations of motion derived from it also respect that symmetry. In this final chapter, we therefore present an introductory account of variational principles and the Lagrangian formalism. Our ultimate aim will be to derive afresh the field equations of general relativity from this new perspective. This will require us to consider some general aspects of classical field theory in flat and curved spacetimes. As a result, this chapter lies somewhat outside the mainstream discussion presented in preceding chapters and may be omitted on a first reading. Nevertheless the variational approach that we shall outline is extremely powerful and provides the basis for most current research into the formulation of classical (and quantum) field theories, including general relativity and other candidate theories of gravitation.

19.1 Hamilton's principle in Newtonian mechanics

To begin, let us remind ourselves of a familiar example of a physical variational principle, namely Hamilton's principle in Newtonian mechanics. Consider a mechanical system whose configuration can be defined uniquely by a number of generalised coordinates q^a, $a = 1, 2, \ldots, n$ (usually distances and angles), together with time t, and which experiences only forces derivable from a potential. Hamilton's principle states that in moving from one configuration at time t_1 to another at time t_2 the motion of such a system is such as to make stationary the *action*

$$\boxed{S = \int_{t_1}^{t_2} L(q^a, \dot{q}^a, t)\, dt.} \tag{19.1}$$

The *Lagrangian* L is defined, in terms of the kinetic energy T and the potential energy V (with respect to some reference situation), by $L = T - V$. Here V is a function of the q^a (and possibly t) only, but not of the $\dot{q}^a$. As discussed in Section 3.19, the coordinates define a *configuration space* with line element $ds^2 = g_{ab}\, dq^a\, dq^b$. For example, the Lagrangian for a particle of mass m can be written as

$$L = T - V = \tfrac{1}{2} m g_{ab} \dot{q}^a \dot{q}^b - V. \tag{19.2}$$

Returning to the general expression (19.1), let us consider an arbitrary variation

$$q^a(t) \to q'^a(t) = q^a(t) + \delta q^a(t)$$

in the trajectory in configuration space and demand that the corresponding variation δS in the action vanishes. Assuming that $\delta q^a(t) = 0$ at the endpoints t_1 and t_2, we know from our discussion of the calculus of variations in Appendix 3C at the end of Chapter 3 that the Lagrangian L must satisfy the Euler–Lagrange (EL) equations

$$\boxed{\frac{\partial L}{\partial q^a} - \frac{d}{dt}\left(\frac{\partial L}{\partial \dot{q}^a}\right) = 0, \qquad a = 1, 2, \ldots, n.}$$

For example, as shown in Section 3.19, the EL equations for the Lagrangian (19.2) are

$$m(\ddot{q}^a + \Gamma^a{}_{bc} \dot{q}^b \dot{q}^c) = -g^{ab} \partial_b V,$$

which corresponds to Newton's second law in an arbitrary coordinate system. If the $q^a(t)$ are taken to be the Cartesian coordinates $x^a(t)$ of the particle, we immediately recover the more familiar form $m\ddot{x}^a = -\delta^{ab} \partial_b V$.

Hamilton's principle is easily extended from the notion of discrete particles to continuous systems. As an example, let us consider a flexible string stretched between two fixed points at $x = 0$ and $x = l$. In this case, we again have one independent time coordinate t, but now in the context of a continuum in which the $q^a(t)$ become the continuous variable $\phi(t, x)$ describing the transverse displacement of the string as a function of position and time (see Figure 19.1). Consequently,

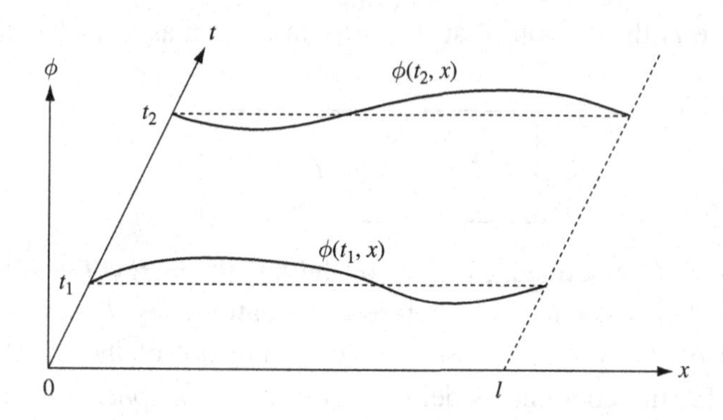

Figure 19.1 The transverse displacement $\phi(t, x)$ of a taut string fixed at two points a distance l apart, viewed as a function in the (t, x)-plane.

the expressions for T and V become integrals over x rather than sums over the label a. If $\rho(x)$ and $\tau(x)$ are the local line density and tension of the string then the kinetic and potential energies of the string for small displacements are given by

$$T = \int_0^l \tfrac{1}{2}\rho\left(\frac{\partial\phi}{\partial t}\right)^2 dx \qquad \text{and} \qquad V = \int_0^l \tfrac{1}{2}\tau\left(\frac{\partial\phi}{\partial x}\right)^2 dx.$$

Thus, the action (19.1) becomes

$$S \equiv \int_{t_1}^{t_2}\int_0^l \mathcal{L}\, dx\, dt = \int_{t_1}^{t_2}\int_0^l \tfrac{1}{2}\left[\rho(\partial_t\phi)^2 - \tau(\partial_x\phi)^2\right] dx\, dt, \tag{19.3}$$

where in the first equality we have defined the Lagrangian density $\mathcal{L}$ and in the final expression we have adopted the shorthand $\partial_t = \partial/\partial t$ and $\partial_x = \partial/\partial x$. Let us now consider an arbitrary variation in the function ϕ of the form

$$\phi(t, x) \to \phi'(t, x) = \phi(t, x) + \delta\phi(t, x). \tag{19.4}$$

This leads to a variation in the action (19.1) given by

$$\delta S = \int_{t_1}^{t_2}\int_0^l \left[\frac{\partial\mathcal{L}}{\partial(\partial_t\phi)}\delta(\partial_t\phi) + \frac{\partial\mathcal{L}}{\partial(\partial_x\phi)}\delta(\partial_x\phi)\right] dx\, dt. \tag{19.5}$$

From (19.4), one immediately notes that $\delta(\partial_t\phi) = \partial_t(\delta\phi)$ and $\delta(\partial_x\phi) = \partial_x(\delta\phi)$. Substituting these expressions in (19.5) and using Leibnitz' rule for the differentiation of a product, we may write

$$\delta S = \delta S_{\mathrm{b}} - \int_{t_1}^{t_2}\int_0^l \left\{\partial_t\left[\frac{\partial\mathcal{L}}{\partial(\partial_t\phi)}\right] + \partial_x\left[\frac{\partial\mathcal{L}}{\partial(\partial_x\phi)}\right]\right\}\delta\phi\, dx\, dt, \tag{19.6}$$

where the 'boundary' (or 'surface') term is given by

$$\delta S_b = \int_{t_1}^{t_2} \int_0^l \left\{ \partial_t \left[\frac{\partial \mathcal{L}}{\partial(\partial_t \phi)} \delta\phi \right] + \partial_x \left[\frac{\partial \mathcal{L}}{\partial(\partial_x \phi)} \delta\phi \right] \right\} dx\, dt$$

$$= \int_0^l \left[\frac{\partial \mathcal{L}}{\partial(\partial_t \phi)} \delta\phi \right]_{t=t_1}^{t=t_2} dx + \int_{t_1}^{t_2} \left[\frac{\partial \mathcal{L}}{\partial(\partial_x \phi)} \delta\phi \right]_{x=0}^{x=l} dt.$$

If we assume that the variation is such that

$$\delta\phi(t_1, x) = 0 = \delta\phi(t_2, x) \qquad \text{and} \qquad \delta\phi(t, 0) = 0 = \delta\phi(t, l)$$

then it vanishes on the entire 'boundary' of the region of interest in the (t, x)-plane, and we have $\delta S_b = 0$. Thus, in this case, by demanding that the total variation (19.6) in the action vanishes ($\delta S = 0$) and using the fact that $\delta\phi$ is arbitrary, we obtain

$$\partial_t \left[\frac{\partial \mathcal{L}}{\partial(\partial_t \phi)} \right] + \partial_x \left[\frac{\partial \mathcal{L}}{\partial(\partial_x \phi)} \right] = \partial_t(\rho \partial_t \phi) - \partial_x(\tau \partial_x \phi) = 0,$$

where, in the first equality, we have evaluated the derivatives of $\mathcal{L}$ with respect to $\partial_t \phi$ and $\partial_x \phi$ using (19.3). If, in addition, ρ and τ do not depend on x or t then

$$\frac{\partial^2 \phi}{\partial x^2} = \frac{1}{c^2} \frac{\partial^2 \phi}{\partial t^2},$$

where $c^2 = \tau/\rho$. This is the wave equation for small transverse oscillations of a taut uniform string.

19.2 Classical field theory and the action

In the above discussion, the function $\phi(t, x)$ may be regarded as a 'field' defined on a two-dimensional space (or manifold) parameterised by the coordinates x and t. To extend the idea of a variational principle to a field theory in spacetime, one therefore needs only to replace $\phi(t, x)$ by a (finite) set of fields $\Phi^a(x^\mu)$ defined on a four-dimensional spacetime parameterised in terms of some (in general) arbitrary set of continuous coordinates x^μ. Alternatively, one could even consider each member of the (finite) set of generalised coordinates $q^a(t)$ in (19.1) as a 'field' defined on a one-dimensional manifold parameterised by the continuous coordinate t, and simply replace the $q^a(t)$ by the set of fields $\Phi^a(x^\mu)$. In either case, the index a acts merely as a label for the individual fields in the theory.

This last point is worth clarifying. If, for example, one were considering a field theory containing a set of M scalar fields $\phi^1, \phi^2, \ldots, \phi^M$ then the set of fields would be simply $\{\Phi^a\} = \{\phi^1, \phi^2, \ldots, \phi^M\}$. Alternatively, one might be interested in a field theory containing a vector field (such as electromagnetism). In this case, the label a would run over the four components of the vector field in the chosen coordinate system, i.e. we would write $\{\Phi^a\} = \{A^0, A^1, A^2, A^3\} = \{A^\mu\}$ and so a would then be a spacetime index. Similar considerations apply to the components of tensor fields. Use of the index a may also be trivially extended to label the components of two or more vector or tensor fields involved in the theory. Indeed, when considering field theories defined on some arbitrary manifold and in arbitrary coordinates, one must always include the metric tensor components in the set of fields. For example, in electromagnetism on an arbitrary manifold, the full set of fields is in fact $\{\Phi^a\} = \{A^\mu, g^{\mu\nu}\}$.

By analogy with (19.3), the action S for a set of fields defined on some general four-dimensional spacetime manifold should take the form of an integral of some function $\mathcal{L}$, called the *Lagrangian density*, of the fields Φ^a and their first (and possibly higher) derivatives over some four-dimensional region $\mathcal{R}$ of the spacetime. Thus, we take the action integral to be

$$\boxed{S = \int_{\mathcal{R}} \mathcal{L}(\Phi^a, \partial_\mu \Phi^a, \partial_\mu \partial_\nu \Phi^a, \ldots)\, d^4x,} \tag{19.7}$$

where d^4x denotes the product of coordinate differentials $dx^0\, dx^1\, dx^2\, dx^3$. It is believed that physical theories should be generally covariant and so this symmetry must be reflected in the action S, which therefore has to be a scalar under general coordinate transformations. From the discussion in Section 2.14, we know that in any arbitrary coordinate system x^μ the invariant volume element (which transforms as a scalar field) is $d^4V = \sqrt{-g}\, d^4x$, where g is the determinant of the metric tensor in that coordinate system (and is negative for the signature of the metric used in this book). It is therefore convenient to write the action (19.7) in the form

$$S = \int_{\mathcal{R}} L\sqrt{-g}\, d^4x,$$

where we have introduced the *field Lagrangian L*, which is clearly related to the Lagrangian density $\mathcal{L}$ by[1]

$$\mathcal{L} = L\sqrt{-g}. \tag{19.8}$$

For the action S to be a scalar, the quantity $L\sqrt{-g}\, d^4x$ must be a scalar field at each point in $\mathcal{R}$. Since the invariant volume element $\sqrt{-g}\, d^4x$ is already a scalar field, then so too must be the Lagrangian L. Taking L to be in general a function of the fields Φ^a and their first (and possibly higher) derivatives, the action for a

[1] Although most authors agree that $\mathcal{L}$ is called the *Lagrangian density*, it is common in field theory for the term *Lagrangian* (and the symbol L) to mean the integral of $\mathcal{L}$ over some three-dimensional spacelike hypersurface, rather than the relationship given in (19.8). We will adopt the convention (19.8) throughout this chapter.

set of classical fields defined on some 4-dimensional spacetime manifold may be written as

$$\boxed{S = \int_{\mathcal{R}} L(\Phi^a, \partial_\mu \Phi^a, \partial_\mu \partial_\nu \Phi^a, \ldots)\sqrt{-g}\, d^4x,}$$

where L is a scalar function of spacetime position. We note finally that the Lagrangian density $\mathcal{L}$ in (19.8) will *not* transform as a scalar field under coordinate transformations; in fact, it is what is known as a *scalar density* of weight unity, although we need not concern ourselves here with the definition of such objects.

19.3 Euler–Lagrange equations

We now derive the form of the field equations for (some subset of) the fields Φ^a by demanding that the action is *stationary*, or invariant, under small variations in (the same subset of) the fields of the form

$$\Phi^a(x) \to \Phi'^a(x) = \Phi^a(x) + \delta\Phi^a(x). \tag{19.9}$$

It is important to note that we are *not* performing any coordinate transformation here; we are considering only variations in the functional forms of the fields Φ^a in a fixed coordinate system. For simplicity, we shall perform our derivation of the field equations under the assumption that the field theory is *local*, which means that second- or higher-order derivatives of the fields do not appear in the action. Thus, we need only consider the consequent variation in the first derivatives of the fields, which, from (19.9), is given by

$$\partial_\mu \Phi^a \to \partial_\mu \Phi'^a = \partial_\mu \Phi^a + \partial_\mu(\delta\Phi^a). \tag{19.10}$$

We also note for later use that, from its definition (19.9), the δ-operator commutes with derivatives since

$$\partial_\mu(\delta\Phi^a) = \partial_\mu(\Phi'^a - \Phi^a) = \partial_\mu \Phi'^a - \partial_\mu \Phi^a = \delta(\partial_\mu \Phi^a). \tag{19.11}$$

The variations (19.9, 19.10) lead to a variation in the action $S \to S + \delta S$, with

$$\delta S = \int_{\mathcal{R}} \delta\mathcal{L}\, d^4x = \int_{\mathcal{R}} \left[\frac{\partial \mathcal{L}}{\partial \Phi^a}\delta\Phi^a + \frac{\partial \mathcal{L}}{\partial(\partial_\mu \Phi^a)}\delta(\partial_\mu \Phi^a)\right] d^4x, \tag{19.12}$$

where, for the time being, it is convenient to work in terms of the Lagrangian density $\mathcal{L}$ defined in (19.8). To derive the field equations, we wish to factor out the variation $\delta\Phi^a$ in the second term of the integrand. Using (19.11), this second term may be written

$$\int_{\mathcal{R}} \frac{\partial\mathcal{L}}{\partial(\partial_\mu\Phi^a)}\partial_\mu(\delta\Phi^a)\,d^4x = \int_{\mathcal{R}} \partial_\mu\left[\frac{\partial\mathcal{L}}{\partial(\partial_\mu\Phi^a)}\delta\Phi^a\right]d^4x - \int_{\mathcal{R}} \partial_\mu\left[\frac{\partial\mathcal{L}}{\partial(\partial_\mu\Phi^a)}\right]\delta\Phi^a\,d^4x,$$

where we have integrated by parts (which corresponds simply to rewriting the integrand using Leibnitz' theorem for the derivative of a product). The first integral on the right-hand side is a total derivative and can therefore be converted into an integral over the bounding surface $\partial\mathcal{R}$ of the region $\mathcal{R}$, by straightforward calculus. If we restrict the permissible variations $\delta\Phi^a$ to those that *vanish* on the boundary $\partial\mathcal{R}$, this integral will also vanish and so (19.12) becomes[2]

$$\delta S \equiv \int_{\mathcal{R}} \frac{\delta\mathcal{L}}{\delta\Phi^a}\delta\Phi^a\,d^4x = \int_{\mathcal{R}} \left\{\frac{\partial\mathcal{L}}{\partial\Phi^a} - \partial_\mu\left[\frac{\partial\mathcal{L}}{\partial(\partial_\mu\Phi^a)}\right]\right\}\delta\Phi^a\,d^4x,$$

where, in the first equality, we define the *variational derivative* $\delta\mathcal{L}/\delta\Phi^a$ of the Lagrangian density with respect to the field Φ^a. If we demand that the action is stationary, so that $\delta S = 0$, under the arbitrary variations $\delta\Phi^a$ we thus require that

$$\boxed{\frac{\delta\mathcal{L}}{\delta\Phi^a} = \frac{\partial\mathcal{L}}{\partial\Phi^a} - \partial_\mu\left[\frac{\partial\mathcal{L}}{\partial(\partial_\mu\Phi^a)}\right] = 0.} \tag{19.13}$$

These are the Euler–Lagrange (EL) equations, which correspond to the field equations of the (local) field theory defined by the action $S = \int_{\mathcal{R}} \mathcal{L}\,d^4x$. If, in addition, the Lagrangian density depends on second- or higher-order derivatives of the fields then the above derivation is straightforwardly generalised. For example, if second-order derivatives also appear then one obtains

$$\frac{\delta\mathcal{L}}{\delta\Phi^a} = \frac{\partial\mathcal{L}}{\partial\Phi^a} - \partial_\mu\left[\frac{\partial\mathcal{L}}{\partial(\partial_\mu\Phi^a)}\right] + \partial_\mu\partial_\nu\left[\frac{\partial\mathcal{L}}{\partial(\partial_\mu\partial_\nu\Phi^a)}\right] = 0, \tag{19.14}$$

provided that the variations $\delta\Phi^a$ *and* their first derivatives vanish on the boundary $\partial\mathcal{R}$.

[2] The restriction that the variation $\delta\Phi^a$ vanishes on the boundary $\partial\mathcal{R}$ is generally allowable, except when discussing topological objects in field theory such as instantons, which are beyond the scope of our discussion.

19.4 Alternative form of the Euler–Lagrange equations

The EL equations in the form (19.13), or generalised to higher-order derivatives of the fields, provide a straightforward means of determing the field equations corresponding to a given action. In particular, these equations still hold if (some of) the fields Φ^a being varied are the components of the metric tensor $g_{\mu\nu}$ (or functions thereof), as will be the case when we derive the Einstein equations from the gravitational action in Section 19.8.

Nevertheless, if the fields Φ^a being varied are not functions of the metric tensor components then the presence of the $\sqrt{-g}$ factor in the Lagrangian density (19.8) makes evaluation of the derivative terms in the EL equations (19.13) unnecessarily cumbersome, although one will nevertheless arrive at the correct field equations. In such cases, however, the Lagrangian L can often be written in terms of the fields Φ^a and their first (and possibly higher-order) *covariant* derivatives $\nabla_\mu \Phi^a$, as opposed to partial derivatives. Indeed, recalling that L should be a scalar function of spacetime position, one might expect this to be the case since scalars are most easily obtained by contracting tensor indices. Let us therefore repeat our derivation of the form of the EL equations, working instead with an action of the form

$$S = \int_{\mathcal{R}} L(\Phi^a, \nabla_\mu \Phi^a, \nabla_\mu \nabla_\nu \Phi^a, \ldots, g_{\mu\nu}, \partial_\sigma g_{\mu\nu}, \ldots)\sqrt{-g}\, d^4x, \tag{19.15}$$

where the fields Φ^a being varied are independent of the metric tensor $g_{\mu\nu}$ but L might still contain $g_{\mu\nu}$, to raise or lower indices, for example (L might also contain the partial derivatives of $g_{\mu\nu}$; recall that the covariant derivatives of the metric vanish identically).

For simplicity, let us again assume that no second- or higher-order covariant derivatives appear in L. The variation (19.9) leads to variations in the first covariant derivatives of the fields given by

$$\nabla_\mu \Phi^a \to \nabla_\mu \Phi'^a = \nabla_\mu \Phi^a + \nabla_\mu(\delta\Phi^a). \tag{19.16}$$

In a similar way to before, we note that the δ-operator commutes with covariant derivatives, so that $\nabla_\mu(\delta\Phi^a) = \delta(\nabla_\mu \Phi^a)$. The variations (19.9–19.16) lead, in turn, to a variation in the action $S \to S + \delta S$, with

$$\delta S = \int_{\mathcal{R}} \delta L \sqrt{-g}\, d^4x = \int_{\mathcal{R}} \left[\frac{\partial L}{\partial \Phi^a}\delta\Phi^a + \frac{\partial L}{\partial(\nabla_\mu \Phi^a)}\delta(\nabla_\mu \Phi^a) \right] \sqrt{-g}\, d^4x, \tag{19.17}$$

where we are now working in terms of the Lagrangian L (as opposed to the Lagrangian density $\mathcal{L}$). The partial derivative appearing in the first term of the integrand on the right-hand side deserves some comment. In (19.17), we are treating Φ^a and $\nabla_\mu \Phi^a$ as independent variables. In general, however, the covariant derivatives $\nabla_\mu \Phi^a$ will contain terms involving the fields Φ^a multiplied by some connection coefficient. If, for some reason, these terms are written out explicitly in the Lagrangian, they must *not* be included when calculating the partial derivative of L with respect to the fields Φ^a.

As in our previous derivation of the EL equations, we must now factor out the variation $\delta\Phi^a$ in the second term of the integrand in (19.17). Using the fact that the δ-operator commutes with the covariant derivative and employing Leibnitz' theorem for the covariant differentiation of a product, this term may be written

$$\int_{\mathcal{R}} \frac{\partial L}{\partial(\nabla_\mu \Phi^a)} \nabla_\mu(\delta\Phi^a)\sqrt{-g}\, d^4x = \int_{\mathcal{R}} \nabla_\mu \left[\frac{\partial L}{\partial(\nabla_\mu \Phi^a)} \delta\Phi^a \right] \sqrt{-g}\, d^4x - \int_{\mathcal{R}} \nabla_\mu \left[\frac{\partial L}{\partial(\nabla_\mu \Phi^a)} \right] \delta\Phi^a \sqrt{-g}\, d^4x. \quad (19.18)$$

We may now use the *divergence theorem* to convert the first integral on the right-hand side to an integral over the boundary $\partial\mathcal{R}$. The divergence theorem reads

$$\int_{\mathcal{R}} (\nabla_\mu V^\mu)\sqrt{|g|}\, d^4x = \int_{\partial\mathcal{R}} n_\mu V^\mu \sqrt{|\gamma|}\, d^3y, \quad (19.19)$$

where V^μ is an arbitrary vector field, γ is the determinant of the induced metric on the boundary in the coordinates y^i (see Section 2.14) and n_μ is a unit normal to the boundary. Applying this theorem to the first integral on the right-hand side of (19.18) and restricting the allowed variations $\delta\Phi^a$ to vanish on $\partial\mathcal{R}$, we see that this integral is zero. Thus (19.18) becomes

$$\delta S \equiv \int_{\mathcal{R}} \frac{\delta L}{\delta \Phi^a} \delta\Phi^a \sqrt{-g}\, d^4x = \int_{\mathcal{R}} \left\{ \frac{\partial L}{\partial \Phi^a} - \nabla_\mu \left[\frac{\partial L}{\partial(\nabla_\mu \Phi^a)} \right] \right\} \delta\Phi^a \sqrt{-g}\, d^4x,$$

where, in the first equality, we define the variational derivative $\delta L/\delta\Phi^a$ of the Lagrangian with respect to the field Φ^a. Thus, demanding stationarity of the action, $\delta S = 0$, we obtain the alternative form for the Euler–Lagrange equations

$$\boxed{\frac{\delta L}{\delta \Phi^a} = \frac{\partial L}{\partial \Phi^a} - \nabla_\mu \left[\frac{\partial L}{\partial(\nabla_\mu \Phi^a)} \right] = 0.} \quad (19.20)$$

We shall make use of this form for the EL equations when we consider the field theories of a real scalar field in Section 19.6 and electromagnetism in Section 19.7.

19.5 Equivalent actions

From the derivation of the EL equations, (19.13), the alert reader will have noticed that there exists an ambigiuity in the definition of the action. This derives from the fact that one can always convert the integral of a total derivative over some region $\mathcal{R}$ into an integral over the bounding surface $\partial\mathcal{R}$. Let us therefore consider the following modification of the Lagrangian density:

$$\boxed{\mathcal{L} \to \bar{\mathcal{L}} = \mathcal{L} + \partial_\mu Q^\mu(\Phi^a),} \tag{19.21}$$

where the Q^μ may, in general, be four arbitrary functions of the fields (but not of their derivatives). The corresponding action thus reads

$$\bar{S} = S + \int_\mathcal{R} \partial_\mu Q^\mu \, d^4x.$$

The variation in this action under the variation in the fields Φ^a (19.9) is given by

$$\delta\bar{S} = \delta S + \int_\mathcal{R} \partial_\mu(\delta Q^\mu) \, d^4x = \delta S + \int_\mathcal{R} \partial_\mu \left(\frac{\partial Q^\mu}{\partial \Phi^a} \delta\Phi^a \right) d^4x,$$

where δS is the variation in the original action given by the equation before (19.13) and we have used the fact that the δ-operator commutes with derivatives. Since the last integral on the right-hand side is a total derivative, it can be converted to a surface integral over the boundary $\partial\mathcal{R}$. Assuming once again that the variations $\delta\Phi^a$ vanish on $\partial\mathcal{R}$, this surface integral is zero and so $\delta\bar{S} = \delta S$. Hence demanding that $\delta\bar{S} = 0$ yields the *same* EL equations as demanding that $\delta S = 0$, and the two actions are said to be *equivalent*. In other words, any two Lagrangian densities related by an expression of the form (19.21) lead to the same EL equations. The above argument is easily extended to the case in which $\mathcal{L}$ contains second- or higher-order derivatives of the fields. For example, if second-order derivatives also appear in $\mathcal{L}$ then the same EL equations (19.14) will be obtained from any Lagrangian density of the form

$$\bar{\mathcal{L}} = \mathcal{L} + \partial_\mu Q^\mu(\Phi^a, \partial_\nu \Phi^a), \tag{19.22}$$

provided that the variations $\delta\Phi^a$ and their first derivatives vanish on the boundary $\partial\mathcal{R}$.

Despite the appealing features of the above mathematical manoeuvre, the very general nature of the allowed transformation (19.21) can lead to problems of principle. In particular, we have not constrained in any way the transformation properties of the four quantities Q^μ. Thus, we have not ensured that the quantity $\partial_\mu Q^\mu \, d^4x$ is a scalar function under coordinate transformations. Strictly speaking, one should ensure that this is true in order that the second term on the right-hand side of (19.21) is a scalar quantity. Without this criterion, the value of this integral (and hence the action $\bar{S}$) is not a scalar, i.e. its value changes depending on the choice of coordinates. We shall see in Section 19.9, however, that the necessary requirements on the quantities Q^μ are not always imposed. A partial defence of such practices is that, as stated earlier, in the variation (19.9) we are *not* performing any coordinate transformation; we are considering only variations in the functional forms of the fields Φ^a in a fixed coordinate system. One might therefore be persuaded that the variational formalism outlined above would survive the introduction of terms in the action that are not scalars under general coordinate transformations. In principle, however, such sleight of hand is best avoided, and one should always aim to construct an action that is a true covariant scalar.

We may also construct equivalent actions when the original action takes the form (19.15), remembering that in this case we are assuming that the fields of interest Φ^a are independent of the components of the metric tensor $g_{\mu\nu}$. Suppose, for example, that no second- or higher-order covariant derivatives of the fields appear in L, and consider the new Lagrangian

$$\boxed{\bar{L} = L + \nabla_\mu Q^\mu(\Phi^a),}$$

where the functions Q^μ depend only on the fields and not on their first covariant derivatives. The corresponding action then reads

$$\bar{S} = S + \int_{\mathcal{R}} \nabla_\mu Q^\mu \sqrt{-g}\, d^4x, \tag{19.23}$$

and its variation is given by

$$\delta\bar{S} = \delta S + \int_{\mathcal{R}} \nabla_\mu(\delta Q^\mu)\sqrt{-g}\, d^4x = \delta S + \int_{\mathcal{R}} \nabla_\mu \left(\frac{\partial Q^\mu}{\partial \Phi^a}\delta\Phi^a\right)\sqrt{-g}\, d^4x,$$

where δS is the variation in the original action and again we have used the result that the δ-operator commutes with covariant derivatives. Using the divergence theorem (19.19), the last integral on the right-hand side can be converted to a surface integral over the boundary $\partial\mathcal{R}$. Assuming once again that the variations $\delta\Phi^a$ vanish on $\partial\mathcal{R}$, we find that $\delta\bar{S} = \delta S$, and so we have obtained the *same* EL equations (19.20) by demanding that $\delta\bar{S} = 0$ as we did by demanding that $\delta S = 0$. We note that, by using the divergence theorem to obtain a surface integral, in the present case we require the Q^μ to be the components of a vector. This also ensures that $\nabla_\mu Q^\mu$ is a scalar field, and so the second term on the right-hand side of (19.23) (and hence the total action $\bar{S}$) is a scalar integral.

19.6 Field theory of a real scalar field

The simplest example of a field theory is that of a single real scalar field $\phi(x^\mu)$ defined on the spacetime. We will also restrict our considerations to a local field theory, so that no second- or higher-order derivatives of the field appear in the Lagrangian L.

As a starting point, we take as inspiration the Lagrangian (19.2) for the classical motion of a mechanical system in Newtonian mechanics. This Lagrangian is expressed in terms of the derivatives of the generalised coordinates $q^a(t)$ with respect to the time parameter t, the metric g_{ab} of the configuration space of the system and a potential $V(q^a)$. Replacing the generalised coordinates by the field $\phi(x^\mu)$ and time derivatives by derivatives with respect to spacetime position, a reasonable choice of Lagrangian is given by

$$\boxed{L = \tfrac{1}{2}g^{\mu\nu}(\nabla_\mu\phi)(\nabla_\nu\phi) - V(\phi),} \tag{19.24}$$

where the first term may be loosely regarded as the 'kinetic energy' of the field and the second term as its 'potential energy'. In the expression (19.24), we have used covariant derivatives rather than partial derivatives since, as stated in Section 19.2, L must itself be a scalar function of spacetime position. However, since the covariant derivative of a scalar quantity reduces to a partial derivative, in this case the latter could be used. Nevertheless, it is usually wiser to retain the manifestly covariant notation in (19.24). In particular, we see immediately that the corresponding action is given by

$$S = \int_{\mathcal{R}} \left[\tfrac{1}{2}g^{\mu\nu}(\nabla_\mu\phi)(\nabla_\nu\phi) - V(\phi)\right]\sqrt{-g}\, d^4x, \tag{19.25}$$

which is of the general form given in (19.15). Varying this action with respect to ϕ, we may therefore use the convenient form of the EL equations given in (19.20).

For the form of Lagrangian (19.24) we have

$$\frac{\partial L}{\partial\phi} = -\frac{dV}{d\phi} \quad \text{and} \quad \frac{\partial L}{\partial(\nabla_\mu\phi)} = \frac{\partial}{\partial(\nabla_\mu\phi)}\left[\tfrac{1}{2}g^{\rho\sigma}(\nabla_\rho\phi)(\nabla_\sigma\phi)\right],$$

where in the second equation we have relabelled the dummy indices in order to make the differentiation more transparent. Evaluating this derivative explicitly gives[3]

$$\frac{\partial L}{\partial(\nabla_\mu\phi)} = \tfrac{1}{2}g^{\rho\sigma}\left[\delta^\mu_\rho(\nabla_\sigma\phi) + (\nabla_\rho\phi)\delta^\mu_\sigma\right] = \tfrac{1}{2}(g^{\mu\sigma}\nabla_\sigma\phi + g^{\rho\mu}\nabla_\rho\phi) = g^{\mu\nu}\nabla_\nu\phi,$$

and so the EL equations (19.20) become

$$-\frac{dV}{d\phi} - \nabla_\mu(g^{\mu\nu}\nabla_\nu\phi) = 0.$$

[3] With a little practice, derivatives of this sort can in fact be evaluated very quickly, without needing to employ the explicit relabelling step used above.

Remembering that the covariant derivative of the metric tensor is zero, and rearranging, we thus find that the dynamical field equation satisfied by ϕ is

$$\boxed{\Box^2\phi + \frac{dV}{d\phi} = 0,} \tag{19.26}$$

where $\Box^2 \equiv \nabla^\mu\nabla_\mu = g^{\mu\nu}\nabla_\mu\nabla_\nu$ is the *covariant* d'Alembertian operator.

A common choice for the potential is $V = \frac{1}{2}m^2\phi^2$, where m is a constant parameter that characterises the dynamics of the scalar field. The field equation (19.26) then becomes

$$\Box^2\phi + m^2\phi = 0,$$

which is known as the *Klein–Gordon* equation. Upon quantisation (which is beyond the scope of our discussion), this field theory describes collections of neutral spinless particles of mass m that do not interact with each other except through their mutual gravitational attraction.

19.7 Electromagnetism from a variational principle

As discussed in Chapter 6, electromagnetism may be described in terms of the vector field A^μ. Thus, using the general description given in Section 19.2, the fields Φ^a $(a = 1, \ldots, 4)$ being varied are the components of this vector field and so a is a spacetime index. To describe the dynamics of the electromagnetic field in terms of the variational principle, again we begin by constructing a Lagrangian L which is a function of A^μ and its first derivatives and which behaves as a scalar field under general coordinate transformations. We will work from the outset assuming arbitrary coordinates.

In the case of electromagnetism, however, we saw in Chapter 6 that the theory also possesses a *gauge invariance*. If A_μ describes the electromagnetic field in some physical situation then the same situation is also described by any other field of the form

$$A'_\mu = A_\mu + \nabla_\mu\psi = A_\mu + \partial_\mu\psi, \tag{19.27}$$

where ψ is any scalar field (the last equality holds because the covariant derivative of the scalar is simply its partial derivative). As discussed earlier, by demanding that the action be invariant under some symmetry one ensures that the resulting equations of motion also respect that symmetry. We must therefore make sure that the action is invariant under the gauge transformation (19.27). This precludes us from forming scalars depending on $A_\mu A^\mu$, since it is easy to show that this expression is not gauge invariant. Nevertheless, the electromagnetic field-strength tensor

$$F_{\mu\nu} = \nabla_\mu A_\nu - \nabla_\nu A_\mu = \partial_\mu A_\nu - \partial_\nu A_\mu \tag{19.28}$$

is easily shown to be gauge invariant; the second equality in (19.28) holds since a convenient cancellation occurs between the terms containing connection coefficients arising from the two covariant derivatives. The most obvious scalar to be constructed from the field-strength tensor is simply $F_{\mu\nu}F^{\mu\nu} = g^{\mu\rho}g^{\nu\sigma}F_{\rho\sigma}F_{\mu\nu}$. Including a factor of $-1/(4\mu_0)$ for later convenience, we shall take the 'free-field' part of the Lagrangian to be

$$L_\text{f} = -\frac{1}{4\mu_0} g^{\mu\rho}g^{\nu\sigma}(\nabla_\rho A_\sigma - \nabla_\sigma A_\rho)(\nabla_\mu A_\nu - \nabla_\nu A_\mu),$$

where, again for later convenience, we have written the expression in terms of covariant derivatives rather than partial derivatives.

So far we have not taken into account that the source of the electromagnetic field is the 4-current density j^μ of any charged matter present. To describe this, we must include an 'interaction term' in the Lagrangian. The most straightforward scalar we may construct from the electromagnetic field and the current density is $j^\mu A_\mu$, and we will take the interaction term to be $L_\text{i} = -j^\mu A_\mu$. Taking the full Lagrangian to be $L = L_\text{f} + L_\text{i}$, the action reads

$$\boxed{S = \int_\mathcal{R} \left[-\frac{1}{4\mu_0} g^{\mu\rho}g^{\nu\sigma}(\nabla_\rho A_\sigma - \nabla_\sigma A_\rho)(\nabla_\mu A_\nu - \nabla_\nu A_\mu) - j^\mu A_\mu \right] \sqrt{-g}\, d^4x.} \tag{19.29}$$

As is immediately apparent, however, the interaction term $-j^\mu A_\mu$ is not automatically gauge invariant. Under the gauge transformation (19.27) the corresponding term in the action becomes

$$-\int_\mathcal{R} [j^\mu A_\mu + j^\mu(\nabla_\mu \psi)]\sqrt{-g}\, d^4x = -\int_\mathcal{R} [j^\mu A_\mu + \nabla_\mu(j^\mu \psi) - (\nabla_\mu j^\mu)\psi]\sqrt{-g}\, d^4x.$$

Using the divergence theorem (19.19), we may write the second term in the integrand on the right-hand side as a surface integral over the boundary $\partial\mathcal{R}$. Taking the source j^μ to vanish on $\partial\mathcal{R}$ (by, for example, taking the boundary to be at spatial infinity), the surface integral is zero. Thus, we see that the part of the action arising from the interaction term is, in fact, gauge invariant, provided that the source j^μ satisfies the covariant continuity equation

$$\nabla_\mu j^\mu = 0,$$

and so the requirement of gauge invariance implies the conservation of charge.

Thus, under the appropriate conditions, the action (19.29) is invariant under the gauge transformation (19.27) and, by construction, is a scalar under general coordinate transformations. Let us now determine the Euler–Lagrange equations resulting from varying the fields A_μ in this action (while keeping the source j^μ fixed). From (19.29), we see that the action has the general form (19.15). Therefore we may once again use the form of the EL equations given in (19.20), which in this case read

$$\frac{\partial L}{\partial A_\nu} - \nabla_\mu \left[\frac{\partial L}{\partial(\nabla_\mu A_\nu)} \right] = 0. \tag{19.30}$$

For the action in (19.29), we have immediately

$$\frac{\partial L}{\partial A_\nu} = -j^\mu \delta^\nu_\mu = -j^\nu, \tag{19.31}$$

but evaluation of the second term on the left-hand side of (19.30) requires more care. Relabelling dummy indices, and writing $\nabla_\mu A_\nu - \nabla_\nu A_\mu = F_{\mu\nu}$ for convenience, we have

$$\begin{aligned}
\frac{\partial L}{\partial(\nabla_\mu A_\nu)} &= \frac{\partial}{\partial(\nabla_\mu A_\nu)} \left[-\frac{1}{4\mu_0} g^{\alpha\rho} g^{\beta\sigma} F_{\rho\sigma} F_{\alpha\beta} \right] \\
&= -\frac{1}{4\mu_0} g^{\alpha\rho} g^{\beta\sigma} \left[\left(\delta^\mu_\rho \delta^\nu_\sigma - \delta^\mu_\sigma \delta^\nu_\rho \right) F_{\alpha\beta} + F_{\rho\sigma} \left(\delta^\mu_\alpha \delta^\nu_\beta - \delta^\mu_\beta \delta^\nu_\alpha \right) \right] \\
&= -\frac{1}{4\mu_0} (g^{\alpha\mu} g^{\beta\nu} - g^{\alpha\nu} g^{\beta\mu}) F_{\alpha\beta} - \frac{1}{4\mu_0} (g^{\mu\rho} g^{\nu\sigma} - g^{\nu\rho} g^{\mu\sigma}) F_{\rho\sigma} \\
&= -\frac{1}{4\mu_0} (F^{\mu\nu} - F^{\nu\mu}) - \frac{1}{4\mu_0} (F^{\mu\nu} - F^{\nu\mu}) = -\frac{1}{\mu_0} F^{\mu\nu},
\end{aligned}$$

where in the last equality we have used the antisymmetry of the field-strength tensor (19.28). Combining this result with (19.31), the EL Lagrange equations (19.30) read

$$\boxed{\nabla_\mu F^{\mu\nu} = \mu_0 j^\nu,}$$

which is the same expression as that for the inhomogeneous Maxwell equations in an arbitrary coordinate system, given in Section 7.7. The remaining homogeneous Maxwell equations are in fact automatically satisfied from the definition (19.28) of the field-strength tensor, since

$$\nabla_\sigma F_{\mu\nu} + \nabla_\nu F_{\sigma\mu} + \nabla_\mu F_{\nu\sigma} = \partial_\sigma F_{\mu\nu} + \partial_\nu F_{\sigma\mu} + \partial_\mu F_{\nu\sigma} = 0.$$

Of course, one may object to the fact that we carefully constructed the action (19.29) (by, for example, including specific factors in L_f and L_i) in such a way that its variation with respect to A_μ led to the field equations for electromagnetism. Nevertheless, the derivation above illustrates the natural way in which the action approach constrains the possible forms for the theory and allows any symmetries in the theory to be made manifest.

19.8 The Einstein–Hilbert action and general relativity *in vacuo*

We now use our experience in expressing scalar field theory and electromagnetism as variational principles to construct an action for gravitation from which the Einstein field equations of general relativity can be derived. For the time being, we will restrict our attention to general relativity *in vacuo*.

To construct an action for general relativity, we must define a Lagrangian L which is a scalar under general coordinate transformations and which depends on the components $g_{\mu\nu}$ of the metric tensor (these are now the dynamical fields), and their first- and possibly higher-order derivatives. The simplest non-trivial scalar that can be constructed from the metric and its derivatives is the Ricci scalar R, which depends on $g_{\mu\nu}$ and its first- and second-order derivatives. In fact, R is the *only* scalar derivable from the metric tensor that depends on derivatives no higher than second order. From our knowledge of gravitation as a manifestation of spacetime curvature, we might also expect L to be derived from the curvature tensor. Thus, in searching for the simplest plausible variational principle for gravitation, one is immediately led to the *Einstein–Hilbert* action

$$\boxed{S_{\text{EH}} = \int_{\mathcal{R}} R\sqrt{-g}\, d^4x.} \tag{19.32}$$

Since the corresponding Lagrangian $L_{\text{EH}} = R$ now depends on the elements of the metric tensor, it is more convenient to work in terms of the Lagrangian density $\mathcal{L}_{\text{EH}} = R\sqrt{-g}$. The resulting EL equations thus take the form (19.14), which in this case reads

$$\frac{\partial \mathcal{L}}{\partial g_{\mu\nu}} - \partial_\sigma\left[\frac{\partial \mathcal{L}}{\partial(\partial_\sigma g_{\mu\nu})}\right] + \partial_\rho\partial_\sigma\left[\frac{\partial \mathcal{L}}{\partial(\partial_\rho\partial_\sigma g_{\mu\nu})}\right] = 0.$$

Unfortunately, the task of evaluating each term in the above equation involves a formidable amount of algebra, albeit straightforward. We shall therefore not pursue this approach any further. Instead, we shall derive the corresponding field equations by considering directly the variation in the action resulting from a variation in the metric tensor.

Let us therefore consider a variation in the metric tensor given by

$$g_{\mu\nu} \to g_{\mu\nu} + \delta g_{\mu\nu},$$

where $\delta g_{\mu\nu}$ *and its first derivative* vanish on the boundary $\partial\mathcal{R}$ of the region $\mathcal{R}$. It will prove useful also to determine the corresponding variation $\delta g^{\mu\nu}$ in the inverse metric components. This is most easily achieved by noting that $g^{\mu\rho}g_{\rho\nu} = \delta^\mu_\nu$ and using the fact that the constant tensor δ^μ_ν does not change under a variation. To first order in the variation, one may therefore write

$$\delta g^{\mu\rho} g_{\rho\nu} + g^{\mu\rho} \delta g_{\rho\nu} = 0. \tag{19.33}$$

Multiplying through by $g^{\nu\sigma}$, relabelling indices and rearranging, one obtains

$$\delta g^{\mu\nu} = -g^{\mu\rho} g^{\nu\sigma} \delta g_{\rho\sigma}.$$

Writing the Ricci scalar as $R = g^{\mu\nu}R_{\mu\nu}$, the first-order variation in the Einstein–Hilbert action (19.32) can be written as

$$\begin{aligned}\delta S_{\text{EH}} &= \int_{\mathcal{R}} \delta g^{\mu\nu} R_{\mu\nu}\sqrt{-g}\, d^4x + \int_{\mathcal{R}} g^{\mu\nu}\delta R_{\mu\nu}\sqrt{-g}\, d^4x + \int_{\mathcal{R}} g^{\mu\nu} R_{\mu\nu}\, \delta(\sqrt{-g})\, d^4x \\ &\equiv \delta S_1 + \delta S_2 + \delta S_3. \end{aligned} \tag{19.34}$$

To derive the field equations, we need to factor out the variation $\delta g^{\mu\nu}$ in the second and third integrals. Let us first focus on the second term and write the variation $\delta R_{\mu\nu}$ in terms of the variation $\delta g^{\mu\nu}$ in the metric tensor. It is in fact more illuminating, and no more work, to determine the variation $\delta R^{\sigma}{}_{\mu\nu\rho}$ in the full curvature tensor, from which the corresponding variation in the Ricci tensor can be obtained immediately by contraction. The curvature tensor is given by

$$R^{\sigma}{}_{\mu\nu\rho} = \partial_\nu \Gamma^{\sigma}{}_{\mu\rho} - \partial_\rho \Gamma^{\sigma}{}_{\mu\nu} + \Gamma^{\tau}{}_{\mu\rho}\Gamma^{\sigma}{}_{\tau\nu} - \Gamma^{\tau}{}_{\mu\nu}\Gamma^{\sigma}{}_{\tau\rho}.$$

Let us first consider the variation in the curvature tensor resulting from an arbitrary variation in the connection coefficients,

$$\Gamma^{\sigma}{}_{\mu\nu} \to \Gamma^{\sigma}{}_{\mu\nu} + \delta\Gamma^{\sigma}{}_{\mu\nu}.$$

It is worth noting that the variation $\delta\Gamma^{\sigma}{}_{\mu\nu}$ is the difference of two connections and is therefore a tensor. As is often the case in proving tensor identities, it is easiest to work in local geodesic coordinates at some arbitrary point P. In such a coordinate system $\Gamma^{\sigma}{}_{\mu\nu}(P) = 0$, and so at the point P we have

$$\delta R^{\sigma}{}_{\mu\nu\rho} = \partial_\nu \left(\delta\Gamma^{\sigma}{}_{\mu\rho}\right) - \partial_\rho \left(\delta\Gamma^{\sigma}{}_{\mu\nu}\right).$$

Moreover, partial derivatives and covariant derivatives coincide at P and so

$$\delta R^{\sigma}{}_{\mu\nu\rho} = \nabla_\nu \left(\delta\Gamma^{\sigma}{}_{\mu\rho}\right) - \nabla_\rho \left(\delta\Gamma^{\sigma}{}_{\mu\nu}\right). \tag{19.35}$$

We now see, however, that the quantities on the right-hand side are tensors, and therefore (19.35) holds not only in geodesic coordinates at P but in any arbitrary coordinate system. Since the point P was chosen arbitrarily, the result (19.35) thus holds generally and is known as the *Palatini equation*. The corresponding variation in the Ricci tensor is obtained by contracting on σ and ρ in (19.35) to give

$$\delta R_{\mu\nu} = \nabla_\nu \left(\delta\Gamma^\sigma{}_{\mu\sigma}\right) - \nabla_\sigma \left(\delta\Gamma^\sigma{}_{\mu\nu}\right). \tag{19.36}$$

We may therefore write the second term on the right-hand side of (19.34) as

$$\begin{aligned}\delta S_2 &= \int_{\mathcal{R}} g^{\mu\nu} \left[\nabla_\nu \left(\delta\Gamma^\sigma{}_{\mu\sigma}\right) - \nabla_\sigma \left(\delta\Gamma^\sigma{}_{\mu\nu}\right)\right] \sqrt{-g}\, d^4x \\ &= \int_{\mathcal{R}} \nabla_\nu \left(g^{\mu\nu}\delta\Gamma^\sigma{}_{\mu\sigma} - g^{\mu\sigma}\delta\Gamma^\nu{}_{\mu\sigma}\right) \sqrt{-g}\, d^4x,\end{aligned}$$

where in the last line we have used the fact that the covariant derivative of the metric vanishes and we have relabelled indices in the second term of the integrand. Using the divergence theorem (19.19), however, we may write δS_2 as a surface integral over the boundary $\partial\mathcal{R}$, which vanishes provided that the variation in the connection vanishes on the boundary. This means that variations in the metric tensor *and* in its first derivatives vanish on $\partial\mathcal{R}$.

Let us now turn our attention to the third term δS_3 in (19.34), in which we must express $\delta\sqrt{-g}$ in terms of the variation $\delta g^{\mu\nu}$. Recalling that $g = \det[g_{\mu\nu}]$, we note that the cofactor of the element $g_{\mu\nu}$ in this determinant is $gg_{\mu\nu}$. It follows that

$$\delta g = g g^{\mu\nu} \delta g_{\mu\nu} = -g g_{\mu\nu} \delta g^{\mu\nu},$$

where in the second equality we have used the result (19.33). Thus, we have

$$\delta\sqrt{-g} = -\tfrac{1}{2}(-g)^{-1/2}\delta g = -\tfrac{1}{2}\sqrt{-g}\, g_{\mu\nu}\delta g^{\mu\nu}. \tag{19.37}$$

Substituting this expression into the third term δS_3 in (19.34) and remembering that $\delta S_2 = 0$, we finally discover that the variation in the Einstein–Hilbert action may be written as

$$\delta S_{\text{EH}} = \int_{\mathcal{R}} \left(R_{\mu\nu} - \tfrac{1}{2} g_{\mu\nu} R\right) \delta g^{\mu\nu} \sqrt{-g}\, d^4x. \tag{19.38}$$

By demanding that $\delta S_{\text{EH}} = 0$ and using the fact that the variation $\delta g^{\mu\nu}$ is arbitrary, we thus recover Einstein's field equations *in vacuo*:

$$\boxed{G_{\mu\nu} \equiv R_{\mu\nu} - \tfrac{1}{2} g_{\mu\nu} R = 0.} \tag{19.39}$$

This is an impressive result, since we have obtained the field equations of general relativity by varying an action (19.32) to which we were led very naturally on the grounds of symmetry and simplicity. This illustrates the power of the variational approach and should be contrasted with the more heuristic approach we had to employ in Section 8.4. Moreover, if one were willing to consider more complicated actions, the variational formalism suggests how Einstein's theory might be modified by adding to the Lagrangian terms proportional to R^2, R^3, etc. The formalism also provides a means for investigating alternative gravitational Lagrangians. For example, the choice $L = R_{\mu\nu\rho\sigma}R^{\rho\mu\nu\sigma}$ leads to an alternative self-consistent theory of gravity considered by Eddington.

19.9 An equivalent action for general relativity *in vacuo*

The Einstein–Hilbert action (19.32) differs from the action (19.25) for scalar field theory and the action (19.29) for electromagnetism in that it depends on second-order derivatives of the dynamical fields. It is therefore of interest to consider whether the empty-space gravitational field equations can be derived from an action that depends only on the metric tensor and its *first* derivatives. As stated in the previous section, however, R is the *only* scalar derivable from the metric tensor that depends on derivatives no higher than second order, so at first our goal appears unattainable. Nevertheless, as we will show, we may use the notion of equivalent actions discussed in Section 19.5 to circumvent this difficulty, albeit in a way that results in a new action that is not a scalar under general coordinate transformations.

The Lagrangian density $\mathcal{L}_{\text{EH}} = \sqrt{-g}R$ in the Einstein–Hilbert action (19.32) may be written as

$$\begin{aligned}\mathcal{L}_{\text{EH}} &= \sqrt{-g}g^{\mu\nu}R_{\mu\nu}\\ &= \sqrt{-g}g^{\mu\nu}\left(\partial_\nu\Gamma^\sigma{}_{\mu\sigma} - \partial_\sigma\Gamma^\sigma{}_{\mu\nu} + \Gamma^\tau{}_{\mu\sigma}\Gamma^\sigma{}_{\tau\nu} - \Gamma^\tau{}_{\mu\nu}\Gamma^\sigma{}_{\tau\sigma}\right)\\ &= \sqrt{-g}g^{\mu\nu}\left(\partial_\nu\Gamma^\sigma{}_{\mu\sigma} - \partial_\sigma\Gamma^\sigma{}_{\mu\nu}\right) - \bar{\mathcal{L}},\end{aligned} \tag{19.40}$$

where in the last line we have defined a new Lagrangian density

$$\boxed{\bar{\mathcal{L}} \equiv \sqrt{-g}\,g^{\mu\nu}\left(\Gamma^\tau{}_{\mu\nu}\Gamma^\sigma{}_{\tau\sigma} - \Gamma^\tau{}_{\mu\sigma}\Gamma^\sigma{}_{\tau\nu}\right),} \tag{19.41}$$

which clearly depends only on the metric and its first derivatives. (Note that the minus sign in (19.40) is for later convenience.) By relabelling indices and using Leibnitz' rule for the differentiation of products, we can write the first term in (19.40) as

$$\begin{aligned}\sqrt{-g}g^{\mu\nu}\left(\partial_\nu\Gamma^\sigma{}_{\mu\sigma} - \partial_\sigma\Gamma^\sigma{}_{\mu\nu}\right) &= \partial_\mu\left(\sqrt{-g}g^{\mu\nu}\Gamma^\sigma{}_{\nu\sigma} - \sqrt{-g}g^{\sigma\nu}\Gamma^\mu{}_{\sigma\nu}\right)\\ &\quad -\partial_\nu(\sqrt{-g}g^{\mu\nu})\Gamma^\sigma{}_{\mu\sigma} + \partial_\sigma(\sqrt{-g}g^{\mu\nu})\Gamma^\sigma{}_{\mu\nu}.\end{aligned} \tag{19.42}$$

To evaluate the last two terms on the right-hand side, we note that

$$\partial_\sigma(\sqrt{-g}g^{\mu\nu}) = \tfrac{1}{2}(-g)^{-1/2}g^{\mu\nu}\partial_\sigma g + \sqrt{-g}\partial_\sigma g^{\mu\nu}. \tag{19.43}$$

Using the result (3.24) derived in Section 3.10, we have $\partial_\sigma g = 2g\Gamma^\rho{}_{\rho\sigma}$ and, since the covariant derivative of the metric (or its inverse) is zero,

$$\nabla_\sigma g^{\mu\nu} = \partial_\sigma g^{\mu\nu} + \Gamma^\mu{}_{\rho\sigma}g^{\rho\nu} + \Gamma^\nu{}_{\rho\sigma}g^{\mu\rho} = 0.$$

Thus, we may write (19.43) as

$$\partial_\sigma(\sqrt{-g}g^{\mu\nu}) = \sqrt{-g}\left(\Gamma^\rho{}_{\rho\sigma}g^{\mu\nu} - \Gamma^\mu{}_{\rho\sigma}g^{\rho\nu} - \Gamma^\nu{}_{\rho\sigma}g^{\mu\rho}\right).$$

Substituting this result into the last two terms on the right-hand side of (19.42) (contracting on ν and σ for the first of these terms), relabelling indices and simplifying, one finds that

$$\sqrt{-g}g^{\mu\nu}\left(\partial_\nu\Gamma^\sigma{}_{\mu\sigma} - \partial_\sigma\Gamma^\sigma{}_{\mu\nu}\right) = \partial_\mu\left(\sqrt{-g}g^{\mu\nu}\Gamma^\sigma{}_{\nu\sigma} - \sqrt{-g}g^{\sigma\nu}\Gamma^\mu{}_{\sigma\nu}\right) + 2\bar{\mathcal{L}}.$$

Thus, we finally discover that the Einstein–Hilbert Lagrangian density (19.40) can be written

$$\boxed{\mathcal{L}_{\text{EH}} = \bar{\mathcal{L}} + \partial_\mu\left(\sqrt{-g}g^{\mu\nu}\Gamma^\sigma{}_{\nu\sigma} - \sqrt{-g}g^{\sigma\nu}\Gamma^\mu{}_{\sigma\nu}\right),} \tag{19.44}$$

where $\bar{\mathcal{L}}$ is given by (19.41).

We see immediately, however, that the second term in (19.44) is a total derivative, and so $\mathcal{L}_{\text{EH}}$ and $\bar{\mathcal{L}}$ are related by an expression of the form (19.22). The two Lagrangian densities are therefore *equivalent*. As discussed in Section 19.5, variation of the new action

$$\boxed{\bar{S} = \int_{\mathcal{R}} g^{\mu\nu}\left(\Gamma^\tau{}_{\mu\nu}\Gamma^\sigma{}_{\tau\sigma} - \Gamma^\tau{}_{\mu\sigma}\Gamma^\sigma{}_{\tau\nu}\right)\sqrt{-g}\,d^4x} \tag{19.45}$$

will thus lead to the *same* field equations as did the Einstein–Hilbert action S_{EH}, provided that the variation in the metric *and* its first derivative vanish on the boundary $\partial\mathcal{R}$. Thus, the variation of (19.45) will again yield Einstein's field equations *in vacuo* (which may be checked directly), but the action depends only on the metric and its first derivatives. There is, however, a price to pay in adopting the above result, since the new action $\bar{S}$ is easily shown *not* to be a scalar with respect to general coordinate transformations (see the discussion in Section 19.5).

19.10 The Palatini approach for general relativity *in vacuo*

A more elegant and illuminating method for obtaining the Einstein field equations from an action depending only on dynamical fields and their first derivatives is provided by the *Palatini approach*, which we now discuss. In this formalism one treats the metric $g_{\mu\nu}$ and the connection $\Gamma^{\sigma}{}_{\mu\nu}$ as *independent* fields. In other words, one does not assume any explicit relationship between the metric and the connection.

We begin again with the Einstein–Hilbert Lagrangian density

$$\mathcal{L}_{\rm EH} = \sqrt{-g}g^{\mu\nu}R_{\mu\nu} = \sqrt{-g}g^{\mu\nu}\left(\partial_\nu\Gamma^{\sigma}{}_{\mu\sigma} - \partial_\sigma\Gamma^{\sigma}{}_{\mu\nu} + \Gamma^{\tau}{}_{\mu\sigma}\Gamma^{\sigma}{}_{\tau\nu} - \Gamma^{\tau}{}_{\mu\nu}\Gamma^{\sigma}{}_{\tau\sigma}\right),$$

which we now consider as a function of the metric, the connection and first derivatives of the connection, i.e. $\mathcal{L}_{\rm EH} = \mathcal{L}_{\rm EH}(g_{\mu\nu}, \Gamma^{\sigma}{}_{\mu\nu}, \partial_\rho\Gamma^{\sigma}{}_{\mu\nu})$. Let us first consider the variation in the action resulting from a variation in the metric alone. This may be written as

$$\delta S_{\rm EH} = \int_{\mathcal{R}} \delta(\sqrt{-g}g^{\mu\nu})R_{\mu\nu}\, d^4x.$$

Demanding that $\delta S_{\rm EH} = 0$ for an arbitrary variation in the metric, we immediately find that

$$\boxed{R_{\mu\nu} = 0,}$$

which gives the Einstein field equations *in vacuo*.

Let us now consider varying the action with respect to the connection, which yields

$$\begin{aligned}\delta S_{\rm EH} &= \int_{\mathcal{R}} \sqrt{-g}g^{\mu\nu}\,\delta R_{\mu\nu}\, d^4x \\ &= \int_{\mathcal{R}} \sqrt{-g}g^{\mu\nu}\left[\nabla_\nu(\delta\Gamma^{\sigma}{}_{\mu\sigma}) - \nabla_\sigma(\delta\Gamma^{\sigma}{}_{\mu\nu})\right] d^4x, \qquad (19.46)\end{aligned}$$

where in the second line we have used the contracted version (19.36) of the Palatini equation. Using Leibnitz' theorem for the differentiation of products and relabelling some dummy indices, we may write (19.46) as

$$\begin{aligned}\delta S_{\rm EH} &= \int_{\mathcal{R}} \nabla_\nu\left(g^{\mu\nu}\delta\Gamma^{\sigma}{}_{\mu\sigma} - g^{\mu\sigma}\delta\Gamma^{\nu}{}_{\mu\sigma}\right)\sqrt{-g}\, d^4x \\ &\quad + \int_{\mathcal{R}} \left[(\nabla_\rho g^{\mu\nu})\,\delta\Gamma^{\rho}{}_{\mu\nu} - (\nabla_\nu g^{\mu\nu})\,\delta\Gamma^{\rho}{}_{\mu\rho}\right]\sqrt{-g}\, d^4x, \qquad (19.47)\end{aligned}$$

where we note that we have *not* assumed that the covariant derivative of the metric vanishes, since we have not (yet) specified any relationship between the connection and the metric. Using the divergence theorem (19.19), we may write the first integral on the right-hand side of (19.47) as a surface integral over the boundary $\partial\mathcal{R}$, which vanishes if we assume that the variation in the connection vanishes on the boundary. Relabelling some dummy indices in the second integral on the right-hand side of (19.47), we thus find

$$\delta S_{\text{EH}} = -\int_{\mathcal{R}} \left(\delta^\nu_\rho \nabla_\sigma g^{\mu\sigma} - \nabla_\rho g^{\mu\nu}\right) \delta\Gamma^\rho{}_{\mu\nu} \sqrt{-g}\, d^4x. \tag{19.48}$$

Since we are assuming that the manifold is torsionless, the variation $\delta\Gamma^\rho{}_{\mu\nu}$ in the connection, although arbitrary, must be symmetric in its lower two indices. As a result, demanding that $\delta S_{\text{EH}} = 0$ only requires the symmetric part of the term in parentheses in (19.48) to vanish; when contracted with $\delta\Gamma^\rho{}_{\mu\nu}$, the antisymmetric part will automatically equal zero. Thus, stationarity of the action requires that

$$\tfrac{1}{2}\delta^\nu_\rho \nabla_\sigma g^{\mu\sigma} + \tfrac{1}{2}\delta^\mu_\rho \nabla_\sigma g^{\nu\sigma} - \nabla_\rho g^{\mu\nu} = 0.$$

We thus deduce that $\nabla_\sigma g^{\mu\nu} = 0$, which in turn implies that $\nabla_\sigma g_{\mu\nu} = 0$. Hence by demanding stationarity of the Einstein–Hilbert action with respect to variations in the (symmetric) connection, we have *derived* that the covariant derivative of the metric must vanish. We may thus write

$$\partial_\rho g_{\mu\nu} = \Gamma^\sigma{}_{\mu\rho} g_{\sigma\nu} + \Gamma^\sigma{}_{\nu\rho} g_{\mu\sigma}.$$

Cyclically permuting the free indices to obtain similiar expressions for $\partial_\nu g_{\rho\mu}$ and $\partial_\mu g_{\nu\rho}$, combining the results and contracting with $g^{\rho\sigma}$ one finds that

$$\boxed{\Gamma^\rho{}_{\mu\nu} = \tfrac{1}{2} g^{\rho\sigma}(\partial_\mu g_{\sigma\nu} + \partial_\nu g_{\mu\sigma} - \partial_\rho g_{\mu\nu}),}$$

and hence the connection must be the metric connection.

19.11 General relativity in the presence of matter

So far we have confined our attention to deriving the gravitational field equations *in vacuo*. We now consider how the full Einstein equations, in the presence of other (non-gravitational) fields, may be obtained by a variational principle. In order to accommodate this generalisation, one simply needs to add an extra term to the action to give

$$S = \frac{1}{2\kappa} S_{\text{EH}} + S_{\text{M}} = \int_{\mathcal{R}} \left(\frac{1}{2\kappa}\mathcal{L}_{\text{EH}} + \mathcal{L}_{\text{M}}\right) d^4x, \tag{19.49}$$

where the Einstein–Hilbert action S_{EH} is considered as a function of the metric and of its first- and second-order derivatives (as in Section 19.8). S_{M} is the 'matter' action for any non-gravitational fields present, and $\kappa = 8\pi G/c^4$. The factor $1/(2\kappa)$ in (19.49) is chosen for later convenience.

Let us now consider varying the action with respect to the (inverse) metric, to obtain

$$\frac{1}{2\kappa}\frac{\delta \mathcal{L}_{\mathrm{EH}}}{\delta g^{\mu\nu}} + \frac{\delta \mathcal{L}_{\mathrm{M}}}{\delta g^{\mu\nu}} = 0.$$

From (19.38), we see that

$$\frac{\delta \mathcal{L}_{\mathrm{EH}}}{\delta g^{\mu\nu}} = \sqrt{-g}\, G_{\mu\nu},$$

where $G_{\mu\nu} = R_{\mu\nu} - \frac{1}{2}g_{\mu\nu}R$ is the Einstein tensor. Thus, if we make the bold assertion that the energy–momentum tensor of the non-gravitational fields (or 'matter') is given by

$$T_{\mu\nu} = \frac{2}{\sqrt{-g}}\frac{\delta \mathcal{L}_{\mathrm{M}}}{\delta g^{\mu\nu}}, \tag{19.50}$$

then we recover the full Einstein equations

$$\boxed{G_{\mu\nu} = -\kappa T_{\mu\nu}.}$$

The definition (19.50) of the 'matter' energy–momentum tensor may appear to be somewhat arbitrary. Nevertheless, as we show in the next section, this tensor has all the properties required of an energy–momentum tensor.

19.12 The dynamical energy–momentum tensor

The quantities $T_{\mu\nu}$ defined in (19.50) are clearly the components of a tensor, which is known more properly as the *dynamical* energy–momentum tensor. From the definition we also see immediately that $T_{\mu\nu}$ is a symmetric tensor, as is required by the full Einstein equations (19.39). Most importantly, however, we now show that it obeys the conservation equation $\nabla_\mu T^{\mu\nu} = 0$.

From the definition (19.50), the variation in the matter action resulting from a variation in the metric is given by

$$\begin{aligned}\delta S_{\mathrm{M}} \equiv \int_{\mathcal{R}} \frac{\delta \mathcal{L}_{\mathrm{M}}}{\delta g^{\mu\nu}}\,\delta g^{\mu\nu}\, d^4x &= \tfrac{1}{2}\int_{\mathcal{R}} T_{\mu\nu}\,\delta g^{\mu\nu}\sqrt{-g}\, d^4x \\ &= -\tfrac{1}{2}\int_{\mathcal{R}} T^{\mu\nu}\,\delta g_{\mu\nu}\sqrt{-g}\, d^4x,\end{aligned} \tag{19.51}$$

where, in the last equality, we have written δS_{M} in terms of the contravariant components $T^{\mu\nu}$ of the energy–momentum tensor for later convenience, using the result (19.33). Let us now consider making an infinitesimal general coordinate transformation

$$x'^{\mu} = x^{\mu} + \xi^{\mu}(x), \tag{19.52}$$

where $\xi^\mu(x)$ is an infinitesimal smooth vector field. Since the action S_M is, by construction, a covariant scalar, then we must have $\delta S_M = 0$ under the coordinate transformation. We know, however, that the metric coefficients must transform as

$$g'_{\mu\nu}(x') = \frac{\partial x^\rho}{\partial x'^\mu}\frac{\partial x^\sigma}{\partial x'^\nu} g_{\rho\sigma}(x) = [\delta^\rho_\mu - \partial_\mu \xi^\rho(x)][\delta^\sigma_\nu - \partial_\nu \xi^\sigma(x)]\, g_{\rho\sigma}(x) \qquad (19.53)$$

$$= g_{\mu\nu}(x) - g_{\rho\nu}(x)\partial_\mu \xi^\rho(x) - g_{\mu\sigma}(x)\partial_\nu \xi^\sigma(x), \qquad (19.54)$$

to first order in ξ^μ, where we have used the expression (17.3) for the transformation matrix corresponding to the infinitesimal coordinate transformation (19.52). We have explicitly included the dependence on x and x' in (19.54), since it is crucial to determining the corresponding variation $\delta g_{\mu\nu}$. As mentioned in Section 19.3, this variation is only of the functional form of the fields $g_{\mu\nu}$. Thus, we have

$$\begin{aligned}\delta g_{\mu\nu}(x) \equiv g'_{\mu\nu}(x) - g_{\mu\nu}(x) &= [g'_{\mu\nu}(x') - g_{\mu\nu}(x)] - [g'_{\mu\nu}(x') - g'_{\mu\nu}(x)] \\ &= [g'_{\mu\nu}(x') - g_{\mu\nu}(x)] - \xi^\sigma(x)\partial_\sigma g'_{\mu\nu}(x) \\ &= [g'_{\mu\nu}(x') - g_{\mu\nu}(x)] - \xi^\sigma(x)\partial_\sigma g_{\mu\nu}(x),\end{aligned}$$

to first order in ξ^μ. Using the expression (19.54) and dropping the explicit dependence on x, we find that

$$\delta g_{\mu\nu} = -g_{\rho\nu}\partial_\mu \xi^\rho - g_{\mu\rho}\partial_\nu \xi^\rho - \xi^\rho \partial_\rho g_{\mu\nu} = -(\nabla_\mu \xi_\nu + \nabla_\nu \xi_\mu),$$

where, in the second equality, we have rewritten the partial derivatives in terms of covariant derivatives, cancelled matching terms involving connection coefficients and used the fact that $\nabla_\rho g_{\mu\nu} = 0$.

Substituting this result into (19.51) and remembering that $\delta S_M = 0$ under a coordinate transformation and that $T^{\mu\nu}$ is symmetric, we have

$$\delta S_M = \int_{\mathcal{R}} T^{\mu\nu}(\nabla_\mu \xi_\nu)\sqrt{-g}\, d^4x = 0.$$

Using Leibnitz' theorem for the covariant differentiation of a product, we write

$$\delta S_M = \int_{\mathcal{R}} \nabla_\mu (T^{\mu\nu}\xi_\nu)\sqrt{-g}\, d^4x - \int_{\mathcal{R}} (\nabla_\mu T^{\mu\nu})\xi_\nu \sqrt{-g}\, d^4x = 0. \qquad (19.55)$$

We may use the divergence theorem (19.19) to write the first integral as a surface integral over the boundary $\partial\mathcal{R}$ in the usual manner. Assuming that the functions $\xi^\nu(x)$ vanish on the boundary $\partial\mathcal{R}$ this surface integral vanishes, leaving only the second integral in (19.55). Since the $\xi^\mu(x)$ are arbitrary, however, one immediately finds that

$$\boxed{\nabla_\mu T^{\mu\nu} = 0,}$$

and so the covariant divergence of the energy–momentum tensor vanishes, as required. Thus, we see that the general covariance of the matter action implies energy–momentum conservation in the same way as the gauge invariance of the action (19.29) for electromagnetism implies charge conservation (see Section 19.7).

Now that we have shown that the tensor $T_{\mu\nu}$ defined by (19.50) has the appropriate properties of an energy–momentum tensor, we may calculate the explicit form of this tensor for some specific 'matter' actions. Let us begin by considering the action (19.25) for a real scalar field ϕ. Varying this action now with respect to the (inverse) metric, rather than the field ϕ, we obtain

$$\begin{aligned}\delta S_\phi &= \int_{\mathcal{R}} \{[\tfrac{1}{2}\delta g^{\mu\nu}(\nabla_\mu\phi)(\nabla_\nu\phi)]\sqrt{-g} \\ &\qquad + [\tfrac{1}{2}g^{\mu\nu}(\nabla_\mu\phi)(\nabla_\nu\phi) - V(\phi)]\,\delta(\sqrt{-g})\}\, d^4x \\ &= \int_{\mathcal{R}} \{\tfrac{1}{2}(\nabla_\mu\phi)(\nabla_\nu\phi) - \tfrac{1}{2}g_{\mu\nu}[\tfrac{1}{2}g^{\rho\sigma}(\nabla_\rho\phi)(\nabla_\sigma\phi) - V(\phi)]\}\,\delta g^{\mu\nu}\sqrt{-g}\, d^4x,\end{aligned}$$

where in the last line we have used the expression (19.37) for $\delta\sqrt{-g}$. Comparing the above expression with that in (19.51), we immediately see that the energy–momentum tensor for a real scalar field is given by

$$\boxed{T^{(\phi)}_{\mu\nu} = (\nabla_\mu\phi)(\nabla_\nu\phi) - g_{\mu\nu}[\tfrac{1}{2}(\nabla_\sigma\phi)(\nabla^\sigma\phi) - V(\phi)],}$$

which agrees with the expression (16.7) adopted in our discussion of inflation in Section 16.3.

We may also obtain the energy–momentum tensor for the electromagnetic field in a similar manner. From (19.29) and (19.28), in the absence of sources we may write the action for electromagnetism as

$$S_{\text{EM}} = -\frac{1}{4\mu_0}\int_{\mathcal{R}} g^{\mu\rho}g^{\nu\sigma}F_{\rho\sigma}F_{\mu\nu}\sqrt{-g}\, d^4x,$$

where $F_{\mu\nu} = \partial_\mu A_\nu - \partial_\nu A_\mu$ and so does not depend on the metric. Varying this action with respect to the (inverse) metric, we have

$$\begin{aligned}\delta S_{\text{EM}} &= -\frac{1}{4\mu_0}\int_{\mathcal{R}} [\delta(g^{\mu\rho}g^{\nu\sigma})F_{\rho\sigma}F_{\mu\nu}\sqrt{-g} + F_{\rho\sigma}F^{\rho\sigma}\delta(\sqrt{-g})]\, d^4x \\ &= -\frac{1}{4\mu_0}\int_{\mathcal{R}} (2g^{\rho\sigma}F_{\mu\rho}F_{\nu\sigma} - \tfrac{1}{2}g_{\mu\nu}F_{\rho\sigma}F^{\rho\sigma})\,\delta g^{\mu\nu}\sqrt{-g}\, d^4x,\end{aligned}$$

where in the second equality we have substituted the expression (19.37) for $\delta\sqrt{-g}$ and relabelled some dummy indices. Comparing the above expression with

(19.51), we find that the energy–momentum tensor for the electromagnetic field is given by

$$T^{(\mathrm{EM})}_{\mu\nu} = -\mu_0^{-1}\left(F_{\mu\rho}F_\nu{}^\rho - \tfrac{1}{4}g_{\mu\nu}F_{\rho\sigma}F^{\rho\sigma}\right),$$

which agrees with the expression derived in Exercise 8.3.

Finally, we note that in field theory it is common to define also a *canonical* energy–momentum tensor, which is based on *Noether's theorem*.[4] This states that for every symmetry of the action there exists a corresponding conserved quantity. In particular, if an action is invariant under a spacetime translation, characterised by a coordinate transformation of the form $x^\mu \to x^\mu + a^\mu$ in which the vector a^μ does not depend on spacetime position, then one can define a tensor $S^{\mu\nu}$ that obeys $\nabla_\mu S^{\mu\nu} = 0$. It is this tensor that is usually called the canonical energy–momentum tensor. Unfortunately, there are some drawbacks in using it, since it is not necessarily symmetric (although it can be made so) or gauge invariant.

Exercises

19.1 If $\rho(x)$ and $\tau(x)$ are the local line density and tension of a string, show that the kinetic and potential energies of the string for small displacements $\phi(t, x)$ are given by

$$T = \int_0^l \tfrac{1}{2}\rho\left(\frac{\partial\phi}{\partial t}\right)^2 dx \quad \text{and} \quad V = \int_0^l \tfrac{1}{2}\tau\left(\frac{\partial\phi}{\partial x}\right)^2 dx.$$

19.2 In classical field theory, the conjugate field momenta are defined in terms of the Lagrangian density $\mathcal{L}$ by

$$\pi_a \equiv \frac{\partial\mathcal{L}}{\partial\dot{\Phi}^a},$$

where $\dot{\Phi}^a \equiv \partial_0\Phi^a$ and x^0 is a timelike coordinate. The Hamiltonian density is then defined as

$$\mathcal{H} \equiv \pi_a\dot{\Phi}^a - \mathcal{L}.$$

Use the Euler–Lagrange equations to show that

$$\dot{\Phi}^a = \frac{\delta\mathcal{H}}{\delta\pi_a} \quad \text{and} \quad \dot{\pi}_a = -\frac{\delta\mathcal{H}}{\delta\Phi^a}.$$

19.3 Consider the quantity

$$E = \int_S \mathcal{H}\, d^3x,$$

[4] See, for example, L. H. Ryder, *Quantum Field Theory*, Cambridge University Press, 1985.

where $\mathcal{H}$ is the Hamiltonian density in Exercise 19.2 and the integral extends over some spacelike hypersurface S for which $x^0 = \text{constant}$. Setting $[x^\mu] \equiv (t, x^i)$ and using a dot to denote ∂_t, show that

$$\frac{dE}{dt} = \int_S \left[\frac{\partial \mathcal{H}}{\partial \Phi^a} \dot{\Phi}^a + \frac{\partial \mathcal{H}}{\partial \pi_a} \dot{\pi}_a + \frac{\partial \mathcal{H}}{\partial(\partial_i \Phi^a)} \partial_i \dot{\Phi}^a + \frac{\partial \mathcal{H}}{\partial t} \right] d^3x.$$

By integrating the third term in the integrand by parts, show that $dE/dt = 0$ provided that $\mathcal{H}$ does not depend explicitly on t.

19.4 Obtain an expression for the Hamiltonian density $\mathcal{H}$ for the string in Exercise 19.1. Hence show that the total energy E of the string is given by

$$E = \int_0^l \mathcal{H}\, dx,$$

and show explicitly that it is a constant of the motion.

19.5 A *relative tensor* of weight w transforms under a coordinate transformation as

$$\mathcal{T}'^{a\cdots}_{\ \ b\cdots} = J^{-w} \frac{\partial x'^a}{\partial x^c} \cdots \frac{\partial x'^b}{\partial x^d} \mathcal{T}^{c\cdots}_{\ \ d\cdots}$$

where J is the Jacobian of the transformation and is given by

$$J = \det \left[\frac{\partial x'^a}{\partial x^b} \right].$$

Show that the product of two relative tensors of weights w_1 and w_2 is a relative tensor of weight $w_1 + w_2$. Show further that $\sqrt{-g}$ is a relative scalar of weight $w = 1$ (called a scalar density).

19.6 For a field theory defined by the action $S = \int_{\mathcal{R}} \mathcal{L}\, d^4x$ show that, if $\mathcal{L}$ depends on first- and second-order derivatives of the fields, the Euler–Lagrange equations take the form

$$\frac{\delta \mathcal{L}}{\delta \Phi^a} = \frac{\partial \mathcal{L}}{\partial \Phi^a} - \partial_\mu \left[\frac{\partial \mathcal{L}}{\partial(\partial_\mu \Phi^a)} \right] + \partial_\mu \partial_\nu \left[\frac{\partial \mathcal{L}}{\partial(\partial_\mu \partial_\nu \Phi^a)} \right] = 0,$$

provided that the variations $\delta\Phi^a$ *and* their first derivatives vanish on the boundary $\partial\mathcal{R}$. How do the Euler–Lagrange equations generalise when $\mathcal{L}$ depends on higher-order derivatives of the fields? What assumptions are required regarding the value of the variation $\delta\Phi^a$ and its derivatives on ∂R?

19.7 Consider a local field theory for which the action has the form

$$S = \int_{\mathcal{R}} \mathcal{L}[\Phi^a(x), \partial_\mu \Phi^a(x)]\, d^4x.$$

Under an infinitesimal general coordinate transformation $x'^\mu = x^\mu + \xi^\mu(x)$, the variation in the action is given by

$$\delta S = \int_{\mathcal{R}'} \mathcal{L}'[\Phi'^a(x'), \partial_\mu \Phi'^a(x')]\, d^4x' - \int_{\mathcal{R}} \mathcal{L}[\Phi^a(x), \partial_\mu \Phi^a(x)]\, d^4x.$$

Adopting the shorthand notation $\delta S = \int_{\mathcal{R}'} \mathcal{L}'(x')\,d^4x' - \int_{\mathcal{R}} \mathcal{L}(x)\,d^4x$, show that

$$\delta S = \int_{\mathcal{R}} [\Delta\mathcal{L}(x) + \mathcal{L}(x)\partial_\mu \xi^\mu(x)]\,d^4x = \int_{\mathcal{R}} \{\delta\mathcal{L}(x) + \partial_\mu[\mathcal{L}(x)\xi^\mu(x)]\}\,d^4x,$$

where $\Delta\mathcal{L}(x) = \mathcal{L}'(x') - \mathcal{L}(x)$ and $\delta\mathcal{L}(x) = \mathcal{L}'(x) - \mathcal{L}(x)$.

19.8 Suppose that the action in Exercise 19.7 is invariant under the given coordinate transformation, so that $\delta S = 0$. Since the range of integration $\mathcal{R}$ can be chosen arbitrarily, show by writing

$$\delta\mathcal{L} = \frac{\partial\mathcal{L}}{\partial\Phi^a}\delta\Phi^a + \frac{\partial\mathcal{L}}{\partial(\partial_\mu\Phi^a)}\delta(\partial_\mu\Phi^a),$$

or otherwise, that

$$\left\{\frac{\partial\mathcal{L}}{\partial\Phi^a} - \partial_\mu\left[\frac{\partial\mathcal{L}}{\partial(\partial_\mu\Phi^a)}\right]\right\}\delta\Phi^a + \partial_\mu\left[\frac{\partial\mathcal{L}}{\partial(\partial_\mu\Phi^a)}\delta\Phi^a + \mathcal{L}\xi^\mu\right] = 0.$$

Hence show that the invariance of the action under the given coordinate transformation implies that $\partial_\mu j^\mu = 0$, where

$$j^\mu = \frac{\partial\mathcal{L}}{\partial(\partial_\mu\Phi^a)}\Delta\Phi^a - \left[\frac{\partial\mathcal{L}}{\partial(\partial_\mu\Phi^a)}\partial_\nu\Phi^a - \delta^\mu_\nu\mathcal{L}\right]\xi^\nu,$$

in which $\Delta\Phi^a(x) = \Phi'^a(x') - \Phi^a(x)$. This result is known as *Noether's theorem*.

19.9 Use your answer to Exercise 19.8 to show that invariance of the action under the infinitesimal translation $x'^\mu = x^\mu + \epsilon^\mu$ implies that $\partial_\mu S^\mu{}_\nu = 0$, where

$$S^\mu{}_\nu = \frac{\partial\mathcal{L}}{\partial(\partial_\mu\Phi^a)}\partial_\nu\Phi^a - \delta^\mu_\nu\mathcal{L},$$

which is known as the *canonical energy–momentum tensor* of the fields Φ^a. Is $S^\mu{}_\nu$ necessarily symmetric in μ and ν?

19.10 For the field theory considered in Exercise 19.7, use the fact that $\mathcal{L}$ does not depend explicitly on the coordinates x^μ to write

$$\partial_\nu\mathcal{L} = \frac{\partial\mathcal{L}}{\partial\Phi^a}\partial_\nu\Phi^a + \frac{\partial\mathcal{L}}{\partial(\partial_\mu\Phi^a)}\partial_\nu\partial_\mu\Phi^a.$$

By multiplying the Euler–Lagrange equations by $\partial_\nu\Phi^a$ and summing over a, use the above result to show directly that $\partial_\mu S^\mu{}_\nu = 0$, where $S^\mu{}_\nu$ is the canonical energy–momentum tensor given in Exercise 19.9.

19.11 Consider the 'modified' energy–momentum tensor

$$\Theta^\mu{}_\nu = S^\mu{}_\nu + \partial_\sigma\psi^{\sigma\mu}{}_\nu,$$

where $S^\mu{}_\nu$ is the canonical energy–momentum tensor given in Exercise 19.9 and $\psi^{\sigma\mu}{}_\nu$ is any tensor that is antisymmetric in σ and μ. Show that $\partial_\mu\Theta^\mu{}_\nu = 0$ and that one can always arrange for $\Theta^\mu{}_\nu$ to be symmetric in μ and ν.

19.12 Consider a local field theory defined on Minkowski spacetime in an arbitrary coordinate system x^μ with metric $g_{\mu\nu}$. The action has the form

$$S = \int_{\mathcal{R}} L[\Phi^a(x), \nabla_\mu \Phi^a(x), g_{\mu\nu}(x)]\sqrt{-g}\, d^4x,$$

where the fields Φ^a are independent of the metric $g_{\mu\nu}$ and L is a scalar under general coordinate transformations. Use the fact that L does not depend explicitly on x^μ to write

$$\nabla_\nu L = \frac{\partial L}{\partial \Phi^a}\nabla_\nu \Phi^a + \frac{\partial L}{\partial(\nabla_\mu \Phi^a)}\nabla_\nu \nabla_\mu \Phi^a.$$

By multiplying the appropriate form of the Euler–Lagrange equations by $\nabla_\nu \Phi^a$, summing over a and noting that covariant derivatives commute in Minkowski spacetime, use the above result to show that $\nabla_\mu S^{\mu\nu} = 0$, where the *covariant* canonical energy–momentum tensor $S^{\mu\nu}$ is given by

$$S^{\mu\nu} = \frac{\partial L}{\partial(\nabla_\mu \Phi^a)}\nabla^\nu \Phi^a - g^{\mu\nu} L.$$

19.13 Consider the 'modified' energy–momentum tensor

$$\Theta^\mu{}_\nu = S^\mu{}_\nu + \nabla_\sigma \psi^{\sigma\mu}{}_\nu$$

where $S^\mu{}_\nu$ is the canonical energy–momentum tensor given in Exercise 19.12 and $\psi^{\sigma\mu}{}_\nu$ is any tensor that is antisymmetric in σ and μ. Show that, in a flat spacetime, $\nabla_\mu \Theta^\mu{}_\nu = 0$ and that one can always arrange for $\Theta^\mu{}_\nu$ to be symmetric in μ and ν.

19.14 In a four-dimensional spacetime, use the divergence theorem to show that

$$\int_{\mathcal{R}} \partial_\mu(\sqrt{-g}v^\mu)\, d^4x = \int_{\partial\mathcal{R}} n_\mu v^\mu \sqrt{-\gamma}\, d^3y,$$

where v^μ is an arbitrary vector field, γ is the determinant of the induced metric on the boundary in the coordinates y^i and n_μ is a unit normal to the boundary.

19.15 Consider a complex scalar field $\phi = (\phi_1 + i\phi_2)/\sqrt{2}$, where $\phi_i (i = 1, 2)$ are real scalar fields with potentials of the form $V = \frac{1}{2}m^2\phi_i^2$. Show that the Lagrangian for ϕ may be written as

$$L = g^{\mu\nu}(\nabla_\mu \phi)(\nabla_\nu \phi^*) - m\phi\phi^*,$$

where the asterisk denotes the complex conjugate. By varying ϕ and ϕ^* independently, show that

$$\Box^2\phi + m^2\phi = 0 \quad \text{and} \quad \Box^2\phi^* + m^2\phi^* = 0,$$

where $\Box^2 \equiv \nabla^\mu \nabla_\mu = g^{\mu\nu}\nabla_\mu \nabla_\nu$ is the *covariant* d'Alembertian operator.

19.16 In the theory of electromagnetism in arbitrary coordinates, the field tensor is defined by $F_{\mu\nu} = \nabla_\mu A_\nu - \nabla_\nu A_\mu$. Show directly that

$$F_{\mu\nu} = \partial_\mu A_\nu - \partial_\nu A_\mu$$

and that

$$\nabla_\sigma F_{\mu\nu} + \nabla_\nu F_{\sigma\mu} + \nabla_\mu F_{\nu\sigma} = \partial_\sigma F_{\mu\nu} + \partial_\nu F_{\sigma\mu} + \partial_\mu F_{\nu\sigma}.$$

Hence show that $F_{\mu\nu}$ automatically satisfies the relation

$$\nabla_\sigma F_{\mu\nu} + \nabla_\nu F_{\sigma\mu} + \nabla_\mu F_{\nu\sigma} = 0.$$

19.17 If $F_{\mu\nu} = \nabla_\mu A_\nu - \nabla_\nu A_\mu$, show that

$$\nabla_\sigma F_{\mu\nu} + \nabla_\nu F_{\sigma\mu} + \nabla_\mu F_{\nu\sigma} = 2(R^\rho{}_{\nu\mu\sigma} + R^\rho{}_{\mu\sigma\nu} + R^\rho{}_{\sigma\nu\mu})A_\rho,$$

where $R^\rho{}_{\nu\mu\sigma}$ is the Riemann tensor. Hence use the cyclic identity (7.17) to show that the above expression is zero.

19.18 An alternative Lagrangian for electromagnetism is given by

$$L = \frac{1}{4\mu_0} F_{\mu\nu} F^{\mu\nu} - \frac{1}{2\mu_0} F^{\mu\nu}(\nabla_\mu A_\nu - \nabla_\nu A_\mu) - j^\mu A_\mu,$$

where $F_{\mu\nu}$ and A^μ are considered as *independent* quantities (i.e. no functional relationship between them is assumed). By varying the corresponding action with respect to $F_{\mu\nu}$ and A^μ independently, show that the Euler–Lagrange equations yield

$$\nabla_\mu F^{\mu\nu} = \mu_0 j^\nu \quad \text{and} \quad F_{\mu\nu} = \nabla_\mu A_\nu - \nabla_\nu A_\mu.$$

19.19 The Lagrangian for a free *massive* vector field A^μ of mass m is

$$L = -\tfrac{1}{4} g^{\mu\rho} g^{\nu\sigma} (\nabla_\rho A_\sigma - \nabla_\sigma A_\rho)(\nabla_\mu A_\nu - \nabla_\nu A_\mu) - \tfrac{1}{2} m^2 A_\mu A^\mu.$$

Show that the field equation for A^μ is given by

$$\nabla_\mu(\nabla^\nu A^\mu - \nabla^\mu A^\nu) + m^2 A^\nu = 0.$$

By making use of the fact that covariant derivatives commute in Minkowski spacetime, show that in this case $\nabla_\nu A^\nu = 0$ and hence that the field equation can be written

$$\Box^2 A^\mu + m^2 A^\mu = 0,$$

where $\Box^2 \equiv \nabla^\mu \nabla_\mu = g^{\mu\nu} \nabla_\mu \nabla_\nu$ is the covariant d'Alembertian operator. These are called the *second-order Proca equations*.

19.20 An alternative Lagrangian for a free massive vector field A^μ of mass m, is

$$L = \tfrac{1}{4} F_{\mu\nu} F^{\mu\nu} - \tfrac{1}{2} F^{\mu\nu}(\nabla_\mu A_\nu - \nabla_\nu A_\mu) - \tfrac{1}{2} m^2 A_\mu A^\mu,$$

where $F_{\mu\nu}$ and A^μ are considered as *independent* quantities. By varying the corresponding action with respect to $F_{\mu\nu}$ and A^μ independently, show that the Euler–Lagrange equations yield

$$\nabla_\mu F^{\nu\mu} + m^2 A^\nu = 0 \quad \text{and} \quad F_{\mu\nu} = \nabla_\mu A_\nu - \nabla_\nu A_\mu,$$

which are called the *first-order Proca equations*.

19.21 The simplest scalar action for gravity *in vacuo* that one can construct from the metric tensor alone is

$$S = \int_{\mathcal{R}} \sqrt{-g}\, d^4x.$$

Show that the corresponding field equations are given by $\sqrt{-g} g_{\mu\nu} = 0$ and clearly do not constitute a viable theory of gravity.

19.22 Under a general infinitesimal coordinate transformation of the form $x'^{\mu} = x^{\mu} + \xi^{\mu}(x)$, show that

$$\delta g_{\mu\nu} \equiv g'_{\mu\nu}(x) - g_{\mu\nu}(x) = -(\nabla_\mu \xi_\nu + \nabla_\nu \xi_\mu).$$

19.23 Consider a general action for gravity *in vacuo* of the form

$$S = \int_{\mathcal{R}} \mathcal{L}(g_{\mu\nu}, \partial_\sigma g_{\mu\nu}, \partial_\rho \partial_\sigma g_{\mu\nu}, \ldots)\, d^4x.$$

By considering a general infinitesimal coordinate transformation of the form $x'^{\mu} = x^{\mu} + \xi^{\mu}(x)$, where the $\xi^{\mu}(x)$ vanish on the boundary $\partial_{\mathcal{R}}$, show that the metric and its derivatives must satisfy the differential constraints

$$\nabla_\mu \left(\frac{\delta \mathcal{L}}{\delta g_{\mu\nu}} \right) = 0,$$

where $\delta \mathcal{L}/\delta g_{\mu\nu}$ is the variational derivative of the Lagrangian density with respect to the metric. Hence show that for the Einstein–Hilbert action these differential constraints lead to the contracted Bianchi identities $\nabla_\mu G^{\mu\nu} = 0$.

19.24 Show explicitly that the quadratic action

$$\bar{S} = \int_{\mathcal{R}} g^{\mu\nu} (\Gamma^{\tau}{}_{\mu\nu} \Gamma^{\sigma}{}_{\tau\sigma} - \Gamma^{\tau}{}_{\mu\sigma} \Gamma^{\sigma}{}_{\tau\nu}) \sqrt{-g}\, d^4x$$

is not a scalar with respect to general coordinate transformations. Show further that varying this action with respect to the metric and its first derivative leads to the Einstein field equations *in vacuo*, provided that the variation in the metric *and* its first derivative vanish on the boundary $\partial\mathcal{R}$.

19.25 Obtain an expression for the dynamical energy–momentum tensor of the complex scalar field considered in Exercise 19.15 and that of the massive vector field considered in Exercise 19.19.

Bibliography

Abramowitz, M. & Stegun, I. A., *Handbook of Mathematical Physics*, Dover, 1972.

Chandrasekhar, S., *An Introduction to the Study of Stellar Structure*, Dover, 1958.

Chandrasekhar, S., *The Mathematical Theory of Black Holes*, Oxford University Press, 1983.

Clarke, C., On the global isometric embedding of pseudo-Riemannian manifolds, *Proceedings of the Royal Society* **A314**, 417–28, 1970.

d'Inverno, R., *An Introduction to Einstein's Relativity*, Oxford University Press, 1992.

Dirac, P. A. M., *General Theory of Relativity*, Princeton Landmarks in Physics Series, Princeton University Press, 1996.

Feynman, R. P., Morinigo, F. B. & Wagner, W. G., *Feynman Lectures on Gravitation*, Addison-Wesley, 1995.

Foster, J. & Nightingale, J. D., *A Short Course in General Relativity*, Springer-Verlag, 1995.

Islam, J. N., *An Introduction to Mathematical Cosmology*, Cambridge University Press, 1992.

Liddle, A. & Lyth, D., *Cosmological Inflation and Large-Scale Structure*, Cambridge University Press, 2000.

Misner, C. W., Thorne, K. S. and Wheeler, J. A., *Gravitation*, Freeman, 1973.

Mukhanov, V. F., Feldman, H. A. & Brandenburger, R. H., Theory of cosmological perturbations, *Physics Reports* **215**, 203–333, 1992.

Nash, J., The imbedding problem for Riemannian manifolds, *Annals of Mathematics* **63**, 20–63, 1956.

Padmanabhan, T., *Structure Formation in the Universe*, Cambridge University Press, 1993.

Padmanabhan, T., *From Gravitons to Gravity: Myths and Reality*, abs/grqc/0409089.

Peacock, J., *Cosmological Physics*, Cambridge University Press, 1999.

Rindler, W., *Relativity: Special, General and Cosmological*, Oxford University Press, 2001.

Ryder, R. H., *Quantum Field Theory*, Cambridge University Press, 1985.

Schutz, B. F., *Geometrical Methods of Mathematical Physics*, Cambridge University Press, 1980.

Schutz, B. F., *A First Course in General Relativity*, Cambridge University Press, 1985.

Tanaka, Y. *et al.*, *Nature* **375**, 659, 1995.

Wald, R. M., *General Relativity*, University of Chicago Press, 1984.

Weinberg, S., *Gravitation and Cosmology*, Wiley, 1972.

Will, C., *Theory and Experiment in Gravitational Physics*, Cambridge University Press, 1981.

Index

An italic page number indicates that there is a figure related to the topic on this page.